Omics-driven Research in Medicinal Plants

Omics-driven Research in Medicinal Plants

Archana Jain

Surendra Sarsaiya

Qihai Gong

Zunyi Medical University, Zunyi, Guizhou, China

CABI

CABI is a trading name of CAB International

CABI
Nosworthy Way
Wallingford
Oxfordshire OX10 8DE
UK

Tel: +44 (0)1491 832111
E-mail: info@cabi.org
Website: www.cabi.org

CABI
200 Portland Street
Boston
MA 02114
USA

T: +1 (617)682-9015
E-mail: cabi-nao@cabi.org

A catalogue record for this book is available from the British Library, London, UK.

ISBN-13: 9781836994183 (hardback)
9781836994190 (ePDF)
9781836994206 (ePub)

DOI: 10.1079/9781836994206.0000

Commissioning Editor: Rebecca Stubbs
Editorial Assistant: Theresa Regueira
Production Editor: Rosie Hayden

Typeset by Exeter Premedia Services Pvt Ltd, Chennai, India
Printed and bound in the UK by CPI Group (UK) Ltd, Croydon, CR0 4YY

Contents

Foreword

It is with immense pleasure and a profound sense of anticipation that I introduce this seminal volume, *Omics-driven Research in Medicinal Plants*, to our global community of scientists, researchers, and students. In an era defined by technological convergence and a pressing need for sustainable healthcare solutions, the exploration of medicinal plants through the lens of modern omics technologies has never been more critical. This book arrives at a pivotal moment, offering a comprehensive roadmap to this dynamic and rapidly evolving field.

For centuries, traditional medicine systems across the world, including the rich heritage of Chinese medicine, have relied on the therapeutic power of plants. While their efficacy is well-documented through empirical evidence, the underlying molecular mechanisms, genetic blueprints, and biosynthetic pathways have often remained elusive. Today, we stand at the frontier of a new age, where genomics, transcriptomics, proteomics, and metabolomics are illuminating these ancient remedies with unprecedented clarity. This integration of traditional knowledge with cutting-edge science is the cornerstone of modern phytomedicine research.

This edited work meticulously charts this exciting journey. The chapters within delve into the foundational principles of omics technologies and progress to explore sophisticated themes such as multi-omics integration, precision breeding, metabolic engineering, and the transcriptional regulation of high-value metabolites. The book wisely extends its scope beyond the laboratory, addressing urgent global challenges such as developing climate-resilient plant varieties and establishing frameworks for the ethical conservation and bioprospecting of medicinal species. Crucially, it peers into the future, examining the powerful convergence of omics with artificial intelligence and synthetic biology—a synergy that promises to accelerate drug discovery and development exponentially.

As the President of Zunyi Medical University, located in the botanically rich province of Guizhou, China, I am deeply committed to advancing research that bridges our natural heritage with scientific innovation. This publication embodies that very mission. I extend my heartfelt congratulations and appreciation to my fellow editors, Dr Archana Jain and Dr Surendra Sarsaiya, for their dedication and scholarly rigor. They have assembled a resource that is both intellectually stimulating and immensely practical.

It is my sincere hope that this book will serve as an invaluable guide and a source of inspiration. May it ignite curiosity, foster interdisciplinary collaboration, and ultimately contribute to the discovery of novel, effective, and sustainable healthcare solutions for all of humanity.

Honorable Prof. Qihai Gong,
President
Zunyi Medical University, Zunyi, Guizhou, China

Preface

The intricate molecular symphony within medicinal plants has captivated scientists for generations. These organisms are not merely sources of therapeutic compounds but are complex, dynamic systems that have evolved sophisticated pathways to produce a vast array of bioactive metabolites. For centuries, we have harnessed their healing potential through traditional knowledge. Today, we are poised to understand and engineer this potential with a depth that was once unimaginable, thanks to the advent of omics technologies.

This book, *Omics-driven Research in Medicinal Plants*, is born from our shared conviction that we are at a transformative juncture in plant science and natural product drug discovery. The convergence of genomics, transcriptomics, proteomics, and metabolomics provides a powerful, holistic toolkit to move beyond descriptive studies to mechanistic and predictive science. Our aim in editing this volume is to synthesize the current state of this rapidly advancing field and to provide a comprehensive resource that bridges fundamental concepts with frontier applications.

The journey through these chapters is designed to be both logical and expansive. We begin by establishing a solid foundation in omics technologies and the critical challenge of integrating multi-omics data to extract biologically meaningful insights. Subsequent chapters delve into the application of this knowledge, exploring how we can precisely edit plant genomes, engineer metabolic pathways, and unravel the transcriptional networks that govern the production of high-value compounds. We then shift our focus to the tangible outcomes of this research, from accelerating drug discovery pipelines to developing climate-resilient medicinal plants for a sustainable future.

Recognizing the broader implications of our work, we have dedicated chapters to the role of medicinal plants in functional foods and nutraceuticals, and to the urgent ethical imperative of conserving biodiversity through omics-guided strategies. Finally, we look ahead to the horizon, where artificial intelligence and synthetic biology are set to amplify the power of omics research, opening new vistas for innovation.

This project has been a collaborative endeavor in the truest sense, and our gratitude extends to the many colleagues and researchers whose pioneering work forms the bedrock of this field. We also wish to acknowledge the unwavering support of our families and our home institution, Zunyi Medical University, for fostering an environment conducive to such scholarly pursuits.

It is our sincere hope that this book will serve as a catalyst—inspiring students, guiding researchers, and providing a strategic overview for policy makers. We invite you to explore the pages that follow and join us in unlocking the profound secrets of medicinal plants, for the betterment of global health and well-being.

Archana Jain, Surendra Sarsaiya, and Qihai Gong
Zunyi Medical University, Zunyi, Guizhou, China

Biographies

Professor Gong Qihai, MD, PhD, is a Second-Level Professor, doctoral supervisor, an honored expert of Guizhou Province, and a recipient of the Special Allowance from the State Council. He also serves as director of the Chinese Pharmacological Society. Currently, he is a member of the Party Committee and president of Zunyi Medical University, China, as well as the leader of an innovation team of the Ministry of Education (including "rolling" support). His main research focus is neuropharmacology and toxicology of traditional Chinese medicine. He has presided over more than 20 projects, including the Key Special Project "Modernization Research of Traditional Chinese Medicine" under the National Key R&D Program, projects from the National Natural Science Foundation, and projects from the Ministry of Education. He holds five national invention patents and has published more than 200 papers domestically and internationally, of which over 100 are included in the Science Citation Index (SCI). He has received the First Prize for Provincial Scientific and Technological Progress and the First Prize for Provincial Graduate Teaching Achievements. He has been selected for the Ministry of Education's New Century Excellent Talent Support Program, honored as an Excellent Young Science and Technology Talent of Guizhou Province, and recognized as a "Hundred"-level High-Level Innovative Talent in Guizhou Province. He is a member of the editorial board of the journals *Pharmacology and Clinical Application of Traditional Chinese Medicine* and *Chinese Journal of New Drugs and Clinical Applications*, and was ranked among the world's top 2% of scientists in the field of pharmacy.

Dr Archana Jain is a distinguished post-doctoral researcher contributing significantly to the field of pharmacology at Zunyi Medical University (ZMU) in China. She is affiliated with the university's prestigious Key Laboratory of Basic Pharmacology and the Joint International Research Laboratory of Ethnomedicine of the Ministry of Education, hubs known for cutting-edge research in drug discovery and traditional medicine. Dr Jain completed her Doctor of Philosophy (PhD) in Education at Mansarovar Global University (MGU) in Sehore, Madhya Pradesh, India, in 2023. With her diverse educational background and research interests in education and microbiology, Dr Jain's research is primarily focused on the pharmacological and phytochemical investigation of natural products, with a special emphasis on exploring the therapeutic potential of medicinal plants. Her work often bridges traditional knowledge and modern scientific validation. This research is evidenced by a substantial body of publications in reputable, high-impact international journals. Her scholarly output, as seen on platforms like Google Scholar, includes numerous articles where she often appears as a leading author. Through her dedicated work at Zunyi Medical University, Dr Jain has established herself as a valuable member of an international research team, actively advancing our understanding of ethnomedicine and contributing to the global pursuit of safer and more effective treatments derived from natural sources.

Dr Surendra Sarsaiya is a distinguished professor and researcher in the field of microbiology, whose work stands at the compelling intersection of microbial ecology, natural product discovery, and sustainable bioresource utilization. He is currently a prominent researcher and investigator at the Zunyi Medical University in Zunyi City, China, where he holds a triple affiliation with three of the institution's most prestigious centers: the Key Laboratory of Basic Pharmacology, the Joint International Research Laboratory of Ethnomedicine of the Ministry of Education, and the Bioresource Institute for Healthy Utilization (BIHU). This unique positioning underscores his multidisciplinary approach, leveraging microbiological expertise to advance pharmacological and ethnomedical research. Dr Sarsaiya's primary research focus lies in harnessing the power of microorganisms, particularly endophytic fungi, for the discovery of novel bioactive compounds. His work delves into the intricate relationships between microbes and medicinal plants, exploring how these symbionts produce therapeutic metabolites with potential applications as anti-cancer, antimicrobial, and antioxidant agents. He possesses extensive expertise in the isolation, identification, and fermentation of microbes, followed by the sophisticated extraction, purification, and structural elucidation of their secondary metabolites. This research is critical for developing new compounds for the pharmaceutical industry. His scholarly contributions are substantial, evidenced by a robust publication record in high-impact, international peer-reviewed journals, where his research articles provide significant insights into the microbial synthesis of valuable compounds and the sustainable exploitation of bioresources. Through his dedicated work at Zunyi Medical University, Dr Sarsaiya has established himself as a key figure in an international research community, passionately committed to unlocking the hidden potential of microbes for plant health and the healthy utilization of our planet's biological wealth.

1 Foundations of Omics Technologies in Medicinal Plant Research

Archana Jain, Surendra Sarsaiya and Qihai Gong*
Key Laboratory of Basic Pharmacology and Joint International Research Laboratory of Ethnomedicine of Ministry of Education, Zunyi Medical University, Zunyi, Guizhou, China

1.1 Introduction

Omics technologies encompass a range of disciplines that collectively enhance our understanding of biological systems, particularly in medicinal plant research (Zhang *et al.*, 2023). Genomics is the foundational study of the complete set of DNA including all genes within an organism. It provides critical insights into the genetic makeup that influences the production of bioactive compounds, essential for drug discovery and development (Zhang *et al.*, 2017). Building on this, transcriptomics focuses on the transcriptome, which includes all RNA molecules transcribed from the genome (Rajawat, 2018). This field is vital for understanding gene expression patterns, helping researchers identify active genes during various developmental stages or in response to environmental stimuli (Lenroot and Giedd, 2011). Proteomics, the large-scale study of proteins, elucidates their functions and structures, revealing roles in metabolic pathways and plant defense mechanisms. This understanding can lead to the identification of novel therapeutic targets (Satrio *et al.*, 2024). Complementing these fields, metabolomics involves the comprehensive analysis of metabolites within biological samples, allowing for the identification and quantification of bioactive compounds that contribute to a plant's medicinal properties (Salem *et al.*, 2020). Epigenomics studies the complete set of epigenetic modifications on genetic material, revealing how environmental factors influence gene expression and metabolite production. This understanding is crucial for optimizing plant-based therapies (Abdul *et al.*, 2017). Together, these omics disciplines provide a holistic view of plant biology, integrating data to inform drug discovery and enhance the therapeutic potential of medicinal plants (Pandita *et al.*, 2021). The foundations of omics technologies—from genomics to epigenomics—are essential for advancing medicinal plant research, offering insights that drive the development of effective natural medicines and therapeutic strategies.

The integration of genomics, transcriptomics, proteomics, metabolomics, and epigenomics—collectively referred to as "omics"—plays a pivotal role in advancing medicinal plant research. Each omics discipline contributes unique insights that enhance our understanding of the complex biology of these plants. Genomic variation analysis is fundamental for identifying genetic variations linked to desirable traits, such as increased production of bioactive

*Corresponding author: gqh@zmu.edu.cn

DOI: 10.1079/9781836994206.0001

compounds, which can inform breeding programs aimed at improving crop yields (Singer *et al.*, 2021). Transcriptome profiling complements this by revealing gene expression patterns under various conditions, thereby identifying genes involved in the biosynthesis and regulation of these compounds (Hoover *et al.*, 2007). Proteomics further enriches this understanding by elucidating protein–protein interactions that are crucial for the biosynthesis of bioactive compounds (Sanches *et al.*, 2024). This knowledge can lead to the identification of new targets for enhancing the properties of medicinal plants. Additionally, metabolomics provides insights into the metabolic profiles of these medicinal plants, allowing researchers to understand how environmental and genetic factors influence the production of bioactive compounds (Wang *et al.*, 2024). Epigenomics adds another layer of complexity by examining how epigenetic modifications, such as DNA methylation and histone modifications, regulate gene expression and compound production (Eichten *et al.*, 2014). This interplay between genetic and epigenetic factors is essential for adapting to environmental changes, which is particularly important for sessile organisms like plants. Finally, the integration of these diverse omics datasets fosters a comprehensive understanding of the intricate regulatory networks governing medicinal plant biology, ultimately leading to improved strategies for crop enhancement and bioactive compound discovery (Zhang *et al.*, 2024). By leveraging these interconnected fields, researchers can unlock new potentials in medicinal plant research, paving the way for advancements in agriculture and pharmacology.

In the realm of omics technologies, each discipline—genomics, transcriptomics, proteomics, metabolomics, and epigenomics—focuses on distinct aspects of biological systems, providing a comprehensive understanding of cellular functions and interactions. Genomics is the foundational study of the complete set of genes within an organism, emphasizing the structure, function, and evolution of genomes. It serves as the basis for personalized medicine, allowing for the prediction and prevention of diseases based on individual genetic profiles (Weissenbach, 2016). Transcriptomics, on the other hand, examines the complete set of RNA transcripts, shedding light on gene expression and regulation. This field is crucial for understanding how genes respond to environmental changes, which is essential for cellular differentiation and morphogenesis in multicellular organisms (Singh *et al.*, 2024). Proteomics focuses on the entire set of proteins produced, analyzing their structure, function, and interactions. This discipline is vital for understanding cellular processes and the complexities of protein networks, which are often influenced by various factors, including post-translational modifications (Miteva *et al.*, 2013). Metabolomics studies the complete set of metabolites, providing insights into metabolic pathways and the biochemical responses of organisms to environmental stimuli. This field is particularly useful in disease diagnostics and monitoring therapeutic responses (Lankadurai *et al.*, 2013). Lastly, epigenomics investigates the chemical modifications to DNA and histones that regulate gene expression without altering the underlying DNA sequence (Abdul *et al.*, 2017). This area is critical for understanding cellular differentiation and the mechanisms of diseases such as cancer. Together, these omics technologies create a holistic view of biological systems, enabling researchers to integrate data and uncover complex interactions within and between cellular networks (Fig. 1.1; Tolani *et al.*, 2021).

1.2 Historical Evolution of Omics in Medicinal Plant Biology

The historical evolution of omics in medicinal plant biology has been marked by several key advancements that have significantly impacted biomedical research (Table 1.1 and Fig. 1.2). The journey began with the sequencing of the *Arabidopsis thaliana* genome in 2000, which was the first plant genome to be sequenced. This milestone not only laid the groundwork for omics research in plants but also facilitated the identification of genes and their functions, leading to improvements in crop yields and the discovery of novel therapeutic compounds (Provart *et al.*, 2016). Following this, the field of plant metabolomics emerged in the 1990s, focusing on the study of plant metabolites and their biological roles (Tian *et al.*, 2016). This advancement has been crucial in uncovering

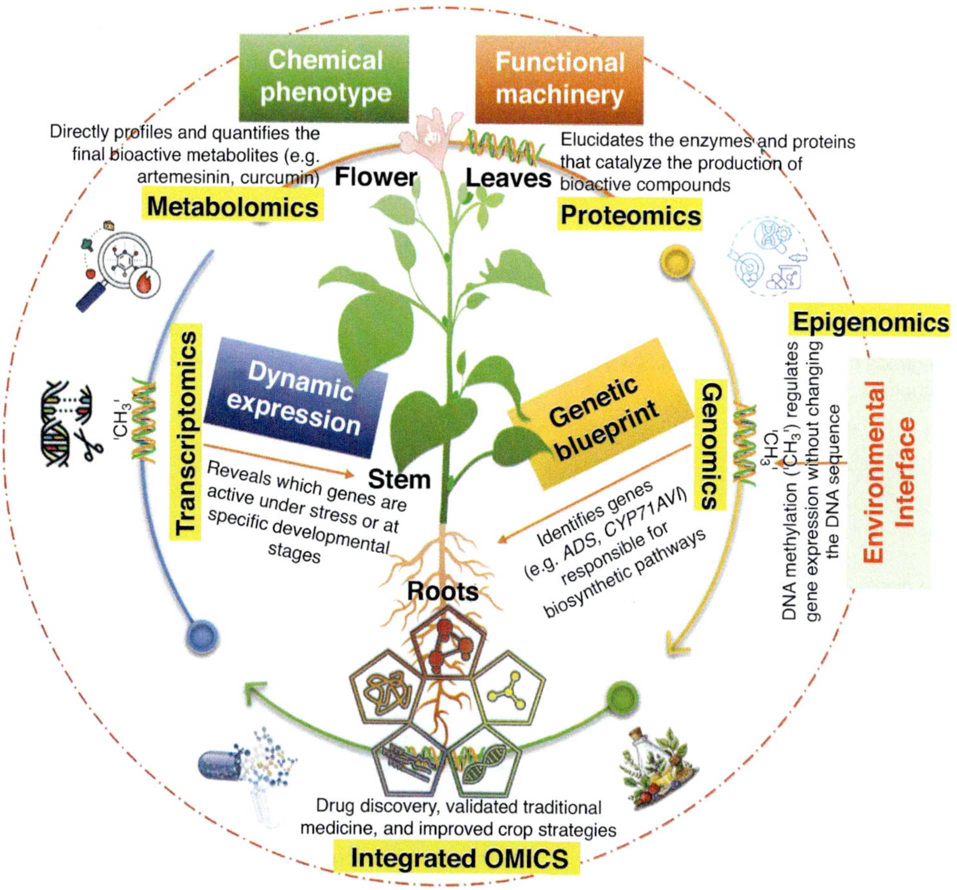

Fig. 1.1. The holistic omics framework in medicinal plant research. (Author's own figure.)

bioactive compounds in medicinal plants, which has direct implications for drug development and therapeutic strategies. Concurrently, expressed sequence tag (EST) sequencing was developed, allowing researchers to identify and characterize gene transcripts, further enhancing our understanding of gene regulation and expression in plants (Hao *et al.*, 2012). The integration of transcriptomics into medicinal plant research has revolutionized our comprehension of gene expression at the transcriptome level, enabling the identification of genes involved in the biosynthesis of bioactive compounds (Zhang *et al.*, 2023). This integration has been complemented by advancements in plant proteomics, which focuses on the functions of plant proteins and their interactions, thereby providing insights into plant defense mechanisms and responses to environmental stimuli (Jabeen *et al.*, 2024). Moreover, the integration of various omics fields has transformed our understanding of medicinal plant biology, facilitated the discovery of novel bioactive compounds, and improved our grasp of gene regulation (Zhang *et al.*, 2023). Collectively, these advancements have not only enriched our knowledge of plant biology but have also paved the way for innovative therapeutic strategies, underscoring the profound biomedical impact of omics research in medicinal plants (Ma *et al.*, 2020).

As the first complete plant genome sequenced, *Arabidopsis* has served as a model organism, allowing researchers to explore complex biological processes and establish foundational knowledge applicable to other plant species (Chang *et al.*, 2016). This genomic

Table 1.1. Historical evolution and impact of omics technologies in medicinal plant biology. Author's own table.

Foundational genomics (2000)	Genomics	The sequencing of the first plant genome (*A. thaliana*) laid the groundwork for plant omics, enabling gene identification, functional studies, and the discovery of novel therapeutic compounds.	(Provart *et al*., 2016)
Emergence of metabolomics (1990s)	Metabolomics	The study of plant metabolites became crucial for uncovering bioactive compounds in medicinal plants, with direct implications for drug development and therapeutic strategies.	(Tian *et al*., 2016)
Gene discovery & profiling	Transcriptomics (ESTs)	Expressed sequence tag (EST) sequencing allowed for the identification and characterization of gene transcripts, enhancing the understanding of gene regulation and expression.	(Hao *et al*., 2012)
Advanced functional analysis	Transcriptomics & proteomics	Revolutionized the comprehension of gene expression and protein functions, enabling the identification of biosynthetic pathways for bioactive compounds and insights into plant defense mechanisms.	(Zhang *et al*., 2023; Jabeen *et al*., 2024)
Systems biology approach	Multi-omics integration	The integration of various omics fields transformed the understanding of medicinal plant biology, facilitating the discovery of novel bioactive compounds and improving the grasp of complex gene regulation networks.	(Ma *et al*., 2020; Zhang *et al*., 2023)

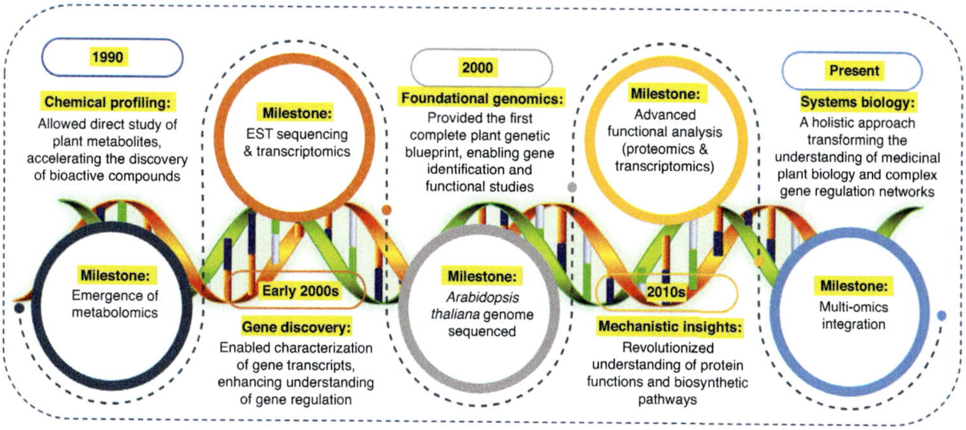

Fig. 1.2. The historical evolution of plant omics. (Author's own figure.)

breakthrough has facilitated advancements in functional genomics, enabling scientists to investigate gene regulation and the interactions between genes and their products (Hilton and Gersbach, 2015). By understanding how genes are expressed in response to environmental cues, researchers can develop crops with improved traits, such as drought tolerance and disease resistance, which are crucial for food security and sustainable agriculture (Fang and Xiong, 2015). Moreover, the insights gained from *Arabidopsis* have propelled comparative genomics, allowing for the identification of conserved genetic elements across different plant species (Yaschenko *et al.*, 2025). This comparative approach not only aids in crop improvement but also enhances our understanding of plant evolution and biodiversity. In the biomedical realm, the knowledge derived from plant genomics has implications for biofuel production, as researchers can optimize plant biomass for renewable energy sources (Mortimer, 2019). Additionally, the understanding of plant–microbe interactions has led to strategies for enhancing disease resistance in crops, which is vital for maintaining agricultural productivity (Wille *et al.*, 2019). Finally, the advent of genome editing technologies, such as CRISPR-Cas9 (clustered regularly interspaced short palindromic repeats-associated protein 9), has been made possible by the foundational knowledge gained from the *Arabidopsis* genome, allowing for precise modifications that can introduce desirable traits into crops (Razzaq *et al.*, 2019). Overall, the sequencing of *A. thaliana* has profoundly influenced both plant biology and its applications in agriculture and biotechnology, paving the way for innovations that address global challenges.

1.3 Core Tools: Sequencing, Mass Spectrometry, and Bioinformatics

The integration of core technologies such as next-generation sequencing (NGS) and liquid chromatography-mass spectrometry (LC-MS) plays a pivotal role in the omics workflows of medicinal plants, facilitating the discovery and characterization of bioactive compounds (Pattnaik, 2023). NGS is a high-throughput sequencing technology that allows for rapid and cost-effective analysis of plant genomes, transcriptomes, and epigenomes (Satam *et al.*, 2023). This technology is essential for identifying genes involved in the biosynthesis of therapeutic agents, thereby enhancing our understanding of plant genomics and supporting the modernization of traditional herbal medicines (Younessi-Hamzekhanlu *et al.*, 2022). On the other hand, LC-MS combines the separation capabilities of liquid chromatography with the detection power of mass spectrometry, making it indispensable for the identification and quantification of bioactive compounds in medicinal plants (Yilmaz, 2020). This technique enables researchers to efficiently analyze complex mixtures of metabolites, which is crucial for discovering new therapeutic leads

(Maekawa and Mano, 2022). Moreover, bioinformatics tools are integral to the analysis of omics data generated from these technologies. Tools such as BLAST (basic local alignment search tool) and KEGG (Kyoto Encyclopedia of Genes and Genomes) allow researchers to interpret vast datasets, identify patterns, and reconstruct metabolic pathways, all of which is vital for understanding the functional roles of genes and metabolites in medicinal plants (Murmu *et al.*, 2024). Together, these technologies create a comprehensive framework for the study of medicinal plants, enabling the identification of novel therapeutic agents and the development of personalized medicine approaches.

By leveraging NGS, LC-MS, and bioinformatics, researchers can enhance the quality control of medicinal plants, ensuring the authenticity and consistency of bioactive compound production (Pattnaik, 2023). This integrated approach not only advances scientific knowledge but also supports the sustainable use of medicinal plants in health care. The analysis of omics data generated from medicinal plants involves a variety of bioinformatics tools and pipelines that facilitate data preprocessing, analysis, and interpretation. One of the foundational steps in this process is the use of Trimmomatic, which is employed for trimming and filtering high-throughput sequencing data to enhance data quality by removing low-quality bases and adapters (Mrozek *et al.*, 2021). Following this, FastQC is often utilized for quality control, providing a comprehensive report on data quality metrics such as base calling accuracy and adapter contamination (de Sena Brandine and Smith, 2021). For RNA-seq data, STAR (spliced transcripts alignment to a reference) is a widely used tool for aligning sequences to a reference genome, ensuring high sensitivity and accuracy in the alignment process (Dobin and Gingeras, 2016). Once aligned, DESeq2 is applied for differential gene expression analysis, allowing researchers to identify genes that are significantly expressed under different conditions (McDermaid *et al.*, 2019). In the realm of metabolomics, Metabo Analyst serves as a comprehensive platform for metabolite identification, statistical analysis, and pathway analysis, which is crucial for understanding the metabolic profiles of medicinal plants (Rai *et al.*, 2017). Additionally, KEGG pathway analysis aids in annotating biological pathways, providing insights into the metabolic processes involved in plant responses (Gu *et al.*, 2018). To integrate various omics data types, Omics Data Integration techniques are employed, utilizing tools like Multi-Omics Factor Analysis to uncover patterns across transcriptomics, proteomics, and metabolomics data (Subramanian *et al.*, 2020). The bioconductor platform further supports this analysis by offering a suite of tools for data preprocessing and interpretation (Sepulveda, 2020). Finally, transcriptome assembly and functional annotation pipelines are essential for reconstructing transcriptomes and assigning biological functions to genes, respectively, using tools like Trinity and Inter Pro Scan (Raghavan *et al.*, 2022). Together, these tools form a robust framework for the comprehensive analysis of omics data in medicinal plant research.

NGS plays a pivotal role in the analysis of medicinal plant genomes, offering several key applications that enhance our understanding and utilization of these valuable resources (Cheng *et al.*, 2021). One of the primary applications is in plant breeding and crop improvement, where NGS facilitates the identification of desirable traits, such as increased yield and disease resistance, thereby addressing global food security challenges (Bhat *et al.*, 2016). Additionally, NGS supports genomic-assisted selection, which employs genetic markers to select for specific traits in medicinal plants. This method increases the efficiency and accuracy of breeding programs, significantly reducing the time and costs associated with traditional breeding methods (Basit *et al.*, 2024). Marker-assisted selection, enabled by NGS, allows breeders to pinpoint genetic markers linked to traits of interest, further accelerating the breeding process and improving program accuracy (Kumar *et al.*, 2024). Moreover, NGS is instrumental in conducting genome-wide association studies (GWAS), which analyze genetic variation across entire genomes to identify associations between specific genes and traits (Bhat *et al.*, 2016). This approach aids in the development of improved crop varieties by uncovering the genetic basis of complex traits. NGS also enhances our understanding of gene expression through transcriptomics analysis, revealing insights into the regulation of genes associated with desirable traits and facilitating the development of novel therapeutic compounds (Abdi *et al.*, 2024). Furthermore, NGS enables metagenomics and epigenomics analyses, which explore microbial communities and epigenetic modifications, respectively, contributing to improved plant health and crop yields (Yadav *et al.*, 2025). Lastly, NGS supports innovative approaches such as gene editing, synthetic biology, and systems biology, which collectively enable the precise modification of genes and the design of novel biological pathways, ultimately leading to the production of new therapeutic compounds and enhanced agricultural strategies (Esvelt and Wang, 2013). These applications underscore the transformative potential of NGS in the field of medicinal plant research (Table 1.2).

1.4 Workflow Design: From Data Generation to Analysis

In medicinal plant research, an effective workflow design encompasses three critical stages: experimental design, data generation, and analysis interpretation. The first step, experimental design, is fundamental for ensuring the reliability and reproducibility of results. It involves careful planning, including the selection of appropriate methodologies, identification of variables, and control of confounding factors, which collectively contribute to high-quality data generation (Sharma and Sarkar, 2013). Following the design phase, data generation is executed through various techniques such as chromatography, spectroscopy, and bioassays. These methods are essential for collecting data on the chemical composition and biological activity of medicinal plants (Salam *et al.*, 2019). High-quality data is crucial for subsequent analysis, as it forms the foundation for understanding the pharmacological effects of the compounds studied. Once data is generated, the next phase involves analysis and interpretation. This is where workflow optimization plays a vital role, as it streamlines the processes involved in data analysis, helping to identify bottlenecks and enhance efficiency (Zhai *et al.*, 2023). Advanced techniques such as machine learning and cheminformatics are increasingly utilized to analyze large datasets, allowing researchers to identify patterns and predict biological activities of compounds (Niazi and Mariam, 2023). Moreover, bioassay-guided fractionation and high-throughput screening techniques facilitate the isolation of bioactive compounds and the rapid assessment of their biological activities, respectively (Mani *et al.*, 2022). Finally, the integration of computational pharmacology aids in predicting drug–target interactions and pharmacokinetics, further enriching the interpretation of the data (Naithani and Guleria, 2024). Well-structured workflow from experimental design through data generation to analysis interpretation is essential for advancing medicinal plant research, ultimately leading to the discovery of new therapeutic agents.

Designing an experimental workflow for medicinal plant research requires careful consideration of several key factors to address the complexities of data generation and analysis. First, adhering to experimental design principles is crucial (Park *et al.*, 2022). These principles, which include careful planning, randomization, and replication, help minimize bias and variability, thereby enhancing the reliability and validity of the findings. Data management is another critical aspect. Implementing robust data management plans ensures that data are systematically collected, stored, and shared, which facilitates integration and reduces errors (Sepanj *et al.*, 2024). This is particularly important given the diverse datasets generated through high-throughput screening and metabolomics analysis, which are essential for identifying bioactive compounds (Foito and Stewart, 2018). Moreover, the use of bioinformatics tools is vital for analyzing large datasets. These tools enable researchers to interpret complex data and identify patterns that may not be evident through traditional methods (Demirbaga *et al.*, 2024). Machine learning algorithms can further enhance data analysis by predicting biological activities and uncovering potential therapeutic agents (Vamathevan *et al.*, 2019). Quality control and assurance processes must also be integrated into the workflow to maintain the integrity of plant materials and data (Kabir *et al.*, 2024). Additionally, data visualization techniques play a significant role in effectively communicating complex findings, allowing researchers to identify trends and patterns that can accelerate the discovery of novel therapeutic agents (Ashiwaju *et al.*, 2023). Finally,

Table 1.2. Core omics tools and workflows in medicinal plant biology. Author's own table.

Technology/ platform	Specific function/ methodology	Key applications in medicinal plant biology	Impact & role in integrated workflow	Citations
NEXT-GENERATION SEQUENCING (NGS)				
Genome sequencing	High-throughput determination of the complete DNA sequence of a plant.	• Establishing reference genomes. • Identifying genes involved in biosynthetic pathways (e.g. for alkaloids, terpenoids). • Serving as a foundation for all other omics analyses.	Provides the fundamental blueprint for understanding the genetic potential of a medicinal plant. Essential for annotating transcripts and proteins.	(Provart *et al.*, 2016; Cheng *et al.*, 2021)
Transcriptomics (RNA-seq)	Sequencing of the complete set of RNA transcripts to measure gene expression.	• Identifying genes differentially expressed under specific conditions (e.g. stress, development). • Uncovering regulatory genes and enzymes in bioactive compound synthesis.	Bridges the gap between the static genome and the dynamic metabolic output. Crucial for linking genes to metabolic pathways.	(Zhang *et al.*, 2023; Abdi *et al.*, 2024)
Genome-wide association studies (GWAS)	Screening genome-wide genetic variants across different accessions to find associations with traits.	• Identifying genetic markers (single nucleotide polymorphisms—SNPs) linked to high yield of bioactive metabolites. • Accelerating the breeding of superior medicinal plant varieties.	Enables marker-assisted selection without prior knowledge of the causative genes, leveraging natural genetic diversity.	(Bhat *et al.*, 2016; Kumar *et al.*, 2024)
Metagenomics	Sequencing the collective genome of microbial communities associated with plants (rhizosphere, endophytes).	• Discovering plant-growth-promoting microbes. • Identifying microbial genes involved in symbiotic synthesis or modulation of plant metabolites.	Expands the understanding of the "holobiont," revealing how associated microbiomes influence medicinal plant health and chemistry.	(Wille *et al.*, 2019; Yadav *et al.*, 2025)

Continued

Table 1.2. Continued

Technology/ platform	Specific function/ methodology	Key applications in medicinal plant biology	Impact & role in integrated workflow	Citations
Epigenomics	Profiling heritable changes in gene expression not involving DNA sequence changes (e.g. DNA methylation).	• Understanding how environmental stresses (drought, pathogen attack) epigenetically regulate biosynthetic gene clusters.	Provides a mechanism for how the environment shapes the medicinal properties of a plant, influencing consistency and quality.	(Yadav *et al.*, 2025)
MASS SPECTROMETRY (MS)				
Liquid chromatography-MS (LC-MS)	Couples chromatographic separation with mass spectrometric detection for identifying and quantifying compounds in complex mixtures.	• Untargeted metabolomics: comprehensive profiling of all detectable metabolites in a plant extract. • Targeted quantification: precise measurement of specific known bioactive compounds. • Metabolite fingerprinting: for quality control and authentication of herbal material.	Directly characterizes the metabolic phenotype (the phytochemicals), which is the ultimate source of bioactivity. The primary tool for discovering novel therapeutic leads.	(Yilmaz, 2020; Maekawa and Mano, 2022; Pattnaik, 2023)
BIOINFORMATICS & DATA INTEGRATION				
Data quality control & preprocessing				
FastQC	Provides quality control reports on raw NGS data (e.g. per-base sequence quality, adapter content).	• Initial assessment of sequencing run quality. • Identifying issues like poor quality ends or adapter contamination.	Ensures that downstream analyses (assembly, alignment) are performed on high-quality, reliable data.	(de Sena Brandine and Smith, 2021)
Trimmomatic	Trims adapter sequences and removes low-quality bases from NGS reads.	• Cleaning raw sequencing reads to improve the accuracy of genome and transcriptome assemblies.	A critical preprocessing step that significantly enhances the quality and interpretability of NGS data.	(Mrozek *et al.*, 2021)

Continued

Table 1.2. Continued

Technology/ platform	Specific function/ methodology	Key applications in medicinal plant biology	Impact & role in integrated workflow	Citations
Genomics/transcriptomics analysis				
STAR	Aligns RNA-seq reads to a reference genome with high speed and accuracy.	• Mapping transcriptomic reads to quantify expression levels of known genes.	Provides the aligned data necessary for downstream differential expression analysis.	(Dobin and Gingeras, 2016)
Trinity	*De novo* assembles RNA-seq reads into a full transcriptome without a reference genome.	• Reconstructing the transcriptome of non-model medicinal plants for which a genome is not available.	Enables transcriptomic studies on a vast number of medicinal species that lack genomic resources.	(Raghavan *et al.*, 2022)
DESeq2	Statistical tool for determining differentially expressed genes from RNA-seq count data.	• Identifying genes that are up- or downregulated in response to elicitors, developmental stages, or environmental stresses.	Pinpoints key candidate genes likely involved in the regulation of valuable metabolic pathways.	(McDermaid *et al.*, 2019)
Functional & pathway annotation				
BLAST	Finds regions of similarity between biological sequences (e.g. genes, proteins).	• Annotating assembled sequences by comparing them to public databases to infer function.	The foundational tool for assigning putative functions to newly discovered genes and transcripts.	(Murmu *et al.*, 2024)
InterProScan	Classifies protein sequences into families and predicts domains and important sites.	• Providing functional annotation for predicted protein-coding genes from genomic or transcriptomic data.	Offers a more robust functional annotation by integrating multiple protein signature databases.	(Raghavan *et al.*, 2022)
KEGG	A database resource for understanding high-level functions and utilities of biological systems.	• Mapping annotated genes and metabolites onto pathway maps (e.g. terpenoid backbone biosynthesis, flavonoid biosynthesis).	Visualizes and interprets the biological context of omics data, revealing complete metabolic pathways.	(Gu *et al.*, 2018; Murmu *et al.*, 2024)

Continued

Table 1.2. Continued

Technology/ platform	Specific function/ methodology	Key applications in medicinal plant biology	Impact & role in integrated workflow	Citations
Metabolomics data analysis				
MetaboAnalyst	A comprehensive web-based platform for metabolomic data analysis, statistics, and functional interpretation.	• Performing statistical analysis (principal component analysis—PCA, partial least squares discriminant analysis—PCA) to find significant metabolites. • Integrating metabolomic data with pathway enrichment analysis.	A user-friendly, all-in-one suite that makes advanced metabolomics data analysis accessible to biologists.	(Rai *et al.*, 2017)
Multi-omics data integration				
Multi-Omics Factor Analysis (MOFA)	A statistical framework for integrating multiple omics datasets to identify the principal sources of variation.	• Simultaneously analyzing transcriptomic, proteomic, and metabolomic data from the same samples to uncover coordinated biological responses.	Moves beyond single-omics to a systems-level understanding, revealing how different molecular layers interact to define plant chemotype.	(Subramanian *et al.*, 2020)
Bioconductor	An open-source software project providing a suite of tools for the analysis and comprehension of high-throughput genomic data.	• Offering a standardized platform for a wide array of omics data preprocessing, statistical analysis, and visualization tasks in R.	Promotes reproducible research by providing a vast, community-vetted collection of bioinformatics packages.	(Sepulveda, 2020)
SYNTHETIC BIOLOGY & ENGINEERING				
Gene editing (e.g. CRISPR-Cas9)	Precise modification of plant genomes to knock out, knock in, or regulate genes.	• Validating the function of genes predicted to be involved in biosynthesis. • Engineering plant metabolism to enhance the production of desired compounds.	Allows for functional validation of omics discoveries and direct engineering of medicinal plants or microbial hosts (e.g. yeast) for compound production.	(Esvelt and Wang, 2013; Razzaq *et al.*, 2019; Wang *et al.*, 2024)

fostering collaborative research networks can enhance resource sharing and expertise, enabling researchers to tackle complex questions more effectively. By considering these elements, researchers can design a comprehensive experimental workflow that addresses the multifaceted challenges of medicinal plant research (Table 1.3 and Fig. 1.3).

1.5 Ethical Foundations: Bioprospecting and Biopiracy

Bioprospecting, particularly in the context of medicinal plants, is underpinned by several ethical foundations that aim to ensure fair and responsible practices (Preetha *et al.*, 2024). Central to these foundations is the concept of access and benefit-sharing (ABS), which mandates equitable sharing of benefits derived from genetic resources, including medicinal plants, between providers and users (Preetha *et al.*, 2024). This principle is crucial for promoting sustainable use and conservation of these resources, aligning with the ethical imperative to respect the rights of Indigenous communities. Prior informed consent (PIC) is another vital ethical component, requiring that Indigenous communities give consent before their genetic resources are accessed (Hudson *et al.*, 2020). This respects their sovereignty and acknowledges their traditional knowledge, which is often the basis for the understanding of the medicinal properties of these plants. However, the risk of biopiracy looms large, as it involves the unauthorized appropriation of these resources without fair compensation, highlighting the need for ethical frameworks to prevent exploitation (Handique and Dubey, 2022). The Nagoya Protocol provides an international framework for ABS, reinforcing ethical principles by ensuring fair benefit-sharing and promoting sustainable use while recognizing the rights of local communities (Escobar-Pemberthy and Calle Saldarriaga, 2020; Nelliyat, 2022). Furthermore, the integration of traditional knowledge through ethnobotany emphasizes the importance of recognizing and valuing Indigenous practices in the sustainable harvesting of medicinal plants (Mbelebele *et al.*, 2024). Ultimately, the ethical foundations of bioprospecting are intertwined with biodiversity conservation efforts, as they promote the sustainable management of medicinal plants while safeguarding the rights and livelihoods of Indigenous communities (Pushpangadan *et al.*, 2018). By adhering to these ethical principles, bioprospecting can contribute to social justice and environmental sustainability, fostering a more equitable relationship between modern science and traditional knowledge.

Biopiracy, particularly in the context of medicinal plants, raises significant ethical concerns regarding the exploitation of Indigenous knowledge and resources (Efferth *et al.*, 2016). Traditional knowledge (TK) is often the foundation for discovering and utilizing plant-based remedies, yet it is frequently appropriated without proper recognition or compensation, leading to biopiracy (Balkrishna *et al.*, 2024). This exploitation undermines the rights of Indigenous peoples, who possess inherent rights to their lands, resources, and knowledge. To combat biopiracy, mechanisms such as ABS are essential (Nelliyat *et al.*, 2022). ABS ensures that benefits derived from genetic resources are shared equitably among stakeholders, including Indigenous communities (Morrison *et al.*, 2021). Fair trade practices can also play a vital role by ensuring that medicinal plant products are sourced in a manner that respects community rights and promotes sustainable development (Raju and Das, 2024). Ultimately, preserving biocultural diversity—an interconnectedness of biological and cultural diversity—is vital for maintaining ecosystem services and preventing biopiracy (Bhattacharyaa and Adhikary, 2025). Cooperation is necessary to create a unified approach to these challenges, ensuring that all stakeholders work together to conserve and sustainably use medicinal plant resources (Shukla, 2023). By addressing these ethical foundations, we can foster a more equitable and sustainable future for both Indigenous communities and the global society that benefits from their knowledge.

Bioprospecting, the search for new species and compounds in nature for medicinal use, presents significant ethical challenges, particularly concerning the rights of Indigenous communities and the potential for biopiracy (Pushpangadan *et al.*, 2018). Central to these challenges is the exploitation of Indigenous knowledge systems (IKS), which encompass

Table 1.3. Integrated workflow for medicinal plant research: From experimental design to discovery. Author's own table.

Workflow phase	Key objectives	Critical methods & tools	Key outputs & deliverables	Major challenges & critical considerations	Citations
PHASE 1: EXPERIMENTAL DESIGN & PLANNING					
1.1 Hypothesis & objective definition	• Formulate a clear, testable hypothesis based on traditional use or prior knowledge. • Define primary and secondary research objectives.	• Literature mining (e.g. ethnobotanical databases, scientific literature). • Defining measurable endpoints (e.g. half maximal inhibitory concentration (IC50) for bioactivity, compound yield).	A detailed research protocol with defined aims, hypotheses, and success metrics.	• Avoiding "fishing expeditions" without a clear hypothesis. • Ensuring objectives are feasible with available resources and technologies.	(Sharma and Sarkar, 2013)
1.2 Sample preparation & standardization	• Ensure biological and technical reproducibility. • Minimize pre-analytical variability.	• Randomization & replication: random sampling of plant populations; multiple biological and technical replicates. • Standardized processing: controlled drying, grinding, and extraction protocols (e.g. using specified solvents, durations). • Quality assurance: botanical authentication (voucher specimens), contamination screening.	Standardized, authenticated, and reproducible plant extracts or tissues. A robust sample metadata tracker.	• Intrinsic variability of plant material due to genotype, environment, and harvest time. • Degradation of labile compounds during processing. • Ensuring consistent starting material quality.	(Park *et al.*, 2022; Kabir *et al.*, 2024)
1.3 Data management planning	• Ensure data integrity, traceability, and future reusability (findable, accessible, interoperable, reusable (FAIR) principles)	• Designing a data management plan (DMP). • Implementing laboratory information management systems (LIMS). • Defining metadata schemas (e.g. MIAME (minimum information about a microarray experiment) and MSI (metabolomics standards initiative) standards for omics).	A formal data management plan (DMP). A structured, searchable digital repository for all raw data and metadata.	• Handling multi-modal, high-volume data (omics, spectral). • Inconsistent metadata reporting. • Long-term data storage and curation costs.	(Sepanj *et al.*, 2024)

Continued

Table 1.3. Continued

Workflow phase	Key objectives	Critical methods & tools	Key outputs & deliverables	Major challenges & critical considerations	Citations
PHASE 2: DATA GENERATION & ACQUISITION					
2.1 Chemical profiling (metabolomics)	• Comprehensively characterize the phytochemical composition.	• Chromatography & spectroscopy: LC-MS (liquid chromatography-mass spectrometry), GC-MS (gas chromatography-mass spectrometry), NMR (nuclear magnetic resonance) for untargeted and targeted metabolomics. • Hyphenated techniques: LC-SPE-NMR (liquid chromatography-solid phase extraction-nuclear magnetic resonance) for compound isolation and identification.	Raw spectral data files (.raw, .d). A list of annotated/identified metabolites. Compound libraries.	• Compound annotation uncertainty (vs identification). • Dynamic range issues (high-abundance compounds masking rare ones). • Co-elution of compounds in complex mixtures.	(Foito and Stewart, 2018; Salam *et al.*, 2019)
2.2 Biological activity screening (bioassays)	• Identify extracts or fractions with desired therapeutic activity.	• High-throughput screening (HTS): robotic assays for rapid activity assessment. • Bioassay-guided fractionation: iterative activity testing of subdivided extracts to isolate active principle. • *In vitro*/*in vivo* models: cell-based assays, enzyme inhibition assays, animal models of disease.	Dose-response curves (e.g. IC50, EC50 (half-maximal effective concentration) values). "Hit" lists of active extracts/fractions.	• False positives/negatives from assay interference (e.g. compound auto-fluorescence, cytotoxicity). • Translating *in vitro* activity to *in vivo* efficacy. • Throughput vs physiological relevance trade-off.	(Vamathevan *et al.*, 2019; Mani *et al.*, 2022)

2.3 Molecular profiling (omics)	• Understand the genetic and functional basis of compound biosynthesis and activity.	• Genomics/transcriptomics: NGS (Illumina, PacBio) for genome sequencing and RNA-Seq. • Proteomics: LC-MS/MS for protein identification and quantification.	Genome assemblies, gene annotations. Differential gene/protein expression tables. Pathway maps.	• Lack of reference genomes for non-model plants. • Integrating data across different omics layers (data fusion). • High computational cost and expertise required.	(Demirbaga *et al.*, 2024)
PHASE 3: DATA ANALYSIS, INTEGRATION, & INTERPRETATION					
3.1 Data preprocessing & quality control	• Clean raw data to ensure high quality for downstream analysis.	• NGS data: FastQC, Trimmomatic for QC and adapter trimming. • MS Data: Peak picking, alignment, normalization (using XCMS, MZmine). • Statistical QC: Principal Component Analysis (PCA) to detect outliers and batch effects.	Quality control reports. Cleaned, normalized, and aligned data matrices ready for statistical analysis.	• Correcting for technical batch effects that can overshadow biological signals. • Choosing appropriate normalization methods for different data types.	(Zhai *et al.*, 2023)

Continued

Table 1.3. Continued

Workflow phase	Key objectives	Critical methods & tools	Key outputs & deliverables	Major challenges & critical considerations	Citations
3.2 Statistical analysis & pattern discovery	• Identify statistically significant features correlated with bioactivity or treatment.	• Univariate statistics: *t*-tests (student's *t*-test), ANOVA (analysis of variance). • Multivariate statistics: PLS-DA, OPLS-DA (orthogonal partial least squares discriminant analysis) for classification and biomarker discovery. • Machine learning: Random Forest, support vector machines (SVM) for predictive model building and feature selection.	Lists of significant biomarkers (genes, proteins, metabolites). Predictive models of bioactivity.	• Overfitting models with high-dimensional data ("curse of dimensionality"). • Distinguishing correlation from causation.	(Vamathevan *et al.*, 2019; Niazi and Mariam, 2023)
3.3 Advanced modeling & mechanistic insight	• Predict mechanisms of action, polypharmacology, and ADMET (absorption, distribution, metabolism, excretion, and toxicity) properties.	• Cheminformatics: molecular docking, pharmacophore modeling. • Network pharmacology: constructing compound–target pathway networks. • Computational pharmacology: physiologically based pharmacokinetic/pharmacodynamic (PBPK/PD) modeling to predict pharmacokinetics and pharmacodynamics.	Hypothesized mechanisms of action. Predicted drug–target interaction networks. ADMET risk assessments.	• Limited accuracy of *in silico* predictions requiring experimental validation. • Accounting for prodrug metabolism and synergistic effects in polyherbal formulations.	(Noor *et al.*, 2022; Naithani and Guleria, 2024)

3.4 Data visualization & knowledge translation	• Communicate complex findings effectively to diverse audiences.	• Interactive dashboards: R/Shiny, Python/Plotly. • Pathway mapping tools: KEGG, Cytoscape for network visualization. • Chemical imaging: MS imaging to visualize spatial distribution of compounds in plant tissue.	Publication-ready figures. Interactive web applications. Comprehensive pathway maps.	• Creating intuitive visualizations from highly complex, multi-layered data. • Ensuring visualizations are accurate and not misleading.	(Ashiwaju *et al.*, 2023)
FINAL OUTCOME	Discovery of novel therapeutic agents, validation of traditional uses, and a systems-level understanding of medicinal plant biology, supported by a reproducible and well-documented data pipeline.				

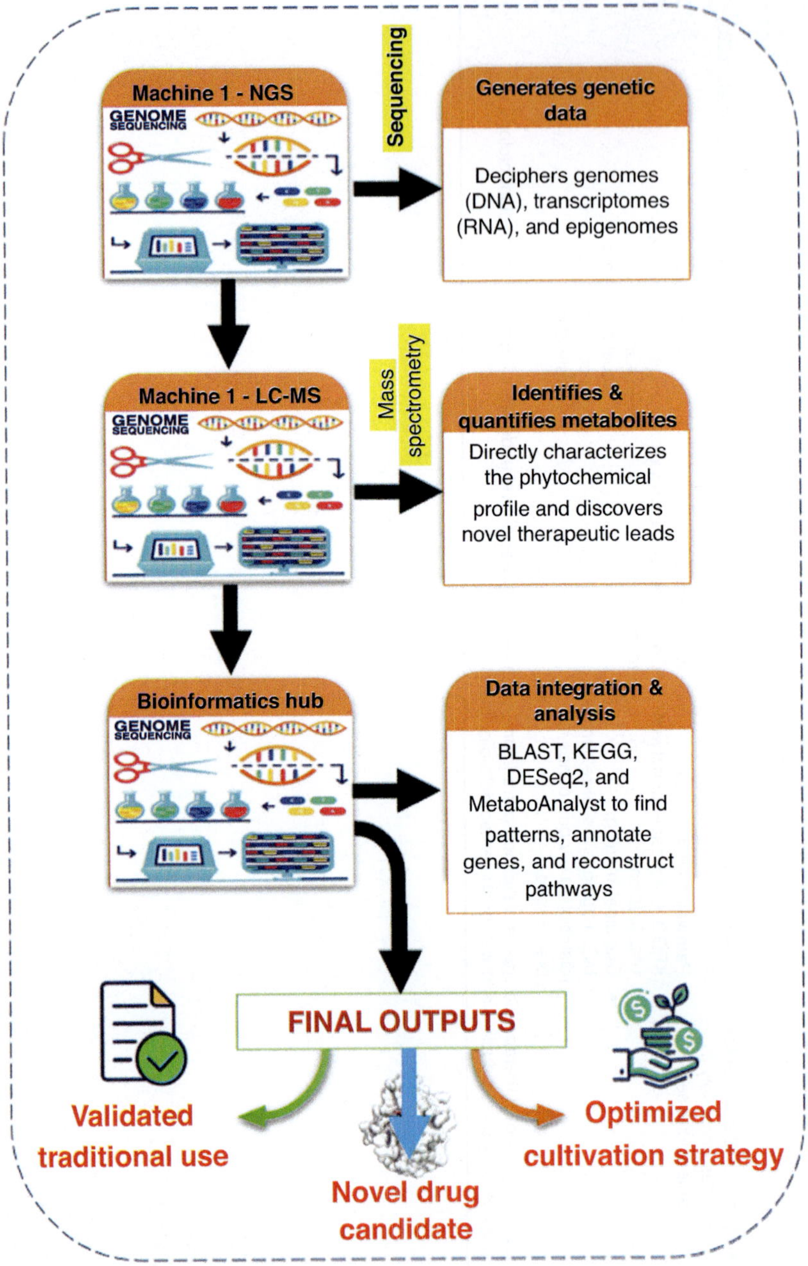

Fig. 1.3. Core omics toolbox from sample to insight. (Author's own figure.)

the traditional knowledge and practices of local communities regarding medicinal plants. Often, this knowledge is overlooked or misappropriated in bioprospecting efforts, raising concerns about the ethical implications of such actions (Pushpangadan *et al.*, 2018). ABS frameworks aim to ensure that benefits derived from genetic resources are shared fairly and equitably with the communities that provide them (Sirakaya, 2022). However, the implementation of ABS is fraught with difficulties, as it requires transparency and prior informed consent (PIC) from Indigenous peoples before accessing their traditional knowledge or genetic resources (Chege Kamau *et al.*, 2015). The lack of adherence to these principles can lead to biopiracy, where companies profit from resources without compensating the communities that have preserved and utilized these resources for generations (Passas, 2017). Intellectual property rights (IPR) can play a dual role in biopiracy. While they can protect the interests of Indigenous communities and prevent biopiracy, they can also create barriers to ABS, complicating the ethical landscape further (Hashimy and Benjamin, 2024). Additionally, the ethical challenges are compounded by the need for fair and equitable benefit-sharing, which is essential for addressing the injustices faced by Indigenous peoples in bioprospecting scenarios. The ethical foundations of bioprospecting and biopiracy in medicinal plants are deeply intertwined with issues of Indigenous rights, knowledge exploitation, and the complexities of legal frameworks designed to protect these interests (Table 1.4 and Fig. 1.4; Efferth *et al.*, 2016).

1.6 Case Study: Omics in Traditional Medicine Validation in Medicinal Plants

The application of omics technologies in validating traditional medicine practices is exemplified through the study of *Artemisia annua* and *Curcuma longa* (Table 1.5 and Fig. 1.5). Both plants have a rich history in traditional medicine, and omics approaches provide a scientific basis for their medicinal properties (Jauhari *et al.*, 2024). For *A. annua*, omics technologies, including genomics, proteomics, and metabolomics, are employed to uncover the molecular mechanisms and bioactive compounds that contribute to its therapeutic effects (Hussain *et al.*, 2023). This integration of omics can lead to the identification of novel drug targets and bioactive compounds, enhancing the development of evidence-based medicinal practices (Jauhari *et al.*, 2024). Similarly, *C. longa* benefits from proteomics and genomics to validate its anti-inflammatory and antioxidant properties (Chakraborty *et al.*, 2021). By studying the interactions of proteins and the genetic basis of its medicinal effects, researchers can confirm the efficacy and safety of *C. longa* in traditional medicine (Witkin and Li, 2013). The integration of these omics approaches not only aids in understanding the plants' bioactive compounds but also facilitates the discovery of novel biomarkers for quality control and efficacy assessment (Odriozola and Corrales, 2015). Moreover, the validation of traditional medicine through omics is crucial for establishing a scientific framework that supports the use of these plants in modern therapeutics. This process involves comprehensive data integration and analysis, which can reveal insights into the complex interactions of metabolites and proteins within these medicinal plants. Ultimately, the omics-based validation of *A. annua* and *C. longa* exemplifies how modern scientific techniques can enhance the credibility and application of traditional medicine (Umaretiya, 2024).

Omics technologies in the validation of traditional medicine, particularly concerning medicinal plants, represent a significant advancement in understanding their therapeutic potential (Jauhari *et al.*, 2024). Omics approaches, including metabolomics, genomics, and proteomics, facilitate the identification and validation of bioactive compounds within these plants, thereby elucidating their molecular mechanisms of action (Singh *et al.*, 2021). In traditional Chinese medicine (TCM), for instance, omics technologies have been employed to validate the efficacy of various medicinal plants, revealing the complex interactions between their bioactive compounds and biological targets (Zhao *et al.*, 2024). Similarly, Ayurvedic and Unani medicine also benefit from these technologies, which help in identifying key compounds that contribute to their therapeutic effects (Mishra, 2025). Phytochemical profiling is a crucial aspect of this validation process, as it standardizes the quality and efficacy of medicinal plants by quantifying their bioactive constituents (Kumar *et al.*, 2023). This profiling, combined with systems biology approaches, allows researchers to integrate diverse omics data, providing a comprehensive understanding of the interactions within biological systems (Pinu *et al.*, 2019). Furthermore, network pharmacology enhances this understanding

Table 1.4. Ethical foundations in medicinal plant research: Navigating bioprospecting and biopiracy. Author's own table.

Ethical concept/ mechanism	Core principle & definition	Key legal & policy frameworks	Positive role/intended function	Critical challenges & ethical gaps	Citations
Access and benefit-sharing (ABS)	The fair and equitable sharing of benefits arising from the utilization of genetic resources and associated traditional knowledge.	• Convention on Biological Diversity (CBD). • Nagoya Protocol (2010). • National ABS legislation (e.g. in provider countries like India, Costa Rica).	Aims to create a legal pathway for research while ensuring communities receive compensation (monetary and non-monetary) for their resources and knowledge. Fosters partnerships instead of exploitation.	• Power imbalance: large corporations often have greater legal and financial resources than Indigenous communities. • Complex negotiations: reaching mutually agreeable terms can be difficult and time-consuming. • Defining "fairness": what constitutes fair compensation? Royalties vs upfront payment vs infrastructure development.	(Nelliyat *et al.*, 2022; Sirakaya, 2022; Preetha *et al.*, 2024)
Prior informed consent (PIC)	The principle that Indigenous and local communities must be fully informed about, and give their authorization for, access to their genetic resources and knowledge.	• Nagoya Protocol (Article 6). • UN Declaration on the Rights of Indigenous Peoples (UNDRIP).	Respects the sovereignty and self-determination of communities. Ensures they have the right to say "no" and to set the terms of engagement.	• Defining the "Community": who legitimately represents the community to give consent? • Informed consent: ensuring understanding of complex scientific and commercial concepts is challenging. • Free consent: consent may not be truly "free" if communities face economic pressure.	(Hudson *et al.*, 2020)

Traditional knowledge (TK) & Indigenous knowledge systems (IKS)	The knowledge, innovations, and practices of Indigenous and local communities developed from experience over centuries and adapted to the local culture and environment.	• CBD (Article 8(j)). • WIPO (World Intellectual Property Organization) Intergovernmental Committee (IGC) on IP & Genetic Resources.	Serves as an invaluable "filter" or "guide" for bioprospecting, dramatically increasing the efficiency of drug discovery. Recognizes the intellectual contribution of communities.	Appropriation vs recognition: TK is often used without acknowledgment or is treated as "public domain" information. • Documentation Paradox: documenting TK in databases can make it more vulnerable to misappropriation if not properly protected. • Erosion: TK is being lost as elders pass away and youth migrate.	(Pushpangadan *et al.*, 2018; Balkrishna *et al.*, 2024; Mbelebele *et al.*, 2024)
Biopiracy	The unauthorized and uncompensated appropriation of biological resources and/or associated traditional knowledge, followed by the monopolization of benefits through intellectual property rights (IPR).	Largely defined *by the absence of* compliance with CBD/Nagoya principles (ABS & PIC).	(Concept is inherently negative.) Highlights the unethical and often illegal nature of "resource grabbing" and knowledge theft.	• IPR as a tool: patents are granted on "isolated" or "modified" compounds without acknowledging the source of the original idea or resource, effectively privatizing collective knowledge. • Legal loopholes: jurisdictional gaps; lack of enforcement in provider countries. • Retroactive justice: difficulty in challenging existing patents (e.g. on turmeric, neem).	(Efferth *et al.*, 2016; Handique and Dubey, 2022)

Continued

Table 1.4. Continued

Ethical concept/ mechanism	Core principle & definition	Key legal & policy frameworks	Positive role/intended function	Critical challenges & ethical gaps	Citations
The Nagoya Protocol	A supplementary international agreement to the CBD that provides a transparent legal framework for the implementation of ABS.	• Nagoya Protocol on Access to Genetic Resources and the Fair and Equitable Sharing of Benefits Arising from their Utilization.	• Creates legal certainty for both users and providers. • Mandates the establishment of National focal points and competent national authorities. • Aims to foster research that contributes to biodiversity conservation.	• Limited ratification: not all countries are parties, creating jurisdictional havens. • Administrative burden: can be complex and costly for academic researchers and small companies to navigate. • Checkpoint monitoring: difficulty in tracking the utilization of genetic resources along the R&D chain.	(Escobar-Pemberthy and Calle Saldarriaga, 2020; Nelliyat, 2022)
Intellectual property rights (IPR)	Legal rights granted to protect creations of the mind, such as inventions (patents), literary works (copyright), and symbols (trademarks).	• World Trade Organization TRIPS Agreement (Agreement on Trade-Related Aspects of Intellectual Property Rights). • National patent laws.	Can theoretically be used to *protect* community rights through trademarks (e.g."Penan Honey") or defensive publishing to prevent third-party patents.	• Systemic misalignment: Western IPR systems are designed for individual, novel, non-obvious inventions, not for collective, ancient, traditional knowledge. This leads to the patenting of "inventions" that are obvious extensions of TK. • Defensive costs: the cost of legally challenging a biopiracy patent is prohibitive for most communities.	(Efferth *et al.*, 2016; Hashimy and Benjamin, 2024)

Biocultural rights & conservation	The rights of communities to steward their territories, resources, and cultural practices, recognizing the inextricable link between cultural and biological diversity.	• UNDRIP. • CBD's ecosystem approach.	Positions communities as essential partners in conservation. Acknowledges that the preservation of medicinal plants is tied to the preservation of the cultures that know how to use them.	• Land dispossession: communities losing access to their traditional lands directly threatens both biodiversity and TK. • Benefit-sharing for conservation: ensuring that benefits from bioprospecting actually reach conservation efforts and local livelihoods.	(Pushpangadan *et al.*, 2018; Raju and Das, 2024; Bhattacharyaa and Adhikary, 2025)
Fair trade & ethical sourcing	A trading partnership based on dialogue, transparency, and respect that seeks greater equity in international trade, offering better trading conditions to marginalized producers.	• Fairtrade international certification. • Union for Ethical BioTrade (UEBT).	Ensures that harvesters of medicinal plants receive a fair price and work in safe conditions. Promotes sustainable harvesting practices directly at the source.	• Greenwashing: companies may engage in superficial ethical sourcing without meaningful ABS or PIC. • Certification costs: can be a barrier for small-scale community enterprises. • Supply chain complexity: difficult to trace ingredients from the final product back to the original harvesters.	(Morrison *et al.*, 2021; Raju and Das, 2024)

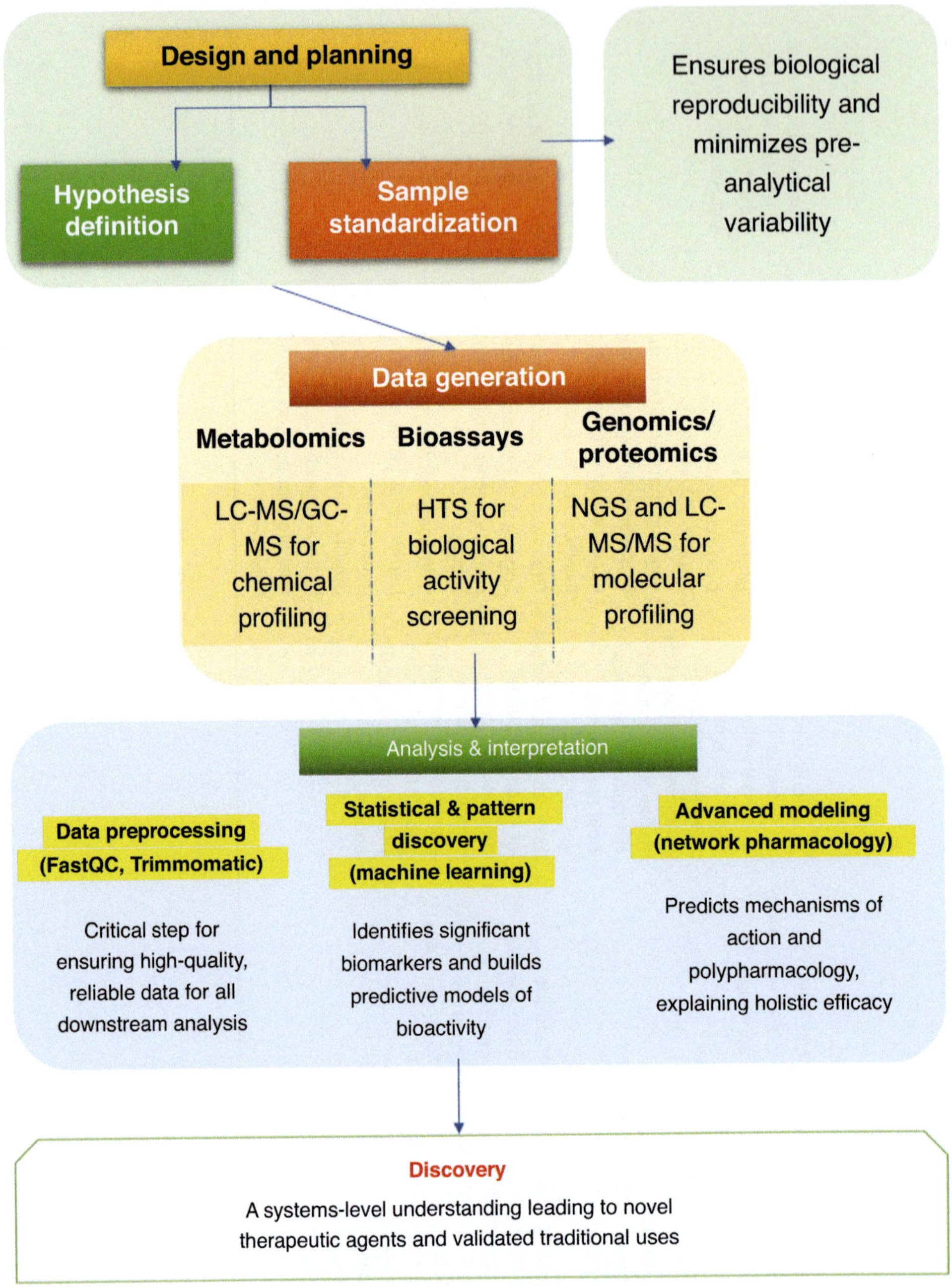

Fig. 1.4. The integrated research workflow pipeline. (Author's own figure.)

by mapping the interactions between bioactive compounds and their targets, thus revealing the underlying mechanisms of action (Noor *et al.*, 2022). The use of bioinformatics tools is essential in analyzing the vast amounts of data generated through these omics studies, enabling the identification of new therapeutic candidates and the development of predictive models for

Table 1.5. Omics-driven validation of traditional medicine: Case studies of *A. annua* and *C. longa*. Author's own table.

Omics technology	Application in *A. annua* (Qinghao)	Key findings & validation for *A. annua*	Application in *C. longa* (turmeric)	Key findings & validation for *C. longa*	Impact on validation & modern application	Citations
Genomics	Sequencing the genome to identify genes involved in the artemisinin biosynthetic pathway.	Identified key genes like *ADS* (amorpha-4,11-diene synthase) and *CYP71AV1* (cytochrome P450) that catalyze critical steps in artemisinin production. This validated the plant's genetic capacity to produce the potent antimalarial compound.	Sequencing the genome and transcriptome to map the genes responsible for curcuminoid biosynthesis.	Identified genes for enzymes like curcumin synthase (CURS) and diketide-CoA synthase (DCS), providing the genetic blueprint for the production of curcumin, demethoxycurcumin, and bisdemethoxycurcumin.	Validates the genetic potential for bioactive compound production. Enables marker-assisted breeding for high-yielding varieties and identifies targets for metabolic engineering.	(Zhao *et al.*, 2023; Jauhari *et al.*, 2024)
Transcriptomics	Analyzing gene expression changes in response to elicitors (e.g. jasmonic acid) or different developmental stages.	Revealed the coordinated upregulation of the entire artemisinin biosynthetic pathway under stress conditions, explaining why traditional harvesting methods (specific growth stages) might optimize potency.	Studying gene expression in rhizomes under different conditions to understand the regulation of curcuminoid biosynthesis.	Uncovered transcription factors that regulate curcuminoid pathways, providing insights into how to boost production and how the plant's environment affects its medicinal chemistry.	Links genetic potential to dynamic expression. Explains variability in compound levels and identifies optimal growth/ harvest conditions to standardize raw material quality.	(Singh *et al.*, 2021; Hussain *et al.*, 2023)
Metabolomics	Profiling the full spectrum of secondary metabolites (including artemisinin, flavonoids, other terpenes) using LC-MS/GC-MS.	Confirmed artemisinin as the primary antimalarial agent but also identified synergistic compounds (e.g. flavonoids) that may enhance efficacy and reduce parasite resistance, validating complex traditional formulations.	Comprehensive profiling of curcuminoids and other bioactive metabolites (e.g. turmerones) in different turmeric chemotypes.	Quantified the complex metabolome, confirming curcuminoids as primary anti-inflammatory agents but also revealing a spectrum of other compounds that contribute to its antioxidant, antimicrobial, and neuroprotective effects.	Provides a chemical fingerprint for authentication and quality control. Identifies synergistic compounds and adulterants, moving beyond a single-molecule view to a holistic phytocomplex profile.	(Kumar *et al.*, 2023; Cardona *et al.*, 2024)

Continued

Table 1.5. Continued

Omics technology	Application in *A. annua* (Qinghao)	Key findings & validation for *A. annua*	Application in *C. longa* (turmeric)	Key findings & validation for *C. longa*	Impact on validation & modern application	Citations
Proteomics	Identifying proteins expressed in response to pathogen attack or involved in the artemisinin pathway.	Elucidated defense-related proteins and the full suite of enzymes in the artemisinin pathway, including their subcellular localization, providing a complete picture of its biosynthesis.	Mapping the protein targets of curcumin in inflammatory pathways (e.g. NF-XB—nuclear factor kappa-light-chain-enhancer of activated B cells, COX-2—cyclooxygenase-2, TNF-α—tumour necrosis factor alpha) in human cells or animal models.	Directly identified how curcuminoids interact with the proteome, inhibiting key proteins in the inflammation cascade and validating the traditional use for treating inflammatory conditions.	Elucidates the mechanism of action (MoA). Shifts validation from "it works" to "how it works" by identifying specific protein targets and affected biological pathways.	(Witkin and Li, 2013; Meissner *et al.*, 2022)
Systems biology & network pharmacology	Integrating genomics, transcriptomics, and metabolomics data to model the entire artemisinin regulatory network.	Created a systems-level model that predicts how genetic or environmental perturbations affect artemisinin yield, guiding bioprocessing and cultivation strategies.	Constructing compound-target-pathway networks to visualize the multi-scale, multi-target action of turmeric's phytocomplex	Mapped curcumin's interactions on to human protein–protein interaction networks, revealing its pleiotropic effects on inflammation, cancer, and neurological pathways, validating its use in diverse traditional indications.	Validates the polypharmacology of traditional remedies. Moves beyond the "one drug, one target" paradigm to show how multiple compounds in a plant interact with multiple targets in a biological network, explaining holistic efficacy	(Pinu *et al.*, 2019; Noor *et al.*, 2022)
Bioinformatics & data integration	Using tools like KEGG for pathway annotation and bioinformatics pipelines for multi-omics data alignment.	Enabled the reconstruction of the complete artemisinin pathway in databases and the correlation of metabolite levels with gene expression data.	Applying bioinformatics for molecular docking (predicting binding of curcumin to targets) and enrichment analysis of omics data.	Provided computational evidence for efficacy, predicting interactions and biological processes affected by curcumin, which can be tested experimentally, accelerating discovery.	Makes sense of big data. Essential for hypothesis generation, identifying novel biomarkers for efficacy, and creating a reproducible, scientific framework for traditional medicine validation.	(Rai *et al.*, 2017; Clark and Lillard, 2024)

Overall impact on validation	Integrating multi-omics approaches.	Scientifically confirms the antimalarial use, reveals the complex phytochemistry, and enables optimized production.	Integrating multi-omics approaches.	Scientifically confirms the anti-inflammatory and antioxidant uses, provides a molecular basis for traditional claims, and enables product standardization.	Bridges traditional and modern medicine by providing molecular evidence for validation, safety, efficacy, and the development of standardized herbal products.	(Younas *et al.*, 2023; Jauhari *et al.*, 2024; Molla and Bitew, 2024; Umaretiya, 2024)

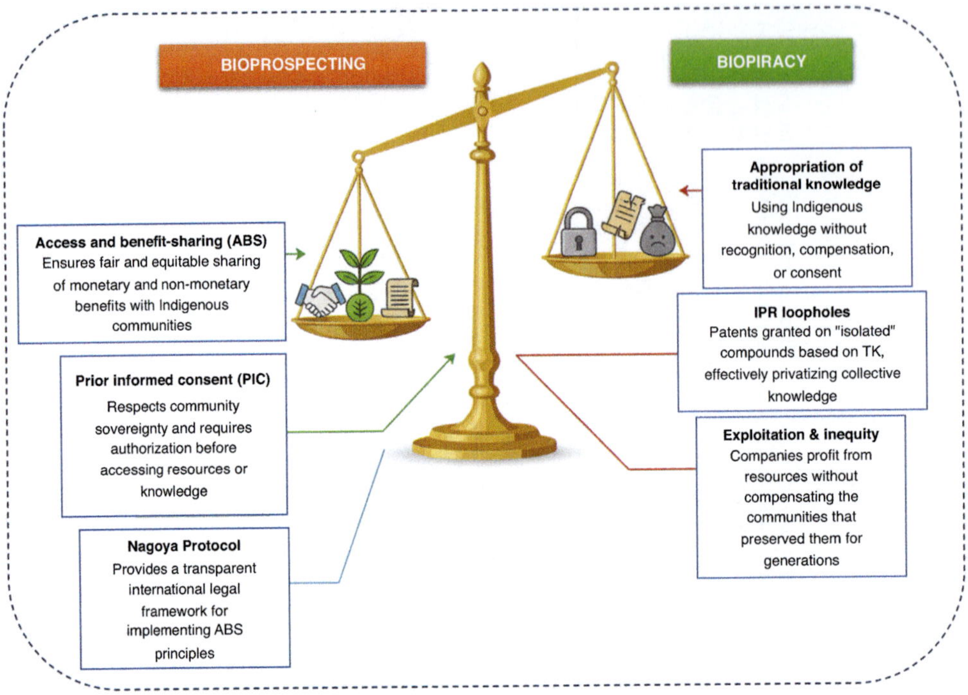

Fig. 1.5. Navigating the ethical landscape bioprospecting vs biopiracy. (Author's own figure.)

efficacy (Clark and Lillard, 2024). Overall, the application of omics technologies in traditional medicine validation not only supports the scientific basis of these practices but also paves the way for the discovery of novel therapeutic agents derived from medicinal plants.

Omics technologies in the study of traditional medicinal plants like *A. annua* and *C. longa* can significantly enhance their efficacy and safety. By employing metabolomics analysis, researchers can identify bioactive compounds responsible for the therapeutic effects of these plants, allowing for targeted extraction and optimization of these compounds to improve their medicinal properties (Cardona *et al.*, 2024). This approach also aids in predicting potential interactions and adverse effects, which is crucial for ensuring safety in medicinal applications. Genomic sequencing provides insights into the genetic basis of these plants' medicinal properties, enabling the identification of genes involved in the biosynthesis of bioactive compounds (Zhao *et al.*, 2023). This knowledge can lead to improved crop yields and disease resistance, ultimately enhancing the quality of the plants used in traditional medicine. Furthermore, proteomics profiling can elucidate the mechanisms of action of these compounds by identifying their protein targets, which can inform the optimization of formulations for better efficacy and safety (Meissner *et al.*, 2022). A systems biology approach integrates data from various omics levels, offering a comprehensive understanding of the biological processes underlying the medicinal properties of these plants (Rai *et al.*, 2017). This holistic view can help identify key regulatory nodes and predict the effects of perturbations, which is essential for developing effective therapeutic strategies. Additionally, omics-based methods can validate traditional medicinal uses by confirming the efficacy and safety of bioactive compounds, thereby bridging the gap between traditional and modern medicine (Younas *et al.*, 2023). Finally, the insights gained from omics technologies can facilitate personalized medicine applications, tailoring treatments based on individual genetic variations that affect responses to these bioactive compounds (Molla and Bitew, 2024). Overall, the integration of omics in the study of *A. annua* and *C. longa* holds great promise for enhancing their therapeutic potential while ensuring safety.

1.7 Conclusion

Omics encompasses a range of disciplines that study the complete set of biological molecules in organisms, including genomics, transcriptomics, proteomics, metabolomics, and epigenomics. Genomics focuses on the structure and function of genomes, providing insights into the genetic basis of bioactive compounds in medicinal plants. Transcriptomics examines RNA transcripts to understand gene expression and regulation, crucial for identifying genes involved in biosynthesis. Proteomics analyzes the complete set of proteins, revealing their functions and interactions, which is essential for understanding the regulation of bioactive compounds. Metabolomics studies metabolites, offering insights into metabolic pathways and networks, thereby aiding in the characterization of bioactive compounds. Finally, epigenomics investigates epigenetic modifications that influence gene expression, highlighting their role in the biosynthesis of bioactive compounds. Historically, the evolution of omics in medicinal plant biology has been marked by significant advancements, such as the first plant genome sequencing, which has had profound biomedical implications by facilitating the identification of therapeutic targets. Core technologies like NGS and LC-MS are pivotal in omics workflows, enabling rapid data generation and analysis of complex biological samples. Bioinformatics plays a critical role in managing and interpreting large-scale omics data, allowing for the identification of relevant genes and pathways. The workflow in omics research involves careful experimental design, data generation, and interpretation, ensuring robust and reliable results. Ethical considerations, such as bioprospecting and biopiracy, present challenges in the field, necessitating a responsible approach to the utilization of medicinal plants. The use of omics technologies in the study of medicinal plants, exemplified by case studies like *A. annua* and *C. longa*, underscores the potential of these approaches in validating traditional medicine and discovering novel therapeutic strategies. The ongoing advancements in omics will continue to enhance our understanding of plant biology and its applications in medicine.

Acknowledgments

The authors express their gratitude for the financial support received through the Distinguished High-Level Talents Research Grant from the Guizhou Science and Technology Corporation Platform Talents Fund (Grant nos.: [2017]5733-001 and CK-1130-002), high-level specialists project fund of Zunyi Medical University (1092/2000006)F-1092, the National Natural Science Foundation of China (82373981), and the support provided by Zunyi Medical University, China. Special appreciation is extended to all laboratory colleagues and research staff members for their valuable insights, constructive guidance, and assistance throughout this study.

References

Abdi, G., Tarighat, M.A., Jain, M., Tendulkar, R., Tendulkar, M. *et al.* (2024) Revolutionizing genomics: Exploring the potential of next-generation sequencing. In: Singh, V. and Kumar, A. (eds) *Advances in Bioinformatics*. Springer Nature, Singapore, pp. 1–33. DOI: 10.1007/978-981-99-8401-5_1.

Abdul, Q.A., Yu, B.P., Chung, H.Y., Jung, H.A. and Choi, J.S. (2017) Epigenetic modifications of gene expression by lifestyle and environment. *Archives of Pharmacal Research* 40(11), 1219–1237. DOI: 10.1007/s12272-017-0973-3.

Ashiwaju, B.I., Orikpete, O.F. and Uzougbo, C.G. (2023) The intersection of artificial intelligence and big data in drug discovery: A review of current trends and future implications. *Matrix Science Pharma* 7(2), 36–42. DOI: 10.4103/mtsp.mtsp_14_23.

Balkrishna, A., Saini, A., Yadav, P., Yadav, P., Kumar, B. *et al.* (2024) Traditional knowledge databases of medicinal plants. In: Nandave, M., Joshi, R. and Upadhyay, J. (eds) *Ethnopharmacology and OMICS Advances in Medicinal Plants*. Springer Nature, Singapore, pp. 75–92. DOI: 10.1007/978-981-97-2367-6_4.

Basit, F., Feng, X. and Bhat, J.A. (2024) Integration of genomics-assisted and speed breeding for enhancement of plant salt tolerance. In: Ganie, S.A. and Wani, S.H. (eds) *Genetics of Salt Tolerance in Plants*. CAB International, Wallingford, UK, pp. 119–135. DOI: 10.1079/9781800623033.0008.

Bhat, J.A., Ali, S., Salgotra, R.K., Mir, Z.A., Dutta, S. *et al.* (2016) Genomic selection in the *era* of next generation sequencing for complex traits in plant breeding. *Frontiers in Genetics* 7, 221. DOI: 10.3389/fgene.2016.00221.

Bhattacharyaa, P. and Adhikary, S. (2025) Sustainable utilization of biodiversity: Legal perspectives on human welfare and intellectual property. *UTTAR PRADESH JOURNAL OF ZOOLOGY* 46(9), 269–276. DOI: 10.56557/upjoz/2025/v46i94941.

Cardona, K.M., Rugeles-Silva, P.A., Tarazona-Pulido, L., Diaz-Ariza, L.A., Muñoz-Florez, J.E. *et al.* (2024) Exploring the potential of omics technologies in medicinal plant research: A review in Colombia. *Boletin Latinoamericano y Del Caribe de Plantas Medicinales y Aromaticas* 23(4), 460–486. DOI: 10.37360/blacpma.24.23.4.32.

Chakraborty, A., Mahajan, S., Jaiswal, S.K. and Sharma, V.K. (2021) Genome sequencing of turmeric provides evolutionary insights into its medicinal properties. *Communications Biology* 4(1), 1193. DOI: 10.1038/s42003-021-02720-y.

Chang, C., Bowman, J.L. and Meyerowitz, E.M. (2016) Field guide to plant model systems. *Cell* 167(2), 325–339. DOI: 10.1016/j.cell.2016.08.031.

Chege Kamau, E., Winter, G. and Stoll, P.-T. (eds) (2015) *Research and Development on Genetic Resources*. Routledge, Abingdon, UK. DOI: 10.4324/9781315717838.

Cheng, Q.-Q., Ouyang, Y., Tang, Z.-Y., Lao, C.-C., Zhang, Y.-Y. *et al.* (2021) Review on the development and applications of medicinal plant genomes. *Frontiers in Plant Science* 12, 791219. DOI: 10.3389/fpls.2021.791219.

Clark, A.J. and Lillard, J.W. (2024) A comprehensive review of bioinformatics tools for genomic biomarker discovery driving precision oncology. *Genes* 15(8), 1036. DOI: 10.3390/genes15081036.

de Sena Brandine, G. and Smith, A.D. (2021) Falco: High-speed fastQC emulation for quality control of sequencing data. *F1000Research* 8, 1874. DOI: 10.12688/f1000research.21142.2.

Demirbaga, Ü., Aujla, G.S., Jindal, A. and Kalyon, O. (2024) Big data analytics in bioinformatics. In: Srirama, S.N., Chun-Wei Lin, J., Bhatnagar, R., Agarwal, S. and Krishna Reddy, P. (eds) *Big Data Analytics*. Springer Nature Switzerland, Cham, Switzerland, pp. 265–284. DOI: 10.1007/978-3-031-55639-5_12.

Dobin, A. and Gingeras, T.R. (2016) Optimizing RNA-seq mapping with STAR. In: Carugo, O. and Eisenhaber, F. (eds) *Data Mining Techniques for the Life Sciences*, Vol. 1415. Springer New York, New York, pp. 245–262.

Efferth, T., Banerjee, M., Paul, N.W., Abdelfatah, S., Arend, J. *et al.* (2016) Biopiracy of natural products and good bioprospecting practice. *Phytomedicine* 23(2), 166–173. DOI: 10.1016/j.phymed.2015.12.006.

Eichten, S.R., Schmitz, R.J. and Springer, N.M. (2014) Epigenetics: Beyond chromatin modifications and complex genetic regulation. *Plant Physiology* 165(3), 933–947. DOI: 10.1104/pp.113.234211.

Escobar-Pemberthy, N. and Calle Saldarriaga, M.A. (2020) International business, trade and the Nagoya Protocol: Best practices and challenges for sustainability in access and benefit-sharing. In: Leal Filho, W., de Brito, P.R. and Frankenberger, F. (eds) *International Business, Trade and Institutional Sustainability*. Springer International Publishing, Cham, Switzerland, pp. 813–830. DOI: 10.1007/978-3-030-26759-9_48.

Esvelt, K.M. and Wang, H.H. (2013) Genome-scale engineering for systems and synthetic biology. *Molecular Systems Biology* 9(1), 641. DOI: 10.1038/msb.2012.66.

Fang, Y. and Xiong, L. (2015) General mechanisms of drought response and their application in drought resistance improvement in plants. *Cellular and Molecular Life Sciences* 72(4), 673–689. DOI: 10.1007/s00018-014-1767-0.

Foito, A. and Stewart, D. (2018) Metabolomics: A high-throughput screen for biochemical and bioactivity diversity in plants and crops. *Current Pharmaceutical Design* 24(19), 2043–2054. DOI: 10.2174/1381612824666180515125926.

Gu, L., Zhang, Z.-Y., Quan, H., Li, M.-J., Zhao, F.-Y. *et al.* (2018) Integrated analysis of transcriptomic and metabolomic data reveals critical metabolic pathways involved in rotenoid biosynthesis in the medicinal plant *Mirabilis himalaica*. *Molecular Genetics and Genomics* 293(3), 635–647. DOI: 10.1007/s00438-017-1409-y.

Handique, P. and Dubey, A. (2022) Legal mechanisms for protecting indigenous knowledge. *Interdisciplinary Studies in Society, Law, and Politics* 1(1), 4–13. DOI: 10.61838/kman.isslp.1.1.2.

Hao, D., Chen, S., Xiao, P. and Liu, M. (2012) Application of high-throughput sequencing in medicinal plant transcriptome studies. *Drug Development Research* 73(8), 487–498. DOI: 10.1002/ddr.21041.

Hashimy, S.Q. and Benjamin, Dr.M. (2024) Intellectual property and biotechnology: A dual driver of agricultural transformation. *SSRN Electronic Journal*. DOI: 10.2139/ssrn.5015935.

Hilton, I.B. and Gersbach, C.A. (2015) Enabling functional genomics with genome engineering. *Genome Research* 25(10), 1442–1455. DOI: 10.1101/gr.190124.115.

Hoover, C.A., Slattery, M. and Marsh, A.G. (2007) A functional approach to transcriptome profiling: Linking gene expression patterns to metabolites that matter. *Marine Biotechnology* 9(4), 411–419. DOI: 10.1007/s10126-007-9008-2.

Hudson, M., Garrison, N.A., Sterling, R., Caron, N.R., Fox, K. *et al.* (2020) Rights, interests and expectations: Indigenous perspectives on unrestricted access to genomic data. *Nature Reviews. Genetics* 21(6), 377–384. DOI: 10.1038/s41576-020-0228-x.

Hussain, S., Sagar, T., Kaur, S., Nipunta, Kapoor, N. *et al.* (2023) Biosynthesis of bioactive compounds in medicinal and aromatic plants. In: *Biosynthesis of Bioactive Compounds in Medicinal and Aromatic Plants: Manipulation by Conventional and Biotechnological Approaches*. Springer Nature Switzerland, Cham, pp. 41–64. DOI: 10.1007/978-3-031-35221-8_2.

Jabeen, F., Khursheed, N., Ahmad, R., Rasool, S., Ali, A. *et al.* (2024) Crosstalk of multiomics approaches with medicinal plants of therapeutic importance. In: Ali, A., Rehman, M.U., Ahmad, S.B. and Arafah, A. (eds) *Biological Insights of Multi-Omics Technologies in Human Diseases*. Academic Press, Cambridge, Massachusetts, pp. 21–51. DOI: 10.1016/B978-0-443-23971-7.00002-X.

Jauhari, S., Jauhari, R., Rupanagunta, G.P., Nandave, M., Upadhyay, J. *et al.* (2024) Insights on the integration of ethnopharmacology and omics in medicinal plant research. In: Nandave, M., Joshi, R. and Upadhyay, J. (eds) *Ethnopharmacology and OMICS Advances in Medicinal Plants*. Springer Nature Singapore, Singapore, pp. 501–514. DOI: 10.1007/978-981-97-4292-9_22.

Kabir, M., Rana, M.R.H. and Debnath, A. (2024) The role of quality assurance in accelerating pharmaceutical research and development: Strategies for ensuring regulatory compliance and product integrity. *Journal of Angiotherapy* 8(12), 1–11. DOI: 10.25163/angiotherapy.81210102.

Kumar, R., Das, S.P., Choudhury, B.U., Kumar, A., Prakash, N.R. *et al.* (2024) Advances in genomic tools for plant breeding: Harnessing DNA molecular markers, genomic selection, and genome editing. *Biological Research* 57(1), 80. DOI: 10.1186/s40659-024-00562-6.

Kumar, S., Jaiswal, A.K., Aggarwal, M., Ekbbal, R. and Gaurav, G.G. (2023) A review on quality control aspects of Indian medicinal plants. *Pharmacognosy Reviews* 17(34), 276–286. DOI: 10.5530/phrev.2023.17.7.

Lankadurai, B.P., Nagato, E.G. and Simpson, M.J. (2013) Environmental metabolomics: An emerging approach to study organism responses to environmental stressors. *Environmental Reviews* 21(3), 180–205. DOI: 10.1139/er-2013-0011.

Lenroot, R.K. and Giedd, J.N. (2011) Annual research review: Developmental considerations of gene by environment interactions. *Journal of Child Psychology and Psychiatry, and Allied Disciplines* 52(4), 429–441. DOI: 10.1111/j.1469-7610.2011.02381.x.

Ma, X., Meng, Y., Wang, P., Tang, Z., Wang, H. *et al.* (2020) Bioinformatics-assisted, integrated omics studies on medicinal plants. *Briefings in Bioinformatics* 21(6), 1857–1874. DOI: 10.1093/bib/bbz132.

Maekawa, M. and Mano, N. (2022) Cutting-edge LC-MS/MS applications in clinical mass spectrometry: Focusing on analysis of drugs and metabolites. *Biomedical Chromatography* 36(5), e5347. DOI: 10.1002/bmc.5347.

Mani, J., Johnson, J., Hosking, H., Hoyos, B.E., Walsh, K.B. *et al.* (2022) Bioassay guided fractionation protocol for determining novel active compounds in selected Australian flora. *Plants (Basel, Switzerland)* 11(21), 2886. DOI: 10.3390/plants11212886.

Mbelebele, Z., Mdoda, L., Ntlanga, S.S., Nontu, Y. and Gidi, L.S. (2024) Harmonizing traditional knowledge with environmental preservation: Sustainable strategies for the conservation of indigenous medicinal plants (IMPs) and their implications for economic well-being. *Sustainability* 16(14), 5841. DOI: 10.3390/su16145841.

McDermaid, A., Monier, B., Zhao, J., Liu, B. and Ma, Q. (2019) Interpretation of differential gene expression results of RNA-seq data: Review and integration. *Briefings in Bioinformatics* 20(6), 2044–2054. DOI: 10.1093/bib/bby067.

Meissner, F., Geddes-McAlister, J., Mann, M. and Bantscheff, M. (2022) The emerging role of mass spectrometry-based proteomics in drug discovery. *Nature Reviews: Drug Discovery* 21(9), 637–654. DOI: 10.1038/s41573-022-00409-3.

Mishra, R. (2025) Unani therapy: Integrating yoga, ayurveda, and traditional biotechnology for holistic health. In: Rani, V., Saxena, B. and Atale, N. (eds) *Advancing Biotechnology: From Science to Therapeutics and Informatics: Technological Advancements in Biosciences and Biotechnology*. Springer Nature Switzerland, Cham, Switzerland, pp. 339–342. DOI: 10.1007/978-3-031-80973-6_28.

Miteva, Y.V., Budayeva, H.G. and Cristea, I.M. (2013) Proteomics-based methods for discovery, quantification, and validation of protein–protein interactions. *Analytical Chemistry* 85(2), 749–768. DOI: 10.1021/ac3033257.

Molla, G. and Bitew, M. (2024) Revolutionizing personalized medicine: Synergy with multi-omics data generation, main hurdles, and future perspectives. *Biomedicines* 12(12), 2750. DOI: 10.3390/biomedicines12122750.

Morrison, C., Humphries, F. and Lawson, C. (2021) A regional review of genetic resource access and benefit sharing – Key issues and research gaps. *Environmental Policy and Law* 51(5), 273–296. DOI: 10.3233/EPL-201028.

Mortimer, J.C. (2019) Plant synthetic biology could drive a revolution in biofuels and medicine. *Experimental Biology and Medicine* 244(4), 323–331. DOI: 10.1177/1535370218793890.

Mrozek, D., Stępień, K., Grzesik, P. and Małysiak-Mrozek, B. (2021) A large-scale and serverless computational approach for improving quality of NGS data supporting big multi-omics data analyses. *Frontiers in Genetics* 12, 699280. DOI: 10.3389/fgene.2021.699280.

Murmu, S., Das, R., Pandey, B., Sharma, S. and Farooqi, M.S. (2024) Exploring secondary metabolites in plants through bioinformatics. In: Chen, J.-T. (ed.) *Bioinformatics for Plant Research and Crop Breeding*. Wiley, Oxford, UK, pp. 435–458. DOI: 10.1002/9781394209965.

Naithani, U. and Guleria, V. (2024) Integrative computational approaches for discovery and evaluation of lead compound for drug design. *Frontiers in Drug Discovery* 4, 1362456. DOI: 10.3389/fddsv.2024.1362456.

Nelliyat, P. (2022) Implementation of the Nagoya Protocol in India: Experiences, challenges and learnings. In: Oommen, O.V., Laladhas, K.P., Nelliyat, P. and Pisupati, B. (eds) *Biodiversity Conservation Through Access and Benefit Sharing (ABS)*. Springer International Publishing, Cham, Switzerland, pp. 17–36.

Nelliyat, P., Pisupati, B., Laladhas, K.P. and Oommen, O.V. (2022) Convention on biological diversity, access and benefit sharing: A glance. In: Oommen, O.V., Laladhas, K.P., Nelliyat, P. and Pisupati, B. (eds) *Biodiversity Conservation Through Access and Benefit Sharing (ABS)*. Springer International Publishing, Cham, Switzerland, pp. 3–16.

Niazi, S.K. and Mariam, Z. (2023) Recent advances in machine-learning-based chemoinformatics: A comprehensive review. *International Journal of Molecular Sciences* 24(14), 11488. DOI: 10.3390/ijms241411488.

Noor, F., Tahir ul Qamar, M., Ashfaq, U.A., Albutti, A., Alwashmi, A.S.S. *et al*. (2022) Network pharmacology approach for medicinal plants: Review and assessment. *Pharmaceuticals* 15(5), 572. DOI: 10.3390/ph15050572.

Odriozola, L. and Corrales, F.J. (2015) Discovery of nutritional biomarkers: Future directions based on omics technologies. *International Journal of Food Sciences and Nutrition* 66 Suppl 1, S31–40. DOI: 10.3109/09637486.2015.1038224.

Pandita, D., Pandita, A., Wani, S.H., Abdelmohsen, S.A.M., Alyousef, H.A. *et al*. (2021) Crosstalk of multi-omics platforms with plants of therapeutic importance. *Cells* 10(6), 1296. DOI: 10.3390/cells10061296.

Park, G., Kim, M.K., Go, S.H., Choi, M. and Jang, Y.P. (2022) Analytical quality by design (AQbD) approach to the development of analytical procedures for medicinal plants. *Plants (Basel, Switzerland)* 11(21), 2960. DOI: 10.3390/plants11212960.

Passas, N. (ed.) (2017) *International Crimes*. Routledge, Oxford, UK. DOI: 10.4324/9781315092591.

Pattnaik, S. (2023) Next-generation sequencing (NGS) for metabolomics study in medicinal plants under stress condition. In: Singh, D., Kumar Mishra, A. and Kumar Srivastava, A. (eds) *Stress-Responsive*

Factors and Molecular Farming in Medicinal Plants. Springer Nature Singapore, Singapore, pp. 323–343.

Pinu, F.R., Beale, D.J., Paten, A.M., Kouremenos, K., Swarup, S. *et al.* (2019) Systems biology and multi-omics integration: Viewpoints from the metabolomics research community. *Metabolites* 9(4), 76. DOI: 10.3390/metabo9040076.

Preetha, N., Balakrishnan, V. and Thomas, G. (2024) Bioprospecting from medicinal plants and benefit sharing. In: Krishna Panicker, L., Nelliyat, P. and Oommen, O.V. (eds) *Biodiversity and Business*. Springer Nature Switzerland, Cham, Switzerland, pp. 247–267.

Provart, N.J., Alonso, J., Assmann, S.M., Bergmann, D., Brady, S.M. *et al.* (2016) 50 years of arabidopsis research: Highlights and future directions. *The New Phytologist* 209(3), 921–944. DOI: 10.1111/nph.13687.

Pushpangadan, P., George, V., Parambil Ijinu, T. and Ambika Chithra, M. (2018) Biodiversity, bioprospecting, traditional knowledge, sustainable development and value added products: A review. *Journal of Traditional Medicine & Clinical Naturopathy* 07(01). DOI: 10.4172/2573-4555.1000256.

Raghavan, V., Kraft, L., Mesny, F. and Rigerte, L. (2022) A simple guide to de novo transcriptome assembly and annotation. *Briefings in Bioinformatics* 23(2), bbab563. DOI: 10.1093/bib/bbab563.

Rai, A., Saito, K. and Yamazaki, M. (2017) Integrated omics analysis of specialized metabolism in medicinal plants. *The Plant Journal* 90(4), 764–787. DOI: 10.1111/tpj.13485.

Rajawat, J. (2018) Transcriptomics. In: Arivaradarajan, P. and Misra, G. (eds) *Omics Approaches, Technologies and Applications*. Springer Singapore, Singapore, pp. 39–56. DOI: 10.1007/978-981-13-2925-8_3.

Raju, S. and Das, M. (2024) Medicinal plants industry in India: Challenges, opportunities and sustainability. *Medicinal Plants - International Journal of Phytomedicines and Related Industries* 16(1), 1–14. DOI: 10.5958/0975-6892.2024.00001.7.

Razzaq, A., Saleem, F., Kanwal, M., Mustafa, G., Yousaf, S. *et al.* (2019) Modern trends in plant genome editing: An inclusive review of the CRISPR/Cas9 toolbox. *International Journal of Molecular Sciences* 20(16), 4045. DOI: 10.3390/ijms20164045.

Salam, A.M., Lyles, J.T. and Quave, C.L. (2019) Methods in the extraction and chemical analysis of medicinal plants. In: Albuquerque, U.P., Farias Paiva de Lucena, R., Fernandes Cruz da Cunha, L.V. and Nóbrega Alves, R.R. (eds) *Methods and Techniques in Ethnobiology and Ethnoecology*. Springer, New York, New York, pp. 257–283. DOI: 10.1007/978-1-4939-8919-5_17.

Salem, M.A., Perez de Souza, L., Serag, A., Fernie, A.R., Farag, M.A. *et al.* (2020) Metabolomics in the context of plant natural products research: From sample preparation to metabolite analysis. *Metabolites* 10(1), 37. DOI: 10.3390/metabo10010037.

Sanches, P.H.G., de Melo, N.C., Porcari, A.M. and de Carvalho, L.M. (2024) Integrating molecular perspectives: Strategies for comprehensive multi-omics integrative data analysis and machine learning applications in transcriptomics, proteomics, and metabolomics. *Biology* 13(11), 848. DOI: 10.3390/biology13110848.

Satam, H., Joshi, K., Mangrolia, U., Waghoo, S., Zaidi, G. *et al.* (2023) Next-generation sequencing technology: Current trends and advancements. *Biology* 12(7), 997. DOI: 10.3390/biology12070997.

Satrio, R.D., Fendiyanto, M.H. and Miftahudin, M. (2024) Tools and techniques used at global scale through genomics, transcriptomics, proteomics, and metabolomics to investigate plant stress responses at the molecular level. In: Shahid, M. and Gaur, R. (eds) *Molecular Dynamics of Plant Stress and Its Management*. Springer Nature Singapore, Singapore, pp. 555–607.

Sepanj, M.H., Moradi, S., Nazemi, A., Preston, C., Lee, A.M.D. *et al.* (2024) From single shot to structure: End-to-end network-based deflectometry for specular free-form surface reconstruction. *Applied Sciences* 14(23), 10824. DOI: 10.3390/app142310824.

Sepulveda, J.L. (2020) Using R and bioconductor in clinical genomics and transcriptomics. *The Journal of Molecular Diagnostics* 22(1), 3–20. DOI: 10.1016/j.jmoldx.2019.08.006.

Sharma, V. and Sarkar, I.N. (2013) Bioinformatics opportunities for identification and study of medicinal plants. *Briefings in Bioinformatics* 14(2), 238–250. DOI: 10.1093/bib/bbs021.

Shukla, S.K. (2023) Conservation of medicinal plants: Challenges and opportunities. *Journal of Medicinal Botany* 7, 5–10. DOI: 10.25081/jmb.2023.v7.8437.

Singer, S.D., Laurie, J.D., Bilichak, A., Kumar, S. and Singh, J. (2021) Genetic variation and unintended risk in the context of old and new breeding techniques. *Critical Reviews in Plant Sciences* 40(1), 68–108. DOI: 10.1080/07352689.2021.1883826.

Singh, R., Singh, P.K., Kumar, R., Kabir, M.T., Kamal, M.A. *et al.* (2021) Multi-omics approach in the identification of potential therapeutic biomolecule for COVID-19. *Frontiers in Pharmacology* 12, 652335. DOI: 10.3389/fphar.2021.652335.

Singh, S., Praveen, A., Dudha, N., Sharma, V.K. and Bhadrecha, P. (2024) Single-cell transcriptomics: A new frontier in plant biotechnology research. *Plant Cell Reports* 43(12), 294. DOI: 10.1007/s00299-024-03383-9.

Sirakaya, A. (2022) Where access and benefit-sharing comes from: A historical overview. *Genetic Resources* 3(6), 74–88. DOI: 10.46265/genresj.PPUF5169.

Subramanian, I., Verma, S., Kumar, S., Jere, A. and Anamika, K. (2020) Multi-omics data integration, interpretation, and its application. *Bioinformatics and Biology Insights* 14, 1177932219899051. DOI: 10.1177/1177932219899051.

Tian, H., Lam, S.M. and Shui, G. (2016) Metabolomics, A powerful tool for agricultural research. *International Journal of Molecular Sciences* 17(11), 1871. DOI: 10.3390/ijms17111871.

Tolani, P., Gupta, S., Yadav, K., Aggarwal, S. and Yadav, A.K. (2021) Big data, integrative omics and network biology. *Advances in Protein Chemistry and Structural Biology* 127, 127–160. DOI: 10.1016/bs.apcsb.2021.03.006.

Umaretiya, V.R. (2024) Network-based drug discovery from ethnomedicinal plants. In: Kumar Srivastava, A., Kumar Ahirwar, R., Yadav, D. and Kumar, D.G. (eds) *Ethnomedicinal Plants for Drug Discovery*. Springer Nature Singapore, Singapore, pp. 511–522. DOI: 10.1007/978-981-97-3405-4_23.

Vamathevan, J., Clark, D., Czodrowski, P., Dunham, I., Ferran, E. *et al.* (2019) Applications of machine learning in drug discovery and development. *Nature Reviews. Drug Discovery* 18(6), 463–477. DOI: 10.1038/s41573-019-0024-5.

Wang, M., Zhang, S., Li, R. and Zhao, Q. (2024) Unraveling the specialized metabolic pathways in medicinal plant genomes: A review. *Frontiers in Plant Science* 15. DOI: 10.3389/fpls.2024.1459533.

Weissenbach, J. (2016) The rise of genomics. *Comptes Rendus Biologies* 339(7–8), 231–239. DOI: 10.1016/j.crvi.2016.05.002.

Wille, L., Messmer, M.M., Studer, B. and Hohmann, P. (2019) Insights to plant–microbe interactions provide opportunities to improve resistance breeding against root diseases in grain legumes. *Plant, Cell & Environment* 42(1), 20–40. DOI: 10.1111/pce.13214.

Witkin, J. and Li, X. (2013) Curcumin, an active constituent of the ancient medicinal herb *Curcuma longa* L.: Some uses and the establishment and biological basis of medical efficacy. *CNS & Neurological Disorders - Drug Targets* 12(4), 487–497. DOI: 10.2174/1871527311312040007.

Yadav, P., Kumar Oraon, P., Malik, G., Pratap Singh, S. and Gupta, A. (2025) Emerging research in nanotechnology, metagenomics, and other omics-based technologies for rhizosphere management and increased plant growth and productivity. In: Mathur, P. and Mukherjee, S. (eds) *Rhizosphere Engineering and Stress Resilience in Plants*. CRC Press, Boca Raton, pp. 175–199. DOI: 10.1201/9781003354109-10.

Yaschenko, A.E., Alonso, J.M. and Stepanova, A.N. (2025) Arabidopsis as a model for translational research. *The Plant Cell* 37(5). DOI: 10.1093/plcell/koae065.

Yilmaz, M.A. (2020) Simultaneous quantitative screening of 53 phytochemicals in 33 species of medicinal and aromatic plants: A detailed, robust and comprehensive LC–MS/MS method validation. *Industrial Crops and Products* 149, 112347. DOI: 10.1016/j.indcrop.2020.112347.

Younas, A., Riaz, N. and Rashid, M. (2023) Multi-omics approaches for breeding in medicinal plants. In: Prakash, C.S., Fiaz, S., Nadeem, M.A., Baloch, F.S. and Qayyum, A. (eds) *Sustainable Agriculture in the Era of the OMICs Revolution*. Springer International Publishing, Cham, Switzerland, pp. 165–191. DOI: 10.1007/978-3-031-15568-0_8.

Younessi-Hamzekhanlu, M., Ozturk, M., Jafarpour, P. and Mahna, N. (2022) Exploitation of next generation sequencing technologies for unraveling metabolic pathways in medicinal plants: A concise review. *Industrial Crops and Products* 178, 114669. DOI: 10.1016/j.indcrop.2022.114669.

Zhai, K., Yousef, M.S., Mohammed, S., Al-Dewik, N.I. and Qoronfleh, M.W. (2023) Optimizing clinical workflow using precision medicine and advanced data analytics. *Processes* 11(3), 939. DOI: 10.3390/pr11030939.

Zhang, M.M., Qiao, Y., Ang, E.L. and Zhao, H. (2017) Using natural products for drug discovery: The impact of the genomics era. *Expert Opinion on Drug Discovery* 12(5), 475–487. DOI: 10.1080/17460441.2017.1303478.

Zhang, W., Zeng, Y., Jiao, M., Ye, C., Li, Y. *et al*. (2023) Integration of high-throughput omics technologies in medicinal plant research: The new era of natural drug discovery. *Frontiers in Plant Science* 14. DOI: 10.3389/fpls.2023.1073848.

Zhang, X., Ibrahim, Z., Khaskheli, M.B., Raza, H., Zhou, F. *et al*. (2024) Integrative approaches to abiotic stress management in crops: Combining bioinformatics educational tools and artificial intelligence applications. *Sustainability* 16(17), 7651. DOI: 10.3390/su16177651.

Zhao, C., Zhang, Z., Sun, L., Bai, R., Wang, L. *et al*. (2023) Genome sequencing provides potential strategies for drug discovery and synthesis. *Acupuncture and Herbal Medicine* 3(4), 244–255. DOI: 10.1097/HM9.0000000000000076.

Zhao, M., Che, Y., Gao, Y. and Zhang, X. (2024) Application of multi-omics in the study of traditional Chinese medicine. *Frontiers in Pharmacology* 15. DOI: 10.3389/fphar.2024.1431862.

2 Multi-omics Integration and Data-driven Insights

Archana Jain, Surendra Sarsaiya and Qihai Gong*
Key Laboratory of Basic Pharmacology and Joint International Research Laboratory of Ethnomedicine of Ministry of Education, Zunyi Medical University, Zunyi, Guizhou, China

2.1 Introduction

Medicinal plants have long been a cornerstone of traditional and modern pharmacopeias, yielding bioactive compounds with significant therapeutic potential, such as artemisinin from *Artemisia annua* for malaria treatment and paclitaxel from *Taxus* species for cancer therapy (Kam and Yap, 2020). However, the biosynthesis of these compounds is governed by intricate molecular networks influenced by genetic, environmental, and developmental factors, posing challenges for optimization and scalable production. Traditional single-omics approaches, while valuable, often fail to capture the dynamic interplay of molecular layers regulating these pathways. The advent of multi-omics integration—synergizing genomics, transcriptomics, proteomics, and metabolomics—has revolutionized medicinal plant research by enabling a systems-level understanding of biosynthetic processes (Satrio *et al.*, 2024). This chapter explores how data-driven multi-omics frameworks are transforming the field. Cross-omics alignment, for instance, synchronizes temporal gene expression with metabolite accumulation in *A. annua*, revealing key regulatory nodes like the ADS and CYP71AV1 genes that drive artemisinin biosynthesis (Lopes *et al.*, 2020). Similarly, normalization frameworks harmonize heterogeneous genomic and metabolomic datasets in *S. miltiorrhiza*, elucidating terpenoid pathways and identifying novel biosynthesis-related genes (Xia *et al.*, 2023a). Validation workflows using CRISPR-edited (clustered regularly interspaced short palindromic repeats) *Taxus* cell lines further demonstrate the power of precision genetic tools, achieving a 25-fold increase in paclitaxel yield by targeting the PAL gene (Brzycki Newton *et al.*, 2023; Das *et al.*, 2024). Systems biology bridges molecular layers, reconstructing gene-metabolite networks in *Hypericum perforatum* to map hypericin biosynthesis and modeling jasmonate-induced alkaloid dynamics in *Catharanthus roseus* (Jain *et al.*, 2024). Network pharmacology extends these insights, linking *Curcuma longa* curcuminoids to multi-target interactions and employing machine learning (ML) to predict anti-inflammatory targets for *Withania somnifera* withanolides (Öztürk *et al.*, 2023).

Despite these advancements, challenges persist in managing the volume, standardization, and reproducibility of omics data. Frameworks like ISA-Tab (investigation/study/assay tab-delimited) can be used for the harmonization of *Ocimum sanctum* data and for FAIR-compliant (findable, accessible, interoperable, and reusable)

*Corresponding author: gqh@zmu.edu.cn

DOI: 10.1079/9781836994206.0002

sharing of *Moringa oleifera* drought-stress datasets (Bhowmick *et al.*, 2024). Artificial intelligence (AI) and ML tools, such as Random Forest models predicting vincristine yield in *C. roseus* (Wang *et al.*, 2024b) and t-SNE-driven (t-distributed stochastic neighbor embedding) chemotype classification in *Cannabis sativa*, transform raw data into actionable insights (Vergara *et al.*, 2020). A case study on *Panax ginseng* exemplifies the translational potential of multi-omics. By integrating genomics, proteomics, and metabolomics, researchers linked CYP450 gene clusters to ginsenoside diversity and identified drought-responsive saponins through transcript-metabolite networks (Huang *et al.*, 2024). These findings underscore the dual role of multi-omics in advancing both fundamental science and industrial applications, paving the way for climate-resilient cultivars and sustainable bioactive compound production. This chapter synthesizes methodologies, challenges, and breakthroughs in multi-omics integration, highlighting its pivotal role in unlocking the therapeutic potential of medicinal plants while addressing global health and agricultural demands.

2.2 Principles of Multi-omics Data Synergy

2.2.1 Cross-omics alignment

Cross-omics alignment in medicinal plants, such as *A. annua*, involves the integration of various omics data to synchronize temporal and spatial aspects of biosynthetic processes. This approach is crucial for understanding the complex biosynthesis of compounds like artemisinin, a potent antimalarial agent. By correlating gene expression peaks with metabolite accumulation, researchers can gain insights into the regulation and optimization of artemisinin production. This synthesis of data from genomics, transcriptomics, proteomics, and metabolomics provides a comprehensive view of the biosynthetic pathways and their regulation, which is essential for both scientific understanding and industrial applications.

The expression of genes involved in artemisinin biosynthesis, such as ADS, CYP71AV1, and CPR, shows temporal variation across different tissues of *A. annua*. For instance, ADS expression peaks in leaves, correlating with the highest artemisinin content, suggesting a temporal synchronization between gene expression and metabolite accumulation (Zeng *et al.*, 2025). Advances in high-throughput sequencing and analytical methods have enabled the integration of omics data, allowing researchers to map the temporal dynamics of biosynthetic pathways. This integration helps identify key regulatory genes and enzymes that control the timing of metabolite production (Singh *et al.*, 2022).

Spatial multi-omics studies reveal that different tissues of *A. annua* contribute variably to artemisinin biosynthesis. Flowers, for example, are identified as primary sites for precursor biosynthesis, while leaves are the main sites for artemisinin accumulation (Yin *et al.*, 2023). The application of single-cell omics technologies allows for the spatial resolution of biosynthetic pathways at the cellular level, providing insights into how different cell types within a tissue contribute to overall metabolite production (Chen *et al.*, 2025). The low natural concentration of artemisinin in *A. annua* poses a challenge for meeting global demand. Metabolic engineering, informed by cross-omics data, aims to enhance artemisinin production by manipulating key biosynthetic genes and pathways (Li *et al.*, 2023a). Cross-omics data also supports synthetic biology approaches, where microbial systems are engineered to produce artemisinin, offering a sustainable alternative to plant extraction (Li *et al.*, 2023a). Cross-omics alignment offers significant advantages in understanding and optimizing the biosynthesis of medicinal compounds. It also presents challenges. The complexity of plant genomes and the need for sophisticated computational tools to integrate and analyze large datasets are significant hurdles. Moreover, the dynamic nature of biosynthetic pathways, influenced by environmental conditions and developmental stages, requires careful experimental design and data interpretation (Flors *et al.*, 2024). Multi-omics data holds great promise for advancing the field of medicinal plant research and the production of valuable natural products (Table 2.1 and Fig. 2.1).

2.2.2 Normalization frameworks

The study of terpenoid pathways in *S. miltiorrhiza* involves scaling genomic (e.g. single nucleotide

Table 2.1. Principles of multi-omics data synergy in medicinal plant research. Author's own table.

Principle	Core concept & methodology	Key application & findings	Identified challenges & future directions	Key citations
Cross-omics alignment	Concept: temporal and spatial integration of disparate omics layers (genomics, transcriptomics, proteomics, metabolomics) to synchronize and map biosynthetic pathways. Methodology: correlation of gene expression peaks (e.g. from RNA-Seq) with metabolite accumulation profiles (e.g. from LC-MS) across different tissues and developmental stages.	Application: elucidating the artemisinin biosynthetic pathway in *A. annua*. Findings: • Temporal alignment revealed ADS gene expression peaks in leaves correlate with highest artemisinin content. • Spatial multi-omics identified flowers as primary sites for precursor synthesis and leaves for final accumulation. • Single-cell omics provide cellular-level resolution of these pathways.	Challenges: • Complexity of plant genomes and dynamic nature of pathways influenced by environment. • Requirement for sophisticated computational tools and careful experimental design. Future directions: • Use insights for metabolic engineering in plants and heterologous production in microbial systems to overcome low natural yields.	(Singh *et al*., 2022; Li *et al*., 2023a; Yin *et al*., 2023; Flors *et al*., 2024; Chen *et al*., 2025; Zeng *et al*., 2025)
Normalization frameworks	Concept: standardization and scaling of heterogeneous data types (e.g. genomic SNPs and metabolomic LC-MS peaks) to enable direct comparison and integration. Methodology: development and application of computational frameworks to normalize data from sources like UHPLC-MS, 2D-LC-MS (two-dimensional liquid chromatography-mass spectrometry), and mass spectrometry imaging for co-analysis.	Application: mapping the terpenoid biosynthesis pathway in *S. miltiorrhiza* (Danshen) for tanshinone production. Findings: • Genomic studies identified 40 terpenoid biosynthesis genes, 27 of which were novel. • Metabolomic profiling characterized numerous terpenoids and phenolic acids. • Spatial imaging revealed specific diterpenoid localization in root tissues.	Challenges: • Complexity of phytochemical pathways and the vast spectrum of metabolites. • Low natural abundance of key metabolites necessitates enhanced production strategies. Future directions: • Advanced analytical techniques for comprehensive metabolite capture. • Leveraging integrated data to identify rate-limiting steps for metabolic engineering in hairy root cultures.	(Ma *et al*., 2012; Shao *et al*., 2016; D'Amelia *et al*., 2017; Surendran *et al*., 2022; Bharti *et al*., 2023; Xia *et al*., 2023a; Sha *et al*., 2024; Zhang *et al*., 2024a; Zhao *et al*., 2024a)

Integration tools	Concept: Using bioinformatics platforms to merge and visualize different omics datasets, establishing direct links between gene expression and metabolite levels. Methodology: Employing tools like MetaboAnalyst, FastQC, TopHat, Cufflinks, and the IMP platform for statistical integration, visualization, and pathway analysis of transcriptomic and metabolomic data.	Application: Deciphering the Terpenoid Indole Alkaloid (TIA) pathway in *C. roseus* (Madagascar periwinkle). Findings: • Transcriptome sequencing identified key genes for TIA biosynthesis, including those for vinblastine. • Integration of data identified differentially expressed genes in specific tissues, correlating with the production of valuable metabolites. • Tools help pinpoint rate-limiting steps for targeted genetic engineering.	Challenges: • Dynamic nature of biosynthetic processes influenced by environmental and developmental factors. • Complexity of plant metabolomes requires advanced computational and analytical methods. Future directions: • Development of more comprehensive integration platforms that incorporate environmental and temporal data. • Enhanced tools for elucidating complex regulatory networks controlling bioactive compound production.	(Verma *et al.*, 2014; She *et al.*, 2019; Surendran *et al.*, 2022; Waris *et al.*, 2022; Li *et al.*, 2023a; Zhao *et al.*, 2024b)
Validation workflows	Concept: using targeted genetic manipulation to functionally confirm hypotheses generated from multi-omics data, closing the loop between prediction and experimental verification. Methodology: applying CRISPR-Cas9 and related technologies (e.g. CRISPR-guided methylation) to edit genes in model systems like *Taxus* cell lines, combined with heterologous expression in systems like *N. benthamiana*.	Application: validating the paclitaxel biosynthesis pathway in *Taxus* species (yew trees). Findings: • Multi-omics identified missing pathway steps, such as a novel C4b-C20 epoxidase. • CRISPR-mediated targeting of the PAL gene led to a 25-fold increase in paclitaxel accumulation. • Heterologous expression validated enzyme function in converting baccatin III to paclitaxel.	Challenges: • Ethical and ecological considerations of genetically modifying slow-growing species. • Regulatory and scalability hurdles for commercial application. Future directions: • Broader application of CRISPR to validate and engineer complex plant pathways. • Integrating synthetic biology with multi-omics to create optimized microbial or plant-based production systems for high-value compounds.	(Xiong *et al.*, 2021; Satish *et al.*, 2022; Brzycki Newton *et al.*, 2023; Zhang *et al.*, 2023b)

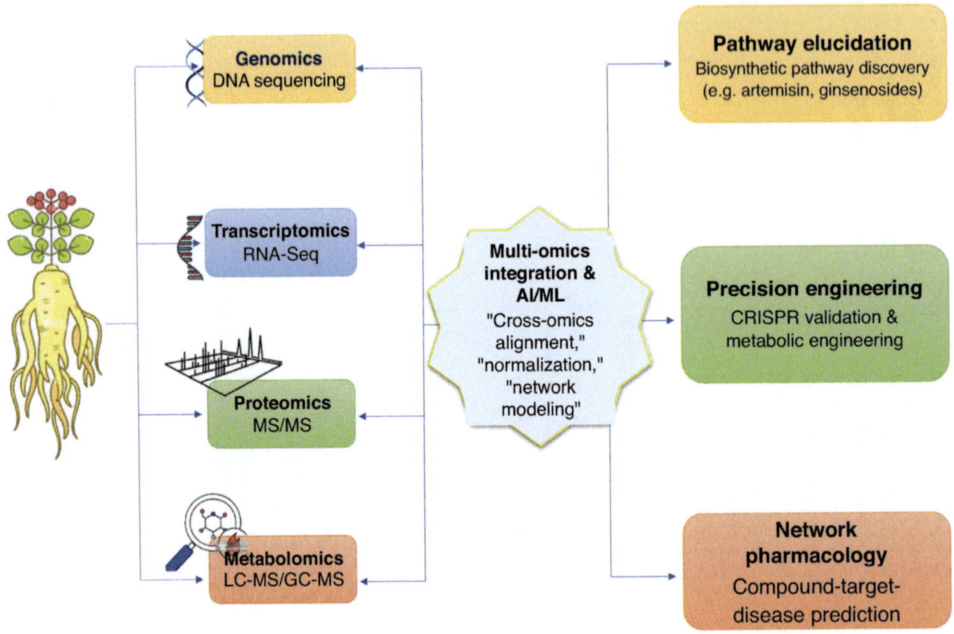

Fig. 2.1. The multi-omics integration workflow in medicinal plant research. (Author's own figure.)

polymorphism (SNP) data) and metabolomic (e.g. liquid chromatography-mass spectrometry (LC-MS) peaks) datasets. The terpenoid pathways in *S. miltiorrhiza*, a medicinal plant known for its therapeutic properties, involves the integration of genomic and metabolomic datasets. This integration is crucial for understanding the biosynthesis of terpenoids, particularly tanshinones, which are significant for their pharmacological activities. The normalization frameworks for scaling genomic data, such as SNP data, and metabolomic data, such as LC-MS peaks, are essential for elucidating these pathways. These frameworks facilitate the comprehensive analysis of the complex biosynthetic networks and the identification of key genes and metabolites involved in terpenoid biosynthesis.

Recent genomic studies have identified 40 terpenoid biosynthesis-related genes in *S. miltiorrhiza*, with 27 being novel. These genes are involved in the biosynthesis of isoprene precursors and labdane-related diterpenoids, which are crucial for tanshinone production (Ma *et al.*, 2012). The Danshen Transcriptional Resource Database (DsTRD) provides a comprehensive collection of transcribed sequences, facilitating gene cloning and functional analysis in *S. miltiorrhiza* (Shao *et al.*, 2016). Advances in genetic engineering, including the use of hairy root cultures and elicitor treatments, have been employed to enhance tanshinone production, highlighting the complexity and potential for metabolic regulation (Bharti *et al.*, 2023).

Techniques such as ultra-high performance liquid chromatography coupled with mass spectrometry (UHPLC-MS) have been used to profile and characterize the multi-components of *S. miltiorrhiza*, identifying numerous terpenoids and phenolic acids (Sha *et al.*, 2024). A polarity-partitioned two-dimensional LC-MS method has been developed to analyze both lipophilic and hydrophilic components, improving the resolution and efficiency of metabolite analysis (Zhang *et al.*, 2024a). Mass spectrometry imaging has revealed the spatial distribution of diterpenoids in *S. miltiorrhiza*, with specific compounds localized in different root tissues, aiding in understanding the biosynthetic pathways (Xia *et al.*, 2023a).

Genomic and metabolomic data allows for the identification of rate-limiting steps in terpenoid biosynthesis pathways. This integration is crucial for metabolic engineering efforts aimed at increasing the production of valuable phytochemicals (Surendran *et al.*, 2022). Combining systems pharmacology, metabolomics, and transcriptomics has been used to reveal the mechanisms of *S. miltiorrhiza* in treating ischemic stroke, demonstrating the potential of integrated approaches in understanding complex biosynthetic pathways (Zhao *et al.*, 2024a). While, this data provides significant insights into the biosynthesis of terpenoids in *S. miltiorrhiza*, challenges remain. The complexity of phytochemical pathways and the need for advanced analytical techniques to capture the full spectrum of metabolites highlight the ongoing need for methodological advancements. Additionally, the low natural abundance of some metabolites necessitates the development of efficient production strategies, such as metabolic engineering and cell culture systems, to meet industrial demands (Table 2.1 and Fig. 2.1; D'Amelia *et al.*, 2017).

2.2.3 Integration tools

The integration of transcriptomic data with metabolite profiles in medicinal plants, such as *C. roseus*, is crucial for understanding the biosynthesis of therapeutic compounds. Platforms like MetaboAnalyst facilitate this integration by providing tools to analyze and visualize complex datasets, thereby linking gene expression with metabolite production. This approach is particularly valuable in elucidating the pathways involved in the biosynthesis of terpenoid indole alkaloids, which are significant for their pharmacological properties. The integration of transcriptomic and metabolomic data can reveal insights into the regulation of these pathways and identify key genes involved in the production of bioactive compounds.

Transcriptome sequencing of *C. roseus* has identified numerous genes involved in the biosynthesis of terpenoid indole alkaloids, with a significant number of transcripts showing homology to known sequences, while others are unique to *C. roseus* (Verma *et al.*, 2014). Tools such as FastQC, Top Hat, and Cufflinks have been used to analyze transcriptomes, identifying genes that are differentially expressed in various tissues, which contribute to the production of specific metabolites like vinblastine (She *et al.*, 2019). Metabolomics provides a comprehensive analysis of the metabolites present in medicinal plants, which is essential for understanding the biosynthetic pathways of bioactive compounds (Waris *et al.*, 2022). The integration of metabolomic data with transcriptomic data can identify rate-limiting steps in metabolic pathways and the corresponding genes, facilitating targeted genetic engineering to enhance metabolite production (Surendran *et al.*, 2022).

MetaboAnalyst and other bioinformatics tools enable the integration of multi-omics data, providing a holistic view of the metabolic pathways and regulatory networks in medicinal plants (Zhao *et al.*, 2024b). The IMP platform offers a comprehensive resource for examining gene annotations and expressions, aiding in the understanding of molecular metabolic pathways in medicinal plants (Li *et al.*, 2023b). While transcriptomic and metabolomic data offers significant insights into the biosynthesis of medicinal compounds, it is important to consider the dynamic nature of these processes. Environmental factors and plant developmental stages can influence gene expression and metabolite production, necessitating a comprehensive approach that includes environmental and temporal data. Additionally, the complexity of plant metabolomes requires advanced analytical techniques and computational tools to fully elucidate the biosynthetic pathways and regulatory networks involved in the production of bioactive compounds (Table 2.1 and Fig. 2.1).

2.2.4 Validation workflows

CRISPR-edited *Taxus* cell lines are used to validate paclitaxel biosynthesis pathways inferred from multi-omics. The validation of paclitaxel biosynthesis pathways using CRISPR-edited *Taxus* cell lines is a cutting-edge approach that leverages advanced genetic and synthetic biology techniques to enhance the understanding and production of this vital anti-cancer compound. Paclitaxel, derived from the *Taxus* species, is a complex diterpenoid with a challenging biosynthetic pathway. The integration of CRISPR technology into this research allows for precise editing of genes involved in paclitaxel biosynthesis, facilitating the validation of pathways inferred from multi-omics studies. This approach not only aids in elucidating the biosynthetic mechanisms but also holds potential for increasing paclitaxel yield through metabolic engineering.

CRISPR-guided DNA methylation has been employed to control the phenylpropanoid pathway in *Taxus chinensis*, which contributes to paclitaxel biosynthesis. This method resulted in a significant 25-fold increase in paclitaxel accumulation by targeting the PAL gene, a key enzyme in the pathway (Brzycki Newton *et al.*, 2023). The use of CRISPR allows for targeted manipulation of specific genes, providing a powerful tool for validating the roles of various enzymes and pathways involved in paclitaxel production (Brzycki Newton *et al.*, 2023).

Multi-omics approaches, including genomics, transcriptomics, and metabolomics, have been instrumental in identifying the complete gene set required for paclitaxel biosynthesis. These studies have uncovered missing steps and enzymes in the pathway, such as a new C4b-C20 epoxidase, which helps overcome bottlenecks in metabolic engineering (Zhang *et al.*, 2023b). The *Taxus* genome has been sequenced to a chromosome-level assembly, revealing unique gene clusters and evolutionary traits that are crucial for understanding and manipulating paclitaxel biosynthesis (Xiong *et al.*, 2021). Heterologous expression systems, such as those using *Nicotiana benthamiana*, have been employed to validate the activity of enzymes involved in paclitaxel biosynthesis. This approach has successfully demonstrated the conversion of intermediates like baccatin III into paclitaxel (Zhang *et al.*, 2023b). The integration of synthetic biology and metabolic engineering strategies, including the use of elicitors and overexpression of biosynthetic genes, has shown promise in enhancing paclitaxel production *in vitro* (Satish *et al.*, 2022).

While CRISPR technology and multi-omics approaches have significantly advanced the understanding and production of paclitaxel, it is important to consider the broader implications of these technologies. The ethical and ecological impacts of genetic modifications, particularly in natural species like *Taxus*, must be carefully evaluated. Additionally, the integration of these advanced techniques into commercial production systems will require addressing regulatory and scalability challenges. Nonetheless, the potential benefits of increased paclitaxel availability for cancer treatment underscore the importance of continued research in this field (Table 2.1 and Fig. 2.1).

2.3 Systems Biology: Bridging Molecular Layers

2.3.1 Network modeling

Systems biology, particularly through network modeling, plays a crucial role in understanding the complex interactions within medicinal plants at a molecular level. This approach allows researchers to bridge various molecular layers, providing insights into the biosynthesis and regulation of secondary metabolites, which are key active ingredients in medicinal plants. Network modeling in systems biology involves the use of molecular networks and dynamic models to represent and understand biological systems, offering a comprehensive view of how different components interact to produce specific phenotypes and behaviors. This methodology is particularly beneficial in the study of medicinal plants, where the complexity of secondary metabolite production and regulation can be daunting.

Systems biology networks are crucial for understanding the complex pathways involved in the biosynthesis of secondary metabolites in medicinal plants (Pathak *et al.*, 2019). Discrete dynamic models, such as Boolean networks, are employed to simulate the behavior of biological systems. These models are particularly useful in large-scale systems where continuous modeling is not feasible, allowing for the prediction of emergent phenomena and system behaviors (Yang and Albert, 2019). The topological properties of these networks are quantified using graph measures, which help in understanding the structural and functional aspects of the networks (Yang and Albert, 2019). Network modeling is applied to explore the biosynthetic pathways of secondary metabolites in medicinal plants. This includes understanding the gene expression and regulation systems that control the production of these metabolites (Rai *et al.*, 2017). The regulation of secondary metabolism in medicinal plants involves complex networks of transcription factors and epigenetic modifications. Network modeling helps in elucidating these multi-layered regulatory mechanisms, providing insights into how gene expression is controlled at various levels (Zhao *et al.*, 2023b). This approach integrates network modeling to identify pharmacological targets from the active ingredients of medicinal plants. It facilitates the systematic investigation of traditional medicines and the discovery of novel compounds for drug development (Noor *et al.*, 2022). Bioinformatics methodologies, combined with network modeling, enhance the identification and characterization of bioactive compounds in medicinal plants. These tools facilitate the analysis of complex chemical compositions and biological activities, advancing drug discovery and development (Kalita *et al.*, 2025).

While network modeling in systems biology provides a powerful framework for understanding the molecular intricacies of medicinal plants, it is important to consider the limitations and challenges associated with this approach. The complexity of biological systems, the vast amount of data generated, and the need for accurate models that can predict real-world behaviors are significant hurdles. Additionally, the integration of various omics data requires sophisticated computational tools and methodologies to ensure meaningful insights. Systems biology approaches, particularly network modeling, are employed to map the complex interactions between genes and metabolites involved in hypericin biosynthesis. This involves integrating various omics data to elucidate the pathways and regulatory mechanisms that govern the production of hypericin, thereby facilitating the development of strategies for enhanced production of this compound (Table 2.2 and Fig. 2.2).

2.3.2 Pathway enrichment

The study of flavonoid pathways in *Ginkgo biloba* through systems biology approaches, particularly pathway enrichment, is crucial for understanding the biosynthesis and regulation of these compounds. Flavonoids are significant due to their medicinal properties, including antioxidant and anti-aging effects, and their role in treating cardiovascular diseases. By integrating transcriptomic and metabolomic data, researchers aim to identify conserved pathways and key regulatory mechanisms that enhance flavonoid production in *G. biloba*. This approach not only aids in understanding the plant's metabolic processes but also supports the development of strategies for the efficient extraction and utilization of flavonoids.

Studies have shown that integrating metabolomic and transcriptomic data can reveal significant changes in flavonoid biosynthesis pathways. For instance, in *G. biloba*, a high-flavonoid mutant (ZY1) exhibited increased levels of flavones and flavanols, while other phenylpropanoid-derived products decreased, suggesting a redirection of metabolic flux toward flavonoid production (Meng *et al.*, 2020). Proteomic and metabolomic analyses have identified stage- and tissue-specific accumulation of flavonoids in *G. biloba*, highlighting the importance of selecting appropriate tissues for flavonoid extraction. Key flavonoids, such as kaempferol and quercetin, were differentially expressed across various tissues and developmental stages (Table 2.2 and Fig. 2.2; Guo *et al.*, 2022).

2.3.2.1 Pathway enrichment and gene regulation

KEGG (Kyoto Encyclopedia of Genes and Genomes) enrichment analysis has identified significant pathways involved in flavonoid biosynthesis, such as the phenylpropanoid pathway. This analysis helps in pinpointing the genes and enzymes that are crucial for flavonoid production under different environmental conditions, such as light exposure (Zhang *et al.*, 2021b). Transcriptome profiling has revealed differentially expressed genes (DEGs) that are involved in flavonoid biosynthesis. For example, genes like GbFLS and GbDFR are upregulated under specific light conditions, promoting the synthesis of flavonoids like quercetin and kaempferol (Zhang *et al.*, 2021a). The integration of transcriptomics and metabolomics is essential for a comprehensive understanding of flavonoid biosynthesis. This approach allows for the identification of rate-limiting steps and key regulatory genes, facilitating metabolic engineering efforts to enhance flavonoid production (Gu *et al.*, 2018). Advances in synthetic biology offer potential for the *in vitro* production of flavonoids, addressing the low natural abundance of these compounds in field-grown plants. This involves the use of elicitors and enzyme inhibitors to modulate

Table 2.2. Systems biology approaches in medicinal plant research. Author's own table.

Section & methodology	Specific application/ model system	Key findings/insights	Strengths of the approach	Limitations/challenges	Key citations
Network modeling	General secondary metabolite biosynthesis in medicinal plants.	• Integrates multi-omics data to map biosynthetic pathways (e.g. for hypericin). • Identifies key regulatory nodes (transcription factors, epigenetic marks). • Topological analysis (graph measures) reveals structural/functional network properties.	• Provides a holistic, systems-level view of complex interactions. • Useful for hypothesis generation and identifying pharmacological targets. • Discrete models (e.g. Boolean) are feasible for large-scale systems where continuous modeling is not.	• High complexity of biological systems is difficult to model accurately. • Generates vast amounts of data requiring sophisticated computational tools for integration. • Models may not always predict real-world behavior accurately.	(Rai *et al.*, 2017; Pathak *et al.*, 2019; Yang and Albert, 2019; Noor *et al.*, 2022; Zhao *et al.*, 2023a; Kalita *et al.*, 2025)
Pathway enrichment	Flavonoid biosynthesis in *G. biloba* (e.g. high-flavonoid mutant ZY1).	• KEGG enrichment pinpoints key pathways (e.g. phenylpropanoid). • Integration of transcriptomic/ metabolomic data reveals metabolic flux redirection (e.g. increased flavones/ flavanols). • Identifies key regulatory genes (e.g. GbFLS, GbDFR) and their responses to environmental cues like light.	• Statistically robust method to identify biologically relevant pathways from omics data. • Reveals tissue and stage-specific accumulation of metabolites, guiding efficient extraction. • Directly supports metabolic engineering by identifying rate-limiting steps.	• Does not fully elucidate the dynamic regulation of the pathways. • The diversity of flavonoids and their derivatives complicates complete analysis. • Potential toxicity and side effects of flavonoids are not addressed by the methodology itself.	(Gu *et al.*, 2018; Meng *et al.*, 2020; Zhang *et al.*, 2021a,b; Guo *et al.*, 2022; Šamec *et al.*, 2022)

Dynamic modeling	Jasmonate-induced alkaloid production in *C. roseus*.	• Dose-dependence is critical; optimal TIA production at 250 μM MeJA. • Transcription factor hierarchy: CrMYC2 (early) activates ORCA3, which regulates biosynthetic genes. • Environmental & developmental factors (light, stage, nutrient) modulate the system.	• Captures temporal and dose-responsive behavior of the system. • Reveals the complex regulatory network (TFs, signaling cross-talk). • Essential for developing effective metabolic engineering strategies (e.g. using cyclodextrins + MeJA).	• Intricate network of factors makes accurate modeling difficult. • Balance between growth and secondary metabolism is a major challenge; excessive JA inhibits growth. • Integrating all influencing factors (environment, development) into a single model is highly complex.	(Zhang *et al.*, 2011; Goklany *et al.*, 2013; Almagro *et al.*, 2014; Goldhaber-Pasillas *et al.*, 2014; Pan *et al.*, 2016; Liu *et al.*, 2021; Song *et al.*, 2022)
Cross-talk mechanisms	Ethylene signaling in sesquiterpene synthesis in *P. ginseng*.	• Ethylene perception initiates a signaling cascade activating specific transcription factors. • Cross-talk with JA and SA pathways fine-tunes sesquiterpene synthesis in response to environment. • Ethylene-responsive elements (EREs) in promoters of sesquiterpene synthase genes directly link signaling to production.	• Elucidates how multiple signaling pathways are integrated to regulate metabolism. • Explains how plants orchestrate complex metabolic responses to stimuli. • Critical for understanding and enhancing the medicinal value (anti-inflammatory, anti-cancer) of the plant.	• The broader context (genetic background, specific stress conditions) significantly influences the outcome of cross-talk. • The interplay between hormones is highly dynamic and context-dependent, making it difficult to generalize. • Isolating the effect of a single pathway (like ethylene) within the network is challenging.	(Cheng *et al.*, 2013; Ghanta *et al.*, 2014; Das *et al.*, 2015; Thakur *et al.*, 2023)

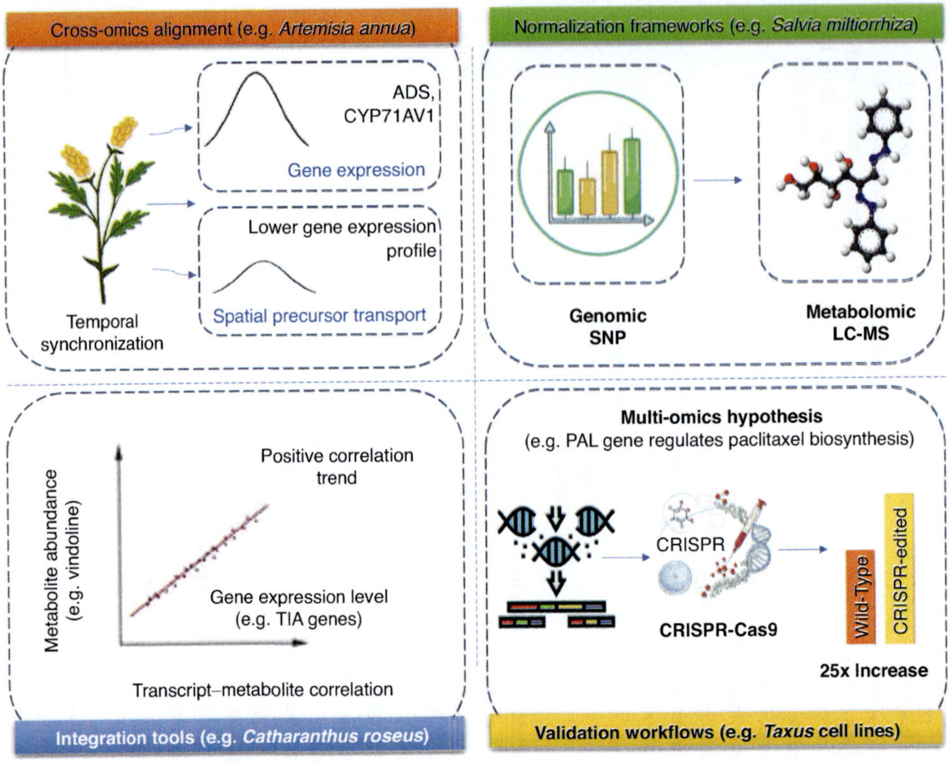

Fig. 2.2. Principles of multi-omics data synergy. (Author's own figure.)

gene expression and metabolic pathways (Amer, 2018). While systems biology approaches provide a detailed understanding of flavonoid pathways in *G. biloba*, challenges remain in fully elucidating these complex networks. The diversity of flavonoids and their derivatives, along with their varied roles in plant physiology and human health, necessitates ongoing research to optimize their extraction and utilization. Additionally, the potential toxicity and side effects of certain flavonoids must be considered in the development of therapeutic applications (Table 2.2 and Fig. 2.2; Šamec *et al.*, 2022).

2.3.3 Dynamic modeling

The dynamic modeling—specifically, the development of kinetic models—of jasmonate-induced alkaloid production in *C. roseus* involves understanding the complex interactions between transcription factors, gene expression, and environmental factors that regulate the biosynthesis of terpenoid indole alkaloids (TIAs). These alkaloids, including vinblastine and vincristine, are of significant pharmaceutical interest due to their anti-cancer properties. The regulation of TIA biosynthesis is primarily mediated by jasmonates, which are plant signaling molecules that activate specific transcription factors and biosynthetic pathways. This dynamic process is influenced by various factors, including the concentration of jasmonates, the developmental stage of the plant, and environmental conditions. The following sections delve into the key aspects of this dynamic modeling.

Jasmonates, such as methyl jasmonate (MeJA), induce the expression of transcription factors like ORCA (Octadecanoid-derivative Responsive *Catharanthus* AP2/ERF-domain) and CrMYC2, which are crucial for TIA biosynthesis in *C. roseus*. ORCA3, in particular, is a key regulator of genes encoding enzymes involved in TIA production (Song *et al.*, 2022). The expression of ORCA and Zct (Zinc finger *Catharanthus*) transcription factors is dose-dependent, with optimal TIA production observed at 250 μM MeJA. At this concentration, ORCA expression is significantly upregulated, while Zct levels remain low, promoting TIA biosynthesis (Goklany *et al.*, 2013). CrMYC2, a basic helix-loop-helix transcription factor, is an immediate-early jasmonate-responsive gene that activates ORCA3 expression, further enhancing the biosynthetic pathway (Zhang *et al.*, 2011).

Environmental factors such as light, temperature, and nutrient availability can modulate TIA biosynthesis by affecting gene expression at transcriptional and post-transcriptional levels (Liu *et al.*, 2021). The developmental stage of the plant also plays a role; for instance, jasmonic acid (JA) has a more pronounced effect on TIA accumulation before flowering compared to during flowering (Goldhaber-Pasillas *et al.*, 2014). The interaction between jasmonate and other signaling molecules, such as ethylene, can further influence TIA production, highlighting the complexity of the regulatory network (Abulfaraj and Al-Andal, 2025).

Metabolic engineering strategies aim to enhance TIA production by manipulating the expression of key regulatory genes and transcription factors. For example, the synergistic use of cyclodextrins and MeJA has been shown to significantly boost alkaloid production in cell cultures (Almagro *et al.*, 2014). Understanding the balance between growth and secondary metabolism is crucial, as excessive JA signaling can inhibit plant growth, thereby reducing overall alkaloid yield. Identifying components like CrJAT1 that modulate JA signaling without compromising growth is essential for optimizing TIA production (Zhang *et al.*, 2024b).

A systems biology approach, integrating metabolomics and other "omics" technologies, is vital for unraveling the complex biosynthetic pathways and regulatory networks involved in TIA production. This approach can help elucidate the cross-talk between different metabolic pathways and the diversion of carbon fluxes (Pan *et al.*, 2016). The use of dynamic modeling and kinetic models can provide insights into the temporal and spatial regulation of TIA biosynthesis, aiding in the development of more effective strategies for enhancing alkaloid production (Pan *et al.*, 2016). While jasmonates are pivotal in regulating TIA biosynthesis, the intricate network of transcription factors and environmental influences presents challenges in fully understanding and optimizing this process. The balance between promoting alkaloid production and maintaining plant growth is a critical consideration in metabolic engineering efforts. Further research into the dynamic interactions within this biosynthetic network is necessary to achieve sustainable and efficient production of these valuable compounds (Table 2.2 and Fig. 2.2).

2.3.4 Cross-talk mechanisms: Ethylene signaling's role in coordinating sesquiterpene synthesis in *P. ginseng*

The role of ethylene signaling in coordinating sesquiterpene synthesis in *P. ginseng* is a complex process that involves multiple molecular layers and cross-talk mechanisms. Ethylene, a plant hormone, is known to play a crucial role in various plant physiological processes, including growth, development, and stress responses. *P. ginseng* ethylene signaling is particularly significant in regulating the synthesis of sesquiterpenes, which are important secondary metabolites with medicinal properties. This synthesis is a result of intricate interactions between ethylene and other signaling pathways, which ultimately influence the plant's metabolic profile. Ethylene is synthesized from methionine through the Yang cycle, and its perception in plants is mediated by ethylene receptors located in the endoplasmic reticulum. Upon binding to ethylene, these receptors undergo a conformational change that initiates a signaling cascade, leading to the activation of transcription factors that regulate gene expression related to sesquiterpene synthesis (Das *et al.*, 2015). Ethylene signaling does not act in isolation; it interacts with other hormonal pathways such as JA and salicylic acid, which are also involved in stress responses and secondary metabolite production. This cross-talk is crucial for fine-tuning the synthesis of sesquiterpenes, as it allows the plant to modulate its metabolic pathways in response to environmental cues (Table 2.2 and Fig. 2.2; Ghanta *et al.*, 2014).

2.3.4.1 *Regulation of sesquiterpene synthase genes*

The expression of sesquiterpene synthase genes is directly influenced by ethylene signaling. These genes encode enzymes that catalyze the formation of sesquiterpenes from farnesyl diphosphate. Ethylene-responsive elements in the promoters of these genes facilitate their transcription in response to ethylene, thereby enhancing sesquiterpene production (Cheng *et al.*, 2013). Sesquiterpenes contribute significantly to the medicinal properties of *P. ginseng*, including anti-inflammatory, antioxidant, and anti-cancer activities. By regulating sesquiterpene synthesis, ethylene signaling indirectly influences the therapeutic efficacy of ginseng, making it a critical area of study for enhancing the plant's medicinal value (Hussain *et al.*, 2023). While ethylene signaling plays a pivotal role in coordinating sesquiterpene synthesis, it is important to consider the broader setting of plant signaling networks. Other factors, such as environmental conditions, genetic background, and developmental stage, also significantly impact sesquiterpene production. Additionally, the interplay between ethylene and other signaling molecules can vary depending on the specific physiological or stress conditions faced by the plant. Understanding these dynamics is essential for developing strategies to optimize the production of valuable secondary metabolites in *P. ginseng* and other medicinal plants (Table 2.2 and Fig. 2.2).

2.4 Network Pharmacology: Mapping Compound–Target Interactions

2.4.1 Compound databases: Annotating *Curcuma longa* metabolites using KNApSAcK and TCM-ID

Network pharmacology has emerged as a crucial tool in understanding the complex interactions between bioactive compounds of medicinal plants and their biological targets. This approach is particularly relevant for annotating metabolites of *C. longa*, such as curcuminoids, using databases like KNApSAcK and TCM-ID (Traditional Chinese Medicine Integrated Database). These databases facilitate the mapping of compound–target interactions, which is essential for exploring the therapeutic potential of *C. longa* in a systematic manner. The integration of traditional knowledge with modern computational techniques allows for a comprehensive analysis of the pharmacological properties of these compounds.

The KNApSAcK database is instrumental in cataloging plant metabolites and their biological activities. It provides a comprehensive list of compounds found in *C. longa*, including curcuminoids, which are known for their diverse therapeutic effects such as anti-inflammatory and anti-cancer properties (Silva *et al.*, 2024). TCM-ID is specifically designed for traditional Chinese medicine and includes detailed information on the bioactive compounds of *C. longa*. It helps in identifying potential targets and pathways that these compounds may influence, thereby supporting the multi-targeted therapeutic approach of network pharmacology (Table 2.3 and Fig. 2.3; Li *et al.*, 2023c).

2.4.1.1 *Network pharmacology and compound–target interactions*

Network pharmacology allows for the analysis of multiple compounds interacting with multiple targets, which is crucial for understanding the complex mechanisms of action of *C. longa*. This approach helps in mapping out the interactions between curcuminoids and various molecular targets, such as transcription factors and protein kinases, which are involved in inflammatory and cancer pathways (Kumari and Subramanya, 2020). By integrating systems biology, network pharmacology provides a holistic view of how curcuminoids interact with biological systems. This integration is essential for predicting the therapeutic outcomes and potential side effects of these compounds, thereby aiding in the development of more effective and safer therapeutic strategies (Nureye, 2024).

Network pharmacology studies the complexity of interactions between multiple compounds and targets. This complexity requires advanced computational models and databases

Table 2.3. Network pharmacology and multi-omics in medicinal plant research. Author's own table.

Section focus	Key databases/tools/ methods	Key findings/ applications	Critical analysis & limitations	Citations
Compound annotation for *C. longa*	KNApSAcK, TCM-ID	• Provides a systematic approach to catalog and annotate curcuminoids and other metabolites. • Facilitates the mapping of compound–target interactions, supporting a multi-target therapeutic understanding. • Integrates traditional knowledge (TCM) with modern computational biology.	• Database limitations: the accuracy and comprehensiveness of predictions are limited by the quality and scope of the underlying databases. They may contain incomplete or unverified entries. • Lack of clinical translation: the approach is heavily predictive; the identified interactions require rigorous experimental and clinical validation to be therapeutically relevant. • Integration challenge: merging traditional knowledge with modern data is promising but non-trivial, as traditional use does not always imply a known mechanistic pathway.	(Mukherjee *et al.*, 2021; Li *et al.*, 2023b; Silva *et al.*, 2024)
Network pharmacology & compound–target interactions	Network pharmacology, systems biology	• Enables a holistic view of how curcuminoids interact with multiple targets (e.g. transcription factors, kinases) simultaneously. • Essential for predicting both therapeutic outcomes and potential side effects of complex natural product mixtures. • Provides a framework for understanding the polypharmacology of herbal medicines.	Computational complexity: modeling complex, non-linear biological networks requires advanced, and often inaccessible, computational models. • Validation gap: the "holistic view" is largely theoretical without subsequent *in vitro*, *in vivo*, and clinical validation. The predictions can be prone to false positives. • Evolving tools: the computational models and databases are "still evolving," meaning conclusions may be transient and require continuous updating.	(Kumari and Subramanya, 2020; Noor *et al.*, 2022; Nureye, 2024)

Continued

Table 2.3. Continued

Section focus	Key databases/tools/methods	Key findings/applications	Critical analysis & limitations	Citations
ML-based interaction prediction for *W. somnifera*	ML (support vector machine—SVM, Random Forest), InflamNat, KNIME, Molecular Docking, QSAR	• ML tools (e.g. InflamNat) can predict anti-inflammatory activity and compound–target relationships with high accuracy. • Identifies specific withanolides (Withaferin A, E, J) as potent modulators of key inflammatory targets (COX-2, NF-XB, AP-1). • Automated workflows (e.g. KNIME) democratize access to these predictive tools.	• Data bias: ML models trained primarily on synthetic compound libraries may perform poorly with the unique structural diversity of natural products like withanolides, leading to inaccurate predictions. • Specialized models needed: highlights the need for developing ML models specifically tailored for natural product chemistry. • Predictive vs confirmatory: docking and QSAR are predictive; their results must be confirmed by experimental binding assays and functional studies.	(Chen *et al.*, 2019; Logie and Vanden Berghe 2020; Niazi and Mariam, 2023; Tian *et al.*, 2023)
ML in predicting interactions	SVMs, Random Forests, Murcko scaffold analysis, K-means clustering	• Advanced ML algorithms are highly effective in screening large datasets for novel COX-2 inhibitors. • Integration of structural analysis techniques (Murcko, clustering) enhances the discovery of novel chemotypes from natural products.	• Algorithmic black box: complex ML models can be "black boxes," making it difficult to interpret the rationale behind their predictions, which is crucial for drug design. • Dependence on data quality: the performance is entirely dependent on the quality, size, and relevance of the training data. "Garbage in, garbage out" is a significant risk.	(Tian *et al.*, 2023; Ghosh *et al.*, 2025)

Multi-omics validation for *E. purpurea*	Proteomics, metabolomics, murine models	• Multi-omics provides a holistic, systems-level understanding of the immunomodulatory effects of *Echinacea* alkamides. • Correlates specific alkamide levels with proteomic changes in immune responses (e.g. increased IL-12, IFN-γ). • Supports the use of *Echinacea* as a potential natural adjuvant for vaccines.	• Causality challenge: while correlations between alkamide levels and proteomic changes are identified, establishing direct causative links is complex due to the multitude of interacting biological variables. • Reproducibility issue: significant variability in plant composition (due to species, growth conditions, extraction methods) leads to inconsistencies in biological outcomes, complicating standardization and clinical translation.	(Spietelun *et al.*, 2013; Swaroop *et al.*, 2021; Liang *et al.*, 2024)
Therapeutic implications of *C. sinensis*	Network pharmacology, epidemiological studies, clinical trials	• Polyphenols (e.g. epigallocatechin gallate—EGCG) exert multi-target effects on MetS by acting as antioxidants, anti-inflammatories, and metabolic regulators (improving insulin sensitivity, lipid profiles, weight). • Epidemiological evidence shows a negative correlation between tea consumption and MetS incidence. • Network pharmacology maps the interaction of tea compounds with multiple biomolecular pathways involved in hypertension and cardiovascular diseases (CVD).	• Clinical evidence gap: the text explicitly states clinical trials are "limited," relying more on preclinical (*in vitro*/*in vivo*) and epidemiological data, which cannot confirm efficacy in humans. • Translational challenges: variability in tea preparation, bioavailability issues, undefined optimal dosing, and potential drug–herb interactions are major hurdles for clinical application. • Complementary, not replacement: polyphenols should be viewed as complementary to, not a replacement for, conventional therapy and lifestyle changes.	(Guo *et al.*, 2020; Hidalgo-Lozada *et al.*, 2024; Chacko *et al.*, 2025; Orsatti *et al.*, 2025)

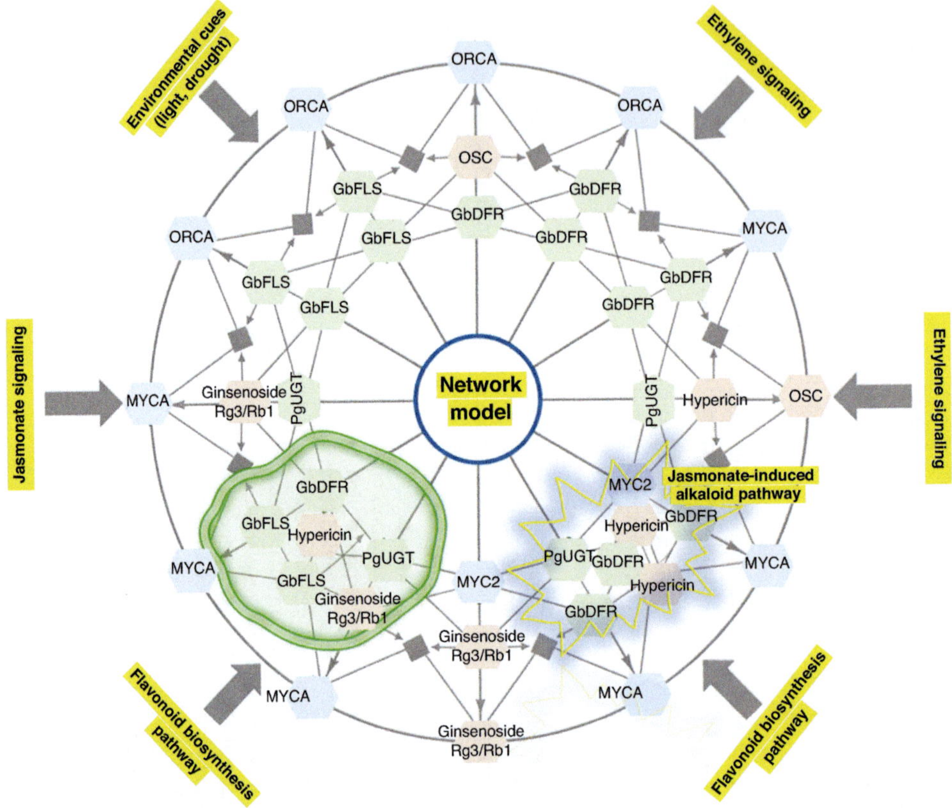

Fig. 2.3. Systems biology bridging molecular layers. (Author's own figure.)

to accurately predict and validate these interactions (Noor *et al.*, 2022). While network pharmacology provides a robust framework for understanding compound–target interactions, there is a need for further validation through clinical trials. This step is crucial for translating the findings from network pharmacology into practical therapeutic applications (Mukherjee *et al.*, 2021). While this pharmacology offers a promising approach to understanding the therapeutic potential of *C. longa*, it is important to consider the limitations of current databases and computational models. These tools, while powerful, are still evolving and require continuous updates and validation to ensure accuracy and reliability. Additionally, the integration of traditional knowledge with modern scientific methods presents both opportunities and challenges in the field of medicinal plant research (Table 2.3 and Fig. 2.3).

2.4.2 Interaction prediction: ML-based tools to link *Withania somniferous* withanolides to human anti-inflammatory targets (e.g. COX-2)

The integration of ML tools in predicting interactions between *W. somniferous* withanolides and human anti-inflammatory targets, such as COX-2, represents a promising frontier in drug discovery. Withanolides, particularly Withaferin A, have shown significant potential in modulating inflammatory pathways, making them attractive candidates for anti-inflammatory drug development. ML-based tools can enhance the prediction of these interactions by leveraging computational models to analyze complex biological data, thus facilitating the identification of effective compounds and their targets.

Withaferin A, a prominent withanolide from *W. somniferous*, has been shown to target key inflammatory pathways, including NF-kB and Nrf2 signaling, which are crucial in managing chronic inflammatory diseases (Logie and Vanden Berghe, 2020). Withanolides E and J have demonstrated inhibitory effects on the AP-1 factor, a significant inflammatory mediator, through molecular docking studies, highlighting their potential as anti-inflammatory agents (Chavadapur *et al.*, 2023). Withaferin A and other withanolides have been effective in suppressing LPS-induced (lipopolysaccharide) pro-inflammatory cytokines in macrophages, indicating their role in modulating innate inflammatory signaling pathways (Table 2.3 and Fig. 2.3; Purushotham *et al.*, 2017).

2.4.2.1 *Machine learning in predicting compound–target interactions*

The use of ML models, such as those developed in the InflamNat platform, allows for the prediction of anti-inflammatory activity and compound–target relationships, achieving high accuracy in identifying potential interactions (Chen *et al.*, 2019). Advanced ML algorithms, including support vector machines and Random Forests, have been employed to predict COX-2 inhibitors, demonstrating the capability of these models to identify novel anti-inflammatory compounds from large datasets (Tian *et al.*, 2023). Automated workflows, like the one developed using KNIME (Konstanz Information Miner), enable the prediction of COX-2 inhibitory potential without requiring extensive coding expertise, thus broadening accessibility to these predictive tools (Ghosh *et al.*, 2025).

Molecular docking and QSAR (quantitative structure–activity relationship) modeling are pivotal in understanding the interaction of phytochemicals with inflammatory targets, providing insights into the structural requirements for effective inhibition (Chihomvu *et al.*, 2024). The integration of ML with traditional computational techniques, such as Murcko scaffold analysis and K-means clustering, enhances the structural analysis of potential inhibitors, facilitating the discovery of novel compounds (Tian *et al.*, 2023). While ML-based tools offer significant advantages in predicting interactions between withanolides and inflammatory targets, challenges remain in ensuring the accuracy and reliability of these predictions. The structural diversity of natural products poses a challenge for models trained on synthetic compounds, necessitating the development of specialized models for natural products (Table 2.3 and Fig. 2.3; Niazi and Mariam, 2023).

2.4.3 Validation via multi-omics: Correlating *Echinacea purpurea* alkamide levels with immune response proteomics in murine models

The validation of medicinal plants like *Echinacea purpurea* through multi-omics approaches offers a comprehensive understanding of their bioactive compounds and their effects on immune responses. *E. purpurea*, known for its immunomodulatory properties, has been studied extensively to correlate its alkamide levels with immune response proteomics in murine models. Multi-omics approaches, including genomics, proteomics, and metabolomics, provide a holistic view of the plant's bioactivity and its potential therapeutic applications. This integration of omics technologies is crucial for unraveling the complex interactions between *E. purpurea*'s compounds and the immune system, particularly in murine models.

E. purpurea is traditionally used for its immune-boosting properties, primarily through the modulation of cytokine and chemokine expression, which enhances the body's defense mechanisms against infections (Trivadila *et al.*, 2025). *E. purpurea* extracts can stimulate phagocytosis and other immune cell functions, acting as both an immunostimulant and an anti-inflammatory agent (Burlou-Nagy *et al.*, 2022). The plant's alkamides, particularly, have been identified as active components that inhibit inflammatory pathways, such as COX-2 activity, and modulate cytokine production, which are crucial for immune response regulation (Table 2.3 and Fig. 2.3; Hou *et al.*, 2010).

2.4.3.1 *Multi-omics approaches in* Echinacea *research*

Multi-omics technologies, including proteomics and metabolomics, are employed to analyze the complex biochemical pathways and interactions of *E. purpurea*'s compounds (Liang *et al.*, 2024). Proteomic analysis helps identify proteins involved in immune responses, while metabolomics provides insights into the metabolic pathways influenced by *Echinacea*'s bioactive compounds (Su *et al.*, 2023). These approaches allow for the identification of key genes and proteins that mediate the plant's immunomodulatory effects, offering a more detailed understanding of its therapeutic potential.

In murine models, *E. purpurea* has been shown to enhance specific immune responses, such as increased production of interleukin-12 and interferon-gamma, which are critical for effective immune function (Swaroop *et al.*, 2021). The alkamide fractions of *E. purpurea* have demonstrated significant immunostimulatory effects, correlating with enhanced antibody production and cytokine expression in response to vaccines (Swaroop *et al.*, 2021). These findings underscore the potential of *E. purpurea* as a natural adjuvant in vaccine formulations, enhancing both humoral and cell-mediated immunity. While the multi-omics approach provides a comprehensive framework for understanding the immunomodulatory effects of *E. purpurea*, it is important to consider the variability in plant composition due to factors such as species, growth conditions, and extraction methods. These factors can lead to inconsistencies in biological outcomes, highlighting the need for standardized extraction and processing methods to ensure reproducibility and efficacy in therapeutic applications (Table 2.3 and Fig. 2.3; Spietelun *et al.*, 2013).

2.4.4 Therapeutic implications: Multi-target effects of *Camellia sinensis* polyphenols in metabolic syndrome networks

C. sinensis, commonly known as tea, is rich in polyphenols, particularly catechins, which have shown significant therapeutic potential in addressing metabolic syndrome (MetS). MetS is a cluster of conditions, including obesity, insulin resistance, hypertension, and dyslipidemia, which collectively increase the risk of cardiovascular diseases and type 2 diabetes. The polyphenols in *C. sinensis* exhibit multi-target effects that can ameliorate various aspects of MetS through their antioxidant, anti-inflammatory, and metabolic regulatory properties. *C. sinensis* polyphenols, such as epigallocatechin gallate (EGCG), are potent antioxidants that reduce oxidative stress, a key factor in MetS pathogenesis. They enhance the body's antioxidant capacity by modulating oxidoreductase enzyme systems and restoring mitochondrial function (Guo *et al.*, 2020). These polyphenols also exhibit anti-inflammatory properties, which help in reducing chronic low-grade inflammation associated with MetS. This is achieved by modulating inflammatory pathways and cytokine production, thereby improving endothelial function and reducing the risk of cardiovascular diseases (Orsatti *et al.*, 2025).

Tea polyphenols contribute to weight management by increasing energy expenditure, enhancing fat oxidation, and stimulating lipolysis. They also improve thermogenesis, which collectively aids in reducing obesity, a core component of MetS (Orsatti *et al.*, 2025). These compounds improve insulin sensitivity and regulate blood glucose levels, thereby reducing the risk of type 2 diabetes. They achieve this by modulating critical metabolic pathways involved in glucose and lipid metabolism (Kang *et al.*, 2019). The consumption of *C. sinensis* has been linked to reduced blood pressure and improved lipid profiles, which are crucial in managing MetS. The polyphenols in tea help in lowering LDL (low-density lipoprotein) cholesterol and triglycerides while increasing HDL (high-density lipoprotein) cholesterol, thus reducing the risk of cardiovascular diseases (Adel Noor *et al.*, 2024). Network pharmacology studies have shown that the secondary metabolites in *C. sinensis* interact with multiple biomolecular pathways, providing a comprehensive approach to managing hypertension and other cardiovascular conditions associated with MetS (Table 2.3 and Fig. 2.3; Chacko *et al.*, 2025).

2.4.4.1 Clinical and epidemiological evidence

Epidemiological studies suggest a negative correlation between polyphenol intake and the incidence of MetS, highlighting the potential of *C. sinensis* as a preventive measure. Clinical trials, although limited, have demonstrated promising results in improving MetS biomarkers

(Hidalgo-Lozada *et al.*, 2024). The therapeutic effects of tea polyphenols are supported by both *in vitro* and *in vivo* studies, which emphasize their role in managing obesity and diabetes, two major components of MetS (Yang *et al.*, 2018). While the therapeutic potential of *C. sinensis* polyphenols in managing MetS is well-supported, it is important to consider the challenges associated with translating these findings into clinical practice. The variability in polyphenol content due to different tea preparations, potential interactions with other medications, and the need for standardized dosing regimens are areas that require further research. Additionally, while polyphenols offer a natural and sustainable approach to managing MetS, they should complement, rather than replace, conventional treatments and lifestyle modifications (Table 2.3 and Fig. 2.3).

2.5 Big Data Challenges: Volume, Standardization, Reproducibility

2.5.1 Cross-omics harmonization: ISA-tab metadata frameworks for *Ocimum sanctum* (holy basil) multi-omics datasets

The integration of multi-omics data for medicinal plants like *O. sanctum* (holy basil) presents significant challenges, particularly in terms of data volume, standardization, and reproducibility. The use of ISA-Tab metadata frameworks is a promising approach to address these challenges by facilitating cross-omics harmonization. This framework can help manage the vast amounts of data generated from various omics platforms, ensuring that datasets are standardized and reproducible, which is crucial for advancing research in medicinal plants. The following sections delve into the specific challenges and the role of ISA-Tab in addressing them.

The advent of high-throughput omics technologies has resulted in the generation of massive datasets, which are often high-dimensional and complex. This is particularly true for medicinal plants like *O. sanctum*, where multi-omics approaches are used to explore genomic, transcriptomic, proteomic, and metabolomic data (Anugraha *et al.*, 2022). The integration of these datasets is essential for a comprehensive understanding of the plant's biosynthetic pathways and therapeutic potential, but the sheer volume of data poses significant challenges in terms of storage, processing, and analysis (Wang *et al.*, 2024a).

Standardization of data is critical for ensuring that datasets from different studies and platforms can be compared and integrated effectively. The ISA-Tab framework provides a structured format for metadata, which helps in standardizing the data across different omics platforms (Krajewski *et al.*, 2015). This standardization is crucial for *O. sanctum*, as it allows researchers to harmonize data from various studies, facilitating cross-comparison and integration of findings (Sharma *et al.*, 2025). Reproducibility is a major concern in multi-omics research due to the variability in experimental conditions and data processing methods. The ISA-Tab framework aids in improving reproducibility by providing detailed metadata that captures the experimental design, protocols, and data processing steps (González-Beltrán *et al.*, 2014). For *O. sanctum*, reproducibility is particularly important as it ensures that findings related to its medicinal properties can be validated and built upon in future research (Table 2.4 and Fig. 2.4; Bhamra *et al.*, 2022).

2.5.1.1 Cross-omics harmonization

Cross-omics harmonization involves integrating data from different omics layers to provide a holistic view of the biological system. The ISA-Tab framework supports this by allowing for the integration of diverse datasets, facilitating the discovery of novel insights into the biosynthetic pathways and regulatory networks of *O. sanctum* (Anugraha *et al.*, 2022). This harmonization is essential for understanding the complex interactions between genes, proteins, and metabolites, which underpin the plant's therapeutic properties (Rahaman *et al.*, 2025). While the ISA-Tab framework offers a robust solution for managing the challenges of volume, standardization, and reproducibility in multi-omics research, it is important to acknowledge the ongoing need for advancements in bioinformatics tools and computational power. These advancements are necessary to further enhance the integration and analysis of large-scale omics data, ultimately leading to more effective utilization of medicinal plants like *O. sanctum* in drug discovery and development (Table 2.4 and Fig. 2.4).

Table 2.4. Critical analysis of big data challenges in medicinal plant multi-omics research. Author's own table.

Medicinal plant	Core challenge	Proposed solution/ framework	Key benefits (as stated)	Critical assessment/ limitations (inferred or stated)	Key citations
O. sanctum (holy basil)	Volume & integration: managing vast, high-dimensional data from genomic, transcriptomic, proteomic, and metabolomic platforms. Standardization & reproducibility: variability in experimental conditions and data processing hinders cross-study comparison and validation.	ISA-Tab metadata framework for cross-omics harmonization.	• Provides a structured format for metadata, enabling standardization across omics platforms. • Improves reproducibility by capturing experimental design and protocols. • Facilitates integration of diverse datasets for a holistic biological view.	• Solution is foundational, not exhaustive: the framework itself is a starting point; its effectiveness is limited by the quality and completeness of the metadata provided by researchers. • Dependent on complementary tools: the text acknowledges a continued need for advanced bioinformatics tools and computational power for effective analysis.	(González-Beltrán *et al.*, 2014; Krajewski *et al.*, 2015; Anugraha *et al.*, 2022)
C. sativa	Scalability: computational and storage demands of terabyte-scale genomics and metabolomics data. Cost-efficiency: managing the economic burden of large-scale data processing.	Cloud-based computing platforms (e.g. Tripal, Phenomenal project).	• Offers near-unlimited, scalable computational resources and flexible payment models. • Significantly reduces data processing times for large datasets. • Facilitates data sharing, collaboration, and reproducibility.	Performance and cost are non-linear: efficiency can decrease with an increase in virtual CPUs, and cost–benefit is highly dependent on the specific scale and platform, leading to potential unpredictability. Introduces new complexities: shifts the challenge from local IT management to cloud-specific issues like data transfer bottlenecks, vendor lock-in, and security configurations.	(Kasson, 2012; Walterbusch *et al.*, 2013; Koppad *et al.*, 2021; Mansueto *et al.*, 2024)

M. oleifera	Reproducibility & reusability: ensuring drought-stress multi-omics data can be found, accessed, integrated, and reused by the global research community.	FAIR data principles (Findable, Accessible, Interoperable, Reusable).	• Enhances findability through standardized metadata and identifiers. • Promotes accessibility via open-access repositories. • Enables interoperability with other datasets through common formats. • Ensures reusability through rich documentation.	High implementation burden: adhering to FAIR principles requires significant effort in data curation, annotation, and management, which may not be fully supported by standard research funding or timelines. Ethical and logistical hurdles: the text mentions but does not resolve challenges like data privacy, consent for genetic resources, and the need for international collaboration on standards.	(Mashamaite *et al.*, 2024; Oyebamiji *et al.*, 2024; Chauhan *et al.*, 2025)
A. vera	Validation & benchmarking: correlating metabolites with genes across independent studies to ensure findings are robust and not study-specific artifacts.	MetaboLights repository for validating cross-study metabolite–gene correlation	• Provides a centralized, public repository for metabolomics data. • Allows for benchmarking and validation of results against independent datasets. • Supports integration with other omics data types.	Data heterogeneity: the value of the repository is contingent on the consistency of data submission practices. Differences in analytical platforms (LC-MS vs NMR), growth conditions, and data processing between studies can complicate direct comparison. Incomplete picture: metabolomics alone is insufficient; full elucidation requires integrated genomics and transcriptomics, which may not be available in the same repository.	(Waris *et al.*, 2022; Cardona *et al.*, 2024)

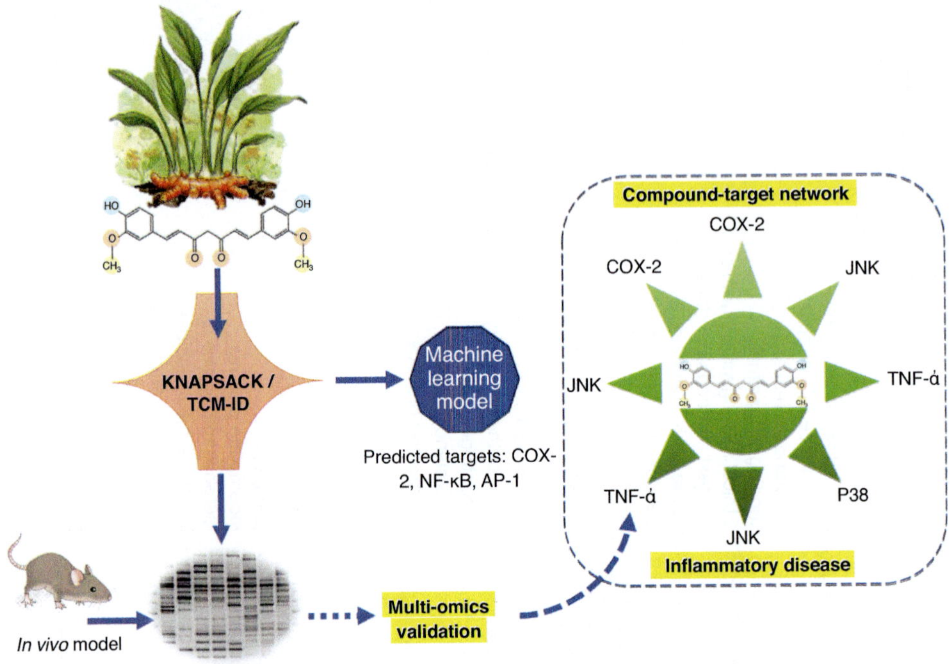

Fig. 2.4. Network pharmacology: from plant to drug target. (Author's own figure.)

2.5.2 Scalability issues: Cloud-based analysis of terabyte-scale *Cannabis sativa* genomics and cannabinoid metabolomics

The scalability issues in cloud-based analysis of terabyte-scale *C. sativa* genomics and cannabinoid metabolomics are multifaceted, involving challenges in data management, computational resources, and integration of diverse datasets. The increasing size and complexity of biological datasets necessitate robust cloud computing solutions to handle the computational and storage demands effectively. Cloud platforms offer potential solutions by providing scalable resources, but they also introduce new challenges related to performance, cost, and data integration. The following sections explore these issues in detail.

The integration of multi-omics data, such as genomics, metabolomics, and proteomics, is crucial for comprehensive *C. sativa* research. Platforms like Tripal facilitate this by hosting public genome assemblies and integrating various data sources, enabling efficient multi-omics analyses (Mansueto *et al.*, 2024). The complexity of *C. sativa*'s chemical composition, which includes cannabinoids and other chemical classes, requires sophisticated data management strategies to avoid biased interpretations and to fully understand the plant's medicinal properties (Myoli *et al.*, 2024).

Cloud computing offers near-unlimited capacity and flexible payment models, which are essential for handling the large-scale datasets typical in genomics and metabolomics research (Kasson, 2012). The PhenoMeNal (Phenome and Metabolome aNalysis) project, for instance, demonstrates how cloud infrastructure can significantly reduce processing times for metabolomics data, although efficiency may decrease as the number of virtual CPUs (central processing units) increases (Koppad *et al.*, 2021). The scalability of cloud platforms is crucial for processing terabyte-scale datasets. However, as datasets grow, performance-limiting factors can change, necessitating new computing paradigms to maintain efficiency and robustness (Kasson, 2012).

While cloud computing can be cost-effective compared to traditional desktop systems, the economic benefits depend on the scale of the data and the specific cloud platform used. Large-scale cloud platforms can offer significant cost savings, but efficiency may decline at higher computational loads (Kasson, 2012). The cost–benefit ratio of different cloud services must be carefully evaluated to ensure that the most suitable option is selected for the specific requirements (Walterbusch *et al.*, 2013).

Cloud platforms facilitate data sharing and reproducibility by making datasets and computational methods easily accessible. This is particularly beneficial in fields like *C. sativa* research, where collaboration and data sharing can accelerate scientific discovery (Kasson, 2012). Tools like the CANDI (Cannabis-derived compound Analysis and Network Discovery Interface) server enhance the ability to predict molecular targets and pathways, providing valuable insights for developing targeted cannabis-based therapies (Ekambaram *et al.*, 2025). Cloud-based solutions offer significant advantages for handling large-scale *C. sativa* genomics and metabolomics data; they also present challenges that must be addressed to fully realize their potential. Issues such as data integration, computational efficiency, and cost management are critical considerations for researchers. Additionally, the rapid evolution of cloud technologies and the increasing complexity of biological datasets necessitate ongoing adaptation and innovation in computational strategies (Table 2.4 and Fig. 2.4).

2.5.3 Reproducibility frameworks: FAIR-compliant sharing of *Moringa oleifera* drought-stress multi-omics data

The FAIR-compliant sharing of *M. oleifera* drought-stress multi-omics data involves the application of the FAIR principles—Findability, Accessibility, Interoperability, and Reusability—to ensure that data related to *M. oleifera*, particularly under drought-stress conditions, is effectively managed and shared. This approach is crucial for advancing research in medicinal plants by facilitating data sharing and reuse, which can lead to new insights and applications.

Findability data should be easily located by both humans and computers. This involves the use of standardized metadata and identifiers. For *M. oleifera*, this means ensuring that datasets related to its drought-stress responses are indexed in searchable databases (Mashamaite *et al.*, 2024). Accessibility data should be accessible under well-defined conditions. This includes the use of open-access repositories where researchers can access *M. oleifera* data without unnecessary barriers (Mashamaite *et al.*, 2024). Interoperability data should be integrated with other datasets and tools. This requires the use of standardized formats and protocols, allowing *M. oleifera* data to be combined with other omics data for comprehensive analyses (Chauhan *et al.*, 2025). Reusability data should be well documented and curated to ensure it can be reused in future research. This involves providing detailed descriptions of the data collection and processing methods used in *M. oleifera* studies (Table 2.4 and Fig. 2.4; Mashamaite *et al.*, 2024).

2.5.3.1 *Multi-omics data in* M. oleifera *research*

Genomics and transcriptomics approaches help in understanding the genetic and transcriptional changes in *M. oleifera* under drought stress, which can reveal key genes involved in stress tolerance and medicinal compound biosynthesis (Oyebamiji *et al.*, 2024). Proteomics and metabolomics techniques provide insights into the protein expression and metabolic pathways affected by drought stress, contributing to the understanding of *M. oleifera*'s adaptive mechanisms and its potential health benefits (Oyebamiji *et al.*, 2024). Integration and analysis of these omics data allows for a holistic view of the plant's response to environmental stressors, facilitating the discovery of novel bioactive compounds and enhancing the plant's medicinal applications (Zhang *et al.*, 2023a). The large scale and high-dimensional nature of multi-omics data present challenges in analysis and interpretation. Advanced computational tools and frameworks are necessary to manage and extract meaningful insights from these datasets (Subramanian *et al.*, 2020).

Ethical and social considerations in compliance with FAIR principles must also address ethical concerns, such as data privacy and consent, especially when dealing with sensitive information. Understanding the genetic and biochemical diversity of *M. oleifera* can lead to the development of improved varieties with enhanced drought tolerance and medicinal properties, addressing global food security and health challenges (Mashamaite *et al.*, 2024). While the FAIR-compliant sharing of *M. oleifera* data offers significant benefits, it is important to consider the broader implications of data sharing, such as the need for international collaboration and the establishment of standardized protocols across different research communities. This approach not only enhances the reproducibility and reliability of research findings but also fosters innovation in the field of medicinal plant research (Table 2.4 and Fig. 2.4).

2.5.4 Benchmarking datasets: Using MetaboLights to validate cross-study *Aloe vera* metabolite–gene correlations

The use of MetaboLights to validate cross-study *Aloe vera* metabolite–gene correlations involves leveraging metabolomics datasets to understand the complex interactions between metabolites and genes in *A. vera*. Metabolomics, as a field, provides a comprehensive analysis of metabolites, which are crucial for understanding the biochemical pathways and therapeutic potential of medicinal plants like *A. vera*. This approach is particularly useful in standardizing and validating the quality of herbal products by identifying and correlating specific metabolites with genetic expressions. The integration of metabolomics with databases like MetaboLights allows for a more robust validation of these correlations across different studies.

Metabolomics offers a detailed analysis of the chemical composition of *A. vera*, which is essential for identifying bioactive compounds that contribute to its medicinal properties. The use of advanced analytical techniques such as LC-MS and nuclear magnetic resonance (NMR) has facilitated the identification and quantification of metabolites, aiding in the understanding of *A. vera*'s healing properties (Waris *et al.*, 2022). Metabolomics helps in the quality assessment of *A. vera* by identifying characteristic quality markers, which are influenced by geographical origins (Yan *et al.*, 2023).

MetaboLights is a repository that stores metabolomics data, providing a platform for researchers to access and compare datasets from different studies (Baby *et al.*, 2022). By using metabolites, researchers can validate metabolite–gene correlations in *A. vera* across various studies, ensuring consistency and reliability in findings (Cardona *et al.*, 2024). This database supports the integration of multi-omics data, which is crucial for a holistic understanding of the metabolic pathways in *A. vera* (Cardona *et al.*, 2024).

Metabolomics involves the analysis of primary and secondary metabolites, a complex task that requires sophisticated analytical techniques (Waris *et al.*, 2022). The integration of genomics, transcriptomics, and metabolomics is necessary to fully elucidate the metabolic pathways and gene expressions in *A. vera* (Pillay *et al.*, 2022). Standardizing protocols and managing large datasets are ongoing challenges, but the use of databases like MetaboLights can help streamline these processes (Elizarraras *et al.*, 2024).

Metabolomics and databases like MetaboLights provide significant insights into *A. vera*'s metabolite–gene correlations and also highlight the broader potential of these technologies in medicinal plant research. The ability to validate findings across studies enhances the reliability of herbal medicine research and supports the development of standardized herbal products. However, the dynamic nature of plant metabolism and the need for large-scale data management remain challenges that require ongoing research and technological advancements (Table 2.4 and Fig. 2.4).

2.6 AI/ML Frameworks for Predictive Omics Modeling

2.6.1 Supervised learning: Random Forest models predicting vincristine yield in *C. roseus* from integrated omics

AI and ML frameworks, particularly supervised learning models like Random Forest, have shown significant promise in predictive omics modeling for medicinal plants. Specifically, these models are being utilized to predict the yield of bioactive compounds such as vincristine in *C. roseus* by leveraging integrated omics data. This approach not only enhances the understanding of the biosynthetic pathways involved but also optimizes the production of these valuable compounds.

These are ensemble learning methods that utilize multiple decision trees to improve prediction accuracy. They are particularly effective in handling high-dimensional data, which is common in omics studies (Barberis *et al.*, 2022). Random Forest models have been applied to predict the yield of vincristine by analyzing genomics, transcriptomics, proteomics, and metabolomics data. This integration allows for a comprehensive understanding of the factors influencing vincristine production (Wang *et al.*, 2024b). The use of Random Forest models provides robustness against overfitting and the ability to handle complex interactions between variables, which are crucial for accurate predictions in biological systems (Barberis *et al.*, 2022).

Omics approaches provide a holistic view of the biological processes involved in vincristine biosynthesis (Wang *et al.*, 2024b). By combining these diverse data types, researchers can identify key genes and enzymes that regulate the biosynthesis of vincristine, leading to more accurate predictions of yield (Guo *et al.*, 2024). Studies have demonstrated that models trained on integrated omics data outperform those using single data types, highlighting the importance of a multi-omics approach in predictive modeling (Guo *et al.*, 2024). While Random Forest models are powerful, their interpretability can be limited, making it difficult to understand the underlying biological mechanisms fully. The application of these models in real-world scenarios must consider regulatory requirements and the practical aspects of scaling up production based on model predictions (Marchese Robinson *et al.*, 2017).

While the use of Random Forest models and integrated omics data in predicting vincristine yield in *C. roseus* is promising, it is essential to consider the broader context of AI and ML in medicinal plant research. The integration of these technologies not only aids in understanding complex biosynthetic pathways but also opens new avenues for optimizing the production of bioactive compounds. However, challenges such as data integration and model interpretability must be addressed to fully realize the potential of these approaches in practical applications. Continued research and interdisciplinary collaboration will be crucial in overcoming these challenges and advancing the field (Table 2.5).

2.6.2 Unsupervised clustering: t-SNE revealing chemotypes in *Cannabis sativa* based on integrated terpenoid and transcriptomic profiles

Unsupervised clustering techniques like t-SNE have shown promise in revealing chemotypes in *C. sativa* by analyzing integrated terpenoid and transcriptomic profiles. This approach allows for a more nuanced understanding of the plant's chemical diversity beyond the traditional focus on cannabinoids. By leveraging these advanced computational techniques, researchers can better classify and differentiate cannabis strains based on their comprehensive chemical profiles, which include terpenoids and other secondary metabolites. This methodology not only enhances the understanding of cannabis chemotypes but also aids in the selection of strains for specific medicinal purposes.

t-SNE (t-distributed stochastic neighbor embedding) is a powerful tool for visualizing high-dimensional data, making it suitable for distinguishing between different cannabis chemotypes based on complex chemical profiles. This technique can effectively cluster cannabis strains by integrating terpenoid and transcriptomic data, revealing distinct chemotypes that may not be apparent through traditional methods (Vergara *et al.*, 2020). By combining terpenoid profiles with transcriptomic data, researchers can gain insights into the biosynthetic pathways and regulatory mechanisms that contribute to the diversity of cannabis chemotypes. This integrated approach allows for a more comprehensive understanding of the plant's chemical space, which is crucial for identifying strains with specific therapeutic properties (Aliferis and Bernard-Perron, 2020).

Terpenoids are a significant class of compounds in cannabis, contributing to its aroma and therapeutic effects. Their biosynthesis involves complex pathways that are influenced by various genetic and environmental factors. Understanding these pathways through transcriptomic analysis can help in identifying

Table 2.5. Critical analysis of AI/ML frameworks for predictive omics in medicinal plants. Author's own table.

Medicinal plant	AI/ML technique	Core application	Key benefits (as stated)	Critical assessment/limitations (inferred or stated)	Key citations
C. roseus (Madagascar periwinkle)	Supervised learning: Random Forest	Predicting vincristine yield from integrated multi-omics data (genomics, transcriptomics, proteomics, metabolomics).	• Effective at handling high-dimensional, complex omics data. • Robust against overfitting due to its ensemble nature. • Models trained on integrated data outperform single-omics models.	"Black box" problem: limited interpretability makes it difficult to extract clear, actionable biological mechanisms from the model, hindering hypothesis generation. Translational gap: the jump from a predictive model to scalable agricultural or bioproduction practices is non-trivial and involves regulatory and practical hurdles not addressed by the model itself.	(Marchese Robinson *et al.*, 2017; Barberis *et al.*, 2022; Wang *et al.*, 2024b)
C. sativa	Unsupervised learning: t-SNE	Revealing chemotypes based on integrated terpenoid and transcriptomic profiles, moving beyond cannabinoid-only classification.	• Powerful visualization for high-dimensional data, revealing hidden patterns and clusters (chemotypes). • Provides a more nuanced, holistic view of the plant's chemical diversity. • Aids in selecting strains for specific medicinal applications based on full profile.	Interpretation is qualitative: t-SNE is primarily a visualization tool; the clusters themselves are not predictive and can be sensitive to parameter tuning. Biological interpretation of the clusters remains a separate, manual step. Data quality dependency: the reliability of the revealed chemotypes is entirely dependent on the quality, completeness, and normalization of the underlying terpenoid and transcriptomic data. Noise or bias in data collection leads to misleading visualizations.	(Aliferis and Bernard-Perron, 2020; Schilling *et al.*, 2020; Vergara *et al.*, 2020)

G. glabra (licorice)	Deep learning: autoencoders	Reconstructing complex gene–metabolite interaction networks under salt stress to understand plant adaptation and metabolite production.	• Excellent for non-linear dimensionality reduction and feature extraction from highly complex omics data. • Can model intricate, non-obvious relationships between genes and metabolites that simpler models might miss. • Useful for generating hypotheses about key regulators in stress response networks.	Ultimate "black box": the internal representations learned by autoencoders are even less interpretable than models like Random Forest. Understanding why a specific network is reconstructed is a major challenge. Data hungry: deep learning models typically require very large datasets to train effectively without overfitting, which may be a limitation for some medicinal plant studies. The "why" behind the predicted networks remains elusive.	(Kumar *et al.*, 2022; Pillay *et al.*, 2022; Sobral *et al.*, 2023)
Annonaceae species (e.g. Soursop family)	Transfer learning: BERT-inspired models	Predicting anti-cancer metabolites in understudied species by leveraging knowledge from data-rich domains (e.g. model plants or well-annotated genomes).	• Addresses the critical problem of data scarcity in understudied species, which are often a source of novel bioactives. • Leverages pre-trained models to achieve meaningful predictions with limited target-specific data. • Can process sequential data (e.g. DNA, protein sequences) to predict biosynthetic potential.	Domain shift risk: performance is highly dependent on the relevance of the source domain (what the model was pre-trained on) to the target domain (Annonaceae). Poor transfer can lead to inaccurate predictions. Compound validation: this is a high-throughput *in silico* screening method. All predictions are hypothetical and require rigorous downstream *in vitro* and *in vivo* validation, which is a significant bottleneck.	(Ahmad *et al.*, 2023; Torabi *et al.*, 2024; Kalita *et al.*, 2025)

key regulatory genes and enzymes involved in terpenoid production (Asiminicesei *et al.*, 2024). Different cannabis strains exhibit unique terpenoid profiles, which can be used to classify them into distinct chemotypes. For instance, some strains may be rich in monoterpenes, while others may have higher concentrations of sesquiterpenes. This diversity is crucial for tailoring cannabis strains to specific medicinal applications (Jin *et al.*, 2021).

The ability to classify cannabis strains based on their comprehensive chemical profiles, including terpenoids, has significant implications for medicinal use. It allows for the selection of strains with specific therapeutic properties, enhancing the efficacy of cannabis-based treatments (Myoli *et al.*, 2024). AI/ML frameworks can also aid in the breeding of cannabis strains by identifying genetic markers associated with desirable chemotypes. This can lead to the development of new strains with optimized cannabinoid and terpenoid profiles, tailored for specific therapeutic purposes (Cembrowska-Lech *et al.*, 2023).

AI/ML frameworks in cannabis research offer numerous benefits. The accuracy of clustering and classification techniques like t-SNE depends on the quality and completeness of the data. Missing data or low signal-to-noise ratios for certain compounds can affect the reliability of the results. Additionally, the complexity of cannabis's chemical space requires careful consideration of various factors, including environmental influences and genetic variability, which can impact the interpretation of chemotype classifications (Table 2.5; Schilling *et al.*, 2020).

2.6.3 Deep learning: Autoencoders reconstructing gene-metabolite networks in *Glycyrrhiza glabra* (licorice) under salt stress

Deep learning, particularly the use of autoencoders, has emerged as a powerful tool in predictive omics modeling, especially for reconstructing gene-metabolite networks in plants like *Glycyrrhiza glabra* (licorice) under stress conditions such as salinity. Autoencoders, a type of neural network, are adept at learning efficient representations of data, which is crucial for understanding complex biological systems. These models can help elucidate the intricate networks of genes and metabolites that respond to salt stress, thereby enhancing our understanding of plant stress responses and potentially improving the therapeutic efficacy of medicinal plants.

These are neural networks designed to learn efficient codings of input data, which is particularly useful in omics for dimensionality reduction and feature extraction. They can reconstruct input data from a compressed representation, making them ideal for modeling complex biological networks (Sobral *et al.*, 2023). Deep learning, including the use of autoencoders, is increasingly applied in plant omics to integrate multi-layered data from fields such as genomics, transcriptomics, and metabolomics. This integration is crucial for understanding the biosynthesis of secondary metabolites under stress conditions (Sobral *et al.*, 2023). *G. glabra* is known for its rich array of metabolites, including terpenoids and flavonoids, which are crucial for its medicinal properties. Metabolomics studies have identified numerous metabolites and their interactions with human proteins, highlighting the plant's potential in neuroprotection (Kumar *et al.*, 2022). Under salt stress, plants like *G. glabra* undergo significant metabolic changes. These changes can be mapped using omics technologies to understand how stress affects metabolite production and gene expression (Alugoju and Tencomnao, 2023). Bioinformatics and systems biology play a crucial role in managing and interpreting the vast amounts of data generated by omics technologies. These fields help in identifying new active compounds and understanding the complex networks of gene–metabolite interactions (Likić *et al.*, 2010).

The biosynthesis of secondary metabolites involves complex networks that are influenced by various environmental factors. Understanding these pathways requires advanced computational models and integrative approaches (Pillay *et al.*, 2022). Environmental stresses, such as salinity, can lead to significant changes in the production of secondary metabolites. These changes are crucial for the plant's adaptation and can be leveraged to enhance the therapeutic potential of medicinal plants (Punetha *et al.*, 2022). While deep learning and autoencoders offer significant potential in modeling gene-metabolite networks, challenges remain in fully understanding the complex interactions within these networks. The integration of multi-omics data and advanced computational models is essential for overcoming these challenges and unlocking the full potential of medicinal plants like *G. glabra* under stress conditions. This approach not only aids in the discovery of new bioactive compounds but also enhances the resilience of plants to environmental stresses, thereby contributing to sustainable agricultural practices and the development of novel therapeutics (Table 2.5).

2.6.4 Transfer learning: BERT-inspired models predicting anti-cancer metabolites in understudied *Annonaceae* species

Transfer learning models inspired by bidirectional encoder representations from transformers (BERT), in predicting anti-cancer metabolites in understudied *Annonaceae* species, represent a promising frontier in medicinal plant research. These models leverage the vast data generated from omics technologies to predict bioactive compounds with potential therapeutic benefits. The integration of AI/ML with omics data facilitates the discovery of novel anti-cancer agents by analyzing complex biological datasets and identifying patterns that may not be apparent through traditional methods. This approach is particularly beneficial for understudied plant species, where data scarcity can be a significant barrier to research.

Transfer learning, particularly models inspired by BERT, can be adapted to predict anti-cancer metabolites by leveraging pre-trained models on large datasets and fine-tuning them on specific tasks related to understudied species. This approach is beneficial in scenarios where data is limited, as it allows the model to apply learned knowledge from related tasks (Torabi *et al.*, 2024). BERT-inspired models can process sequential data, such as genomic sequences, to predict the presence of bioactive compounds, offering a novel method to explore the therapeutic potential of *Annonaceae* species (Torabi *et al.*, 2024). This comprehensive data integration supports the identification of candidate genes and metabolites involved in anti-cancer activity (Ahmad and Asif, 2023). High-throughput sequencing and omics technologies have advanced the field of herbal genomics, providing insights into the genetic and metabolic pathways that govern the production of bioactive compounds in medicinal plants (Ahmad and Asif, 2023).

The democratization of ML in biology and the development of accessible ML tools can enhance the exploration of plant biosynthetic pathways, ultimately improving the discovery and optimization of anti-cancer agents (Kalita *et al.*, 2025). While the application of AI/ML frameworks in predicting anti-cancer metabolites in understudied *Annonaceae* species holds significant promise, it is essential to consider the broader context of medicinal plant research. The integration of omics data and AI/ML technologies can lead to breakthroughs in drug discovery, but it also necessitates careful consideration of ethical, regulatory, and practical challenges. Continued research and collaboration across disciplines will be crucial to fully realizing the potential of these technologies in improving human health outcomes (Table 2.5).

2.7 Case Study: Multi-Omics in *P. ginseng* Research

2.7.1 Multi-omic profiling: Genomics (CYP450 gene clusters), proteomics (OSC enzyme activity), and metabolomics (ginsenoside Rg3/Rb1 ratios)

P. ginseng research has significantly advanced the understanding of its complex biological systems, particularly in the biosynthesis of ginsenosides, which are the primary active compounds. Multi-omics profiling, encompassing genomics, proteomics, and metabolomics, provides a comprehensive view of the molecular mechanisms underlying ginseng's medicinal properties. This approach has been instrumental in identifying key genes, enzymes, and metabolites involved in ginsenoside biosynthesis, offering insights into the plant's adaptation to environmental stresses and its pharmacological potential. The following sections detail the contributions of each omics discipline to *P. ginseng* research (Table 2.6 and Fig. 2.5).

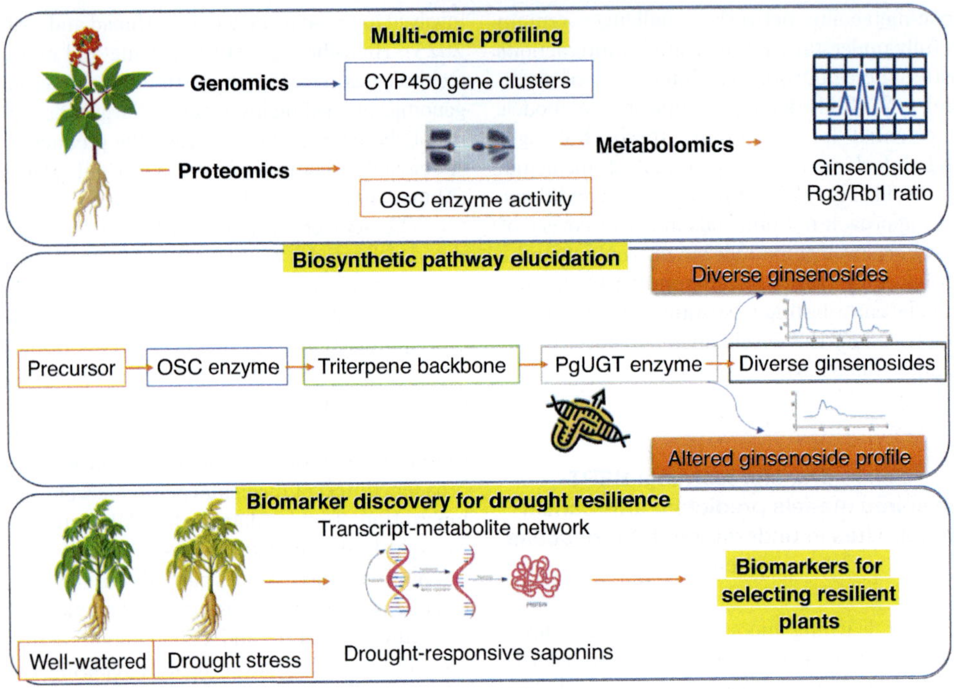

Fig. 2.5. Case study—multi-omics in *P. ginseng*. (Author's own figure.)

2.7.1.1 Genomics: CYP450 gene clusters

Genomic studies have identified the cytochrome P450 (CYP450) gene family as crucial in the biosynthesis of ginsenosides. These enzymes are involved in the hydroxylation steps of ginsenoside synthesis, which are essential for the structural diversity of these compounds (Li *et al.*, 2023d). The telomere-to-telomere reference genome of *P. ginseng* has revealed the expansion of gene families associated with ginsenoside biosynthesis, highlighting the evolutionary adaptations of this plant (Song *et al.*, 2024). Functional validation of specific CYP450 genes, such as PgCYP309, has demonstrated their role in enhancing ginsenoside production, particularly under stress conditions induced by methyl jasmonate (Jiang *et al.*, 2024).

Proteomic analyses have focused on the oxidosqualene cyclase (OSC) enzymes, which catalyze the cyclization of 2,3-oxidosqualene to form the triterpene backbone of ginsenosides (Wang *et al.*, 2022). Comparative proteomics has identified stress-responsive proteins that may influence OSC activity and, consequently, ginsenoside biosynthesis. These include heat shock proteins and enzymes involved in redox homeostasis (Yin *et al.*, 2021). The integration of proteomics with other omics data has provided insights into the regulation of secondary metabolite pathways and the plant's response to environmental challenges (Majumdar and Keller, 2021).

Metabolomic profiling has been pivotal in quantifying and comparing the ratios of different ginsenosides, such as Rg3 and Rb1, which are indicative of the plant's medicinal quality (Xia *et al.*, 2023b). Advanced metabolomics techniques, such as UHPLC/IM-QTOF-MS (ultra-high performance liquid chromatography/ion mobility/quadrupole time-of-flight mass spectrometry), have enabled the identification of chemical markers that differentiate between *Panax* species, aiding in the authentication and quality control of ginseng products (Jia *et al.*, 2019). The dynamic changes in metabolite profiles during ginseng development and under stress conditions have been mapped, providing a deeper understanding of the biosynthetic pathways and their regulation (Liu *et al.*, 2025).

For *P. ginseng*, challenges remain in integrating these diverse datasets to form a cohesive picture of the plant's biology. The complexity of ginseng's genetic background and the intricate interactions between its metabolic pathways necessitate further research to fully elucidate the mechanisms of ginsenoside biosynthesis and their pharmacological implications. Additionally, the development of more sophisticated bioinformatics tools and analytical techniques will be crucial in overcoming these challenges and unlocking the full potential of ginseng as a medicinal resource (Table 2.6 and Fig. 2.5).

2.7.2 Biosynthetic pathway elucidation: Linking *PgUGT* gene variants to ginsenoside diversity via CRISPR-omics

The elucidation of the biosynthetic pathway of ginsenosides in *P. ginseng*, particularly through the lens of multi-omics and CRISPR-omics, is a significant area of research. This approach aims to link *P. ginseng* UDP-glycosyltransferase (PgUGT) gene variants to the diversity of ginsenosides, which are the primary bioactive compounds in ginseng. The use of CRISPR technology allows for precise editing of genes involved in the biosynthesis pathway, potentially enhancing the production and diversity of ginsenosides. This research is crucial for understanding the genetic and enzymatic mechanisms underlying ginsenoside biosynthesis and for developing strategies to increase their yield and diversity.

UGTs, particularly the UGT94 family, play a critical role in the glycosylation of ginsenosides, which is essential for their structural diversity and pharmacological activity. The UGT94 family has been shown to catalyze sugar chain elongation at various positions on ginsenoside molecules, contributing to the diversity of these compounds (Yang *et al.*, 2020). The identification and functional characterization of UGTs has been a focus of recent research, with significant progress made in understanding their role in ginsenoside biosynthesis in both plants and microorganisms (Yuan *et al.*, 2024).

CRISPR-Cas9 technology has been applied to target and edit specific genes within the ginsenoside biosynthesis pathway. This approach allows for the precise manipulation of genes, such as those encoding UGTs, to enhance ginsenoside production and diversity (Wang *et al.*, 2025a). The application of CRISPR-Cas9 in ginseng has provided insights into the regulation of the biosynthesis pathway and has the potential to improve the medicinal value of ginseng by increasing the yield of specific ginsenosides (Liu, 2023).

Multi-omics approaches facilitate the identification of key genes and enzymes involved in the pathway and provide a comprehensive understanding of the molecular mechanisms underlying ginsenoside biosynthesis (Borah *et al.*, 2024). The integration of omics data has led to the discovery of novel genes and pathways, enhancing the understanding of ginsenoside biosynthesis and enabling the development of strategies for metabolic engineering (Yin *et al.*, 2021). While CRISPR-omics and multi-omics approaches in ginsenoside research hold great promise, it is important to consider the potential limitations and ethical implications of genetic editing in medicinal plants. The integration of these technologies must be carefully managed to ensure sustainable and ethical advancements in the field. Additionally, the complexity of the ginsenoside biosynthesis pathway presents ongoing challenges that require further research and innovation (Table 2.6 and Fig. 2.5).

2.7.3 Biomarker discovery: Integrated transcript-metabolite networks identifying drought-responsive saponins

The integration of multi-omics approaches in *P. ginseng* research has significantly advanced the understanding of biomarker discovery, particularly in identifying drought-responsive saponins. Multi-omics technologies, which include transcriptomics, metabolomics, and proteomics, have been pivotal in elucidating the complex biosynthetic pathways and regulatory networks involved in the production of bioactive compounds such as saponins. These approaches have enabled researchers to identify key genes and metabolites that respond to environmental

Table 2.6. Critical analysis of multi-omics applications in P. ginseng research. Author's own table.

Research focus	Omics technologies employed	Key findings & strengths	Critical gaps & limitations	Key citations
Multi-omic profiling of ginsenoside biosynthesis	Genomics: CYP450 gene family analysis. Proteomics: OSC enzyme activity profiling. Metabolomics: Ginsenoside Rg3/Rb1 ratio quantification.	• Comprehensive parts list: identifies key players (genes, enzymes, metabolites) in the ginsenoside pathway. • High-resolution data: advanced techniques (e.g. UHPLC/IM-QTOF-MS, telomere-to-telomere genome) provide precise chemical and genetic data for authentication and quality control. • Stress response insight: reveals how environmental factors (e.g. methyl jasmonate) dynamically influence the pathway.	Correlative vs causal: primarily generates lists of correlated components (e.g. a CYP gene is present when a ginsenoside is produced). It often lacks direct mechanistic proof of function in planta. Data silos: the text explicitly states that integrating genomic, proteomic, and metabolomic datasets into a "cohesive picture" remains a major challenge. The relationships between a genomic variant, its protein product's activity, and the final metabolite output are not fully resolved.	(Wang *et al.*, 2022; Li *et al.*, 2023d; Xia *et al.*, 2023b; Song *et al.*, 2024)
Biosynthetic pathway elucidation via CRISPR-omics	Functional genomics: CRISPR-Cas9 gene editing of PgUGT genes. Multi-omics: analysis of edited lines to link gene function to metabolite output.	Establishes causality: CRISPR moves beyond correlation by enabling direct functional validation. Knocking out a PgUGT gene and observing a change in ginsenoside profile proves its role. Precision engineering potential: offers a direct path to rationally engineer ginseng for enhanced or altered ginsenoside profiles (e.g. increasing high-value Rg3).	Technical hurdles in plants: CRISPR efficiency in a complex, polyploid plant like ginseng can be low, and regeneration of edited plants is challenging and time-consuming. Pleiotropic effects: editing a single gene (e.g. in UGT family) may have unintended consequences on other pathways or plant fitness, which are not always assessed. Ethical & regulatory barriers: the text mentions but does not resolve the significant ethical and regulatory challenges of commercializing genetically edited medicinal plants.	(Yang *et al.*, 2020; Liu, 2023; Wang *et al.*, 2025b)

Biomarker discovery for drought resilience	Transcriptomics & metabolomics: integrated network analysis (e.g. WGCNA) to link gene expression to saponin accumulation under drought.	Systems-level view: identifies co-expressed gene modules and associated metabolites, revealing the broader regulatory network, not just single genes. Applied potential for breeding: discovers candidate biomarker genes/metabolites that can be used to screen and breed for drought-tolerant ginseng varieties with stable medicinal compound production.	Predictive vs descriptive: the network models are often descriptive of the observed response. Their predictive power for how the plant will respond to new, complex environmental scenarios is limited without further validation. Post-transcriptional blind spot: as noted, this approach misses crucial post-transcriptional and post-translational regulation (e.g. protein phosphorylation). A drought-responsive transcript may not lead to a corresponding change in enzyme activity.	(Kim *et al.*, 2017; Di *et al.*, 2023; Wang *et al.*, 2025b)

stressors like drought, thereby enhancing the potential for developing more resilient ginseng varieties.

Transcriptomic and metabolomic data has been instrumental in understanding the biosynthesis of ginsenosides, the primary bioactive compounds in ginseng. For instance, studies have shown that different light intensities can significantly affect the accumulation of ginsenosides in *P. ginseng* leaves, with higher light intensities promoting increased synthesis of these compounds (Di *et al.*, 2023). This highlights the importance of environmental factors in modulating the expression of genes involved in ginsenoside biosynthesis. Proteomic analyses have provided insights into the protein profiles of ginseng under various stress conditions, identifying stress-responsive proteins that play crucial roles in the plant's adaptation to environmental changes (Kim *et al.*, 2017). This information complements transcriptomic data, offering a more comprehensive view of the molecular mechanisms underlying saponin biosynthesis.

Studies have identified numerous genes associated with the biosynthesis of saponins, particularly under drought conditions. For example, weighted gene co-expression network analysis (WGCNA) has been used to link environmental factors with saponin accumulation, revealing specific transcript modules that respond to drought stress (Wang *et al.*, 2025a). This approach has led to the identification of biomarkers that can be used to select for drought-resistant ginseng varieties. The integration of transcriptomic and metabolomic data has facilitated the construction of integrated networks that map the interactions between genes and metabolites involved in saponin biosynthesis. These networks have been crucial in identifying key regulatory nodes that can be targeted for genetic improvement (Zhang *et al.*, 2020). Despite advances in the understanding of ginsenoside biosynthesis, challenges remain in fully elucidating the complex interactions between different omics layers. The integration of proteomics with transcriptomics and metabolomics is particularly challenging due to the post-transcriptional and post-translational modifications that proteins undergo, which can complicate data interpretation (Kim *et al.*, 2017). Future research should focus on developing more sophisticated integrative models that can accurately predict the effects of environmental stressors on saponin biosynthesis, thereby enhancing the potential for breeding more resilient ginseng varieties. *P. ginseng* research has provided valuable insights into the molecular mechanisms underlying saponin biosynthesis and environmental adaptation. These approaches have not only facilitated the discovery of drought-responsive biomarkers but also highlighted the potential for developing improved ginseng varieties through targeted breeding strategies. However, the complexity of integrating different omics data remains a significant challenge that requires further methodological advancements (Table 2.6 and Fig. 2.5).

2.8 Conclusion

The integration of multi-omics technologies has fundamentally transformed the study of medicinal plants, unlocking unprecedented insights into their biosynthetic pathways, regulatory networks, and therapeutic potential. By harmonizing genomics, transcriptomics, proteomics, and metabolomics, researchers can decode the spatiotemporal dynamics of metabolite production, as exemplified by cross-omics alignment in *A. annua* to optimize artemisinin synthesis and CRISPR-validated workflows in *Taxus* cell lines to elucidate paclitaxel biosynthesis. These approaches not only accelerate the identification of rate-limiting enzymes and regulatory genes but also enable precision metabolic engineering to enhance yields of high-value compounds. Furthermore, systems biology bridges molecular layers, revealing intricate gene-metabolite networks in *H. perforatum* for hypericin biosynthesis and multi-target interactions of *C. longa* curcuminoids through network pharmacology. Such holistic frameworks are critical for translating traditional medicinal knowledge into validated, mechanism-driven therapies. Despite these advancements, significant challenges persist in managing the sheer volume and heterogeneity of omics data. Standardization remains a hurdle, as seen in the adoption of ISA-Tab metadata frameworks for *O. sanctum* datasets, which aim to harmonize disparate data types while ensuring interoperability. Reproducibility, another critical concern, is addressed through FAIR-compliant sharing of *M. oleifera* drought-stress omics data, yet variability in experimental designs and environmental factors complicates cross-study validation. AI/ML frameworks, such as Random Forest models predicting vincristine yield in *C. roseus* and autoencoders reconstructing stress-responsive gene-metabolite networks in *G. glabra*, offer powerful solutions to these challenges. These tools not only enhance predictive accuracy but also uncover non-obvious patterns in terabyte-scale datasets, as demonstrated by t-SNE-driven chemotype classification in *C. sativa*. The *P. ginseng* case study epitomizes the transformative potential of multi-omics integration. By linking CYP450 gene clusters to ginsenoside diversity via CRISPR-omics and identifying drought-responsive saponins through transcript-metabolite networks, researchers are paving the way for climate-resilient cultivars and tailored bioactive compound production. Such discoveries underscore the dual role of multi-omics in both advancing fundamental science and addressing industrial and agricultural demands.

Acknowledgments

The authors express their gratitude for the financial support received through the Distinguished High-Level Talents Research Grant from the Guizhou Science and Technology Corporation Platform Talents Fund (Grant nos.: [2017]5733-001 and CK-1130-002), high-level specialists project fund of Zunyi Medical University (1092/2000006)F-1092, the National Natural Science Foundation of China (82373981), and the support provided by Zunyi Medical University, China. Special appreciation is extended to all laboratory colleagues and research staff members for their valuable insights, constructive guidance, and assistance throughout this study.

References

Abulfaraj, A.A. and Al-Andal, A. (2025) Regulation of alternative splicing in *Catharanthus roseus* in response to methyl jasmonate modulation during development and stress resilience. *Functional Plant Biology* 52(4), FP25017. DOI: 10.1071/FP25017.

Adel Noor, S., El-Mallah, M.M. and EL-Masry, H.G. (2024) The effect of matcha (camellia sinensis) and yerba mate (ilex paraguariensis) aqueous extract on oxidative stress status in obese rats. مجلة الاقتصاد المنزلي 40(2), 23–54. DOI: 10.21608/jhe.2024.372305.

Ahmad, A. and Asif, A. (2023) *Omics Studies of Medicinal Plants*. CRC Press, New York.

Ahmad, A., Mustafa, G., Rana, A. and Zia, A.R. (2023) Improvements in bioremediation agents and their modified strains in mediating environmental pollution. *Current Microbiology* 80(6), 208. DOI: 10.1007/s00284-023-03316-x.

Aliferis, K.A. and Bernard-Perron, D. (2020) Cannabinomics: Application of metabolomics in cannabis (*Cannabis sativa* L.) research and development. *Frontiers in Plant Science* 11, 554. DOI: 10.3389/fpls.2020.00554.

Almagro, L., Gutierrez, J., Pedreño, M.A. and Sottomayor, M. (2014) Synergistic and additive influence of cyclodextrins and methyl jasmonate on the expression of the terpenoid indole alkaloid pathway genes and metabolites in *Catharanthus roseus* cell cultures. *Plant Cell, Tissue and Organ Culture (PCTOC)* 119(3), 543–551. DOI: 10.1007/s11240-014-0554-9.

Alugoju, P. and Tencomnao, T. (2023) Medicinal plants proteomics in response to abiotic stresses. In: Husen, A. and Iqbal, M. (eds) *Medicinal Plants*. Springer Nature Singapore, Singapore, pp. 35–107.

Amer, A. (2018) Biotechnology approaches for *in vitro* production of flavonoids. *Journal of Microbiology, Biotechnology and Food Sciences* 7(5), 457–468. DOI: 10.15414/jmbfs.2018.7.5.457-468.

Anugraha, A.C., Thomas, T. and Dennis Thomas, T. (2022) Multiomics approach in medicinal plants. In: Rout, G.R. and K.v, P. (eds) *Omics in Horticultural Crops*. Academic Press, Cambridge, Massachusetts, pp. 589–602.

Asiminicesei, D.-M., Fertu, D.I. and Gavrilescu, M. (2024) Impact of heavy metal pollution in the environment on the metabolic profile of medicinal plants and their therapeutic potential. *Plants (Basel, Switzerland)* 13(6), 913. DOI: 10.3390/plants13060913.

Baby, J., Thomas, T. and Dennis Thomas, T. (2022) Metabolomics of important medicinal plants. In: Swamy, M.K. and Kumar, A. (eds) *Phytochemical Genomics*. Springer Nature Singapore, Singapore, pp. 285–317.

Barberis, E., Khoso, S., Sica, A., Falasca, M., Gennari, A. *et al.* (2022) Precision medicine approaches with metabolomics and artificial intelligence. *International Journal of Molecular Sciences* 23(19), 11269. DOI: 10.3390/ijms231911269.

Bhamra, S.K., Heinrich, M., Johnson, M.R.D., Howard, C. and Slater, A. (2022) The cultural and commercial value of tulsi (*Ocimum tenuiflorum* L.): Multidisciplinary approaches focusing on species authentication 11(22), 3160. DOI: 10.3390/plants11223160.

Bharti, P.K., Santhosh, B., Singh, S.K. and Kumari, S. (2023) Genetic manipulation in medicinal plants for enhancement of secondary metabolites. In: Singh, R. and Kumar, N. (eds) *Genetic Manipulation of Secondary Metabolites in Medicinal Plant*. Springer Nature Singapore, Singapore, pp. 129–145.

Bhowmick, S., Singh, T. and Chauhan, P.S. (2024) Harnessing the power of aromatic and medicinal plants for natural product innovation. In: Kumar, L., Bharadvaja, N., Singh, R. and Anand, R. (eds) *Medicinal and Aromatic Plants: Current Research Status, Value-Addition to Their Waste, and Agro-Industrial Potential (Vol I)*. Springer Nature Switzerland, Cham, Switzerland, pp. 211–222.

Borah, A., Singh, S., Chattopadhyay, R., Kaur, J. and Bari, V.K. (2024) Integration of CRISPR/Cas9 with multi-omics technologies to engineer secondary metabolite productions in medicinal plant: Challenges and prospects. *Functional & Integrative Genomics* 24(6), 207. DOI: 10.1007/s10142-024-01486-w.

Brzycki Newton, C., Young, E.M. and Roberts, S.C. (2023) Targeted control of supporting pathways in paclitaxel biosynthesis with CRISPR-guided methylation. *Frontiers in Bioengineering and Biotechnology* 11, 1272811. DOI: 10.3389/fbioe.2023.1272811.

Burlou-Nagy, C., Bănică, F., Jurca, T., Vicaș, L.G., Marian, E. *et al.* (2022) *Echinacea purpurea* (L.) Moench: Biological and pharmacological properties. A Review. *Plants* 11(9), 1244. DOI: 10.3390/plants11091244.

Cardona, K.M., Rugeles-Silva, P.A., Tarazona-Pulido, L., Diaz-Ariza, L.A., Muñoz-Florez, J.E. *et al.* (2024) Exploring the potential of omics technologies in medicinal plant research: A review in Colombia. *Boletin Latinoamericano y Del Caribe de Plantas Medicinales y Aromaticas* 23(4), 460–486. DOI: 10.37360/blacpma.24.23.4.32.

Cembrowska-Lech, D., Krzemińska, A., Miller, T., Nowakowska, A., Adamski, C. *et al.* (2023) An integrated multi-omics and artificial intelligence framework for advance plant phenotyping in horticulture. *Biology* 12(10), 1298. DOI: 10.3390/biology12101298.

Chacko, A.M., Bylappa, Y. and Nag, A. (2025) Bioinformatics applications for evaluating health and pharmacological properties of tea. In: Preedy, V.R. (ed.) *Tea in Health and Disease Prevention*. Academic Press, Cambridge, Massachusetts, pp. 379–390.

Chauhan, D., sahu, S., Balley, P., Rashmi, M., Murmu, S. *et al.* (2025) Moringagdb: A bioinformatics approach to unravel *Moringa oleifera*'s nutritional and medicinal benefits. DOI: 10.2139/ssrn.5097948.

Chavadapur, S.S., Biradar, S., Joga, R. and Lamani, B. (2023) *Withania somnifera*: The wonder plant with withanolide E and J inhibit inflammatory mediator AP-1 factor. *Current Bioactive Compounds* 19(2). DOI: 10.2174/1573407218666220601121738.

Chen, Y., Bruyn Kops, C. and Kirchmair, J. (2019) Resources for chemical, biological, and structural data on natural products. In: *Progress in the Chemistry of Organic Natural Products 110: Cheminformatics in Natural Product Research*, Vol. 110. pp. 37–71. DOI: 10.1007/978-3-030-14632-0_2.

Chen, S., Fei, Y.-R., Cai, X., Wang, C., Tong, S. *et al.* (2025) Exploring the role of metabolic pathways in TNBC immunotherapy: Insights from single-cell and spatial transcriptomics. *Frontiers in Endocrinology* 15, 1528248. DOI: 10.3389/fendo.2024.1528248.

Cheng, M.-C., Liao, P.-M., Kuo, W.-W. and Lin, T.-P. (2013) The *Arabidopsis* ETHYLENE RESPONSE FACTOR 1 regulates abiotic stress-responsive gene expression by binding to different *cis*-acting elements in RESPONSE to different stress signals. *Plant Physiology* 162(3), 1566–1582. DOI: 10.1104/pp.113.221911.

Chihomvu, P., Ganesan, A., Gibbons, S., Woollard, K. and Hayes, M.A. (2024) Phytochemicals in drug discovery–a confluence of tradition and innovation. *International Journal of Molecular Sciences* 25(16), 8792. DOI: 10.3390/ijms25168792.

D'Amelia, V., Ruggiero, A., Tranchida-Lombardo, V., Leone, A., Tucci, M. *et al.* (2017) Biosynthesis of *Salvia* specialized metabolites and biotechnological approaches to increase their production. In: Georgiev, V. and Pavlov, A. (eds) *Salvia Biotechnology*. Springer International Publishing, Cham, Switzerland, pp. 241–270.

Das, S., Dutta, S.S., Chowdhury, S. and Das, K. (2015) Ethylene signal transduction and signaling roles–a review. *Agricultural Reviews* 36(2), 133. DOI: 10.5958/0976-0741.2015.00015.X.

Das, S., Kwon, M. and Kim, J.-Y. (2024) Enhancement of specialized metabolites using CRISPR/Cas gene editing technology in medicinal plants. *Frontiers in Plant Science* 15, 1279738. DOI: 10.3389/fpls.2024.1279738.

Di, P., Yang, X., Wan, M., Han, M., Zhang, Y. *et al.* (2023) Integrative metabolomic and transcriptomic reveals potential mechanism for promotion of ginsenoside synthesis in *Panax ginseng* leaves under different light intensities. *Frontiers in Bioengineering and Biotechnology* 11, 1298501. DOI: 10.3389/fbioe.2023.1298501.

Ekambaram, S., Wang, J. and Dokholyan, N.V. (2025) CANDI: A web server for predicting molecular targets and pathways of cannabis-based therapeutics. *Journal of Cannabis Research* 7(1), 13. DOI: 10.1186/s42238-025-00268-w.

Elizarraras, J.M., Liao, Y., Shi, Z., Zhu, Q., Pico, A.R. *et al.* (2024) WebGestalt 2024: Faster gene set analysis and new support for metabolomics and multi-omics. *Nucleic Acids Research* 52(W1), W415–W421. DOI: 10.1093/nar/gkae456.

Flors, V., Kyndt, T., Mauch-Mani, B., Pozo, M.J., Ryu, C.-M. *et al.* (2024) Enabling sustainable crop protection with induced resistance in plants. *Frontiers in Science* 2. DOI: 10.3389/fsci.2024.1407410.

Ghanta, S., Datta, R., Bhattacharyya, D., Sinha, R., Kumar, D. *et al.* (2014) Multistep involvement of glutathione with salicylic acid and ethylene to combat environmental stress. *Journal of Plant Physiology* 171(11), 940–950. DOI: 10.1016/j.jplph.2014.03.002.

Ghosh, P., Kumar, A. and Singh, S.K. (2025) COX-2 inhibitor prediction with KNIME: A codeless automated machine learning-based virtual screening workflow. *Journal of Computational Chemistry* 46(2), e70030. DOI: 10.1002/jcc.70030.

Goklany, S., Rizvi, N.F., Loring, R.H., Cram, E.J. and Lee-Parsons, C.W.T. (2013) Jasmonate-dependent alkaloid biosynthesis in *Catharanthus roseus* hairy root cultures is correlated with the relative expression of *Orca* and *Zct* transcription factors. *Biotechnology Progress* 29(6), 1367–1376. DOI: 10.1002/btpr.1801.

Goldhaber-Pasillas, G.D., Mustafa, N.R. and Verpoorte, R. (2014) Jasmonic acid effect on the fatty acid and terpenoid indole alkaloid accumulation in cell suspension cultures of *Catharanthus roseus*. *Molecules (Basel, Switzerland)* 19(7), 10242–10260. DOI: 10.3390/molecules190710242.

González-Beltrán, A., Maguire, E., Sansone, S.-A. and Rocca-Serra, P. (2014) LinkedISA: Semantic representation of ISA-tab experimental metadata. *BMC Bioinformatics* 15, S4. DOI: 10.1186/1471-2105-15-S14-S4.

Gu, L., Zhang, Z., Quan, H., Li, M., Zhao, F. *et al.* (2018) Integrated analysis of transcriptomic and metabolomic data reveals critical metabolic pathways involved in rotenoid biosynthesis in the medicinal plant *Mirabilis himalaica*. *Molecular Genetics and Genomics* 293(3), 635–647. DOI: 10.1007/s00438-017-1409-y.

Guo, Q., Li, F., Duan, Y., Wen, C., Wang, W. *et al.* (2020) Oxidative stress, nutritional antioxidants and beyond. *Science China Life Sciences* 63(6), 866–874. DOI: 10.1007/s11427-019-9591-5.

Guo, J., Wu, Y., Guo, F. and Wang, G. (2022) Proteomic and metabolomic analyses reveal stage-and tissue-specific flavonoid accumulation in *Ginkgo biloba*. *LWT* 171, 114111. DOI: 10.1016/j.lwt.2022.114111.

Guo, M., Lv, H., Chen, H., Dong, S., Zhang, J. *et al.* (2024) Strategies on biosynthesis and production of bioactive compounds in medicinal plants. *Chinese Herbal Medicines* 16(1), 13–26. DOI: 10.1016/j.chmed.2023.01.007.

Hidalgo-Lozada, G.M., Villarruel-López, A., Nuño, K., García-García, A., Sánchez-Nuño, Y.A. *et al.* (2024) Clinically effective molecules of natural origin for obesity prevention or treatment. *International Journal of Molecular Sciences* 25(5), 2671. DOI: 10.3390/ijms25052671.

Hou, C.-C., Chen, C.-H., Yang, N.-S., Chen, Y.-P., Lo, C.-P. *et al.* (2010) Comparative metabolomics approach coupled with cell- and gene-based assays for species classification and anti-inflammatory bioactivity validation of *Echinacea* plants. *The Journal of Nutritional Biochemistry* 21(11), 1045–1059. DOI: 10.1016/j.jnutbio.2009.08.010.

Huang, X., Zhang, W., Liao, Y., Ye, J. and Xu, F. (2024) Contemporary understanding of transcription factor regulation of terpenoid biosynthesis in plants. *Planta* 259(1), 2. DOI: 10.1007/s00425-023-04268-z.

Hussain, S., Sagar, T., Kaur, S., Nipunta, S.S., Kapoor, N. *et al.* (2023) Biosynthesis of bioactive compounds in medicinal and aromatic plants. In: *Biosynthesis of Bioactive Compounds in Medicinal and Aromatic Plants: Manipulation by Conventional and Biotechnological Approaches*. Springer Nature Switzerland, Cham, Switzerland, pp. 41–64.

Jain, D., Bisht, S., Parvez, A., Singh, K., Bhaskar, P. *et al.* (2024) Effective biotic elicitors for augmentation of secondary metabolite production in medicinal plants. *Agriculture* 14(6), 796. DOI: 10.3390/agriculture14060796.

Jia, L., Zuo, T., Zhang, C., Li, W., Wang, H. *et al.* (2019) Simultaneous profiling and holistic comparison of the metabolomes among the flower buds of *Panax ginseng*, *Panax quinquefolius*, and *Panax notoginseng* by UHPLC/IM-QTOF-HDMSE-based metabolomics analysis. *Molecules* 24(11), 2188. DOI: 10.3390/molecules24112188.

Jiang, Y., He, G., Li, R., Wang, K., Wang, Y. *et al.* (2024) Functional validation of the cytochrome P450 family PgCYP309 gene in *Panax ginseng*. *Biomolecules* 14(6), 715. DOI: 10.3390/biom14060715.

Jin, D., Henry, P., Shan, J. and Chen, J. (2021) Identification of chemotypic markers in three chemotype categories of cannabis using secondary metabolites profiled in inflorescences, leaves, stem bark, and roots. *Frontiers in Plant Science* 12, 699530. DOI: 10.3389/fpls.2021.699530.

Kalita, I., Bhattacharjee, S. and Saharia, M. (2025) Advancements in medicinal plant research. In: Chen, J.-T. (ed.) *Biotechnology, Multiple Omics, and Precision Breeding in Medicinal Plants*. CRC Press, Boca Raton, pp. 135–145.

Kam, M.Y.Y. and Yap, W.S.P. (2020) An oxidatively stressful situation: A case of *Artemisia annua* L. *Biotechnology & Genetic Engineering Reviews* 36(1), 1–31. DOI: 10.1080/02648725.2020.1749818.

Kang, G.G., Francis, N., Hill, R., Waters, D., Blanchard, C. *et al.* (2019) Dietary polyphenols and gene expression in molecular pathways associated with type 2 diabetes mellitus: A review. *International Journal of Molecular Sciences* 21(1), 140. DOI: 10.3390/ijms21010140.

Kasson, P.M. (2012) Computational biology in the cloud: Methods and new insights from computing at scale. In: Altman, R.B., Dunker, A.K., Hunter, L., Murray, T.A. and Klein, T.E. (eds) *Biocomputing 2013: Proceedings of the Pacific Symposium, Pacific Symposium on Biocomputing 2013*, Kohala Coast, Hawaii, USA, January 3–7, 2013, pp. 451–453.

Kim, S.W., Lee, S.H., Min, C.W., Jo, I.H., Bang, K.H. *et al.* (2017) Ginseng (*Panax* sp.) proteomics: An update. *Applied Biological Chemistry* 60(3), 311–320. DOI: 10.1007/s13765-017-0283-y.

Koppad, S., B, A., Gkoutos, G.V. and Acharjee, A. (2021) Cloud computing enabled big multi-omics data analytics. *Bioinformatics and Biology Insights* 15. DOI: 10.1177/11779322211035921.

Kumar, S., Saini, R., Suthar, P., Kumar, V. and Sharma, R. (2022) Plant secondary metabolites: Their food and therapeutic importance. In: Kumar Sharma, A. and Sharma, A. (eds) *Plant Secondary Metabolites*. Springer Nature Singapore, Singapore, pp. 371–413.

Krajewski, P., Chen, D., Ćwiek, H., Van Dijk, A.D.J., Fiorani, F. *et al.* (2015) Towards recommendations for metadata and data handling in plant phenotyping. *Journal of Experimental Botany* 66(18), 5417–5427. DOI: 10.1093/jxb/erv271.

Kumari, S. and Subramanya, H.S. (2020) Network pharmacology study of *Curcuma longa* L.: Potential target proteins and their functional enrichment analysis. *BMC Research Notes* 13(1), 468. DOI: 10.1186/s13104-020-05301-0.

Li, M., Cai, S., You, S. and Liu, Y. (2023a) Multiomics techniques for plant secondary metabolism engineering: Pathways to shape the bioeconomy. In: Lopez-Correa, C. and Suarez-Gonzalez, A. (eds) *Genomics and the Global Bioeconomy*. Academic Press, Cambridge, Massachusetts, pp. 205–252.

Liu, Y., Patra, B., Singh, S.K., Paul, P., Zhou, Y. *et al.* (2021) Terpenoid indole alkaloid biosynthesis in *Catharanthus roseus*: Effects and prospects of environmental factors in metabolic engineering. *Biotechnology Letters* 43(11), 2085–2103. DOI: 10.1007/s10529-021-03179-x.

Li, Z., Wang, Y., Xu, M., Liu, H., Li, L. *et al.* (2023b) Molecular mechanism overview of metabolite biosynthesis in medicinal plants. *Plant Physiology and Biochemistry* 204, 108125. DOI: 10.1016/j.plaphy.2023.108125.

Li, J., Huang, H.-Y., Lin, Y.-C.-D., Zuo, H., Tang, Y. *et al.* (2023c) *Cinnamomi ramulus* inhibits cancer cells growth by inducing G2/M arrest. *Frontiers in Pharmacology* 14, 1121799. DOI: 10.3389/fphar.2023.1121799.

Li, M., Ma, M., Wu, Z., Liang, X., Zheng, Q. *et al.* (2023d) Advances in the biosynthesis and metabolic engineering of rare ginsenosides. *Applied Microbiology and Biotechnology* 107(11), 3391–3404. DOI: 10.1007/s00253-023-12549-6.

Liang, S., Zhao, Y., Luo, Z., Liu, J., Jiang, X. *et al.* (2024) Integrated metabolomics and proteomics analysis reveals the accumulation mechanism of bioactive components in *Polygonatum odoratum*. *Frontiers in Plant Science* 15, 1487613. DOI: 10.3389/fpls.2024.1487613.

Likić, V.A., McConville, M.J., Lithgow, T. and Bacic, A. (2010) Systems biology: The next frontier for bioinformatics. *Advances in Bioinformatics* 2010, 1–10. DOI: 10.1155/2010/268925.

Liu, C. (2023) Targeted editing of the ginsenoside biosynthesis pathway in ginseng using CRISPRCas9 technology and its implications. *Medicinal Plant Research* 13. DOI: 10.5376/mpr.2023.13.0003.

Liu, C., Xu, H., Li, Z., Wang, Y., Qiao, S. *et al.* (2025) Application and progress of genomics in deciphering the genetic regulation mechanisms of plant secondary metabolites. *Plants* 14(9), 1316. DOI: 10.3390/plants14091316.

Logie, E. and Vanden Berghe, W. (2020) Tackling chronic inflammation with withanolide phytochemicals–A withaferin a perspective. antioxidants. *Antioxidants (Basel, Switzerland)* 9(11), 1107. DOI: 10.3390/antiox9111107.

Lopes, E.M., Guimarães-Dias, F., Gama, T. do S.S., Macedo, A.L., Valverde, A.L. *et al.* (2020) *Artemisia annua* L. and photoresponse: From artemisinin accumulation, volatile profile and anatomical modifications to gene expression. *Plant Cell Reports* 39(1), 101–117. DOI: 10.1007/s00299-019-02476-0.

Ma, Y., Yuan, L., Wu, B., Li, X., Chen, S. *et al.* (2012) Genome-wide identification and characterization of novel genes involved in terpenoid biosynthesis in *Salvia miltiorrhiza*. *Journal of Experimental Botany* 63(7), 2809–2823. DOI: 10.1093/jxb/err466.

Majumdar, S. and Keller, A.A. (2021) Omics to address the opportunities and challenges of nanotechnology in agriculture. *Critical Reviews in Environmental Science and Technology* 51(22), 2595–2636. DOI: 10.1080/10643389.2020.1785264.

Mansueto, L., Kretzschmar, T., Mauleon, R. and King, G.J. (2024) Building a community-driven bioinformatics platform to facilitate *Cannabis sativa* multi-omics research. *GigaByte (Hong Kong, China)* 2024. DOI: 10.46471/gigabyte.137.

Marchese Robinson, R.L., Palczewska, A., Palczewski, J. and Kidley, N. (2017) Comparison of the predictive performance and interpretability of random forest and linear models on benchmark data sets. *Journal of Chemical Information and Modeling* 57(8), 1773–1792. DOI: 10.1021/acs.jcim.6b00753.

Mashamaite, C.V., Ramatsitsi, M.N. and Manyevere, A. (2024) *Moringa oleifera* Lam.: A versatile climate-smart plant for nutritional security and therapeutic usage in semi-arid regions. *Journal of Agriculture and Food Research* 16, 101217. DOI: 10.1016/j.jafr.2024.101217.

Meng, X., Li, G., Gu, L., Sun, Y., Li, Z. *et al.* (2020) Comparative metabolomic and transcriptome analysis reveal distinct flavonoid biosynthesis regulation between petals of white and purple *Phalaenopsis amabilis*. *Journal of Plant Growth Regulation* 39(2), 823–840. DOI: 10.1007/s00344-019-10025-y.

Mukherjee, P.K., Banerjee, S. and Kar, A. (2021) Molecular combination networks in medicinal plants: Understanding synergy by network pharmacology in indian traditional medicine. *Phytochemistry Reviews* 20(4), 693–703. DOI: 10.1007/s11101-020-09730-4.

Myoli, A., Choene, M., Kappo, A.P., Madala, N.E., Van der Hooft, J.J.J. *et al.* (2024) Charting the cannabis plant chemical space with computational metabolomics. *Metabolomics* 20(3), 62. DOI: 10.1007/s11306-024-02125-y.

Niazi, S.K. and Mariam, Z. (2023) Recent advances in machine-learning-based chemoinformatics: A comprehensive review. *International Journal of Molecular Sciences* 24(14), 11488. DOI: 10.3390/ijms241411488.

Noor, F., Tahir ul Qamar, M., Ashfaq, U.A., Albutti, A., Alwashmi, A.S.S. *et al.* (2022) Network pharmacology approach for medicinal plants: Review and assessment. *Pharmaceuticals* 15(5), 572. DOI: 10.3390/ph15050572.

Nureye, D. (2024) Reverse pharmacology and network pharmacology: Principles and applications in herbal medicine. *Journal of Traditional Medicine & Applications* 3(2), 1–24. DOI: 10.33140/JTMA.03.02.06.

Orsatti, F.L., de Queiroz Freitas, A.C., Borges, A.V.B. e, Santato, A.S., de Oliveira Assumpção, C. *et al.* (2025) Unveiling the role of exercise in modulating plasma heat shock protein 27 levels: Insights for exercise immunology and cardiovascular health. *Molecular and Cellular Biochemistry* 480(3), 1381–1401. DOI: 10.1007/s11010-024-05089-8.

Oyebamiji, Y.O., Adigun, B.A., Shamsudin, N.A.A., Ikmal, A.M., Salisu, M.A. *et al.* (2024) Recent advancements in mitigating abiotic stresses in crops. *Horticulturae* 10(2), 156. DOI: 10.3390/horticulturae10020156.

Öztürk, M., Sridhar, K.R., Sarwat, M., Altay, V. and Huerta-Martínez, F.M. (2023) *Ethnic Knowledge and Perspectives of Medicinal Plants*. Apple Academic Press, New York.

Pan, Q., Mustafa, N.R., Tang, K., Choi, Y.H. and Verpoorte, R. (2016) Monoterpenoid indole alkaloids biosynthesis and its regulation in *Catharanthus roseus*: A literature review from genes to metabolites. *Phytochemistry Reviews* 15(2), 221–250. DOI: 10.1007/s11101-015-9406-4.

Pathak, S., Agarwal, A.V., Agarwal, P. and Trivedi, P.K. (2019) Secondary metabolite pathways in medicinal plants: Approaches in reconstruction and analysis. In: Singh, S.P., Kumar Upadhyay, S., Pandey, A. and Kumar, S. (eds) *Molecular Approaches in Plant Biology and Environmental Challenges*. Springer, Singapore, pp. 339–364.

Pillay, L.C., Nekati, L., Makhwitine, P.J. and Ndlovu, S.I. (2022) Epigenetic activation of silent biosynthetic gene clusters in endophytic fungi using small molecular modifiers. *Frontiers in Microbiology* 13, 815008. DOI: 10.3389/fmicb.2022.815008.

Punetha, A., Kumar, D., Suryavanshi, P., Padalıa, R. and Venkatesha, K.T. (2022) Environmental abiotic stress and secondary metabolites production in medicinal plants: A review. *Journal of Agricultural Sciences* 28(3), 351–362. DOI: 10.15832/ankutbd.999117.

Purushotham, P.M., Kim, J., Jo, E. and Senthil, K. (2017) Withanolides against TLR4-activated innate inflammatory signalling pathways: A comparative computational and experimental study. *Phytotherapy Research* 31(1), 152–163. DOI: 10.1002/ptr.5746.

Rahaman, M.M., Islam, M.R., Bhuiyan, M.M.R., Noman, I.R., Aziz, M.M. *et al.* (2025) Harnessing big data in biotechnology: A machine learning approach to multi-omics. In: *2025 International Conference on Machine Learning and Autonomous Systems (ICMLAS)*, IEEE, Prawet, Thailand, March 10, 2025, pp. 1391–1401.

Rai, A., Saito, K. and Yamazaki, M. (2017) Integrated omics analysis of specialized metabolism in medicinal plants. *The Plant Journal* 90(4), 764–787. DOI: 10.1111/tpj.13485.

Šamec, D., Karalija, E., Dahija, S. and Hassan, S.T.S. (2022) Biflavonoids: Important contributions to the health benefits of ginkgo (*Ginkgo biloba* L.). *Plants* 11(10), 1381. DOI: 10.3390/plants11101381.

Satish, L., Seher, Y., Rakkammal, K., Muthuramalingam, P., Lakshmi, C.R. *et al.* (2022) Metabolic engineering strategies to enhance the production of anticancer drug, paclitaxel. In: Kumara Swamy, M., Pullaiah, T. and Chen, Z.-S. (eds) *Paclitaxel*. Academic Press, Cambridge, Massachusetts, pp. 229–250.

Satrio, R.D., Fendiyanto, M.H. and Miftahudin, M. (2024) Tools and techniques used at global scale through genomics, transcriptomics, proteomics, and metabolomics to investigate plant stress responses at the molecular level. In: Shahid, M. and Gaur, R. (eds) *Molecular Dynamics of Plant Stress and Its Management*. Springer Nature Singapore, Singapore, pp. 555–607.

Schilling, S., Dowling, C.A., Shi, J., Ryan, L., Hunt, D. *et al.* (2020) The cream of the crop: Biology, breeding and applications of *Cannabis sativa*. *Authorea Preprints*. DOI: 10.22541/au.160139712.25104053/v2.

Sha, F., Zhang, J., Yang, H., Hu, Y., Wei, W. *et al.* (2024) Systematical targeted multicomponent characterization and comparison of *Arnebiae radix* and its three confusing species by offline two-dimensional liquid chromatography/LTQ-orbitrap mass spectrometry. *Analytical and Bioanalytical Chemistry* 416(2), 583–595. DOI: 10.1007/s00216-023-05067-x.

Shao, Y., Wei, J., Wu, F., Zhang, H., Yang, D. *et al.* (2016) DsTRD: Danshen transcriptional resource database. *PLoS ONE* 11(2), e0149747. DOI: 10.1371/journal.pone.0149747.

Sharma, S., Naman, S., Dwivedi, J., Ashawat, M.S., Chandan, A. *et al.* (2025) Compliance with regulatory standards: A comprehensive review on *Ocimum sanctum* (tulsi). *Applied Drug Research, Clinical Trials and Regulatory Affairs* 10, e26673371313587. DOI: 10.2174/0126673371313587240925051006.

She, J., Yan, H., Yang, J., Xu, W. and Su, Z. (2019) CroFGD: *Catharanthus roseus* functional genomics database. *Frontiers in Genetics* 10, 238. DOI: 10.3389/fgene.2019.00238.

Silva, A.C. de J., Fassio, A.V., Barcelos, M.P. and Hage-Melim, L.I. da S. (2024) Herbal medicines: From history to current research—a comprehensive survey. In: Taft, C.A. and Lazaro, S.R. (eds) *Progress in Hydrogen Energy, Fuel Cells, Nano-Biotechnology and Advanced, Bioactive Compounds*. Springer Nature Switzerland, Cham, Switzerland, pp. 315–351.

Singh, K.S., Van der Hooft, J.J.J., Van Wees, S.C.M. and Medema, M.H. (2022) Integrative omics approaches for biosynthetic pathway discovery in plants. *Natural Product Reports* 39(9), 1876–1896. DOI: 10.1039/D2NP00032F.

Sobral, P.S., Luz, V.C.C., Almeida, J.M.G.C.F., Videira, P.A. and Pereira, F. (2023) Computational approaches drive developments in immune-oncology therapies for PD-1/PD-L1 immune checkpoint inhibitors. *International Journal of Molecular Sciences* 24(6), 5908. DOI: 10.3390/ijms24065908.

Song, C., Cao, Y., Dai, J., Li, G., Manzoor, M.A. *et al.* (2022) The multifaceted roles of MYC2 in plants: Toward transcriptional reprogramming and stress tolerance by jasmonate signaling. *Frontiers in Plant Science* 13, 868874. DOI: 10.3389/fpls.2022.868874.

Song, Y., Zhang, Y., Wang, X., Yu, X., Liao, Y. *et al.* (2024) Telomere-to-telomere reference genome for *Panax ginseng* highlights the evolution of saponin biosynthesis. *Horticulture Research* 11(6). DOI: 10.1093/hr/uhae107.

Spietelun, A., Kloskowski, A., Chrzanowski, W. and Namieśnik, J. (2013) Understanding solid-phase microextraction: Key factors influencing the extraction process and trends in improving the technique. *Chemical Reviews* 113(3), 1667–1685. DOI: 10.1021/cr300148j.

Su, Z.-W., Yan, T.-Y., Feng, J., Zhang, M.-Y., Han, L. *et al.* (2023) Protective effects and mechanism of polysaccharides from edible medicinal plants in alcoholic liver injury: A review. *International Journal of Molecular Sciences* 24(22), 16530. DOI: 10.3390/ijms242216530.

Subramanian, I., Verma, S., Kumar, S., Jere, A. and Anamika, K. (2020) Multi-omics data integration, interpretation, and its application. *Bioinformatics and Biology Insights* 14, 1177932219899051. DOI: 10.1177/1177932219899051.

Surendran, K., Ranjisha, K.R., Aswati Nair, R. and Pillai, P.P. (2022) Genomics and metabolomics: A strategy for elucidation of metabolic pathways in medicinal plants. In: Kumara Swamy, M. and Kumar, A. (eds) *Phytochemical Genomics*. Springer Nature Singapore, Singapore, pp. 343–360.

Swaroop, A.K., Motamarri Venkata Naga Lalitha, C., Shanmugam, M., Subramanian, G., Natarajan, J. *et al.* (2021) Plant derived immunomodulators; A critical review. *Advanced Pharmaceutical Bulletin* 12(4), 712–729. DOI: 10.34172/apb.2022.074.

Thakur, N., Nigam, M., Mann, N.A., Gupta, S., Hussain, C.M. *et al.* (2023) Host-mediated gene engineering and microbiome-based technology optimization for sustainable agriculture and environment. *Functional & Integrative Genomics* 23(1), 57. DOI: 10.1007/s10142-023-00982-9.

Tian, Y., Yang, Z., Wang, H. and Yan, A. (2023) Prediction of bioactivities of microsomal prostaglandin. *Chemical Biology & Drug Design* 101(6), 1307–1321. DOI: 10.1111/cbdd.14214.

Torabi, M., Haririan, I., Foroumadi, A., Ghanbari, H. and Ghasemi, F. (2024) A deep learning model based on the BERT pre-trained model to predict the antiproliferative activity of anti-cancer chemical compounds. *SAR and QSAR in Environmental Research* 35(11), 971–992. DOI: 10.1080/1062936X.2024.2431486.

Trivadila, T., Iswantini, D., Rahminiwati, M., Rafi, M., Salsabila, A.P. *et al.* (2025) Herbal immunostimulants and their phytochemicals: Exploring *Morinda citrifolia*, *Echinacea purpurea*, and *Phyllanthus niruri*. *Plants* 14(6), 897. DOI: 10.3390/plants14060897.

Vergara, D., Gaudino, R., Blank, T. and Keegan, B. (2020) Modeling cannabinoids from a large-scale sample of *Cannabis sativa* chemotypes. *PLOS ONE* 15(9), e0236878. DOI: 10.1371/journal.pone.0236878.

Verma, M., Ghangal, R., Sharma, R., Sinha, A.K. and Jain, M. (2014) Transcriptome analysis of *Catharanthus roseus* for gene discovery and expression profiling. *PLoS ONE* 9(7), e103583. DOI: 10.1371/journal.pone.0103583.

Walterbusch, M., Martens, B. and Teuteberg, F. (2013) Evaluating cloud computing services from a total cost of ownership perspective. *Management Research Review* 36(6), 613–638. DOI: 10.1108/01409171311325769.

Wang, J., Guo, Y., Yin, X., Wang, X., Qi, X. *et al*. (2022) Diverse triterpene skeletons are derived from the expansion and divergent evolution of 2,3-oxidosqualene cyclases in plants. *Critical Reviews in Biochemistry and Molecular Biology* 57(2), 113–132. DOI: 10.1080/10409238.2021.1979458.

Wang, C., Meng, L., Zhang, G., Yang, X., Pang, B. *et al*. (2024a) Unraveling crop enzymatic browning through integrated omics. *Frontiers in Plant Science* 15, 1342639. DOI: 10.3389/fpls.2024.1342639.

Wang, M., Zhang, S., Li, R. and Zhao, Q. (2024b) Unraveling the specialized metabolic pathways in medicinal plant genomes: A review. *Frontiers in Plant Science* 15, 1459533. DOI: 10.3389/fpls.2024.1459533.

Wang, C., Sun, H., Yang, Y., Peng, C., Liu, Y. *et al*. (2025a) Coordinated gene expression and hormonal fluxes dictating ginsenoside Rb3 biosynthesis in floral development of *Panax notoginseng*. *BMC Plant Biology* 25(1), 177. DOI: 10.1186/s12870-025-06149-x.

Wang, M., Chen, L., Zhang, Z. and Wang, Q. (2025b) Recent advances in genome mining and synthetic biology for discovery and biosynthesis of natural products. *Critical Reviews in Biotechnology* 45(1), 236–256. DOI: 10.1080/07388551.2024.2383754.

Waris, M., Koçak, E., Gonulalan, E.M., Demirezer, L.O., Kır, S. *et al*. (2022) Metabolomics analysis insight into medicinal plant science. *TrAC Trends in Analytical Chemistry* 157(2), 116795. DOI: 10.1016/j.trac.2022.116795.

Xia, J., Lou, G., Zhang, L., Huang, Y., Yang, J. *et al*. (2023a) Unveiling the spatial distribution and molecular mechanisms of terpenoid biosynthesis in *Salvia miltiorrhiza* and *S. grandifolia* using multi-omics and DESI-MSI. *Horticulture Research* 10(7), uhad109. DOI: 10.1093/hr/uhad109.

Xia, Z., Sun, B., Wen, J.-F., Ma, R.-X., Wang, F.-Y. *et al*. (2023b) Research progress on metabolomics in the quality evaluation and clinical study of *Panax ginseng*. *Biomedical Chromatography* 37(7), e5546. DOI: 10.1002/bmc.5546.

Xiong, X., Gou, J., Liao, Q., Li, Y., Zhou, Q. *et al*. (2021) The *Taxus* genome provides insights into paclitaxel biosynthesis. *Nature Plants* 7(8), 1026–1036. DOI: 10.1038/s41477-021-00963-5.

Yan, Y., Abdulla, R., Ma, Q. and Aisa, H.A. (2023) Comprehensive identification of chemical fingerprint and screening of potential quality markers of *Aloe vera* (L.) Burm. f. from different geographical origins via ultra-high-performance liquid chromatography hyphenated with quadrupole–orbitrap-high-resolution mass spectrometry combined with chemometrics. *Journal of Chromatographic Science* 61(4), 312–321. DOI: 10.1093/chromsci/bmad009.

Yang, G. and Albert, R. (2019) Modeling of molecular networks. In: Bianchi, A., Hillen, T., Lewis, M.A. and Yi, Y. (eds) *The Dynamics of Biological Systems*. Springer International Publishing, Cham, Switzerland, pp. 35–62.

Yang, C.S., Wang, H. and Sheridan, Z.P. (2018) Studies on prevention of obesity, metabolic syndrome, diabetes, cardiovascular diseases and cancer by tea. *Journal of Food and Drug Analysis* 26(1), 1–13. DOI: 10.1016/j.jfda.2017.10.010.

Yang, C., Li, C., Wei, W., Wei, Y., Liu, Q. *et al*. (2020) The unprecedented diversity of UGT94-family UDP-glycosyltransferases in Panax plants and their contribution to ginsenoside biosynthesis. *Scientific Reports* 10(1), 15394. DOI: 10.1038/s41598-020-72278-y.

Yin, X., Hu, H., Shen, X., Li, X., Pei, J. *et al*. (2021) Ginseng omics for ginsenoside biosynthesis. *Current Pharmaceutical Biotechnology* 22(5), 570–578. DOI: 10.2174/1389201021666200807113723.

Yin, Q., Wu, T., Gao, R., Wu, L., Shi, Y. *et al*. (2023) Multi-omics reveal key enzymes involved in the formation of phenylpropanoid glucosides in *Artemisia annua*. *Plant Physiology and Biochemistry* 201, 107795. DOI: 10.1016/j.plaphy.2023.107795.

Yuan, X., Li, R., He, W., Xu, W., Xu, W. *et al*. (2024) Progress in identification of UDP-glycosyltransferases for ginsenoside biosynthesis. *Journal of Natural Products* 87(4), 1246–1267. DOI: 10.1021/acs.jnatprod.3c00630.

Zeng, T., Su, H., Wang, M., He, J., Gu, L. *et al*. (2025) The role of MYC2 transcription factors in plant secondary metabolism and stress response mechanisms. *Plants* 14(8), 1255. DOI: 10.3390/plants14081255.

Zhang, H., Hedhili, S., Montiel, G., Zhang, Y., Chatel, G. *et al*. (2011) The basic helix-loop-helix transcription factor CrMYC2 controls the jasmonate-responsive expression of the *ORCA* genes that regulate alkaloid biosynthesis in *Catharanthus roseus*. *The Plant Journal* 67(1), 61–71. DOI: 10.1111/j.1365-313X.2011.04575.x.

Zhang, J.-Y., Cun, Z., Wu, H.-M. and Chen, J.-W. (2020) Integrated analysis on biochemical profiling and transcriptome revealed nitrogen-driven difference in accumulation of saponins in a medicinal plant *Panax notoginseng*. *Plant Physiology and Biochemistry* 154, 564–580. DOI: 10.1016/j.plaphy.2020.06.049.

Zhang, L., Wang, G., Wang, G. and Cao, F. (2021a) *Ginkgo biloba* L. responds to red and blue light: Via phenylpropanoid and flavonoid biosynthesis pathway. *Forests* 12(8), 1079. DOI: 10.3390/f12081079.

Zhang, L., Zhang, Z., Fang, S., Liu, Y. and Shang, X. (2021b) Integrative analysis of metabolome and transcriptome reveals molecular regulatory mechanism of flavonoid biosynthesis in *Cyclocarya paliurus* under salt stress. *Industrial Crops and Products* 170, 113823. DOI: 10.1016/j.indcrop.2021.113823.

Zhang, Y., Wiese, L., Fang, H., Alseekh, S., Perez de Souza, L. *et al*. (2023b) Synthetic biology identifies the minimal gene set required for paclitaxel biosynthesis in a plant chassis. *Molecular Plant* 16(12), 1951–1961. DOI: 10.1016/j.molp.2023.10.016.

Zhang, L., Cai, Y., Zhong, Q., Zhang, S., Shen, L. *et al*. (2024a) Comprehensive characterization of active components in *Salvia miltiorrhiza* using polarity-partitioned two-dimensional liquid chromatography coupled with mass spectrometry. *Journal of Chromatography A* 1736, 465424. DOI: 10.1016/j.chroma.2024.465424.

Zhang, M., Yang, B., Wang, Y. and Yu, F. (2024b) CrJAT1 regulates endogenous JA signaling for modulating monoterpenoid indole alkaloid biosynthesis in *Catharanthus roseus*. *Genes* 15(3), 324. DOI: 10.3390/genes15030324.

Zhang, W., Zeng, Y., Jiao, M., Ye, C., Li, Y. *et al*. (2023a) Integration of high-throughput omics technologies in medicinal plant research: The new era of natural drug discovery. *Frontiers in Plant Science* 14, 1073848. DOI: 10.3389/fpls.2023.1073848.

Zhao, C., Zhang, Z., Sun, L., Bai, R., Wang, L. *et al*. (2023a) Genome sequencing provides potential strategies for drug discovery and synthesis. *Acupuncture and Herbal Medicine* 3(4), 244–255. DOI: 10.1097/HM9.0000000000000076.

Zhao, Y., Liu, G., Yang, F., Liang, Y., Gao, Q. *et al*. (2023b) Multilayered regulation of secondary metabolism in medicinal plants. *Molecular Horticulture* 3(1), 11. DOI: 10.1186/s43897-023-00059-y.

Zhao, C., Bai, X., Ding, Y., Wen, A. and Fu, Q. (2024a) Combining systems pharmacology, metabolomics, and transcriptomics to reveal the mechanism of *Salvia miltiorrhiza-Cortex moutan* herb pair for the treatment of ischemic stroke. *Frontiers in Pharmacology* 15, 1431692. DOI: 10.3389/fphar.2024.1431692.

Zhao, M., Che, Y., Gao, Y. and Zhang, X. (2024b) Application of multi-omics in the study of traditional Chinese medicine. *Frontiers in Pharmacology* 15, 1431862. DOI: 10.3389/fphar.2024.1431862.

3 Precision Breeding and Genome Editing in Medicinal Plants

Archana Jain, Surendra Sarsaiya and Qihai Gong*
Key Laboratory of Basic Pharmacology and Joint International Research Laboratory of Ethnomedicine of Ministry of Education, Zunyi Medical University, Zunyi, Guizhou, China

3.1 Introduction

Medicinal plants are pivotal to both traditional and modern medicine due to their ability to synthesize bioactive secondary metabolites such as alkaloids, terpenoids, and flavonoids. These compounds are indispensable for pharmaceuticals, with over 25% of modern drugs derived from plant sources (Roy *et al.*, 2022). Secondary metabolites exhibit diverse therapeutic properties, including antimicrobial, anti-inflammatory, and anti-cancer activities, forming the basis of drugs like aspirin and quinine (Safdar *et al.*, 2023). Beyond their medicinal value, these metabolites play critical roles in plant defense mechanisms, protecting against pathogens and environmental stressors (Kaur and Ahmed, 2021). However, their natural production is inherently limited by ecological variability, genetic constraints, and agronomic challenges, which are further exacerbated by climate change, habitat degradation, and unsustainable harvesting practices (Ona *et al.*, 2024).

Climate change intensifies these challenges by destabilizing ecosystems, increasing pest invasions, and reducing the availability of key species, thereby threatening both biodiversity and global pharmaceutical supply chains. Unsustainable harvesting further compounds these risks, risking extinction (Ona *et al.*, 2024). To address these limitations, precision-breeding and genome-editing technologies, particularly CRISPR-Cas9 (clustered regularly interspaced short palindromic repeats/associated protein 9), have emerged as revolutionary tools. These approaches enable targeted modifications in biosynthetic pathways, stress-responsive genes, and regulatory elements, surpassing traditional mutagenesis in precision and efficiency (Asha and Sunita, 2024). For instance, CRISPR-mediated knockout of repressor genes in alkaloid pathways or fine-tuning of transcription factors can amplify metabolite yields without compromising plant vitality (Arun *et al.*, 2023). Innovations like base editing and prime editing further enhance precision, enabling single-nucleotide changes without double-strand breaks, thereby minimizing off-target effects (Das *et al.*, 2024).

CRISPR-Cas9 also enhances biotic and abiotic stress tolerance—critical for medicinal plants facing salinity, extreme temperatures, and pathogen attacks (Shah *et al.*, 2023). Concurrently, synthetic biology complements genome editing by constructing artificial pathways or rewiring existing ones to optimize

*Corresponding author: gqh@zmu.edu.cn

DOI: 10.1079/9781836994206.0003

high-value metabolite production (e.g. terpenoids, phenolics) and even synthesize novel compounds (Asha and Sunita, 2024). Industrial scalability is exemplified by CRISPR-driven innovations in *Mentha* spp., where gene editing combined with growth promoters boosted menthol yields by over 30%, demonstrating the synergy of genetic and agronomic strategies (Ray *et al.*, 2023).

However, translating laboratory success into market viability faces hurdles. Limited genomic data and inefficient transformation/regeneration protocols hinder CRISPR applications in non-model medicinal species (Arun *et al.*, 2023). Regulatory fragmentation—such as the European Union's (EU) process-focused vs the US's product-focused frameworks—complicates commercialization (Asghar *et al.*, 2025). Ethical concerns, including gene flow risks and corporate monopolization of CRISPR technologies, alongside socio-economic disparities and public skepticism fueled by anti-genetically modified organism (GMO) sentiments, pose additional barriers (Ray *et al.*, 2023). Harmonizing international policies to balance biosafety with innovation, coupled with transparent public engagement, is critical for equitable adoption (Asghar *et al.*, 2025).

Ultimately, integrating CRISPR and synthetic biology with sustainable practices—such as ecological cultivation, habitat conservation, and leveraging local genetic resources—is essential to preserve biodiversity, maintain genetic diversity, and ensure the long-term viability of medicinal plants (Ona *et al.*, 2024). This holistic approach promises to revolutionize pharmaceutical innovation while addressing global health and environmental challenges (Fig. 3.1).

3.2 CRISPR-Cas9 and Synthetic Biology Tools

The CRISPR-Cas9 system and synthetic biology tools have revolutionized the field of genome editing, particularly in medicinal plants, by enabling precise modifications to enhance the production of valuable secondary metabolites (Table 3.1 and Fig. 3.2). These metabolites, such as alkaloids, terpenoids, and flavonoids, are crucial for the pharmaceutical industry due to their therapeutic properties. CRISPR-Cas9 offers a targeted approach to manipulate genes involved in the biosynthetic pathways of these compounds, thus improving their yield and quality. This technology surpasses traditional breeding methods by providing a more efficient and precise means of genetic modification. The following sections delve into the specifics of CRISPR-Cas9 applications and the role of synthetic biology in medicinal plants.

CRISPR-Cas9 facilitates the targeted mutagenesis of genes that are critical in the biosynthetic pathways of secondary metabolites. This allows for the enhancement of specific compounds by knocking out genes that negatively regulate their production (Naik *et al.*, 2022). Gene regulation and transcription control systems can also be used to regulate gene expression, thereby optimizing the production of desired metabolites. This is achieved by modifying transcription factors or other regulatory elements within the plant genome (Naik *et al.*, 2022). CRISPR-Cas9 has been employed to improve stress tolerance in medicinal plants, which is crucial for maintaining high yields of secondary metabolites under adverse environmental conditions (Insaf *et al.*, 2023). Recent developments include engineered Cas9 proteins and newly identified CRISPR systems like Cas9-NG and CRISPR-Cas9, which offer enhanced specificity and efficiency for genome editing in plants (Yaseen *et al.*, 2023). Advanced base and prime editing techniques allow for precise nucleotide changes without introducing double-strand breaks, reducing the risk of off-target effects and improving the accuracy of genome modifications (Yaseen *et al.*, 2023). Synthetic biology tools, in conjunction with CRISPR-Cas9, enable the construction of synthetic pathways for the production of novel compounds or the enhancement of existing ones. This approach can lead to the development of medicinal plants with optimized profiles for pharmaceutical use (Jamwal *et al.*, 2023). CRISPR-Cas9 aids in functional genomics studies by allowing researchers to dissect the roles of specific genes in secondary metabolite biosynthesis, thereby facilitating the design of plants with improved therapeutic properties (Hu *et al.*, 2016).

While CRISPR-Cas9 and synthetic biology tools offer significant advantages in the genetic

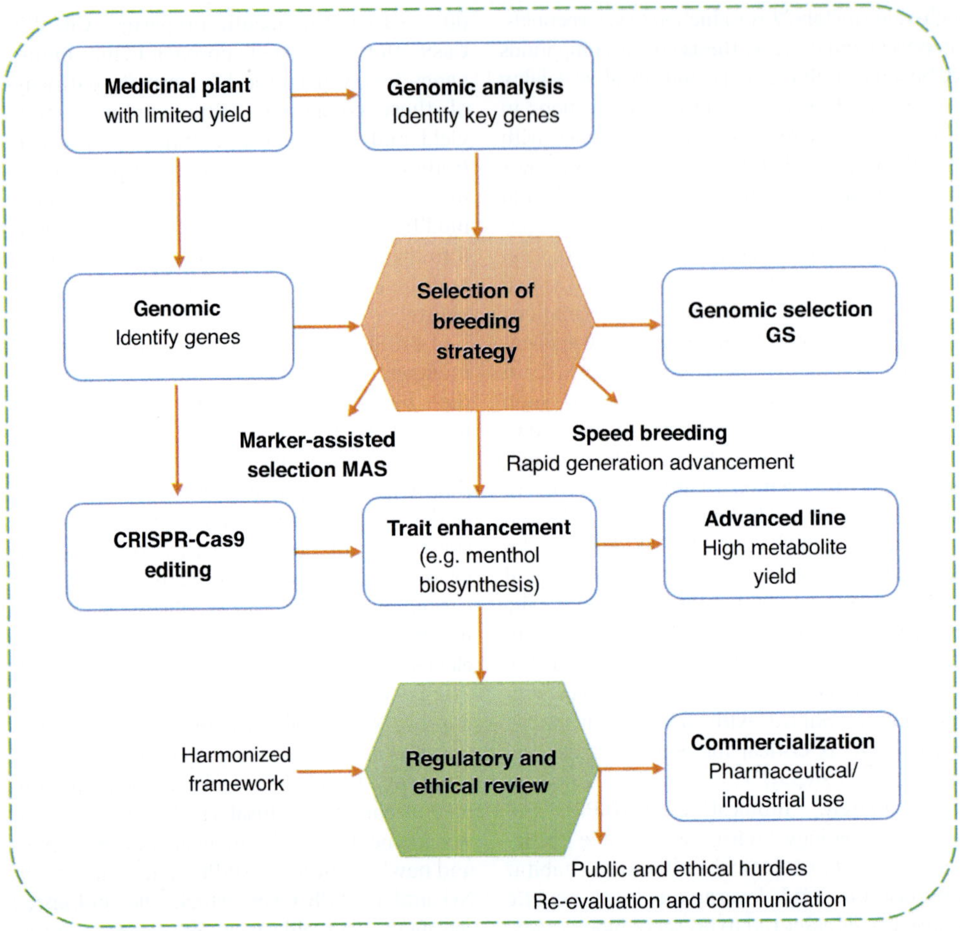

Fig. 3.1. Overall workflow for precision breeding in medicinal plants. (Author's own figure.)

engineering of medicinal plants, challenges remain. The application of these technologies is still in its nascent stages for many medicinal plant species due to limited genomic information and transformation techniques. Additionally, ethical and regulatory considerations must be addressed to ensure the safe and responsible use of genome-editing technologies in agriculture and medicine (Insaf *et al.*, 2023). CRISPR-Cas9 technology has emerged as a transformative tool in precision breeding and genome editing of medicinal plants, particularly for trait-specific modifications such as disease resistance and metabolite enhancement. This technology allows for precise genetic modifications, enabling the development of plants with improved traits by targeting specific genes involved in biosynthetic pathways. The application of CRISPR-Cas9 in medicinal plants is still in its nascent stages, but it holds significant promise for enhancing the production of valuable secondary metabolites and improving plant resilience to biotic and abiotic stresses.

CRISPR-Cas9 facilitates the manipulation of genes involved in the biosynthetic pathways of secondary metabolites, such as alkaloids, terpenoids, flavonoids, and phenolics, which are crucial for the medicinal properties of plants (Naik *et al.*, 2022). By targeting specific genes, CRISPR-Cas9 can enhance the production of these metabolites, thereby increasing the medicinal value of plants and reducing production costs (Das *et al.*, 2024). The technology's ability to create minor heritable mutations is

Table 3.1. Applications and challenges of CRISPR-Cas9 and synthetic biology in medicinal plants. Author's own table.

Application domain	Specific action/ strategy	Key outcome/goal	Associated challenge/limitation	Citations
Enhancement of secondary metabolites	Knockout of repressor genes or negative regulators in biosynthetic pathways (e.g. for alkaloids, terpenoids).	Amplified yield of high-value compounds (e.g. boosted menthol yields by >30% in *Mentha* spp.).	Limited genomic data for non-model medicinal plant species hinders target identification.	(Arun *et al.*, 2023; Ray *et al.*, 2023)
	Fine-tuning of transcription factors and regulatory elements to control gene expression.	Optimization of metabolic flux toward desired compounds without compromising plant vitality.	Inefficient transformation and regeneration protocols for many recalcitrant medicinal species.	(Naik *et al.*, 2022; Arun *et al.*, 2023)
Precision genome editing	Use of base editing and prime editing techniques.	Introduction of precise nucleotide changes without double-strand breaks, minimizing off-target effects.	Technology is still nascent in medicinal plants; requires highly specific guide RNA design and efficient delivery.	(Yaseen *et al.*, 2023; Das *et al.*, 2024)
	Deployment of engineered systems (e.g. Cas9-NG, CRISPR-Cas9).	Enhanced specificity and efficiency of edits, expanding the range of editable genomic sites.	Discovery and adaptation of new systems for diverse plant species is an ongoing process.	(Yaseen *et al.*, 2023)
Stress resilience	Editing of stress-responsive genes (biotic and abiotic).	Improved tolerance to salinity, extreme temperatures, and pathogen attacks, securing metabolite yield.	Complexity of polygenic stress traits makes multi-gene editing necessary.	(Insaf *et al.*, 2023; Shah *et al.*, 2023)
Synthetic pathway engineering	Rewiring existing pathways or constructing artificial biosynthetic pathways using synthetic biology.	Production of novel compounds or optimized profiles of high-value metabolites (e.g. terpenoids, phenolics).	Integration and stable expression of complex synthetic gene circuits in plants remains challenging.	(Jamwal *et al.*, 2023; Asha and Sunita, 2024)
Functional genomics	Use of CRISPR-Cas9 for gene knockout or modulation to dissect gene function.	Elucidation of roles of specific genes in secondary metabolite biosynthesis for informed engineering.	Dependent on the availability of high-quality genome annotations.	(Hu *et al.*, 2016)
Disease resistance	Precise insertions/ deletions in genes associated with disease resistance.	Development of disease-resistant plant varieties, reducing reliance on chemical pesticides.	Potential fitness trade-offs between resistance and other agronomic or metabolic traits.	(Lemke, 2022; Sharma *et al.*, 2024b)
Commercialization & regulation	Navigating product-focused (e.g. US) vs process-focused (e.g. EU) regulatory frameworks.	Aiming for market viability and commercialization of edited medicinal plants.	Regulatory fragmentation and lack of international harmonization.	(Asghar *et al.*, 2025)

Continued

Table 3.1. Continued

Application domain	Specific action/ strategy	Key outcome/goal	Associated challenge/limitation	Citations
Socio-ethical adoption	Public engagement and transparent communication about technology.	Equitable adoption and acceptance of CRISPR-edited plants.	Public skepticism fueled by anti-GMO sentiments; ethical concerns over gene flow and corporate monopolization.	(Ray *et al.*, 2023; Asghar *et al.*, 2025)

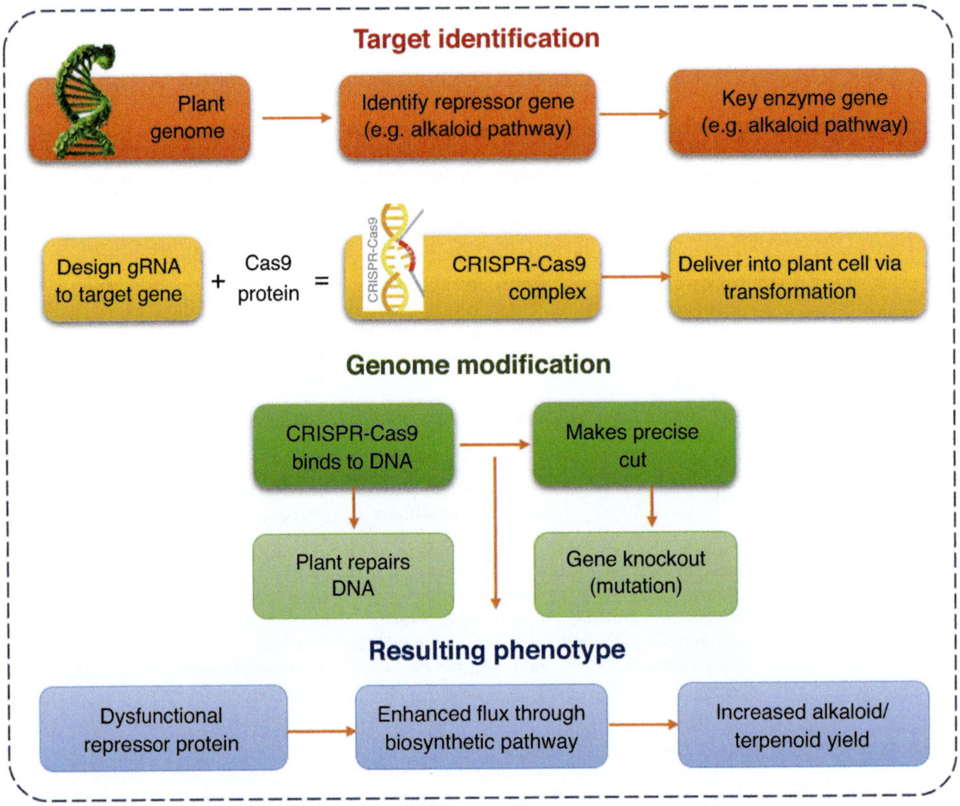

Fig. 3.2. The CRISPR-Cas9 gene-editing process for metabolic engineering. (Author's own figure.)

significant for elucidating and optimizing secondary metabolite pathways, which can lead to the development of plants with enhanced nutritional and pharmaceutical profiles (Arun *et al.*, 2023).

CRISPR-Cas9 has been effectively used to enhance disease resistance in plants by enabling precise genetic insertions and deletions that improve plant resilience against phytopathogens. This is crucial for reducing reliance on chemical pesticides, which pose environmental and health risks (Sharma *et al.*, 2024b). The technology allows for the rapid generation of disease-resistant plant varieties, which is essential for maintaining global food security and agricultural productivity (Sharma *et al.*, 2024b). CRISPR-Cas9's precision in editing specific genes associated with disease resistance makes it a valuable tool in modern plant breeding programs, offering an alternative to traditional breeding methods (Lemke, 2022).

3.3 Marker-assisted vs Genomic Selection Strategies

Marker-assisted selection (MAS) and genomic selection (GS) are two prominent strategies in plant and animal breeding, each with distinct methodologies and applications. MAS focuses on using molecular markers linked to specific traits for selection, while GS employs genome-wide marker data to predict the genetic merit of individuals. Both strategies aim to enhance breeding efficiency and genetic gain, but they differ in their approach to handling complex traits and the scale of marker data used. The integration of MAS and GS can potentially leverage the strengths of both methods to improve breeding outcomes.

3.3.1 MAS

MAS is particularly effective for traits controlled by a few major genes, where markers are closely linked to the trait of interest. This allows for precise selection of desirable alleles, especially in cases of simply inherited traits (Nayak *et al.*, 2017).

MAS has been successfully used in various breeding programs, including the development of disease-resistant varieties and the improvement of specific traits like nematode resistance in peanuts (Chu *et al.*, 2011). The effectiveness of MAS is limited for complex traits governed by multiple genes with small effects. The requirement for prior knowledge of trait-linked markers and the challenges in detecting quantitative trait loci (QTL) for polygenic traits restrict its broader application (Varshney *et al.*, 2017).

3.3.2 GS

GS uses genome-wide marker data to predict the genetic value of individuals, capturing the effects of all markers simultaneously. This approach is particularly advantageous for complex traits influenced by numerous small-effect genes (Varshney *et al.*, 2017). GS can significantly reduce the breeding cycle duration and increase selection accuracy, leading to higher genetic gains per unit time. It is effective in improving crop yield and other complex traits by utilizing high-density single nucleotide polymorphism (SNP) markers distributed across the genome (Unêda-Trevisoli *et al.*, 2017). While GS offers substantial benefits, it requires large training populations and sophisticated statistical models to accurately estimate breeding values. The potential for increased inbreeding and reduced genetic diversity is a concern that needs to be managed (Degen and Müller, 2023).

3.3.3 Integration of MAS and GS

Combining MAS and GS can exploit the advantages of both methods, providing a more comprehensive selection strategy. This integration can enhance the precision of selection for both simple and complex traits, improving breeding efficiency and accelerating the development of new varieties (Verma *et al.*, 2024). The integration is particularly beneficial in scenarios where both major gene effects and polygenic traits need to be considered, such as in the development of crops with improved functional traits and resistance to biotic and abiotic stresses (Moniruzzaman *et al.*, 2015). MAS and GS each have their strengths and limitations; the choice between them often depends on the specific breeding objectives and the nature of the traits being targeted. MAS remains a valuable tool for traits with known genetic markers, while GS is increasingly favored for its ability to handle complex traits and deliver rapid genetic gains. The ongoing advancements in genomics and sequencing technologies continue to enhance the feasibility and effectiveness of both strategies, offering promising prospects for future breeding programs.

3.3.4 Comparative analysis of breeding methods for precision and scalability

The comparative analysis of breeding methods for precision and scalability reveals a dynamic landscape where traditional and modern techniques each offer unique advantages and challenges (Table 3.2). Traditional methods, such as mass selection and hybridization, have been foundational in plant and animal breeding, providing cost-effective and straightforward approaches. However, modern techniques, including CRISPR-Cas9 and genomic selection, offer precision and scalability that are transforming the field. This analysis will explore the distinct characteristics of these breeding strategies, focusing on their precision, scalability, and unique contributions to breeding practices.

Mass selection and hybridization involve breeding individuals with desirable traits to enhance those characteristics in subsequent generations. These methods are cost-effective and require less technical expertise than advanced biotechnological approaches like marker-assisted selection or genetic engineering, making them widely accessible to breeders (Ugandhar, 2025). Backcross breeding and pure-line selection techniques are used for self-pollinated crops, focusing on stabilizing desirable traits over generations. They are reliable but can be time-consuming and less precise compared to modern methods (Ugandhar, 2025). The CRISPR-Cas9 genome-editing tool allows for precise modifications at specific genetic loci, enabling the development of crops with enhanced traits such as disease resistance and stress tolerance. It is highly scalable and can be applied across various plant species (Bhuvaneswari *et al.*, 2020).

GS and MAS methods use genetic markers to predict and select for desirable traits, significantly speeding up the breeding process and improving accuracy. They are particularly effective for complex traits influenced by multiple genes (Bhuvaneswari *et al.*, 2020). High-throughput phenotyping technology allows for the rapid assessment of large plant populations, facilitating the selection of superior traits at an unprecedented scale (Bhuvaneswari *et al.*, 2020). Modern techniques like CRISPR-Cas9 and genomic selection offer unparalleled precision by targeting specific genes and using genetic markers to guide selection. This precision reduces the time and resources needed to achieve desired traits (Kumar *et al.*, 2024). The integration of big data and artificial intelligence in modern breeding methods enhances scalability by enabling the analysis of vast datasets to predict breeding outcomes and optimize selection processes (Fang, 2024).

3.4 Speed Breeding for Rapid Trait Development

Speed breeding is a transformative approach in crop improvement that significantly accelerates the development of new plant varieties by optimizing environmental conditions to hasten plant growth and reproduction cycles (Table 3.3 and Fig. 3.3). This method allows for the production of multiple generations per year, thereby reducing the time required for breeding cycles and enabling rapid trait development. By manipulating factors such as photoperiod, temperature, light intensity, and nutrition, speed breeding facilitates the rapid advancement of crop varieties with enhanced yield potential, stress resilience, and improved nutritional quality. The integration of modern technologies such as marker-assisted selection, high-throughput phenotyping, and genomic tools further enhances the efficiency and precision of speed breeding programs.

Speed breeding involves controlling environmental factors such as light spectrum, photoperiod, temperature, humidity, and CO_2 levels to accelerate plant growth and reproduction cycles. This manipulation allows for the production of up to six generations per year for certain crops like spring wheat and barley, compared to one to two generations in traditional breeding methods (Shyamala and Sasipriya, 2023; Wattoo *et al.*, 2024). The use of growth chambers and greenhouses with controlled conditions enables precise regulation of plant growth factors, facilitating rapid generation turnover and trait development (Kumar and Walia, 2024). Speed breeding is often combined with advanced breeding technologies such as marker-assisted selection, genomic selection, and CRISPR to enhance the precision and efficiency of breeding programs (Watson *et al.*, 2018). By accelerating the breeding cycle, speed breeding allows for quicker genetic improvements and the development of crop varieties with desirable traits such as higher yield, disease

Table 3.2. Comparative analysis of MAS vs GS. Author's own table.

Feature	MAS	GS	Citations
Core principle	Uses a few molecular markers known to be linked to specific genes or quantitative trait loci (QTL).	Uses genome-wide, high-density markers (e.g. SNPs) to predict the total genetic value, capturing all small-effect loci.	(Varshney *et al.*, 2017)
Trait suitability	Best for simple traits controlled by a few major genes (e.g. specific disease resistance, nematode resistance).	Best for complex, polygenic traits (e.g. crop yield, stress tolerance) influenced by many genes with small effects.	(Nayak *et al.*, 2017; Varshney *et al.*, 2017)
Prerequisite knowledge	Requires prior, precise knowledge of the trait's genetics and the identification of tightly linked markers.	Requires a large, representative training population to build a robust prediction model, but no prior gene mapping is needed.	(Chu *et al.*, 2011; Varshney *et al.*, 2017)
Precision & accuracy	High accuracy for selecting specific, known alleles. Accuracy drops significantly for complex traits where it fails to capture all genetic variance.	High prediction accuracy for the overall genetic merit of an individual for complex traits by summing all small-effect markers.	(Unêda-Trevisoli *et al.*, 2017; Varshney *et al.*, 2017)
Scalability & speed	Can be integrated into breeding cycles but is limited by the need to develop and validate markers for each major gene.	Significantly reduces the breeding cycle time, enabling faster genetic gains per unit time, especially when combined with high-throughput phenotyping.	(Unêda-Trevisoli *et al.*, 2017; Bhuvaneswari *et al.*, 2020)
Key advantage	Cost-effective and precise for introgressing or selecting a few well-characterized, high-impact genes.	Superior for improving polygenic traits, accelerates breeding, and does not require prior functional knowledge of genes.	(Chu *et al.*, 2011; Varshney *et al.*, 2017)
Primary limitation	Ineffective for polygenic traits; its utility is constrained by the genetic architecture of the target trait.	Requires large, initial investment in genotyping and phenotyping to train models; risk of increased inbreeding if not managed.	(Varshney *et al.*, 2017; Degen and Müller, 2023)
Integration potential	Can be effectively combined with GS to simultaneously select for major genes (via MAS) and polygenic background (via GS).	Serves as the foundational selection strategy, with MAS integrated to fix critical major alleles within a GS framework.	(Moniruzzaman *et al.*, 2015; Verma *et al.*, 2024)

Continued

Table 3.2. Continued

Feature	MAS	GS	Citations
Role in modern breeding	A valuable tool for precision introgression of specific traits (e.g. disease resistance genes) into elite lines.	A transformative, scalable strategy for overall genetic improvement and accelerating the development of new varieties.	(Bhuvaneswari *et al.*, 2020; Kumar *et al.*, 2024)

resistance, and climate adaptations (Abdul Fiyaz *et al.*, 2020).

The ability to conduct breeding programs throughout the year without being limited by seasonal constraints is a significant advantage of speed breeding (Patel and Gurjar, 2024). Speed breeding has been successfully applied to a wide range of crops, including cereals, legumes, and vegetables, enabling the rapid introgression of desirable traits across different species (Kumar and Walia, 2024). Implementing speed breeding requires specialized infrastructure and can be cost-intensive, posing challenges for widespread adoption, particularly in developing countries (Shinde *et al.*, 2023). The accelerated conditions may impact genetic diversity and induce stress responses in plants, necessitating careful management and optimization of protocols (Kumar and Walia, 2024). The integration of high-throughput phenotyping and genotyping generates large volumes of data, requiring efficient data management systems (Patel and Gurjar, 2024).

While speed breeding offers significant advantages in accelerating crop improvement, it is essential to consider the potential challenges and limitations associated with its implementation. The need for specialized infrastructure and the impact on genetic diversity are critical factors that must be addressed to fully realize the potential of speed breeding. Ongoing advancements in technology and collaborative efforts can help overcome these challenges, making speed breeding an increasingly valuable tool for addressing global food security and agricultural sustainability in the face of climate change. The acceleration of breeding cycles is a critical advancement in agricultural science, aimed at meeting the growing global food demand and adapting to climate change. Techniques such as speed breeding, controlled environments, and light regime manipulation have been developed to shorten the time required to develop new crop varieties. These methods enhance the efficiency of breeding programs by allowing multiple generations per year, thus expediting the selection and improvement of desirable traits.

Growth chambers and glasshouse facilities allow for precise control over environmental conditions such as temperature, humidity, and light, which are crucial for accelerating plant growth cycles. By optimizing these conditions, plants can be grown year-round, independent of external weather conditions, thus enabling multiple generations per year (Shanmugavel *et al.*, 2023). By manipulating light systems to create shorter artificial days, the overall growth cycle of plants can be reduced. This method maximizes growing time and accelerates the transition from seed to seed. Increasing the duration of daily light exposure can significantly speed up plant development. This technique, initially developed by NASA, involves continuous light exposure to enhance photosynthesis and accelerate growth and reproduction (Swami *et al.*, 2023). Adjusting the intensity and spectrum of light can influence plant physiological processes, such as flowering and seed maturation, thereby shortening the breeding cycle. The single-seed descent (SSD) method involves selecting a single seed from each plant to advance to the next generation, allowing for rapid cycling and the development of near-homozygous lines in a short time frame (Sharma *et al.*, 2024a). Utilizing advanced phenotyping platforms in controlled environments enables detailed monitoring of plant traits, facilitating faster selection and breeding of superior varieties (Sharma *et al.*, 2024a).

Table 3.3. Speed breeding: Techniques, benefits, and limitations. Author's own table.

Aspect	Key findings & methodologies	Applications & outcomes	Key challenges & limitations	Citations
Core principle	Manipulation of environmental factors (photoperiod, light spectrum, temperature, CO_2) in controlled environments (growth chambers, greenhouses) to accelerate plant growth and reproduction.	Enables multiple generations per year (e.g. up to six for wheat/ barley vs one to two traditionally), drastically shortening the breeding cycle.	Requires specialized, cost-intensive infrastructure, limiting widespread adoption, especially in developing countries.	(Kumar and Walia, 2024; Wattoo *et al.*, 2024)
Key techniques	• Extended photoperiod (up to 22 hour light) to enhance photosynthesis. • Single-seed descent (SSD) for rapid generation of homozygous lines. • Precise control of temperature, humidity, and light intensity.	Allows year-round breeding, independent of external seasons. Rapid development of genetically stable lines for testing.	Less effective for short-day plants or species with vernalization requirements. Artificial conditions may not fully reflect field performance.	(Shanmugavel *et al.*, 2023; Swami *et al.*, 2023; Sharma *et al.*, 2024a)
Integration with modern technologies	Combined with MAS, GS, and CRISPR-Cas9 gene editing for precision breeding.	Accelerates genetic gain by allowing rapid cycling and precise selection of complex traits (yield, disease resistance, climate adaptation).	Generates large volumes of phenotypic and genotypic data, requiring efficient data management and analysis systems.	(Alahmad *et al.*, 2022; Temesgen, 2022; Patel and Gurjar, 2024)
Crop applicability	Successfully applied to a wide range of crops including cereals (wheat, barley), legumes, and vegetables.	Enables rapid introgression of desirable traits (e.g. disease resistance, stress tolerance) across different species.	Protocols need to be optimized species-by-species. Accelerated conditions may induce unintended stress responses.	(Abdul Fiyaz *et al.*, 2020; Kumar and Walia, 2024)

Continued

Table 3.3. Continued

Aspect	Key findings & methodologies	Applications & outcomes	Key challenges & limitations	Citations
Impact on genetic diversity	The SSD method and rapid cycling can quickly fix alleles and develop homozygous lines.	Efficient for developing uniform breeding lines and purging undesirable alleles.	Risk of reduced genetic diversity and increased inbreeding if not carefully managed within a broader breeding strategy.	(Kumar and Walia, 2024; Sharma *et al.*, 2024a)
Overall goal & potential	A transformative tool to revolutionize crop improvement timelines and address global food security challenges under climate change.	Facilitates the rapid development of resilient, high-yielding varieties with improved nutritional quality.	Bridging the gap between controlled environment performance and field-level adaptability remains a critical hurdle for validation.	(Watson *et al.*, 2018; Shyamala and Sasipriya, 2023)

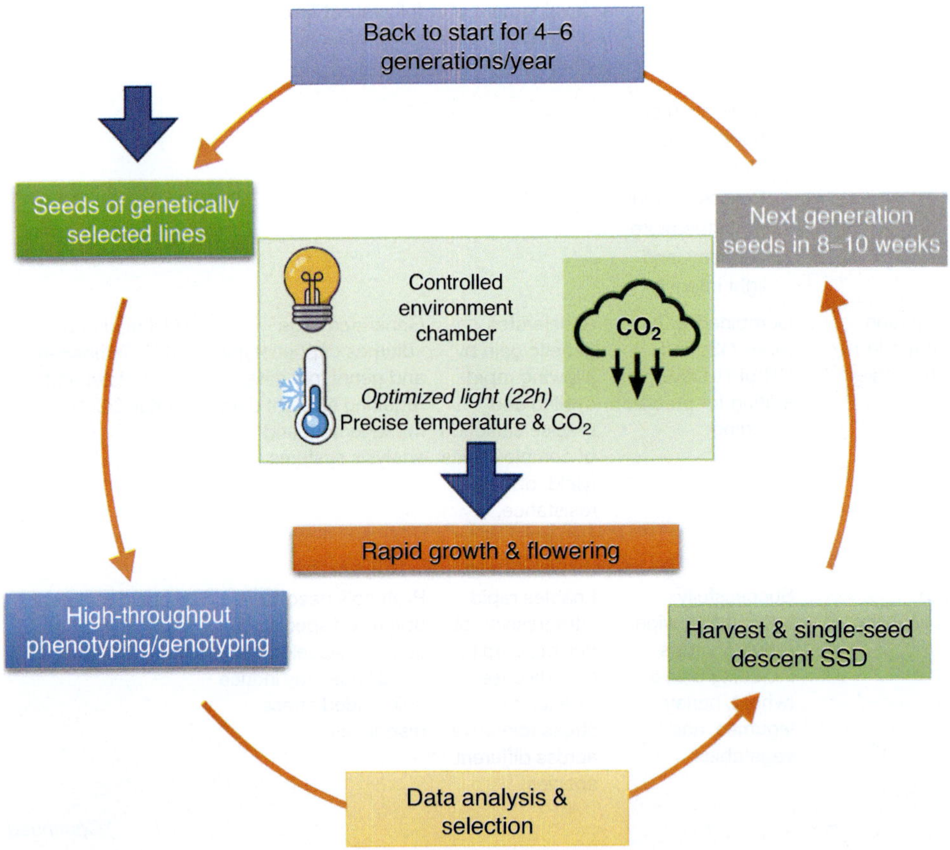

Fig. 3.3. The speed breeding cycle for accelerated trait development. (Author's own figure.)

Marker-assisted selection and genomic selection techniques allow for the precise selection of desirable traits at the genetic level, accelerating the breeding process by reducing the time needed for phenotypic evaluations (Alahmad *et al.*, 2022). Gene editing and CRISPR-Cas systems are cutting-edge technologies that enable the direct modification of plant genomes to incorporate beneficial traits, further reducing the time required to develop new crop varieties (Temesgen, 2022). While these techniques offer significant advantages in shortening breeding cycles, they also present challenges. For instance, speed breeding is less effective for short-day plants or those requiring vernalization, and the artificial environments used may not perfectly replicate field conditions, potentially affecting the adaptability of developed varieties (Swami *et al.*, 2023). Nonetheless, the integration of these advanced techniques holds promise for revolutionizing crop improvement and addressing future food security challenges.

3.5 Case Study: Enhancing Essential Oil Yield in *Mentha* spp.

Enhancing the yield of essential oils in *Mentha* spp. involves a multifaceted approach that includes optimizing growth conditions, genetic manipulation, and the use of growth promoters. Various studies have explored these strategies, demonstrating significant improvements in essential oil yield and composition. The following analysis synthesizes findings from multiple research papers to provide a comprehensive overview of methods to enhance essential oil yield in *Mentha* spp.

Foliar application of plant-derived smoke water (PDSW) and karrikinolide (KAR1) has been shown to significantly enhance the growth and essential oil yield of *Mentha arvensis* L. Optimal concentrations of 1:500 v/v PDSW and 10^{-8} M KAR1 increased plant height, fresh and dry weight, leaf area, and essential oil content by up to 37.09% compared to controls (Singh *et al.*, 2023). These treatments also improved photosynthetic parameters, such as chlorophyll content and carotenoid content, contributing to increased essential oil productivity per plant (Singh *et al.*, 2023). Genetic manipulation of biosynthetic pathways in *Mentha* spp. can enhance essential oil yield by altering carbon flux through the monoterpenoid biosynthesis pathway. Techniques include overexpression and RNA interference (RNAi) suppression of key biosynthetic enzymes and transcription factors (Fuchs *et al.*, 2022). This approach targets the methylerythritol phosphate (MEP) pathway and menthol biosynthetic pathways, aiming to increase the yield and improve the composition of essential oils (Fuchs *et al.*, 2022).

Selecting high-yielding cultivars, such as CIM-Mohak and MSS-5, can significantly increase essential oil yield in *Mentha spicata*. These cultivars have shown higher herb and essential oil yields under subtropical conditions (Kumar *et al.*, 2022). Environmental factors, such as full sun exposure, have been found to enhance biomass and essential oil content compared to shaded conditions, indicating the importance of optimizing cultivation environments. The application of organic and nitrogen fertilizers has been shown to improve growth parameters and essential oil yield in *Mentha canadensis*. A combination of organic and nitrogen fertilizers increased menthol content and essential oil yield (Ardalani *et al.*, 2017). Growth promoters like methyl jasmonate (MJ) and oligomeric sodium alginate with chitosan have also been effective. MJ increased essential oil content and glandular trichome density, while sodium alginate and chitosan improved growth and essential oil composition in *M. arvensis* (Fazili *et al.*, 2017; Huang *et al.*, 2025). Path analysis of essential oil-related characters in *Mentha* spp. suggests that optimizing morphological traits such as stem diameter, leaf length, and flowering time can indirectly enhance essential oil yield. These traits can be targeted for genetic improvement to increase yield. While these strategies have shown promise in enhancing essential oil yield in *Mentha* spp., it is important to consider the economic and practical feasibility of each method. For instance, while genetic manipulation offers precise control over biosynthetic pathways, it may require significant investment and technical expertise. Conversely, simpler methods, like the use of PDSW and cultivar selection, may offer more immediate and cost-effective benefits for farmers. Each approach should be evaluated in the context of specific environmental conditions and resource availability to maximize the potential benefits.

The integration of CRISPR technology with traditional breeding methods presents a promising synergy for enhancing industrial-scale essential oil production. CRISPR-Cas systems offer precise genome-editing capabilities that can significantly improve oil yield, quality, and composition, which are critical for meeting industrial demands. This synergy is particularly evident in the context of oil crops like oilseed rape and pennycress, where CRISPR has been used to target specific genes involved in oil biosynthesis pathways, thereby accelerating the breeding process and enhancing oil characteristics. CRISPR-Cas technology allows for precise modifications in the genome, enabling the alteration of multiple traits simultaneously with high accuracy. This is particularly beneficial for improving oil yield and composition in crops like oilseed rape and pennycress, where specific genes such as FAD2 and ROD1 have been targeted to increase oleic acid content, a desirable trait for industrial applications (Li *et al.*, 2025). Traditional breeding methods are often slow and labor-intensive. CRISPR accelerates this process by enabling rapid and targeted genetic modifications, which is crucial for crops with complex genomes like *Brassica napus* (oilseed rape) (Ali and Zhang, 2023). By targeting genes involved in fatty acid biosynthesis, CRISPR can improve the nutritional and industrial quality of oils. For instance, modifications in pennycress have led to increased oleic acid content, which offers better oxidative stability and cold flow properties, essential for biofuels and other industrial uses (Jarvis *et al.*, 2021; Smajlovic *et al.*, 2021).

Combining CRISPR with traditional breeding techniques can enhance the overall efficiency of developing new crop varieties. While CRISPR provides the precision needed for specific trait improvements, traditional breeding ensures the stability and adaptability of these traits in different environmental conditions (Ali and Zhang, 2023). Recent advancements have demonstrated the successful application of CRISPR-edited crops in field conditions, highlighting the practical potential of this technology in real-world agricultural settings. This is crucial for scaling up production to meet industrial demands (Boniecka, 2024). The enhanced oil yield and quality achieved through CRISPR and breeding synergies can significantly contribute to biofuel production, offering a sustainable alternative to fossil fuels. This aligns with global sustainability goals and the transition to a bio-based economy (Hajinajaf *et al.*, 2024).

Improved oil crops can reduce reliance on petroleum-based products, offering economic benefits through increased crop value and environmental benefits by reducing carbon emissions (Hajinajaf *et al.*, 2024). While CRISPR technology offers significant advantages in improving oil crop production, it is essential to consider potential challenges such as off-target effects and regulatory hurdles. Additionally, the integration of CRISPR with traditional breeding requires careful management to ensure the stability and adaptability of new crop varieties. Despite these challenges, the synergy between CRISPR and breeding holds great promise for advancing industrial-scale essential oil production, contributing to both economic growth and environmental sustainability.

3.6 Regulatory Frameworks for Genetically Modified Medicinal Plants

The regulatory frameworks for genetically modified (GM) medicinal plants are complex and vary significantly across different regions (Table 3.4 and Fig. 3.4). These frameworks are designed to ensure the safety, efficacy, and environmental sustainability of GM medicinal plants, which are increasingly used for pharmaceutical purposes. The regulatory landscape is shaped by a combination of international agreements, national laws, and regional policies, each with its own set of challenges and opportunities for harmonization. The following sections provide a detailed overview of the key aspects of these regulatory frameworks.

The Cartagena Protocol on Biosafety is a key international agreement that guides the regulation of GMOs, including GM medicinal plants, by emphasizing the need for biosafety measures to protect human health and the environment (Glover *et al.*, 2018). The World Health Organization (WHO) plays a significant role in promoting the harmonization of regulations for medicinal plants globally, aiming to ensure

Table 3.4. Regulatory and ethical landscape of genome-edited medicinal plants. Author's own table.

Regulatory aspect	Key features & approaches	Regional examples & applications	Major challenges & criticisms	Citations
International guidance	Cartagena Protocol on Biosafety: provides a foundational framework for safe handling, transport, and use of living modified organisms (LMOs).	Guides national biosafety laws globally, particularly influencing developing countries.	Implementation is uneven; creates a complex patchwork of international requirements.	(Glover *et al.*, 2018)
	World Health Organization (WHO): promotes harmonization of regulations for medicinal plants to ensure safety and efficacy.	Aims for global standards for plant-derived pharmaceuticals, including those from GM plants.	Lack of binding authority; recommendations may be adopted inconsistently.	(Picking, 2017, 2024)
Regional regulatory philosophies	Process-based (e.g. EU): regulation is triggered by the process of genetic modification itself, regardless of the final product.	Stringent, case-by-case risk assessment. Often criticized for being complex, costly, and hindering innovation.	Creates high regulatory burdens and barriers to commercialization; criticized for not being scientifically proportional to risk.	(Sparrow *et al.*, 2013; Kopertekh and Wilhelm, 2023)
	Product-based (e.g. USA): regulation focuses on the characteristics of the final product, not the method used to create it.	Considered more flexible and pragmatic. Facilitates commercialization of products deemed "substantially equivalent" to conventional ones.	May be perceived as less rigorous for novel traits; public trust issues can arise.	(Sparrow *et al.*, 2013)
Regional harmonization efforts	Africa (COMESA): aims to harmonize biotechnology and biosafety regulations across member states to facilitate trade and development.	Draft policy on GM crop cultivation and trade developed through stakeholder consultation.	Many African nations lack functional national biosafety frameworks due to policy, legal, and operational capacity issues.	(Belay *et al.*, 2014; Nang'ayo *et al.*, 2014)
	Central America: countries like Honduras and Costa Rica have developed governance structures to support trade of GM crops.	Supports regional development of agricultural biotechnology with traits like herbicide tolerance.	Requires continuous capacity building and alignment with international market demands.	(Garro Monge, 2012)

Continued

Table 3.4. Continued

Regulatory aspect	Key features & approaches	Regional examples & applications	Major challenges & criticisms	Citations
Emerging challenges	New genetic techniques (nGMs): technologies like gene editing (e.g. CRISPR) challenge existing regulatory definitions.	Ongoing debate on whether nGMs should be regulated as stringently as traditional GMOs, especially if no foreign DNA is inserted.	Regulatory uncertainty for new techniques; need for updated, science-based definitions and frameworks.	(Eckerstorfer *et al*., 2019b)
	Molecular farming: using GM plants to produce pharmaceuticals (e.g. vaccines, therapeutic proteins).	Recognized for its potential in low-cost, scalable production of high-value drugs.	Significant regulatory uncertainties and a lack of specific guidelines tailored to pharmaceutical production in plants.	(Guimarães *et al*., 2014)
Socio-political factors	Public perception & legitimacy: consumer acceptance is a major hurdle, influenced by awareness, trust, and values.	EU regions declaring 'GM-free zones'; public resistance affecting market access globally.	Overcoming deep-seated public skepticism and anti-GMO sentiments requires transparent communication and education.	(Hunt, 2011; Krishna *et al*., 2023)
	Subnational governance: regions/states within a country can enact their own policies, creating internal market fragmentation.	Example of EU regions exerting significant influence, sometimes contradicting broader EU or national stances.	Creates a complex, multi-level governance landscape that is challenging for developers to navigate.	(Hunt, 2011)

patient safety and product efficacy (Picking, 2017, 2024). The EU has a stringent regulatory framework for GM products, including GM medicinal plants, which is often criticized for its complexity and high costs. The EU's regulatory approach is primarily process-oriented, focusing on the methods used to create GMOs rather than the final product characteristics. In contrast, the US regulatory framework is more product-oriented, focusing on the safety and efficacy of the final product. This approach is considered more flexible and less burdensome for the commercialization of GM medicinal plants (Sparrow *et al.*, 2013).

The African Union (AU) is working toward developing a cohesive regulatory framework for GM products, leveraging the experience of the New Partnership for Africa's Development (NEPAD) in establishing regulatory systems for GM crops (Glover *et al.*, 2018). Achieving international harmonization of regulatory frameworks remains a significant challenge due to the diverse approaches adopted by different countries and regions. This lack of harmonization can hinder international trade and the global commercialization of GM medicinal plants (Eckerstorfer *et al.*, 2019b). The high regulatory burden associated with GM medicinal plants, particularly in the EU, is a barrier to innovation and commercialization. There is a call for amendments to be made to existing regulations to reduce this burden and facilitate the development of new GM medicinal products (Kopertekh and Wilhelm, 2023). Public perception and the legitimacy of regulatory frameworks are critical issues, especially in regions like the EU, where there is significant public skepticism toward GM products. Ensuring transparency and public accountability in the regulatory process is essential for gaining public trust.

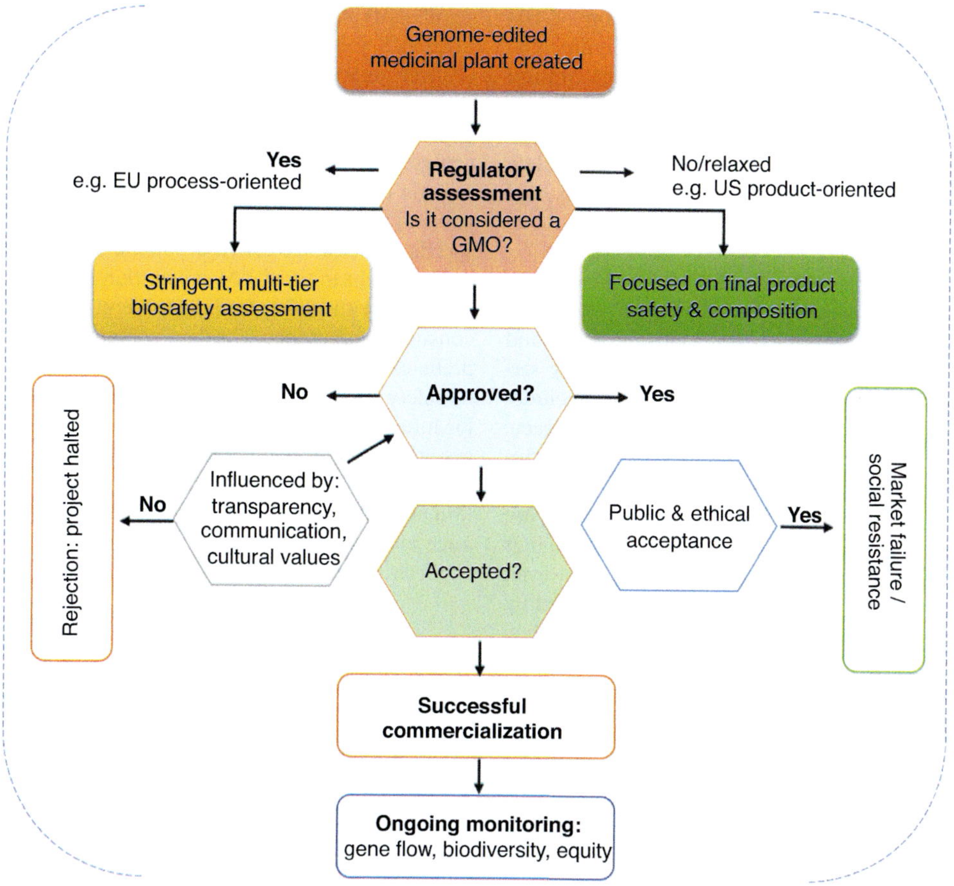

Fig. 3.4. Navigating the regulatory and ethical landscape. (Author's own figure.)

The development of new genetic modification techniques (nGMs) presents new regulatory challenges and opportunities. There is ongoing debate about whether existing regulatory frameworks are adequate for these new technologies or if new, specific regulations are needed (Eckerstorfer *et al.*, 2019b). The use of GM plants for molecular farming, which involves producing pharmaceuticals in plants, is gaining traction. However, regulatory uncertainties and the need for specific guidelines for molecular farming remain significant hurdles (Guimarães *et al.*, 2014). While the regulatory frameworks for GM medicinal plants are evolving, they face several challenges, including the need for international harmonization, reducing regulatory burdens, and addressing public concerns. The development of new genetic modification techniques and molecular farming further complicates the regulatory landscape, necessitating ongoing adaptation and refinement of existing frameworks. As these technologies advance, it is crucial for regulatory bodies to balance innovation with safety and public trust.

The commercialization of GM crops is a complex process that requires region-specific policies to address diverse regulatory, economic, and social factors. These policies are crucial for ensuring that GM crops can be safely and effectively integrated into local agricultural systems while addressing public concerns and trade considerations. Different regions have adopted various approaches to GM crop commercialization, reflecting their unique agricultural, economic, and regulatory contexts. In sub-Saharan Africa, the authorization of GM crops for environmental release and commercial cultivation is aligned with broader agricultural development policies. These policies emphasize the need for harmonization across sectors, particularly in plant breeding, testing, and variety release, to support seed system reforms and agricultural productivity growth (Komen and Wafula, 2021).

The Common Market for Eastern and Southern Africa (COMESA) has developed a regional approach to biotechnology and biosafety, aiming to harmonize regulatory services and remove trade barriers. This includes a draft policy on GM crop cultivation and trade, which has been developed through extensive stakeholder consultations (Belay *et al.*, 2014).

In Africa, regulatory frameworks are influenced by international agreements such as the Cartagena Protocol on Biosafety. However, many countries face challenges in implementing functional national biosafety frameworks due to policy, legal, and operational issues (Nang'ayo *et al.*, 2014). The commercialization of GM crops in Africa is seen as a potential driver for enhancing agricultural productivity, particularly for smallholder farmers. However, challenges, such as building robust seed systems, providing effective extension services, and improving market access, need to be addressed to realize these benefits (Endale *et al.*, 2022).

In Central America, countries like Honduras and Costa Rica have developed governance structures for biosafety regulation, which support the trade of GM crops with traits like herbicide tolerance and pest resistance. These structures are crucial for facilitating the regional development of agricultural biotechnology (Garro Monge, 2012). In Indonesia, local certification of GM products is emphasized to support food sovereignty and economic activities. This involves establishing regulatory frameworks at the local level to ensure the quality and safety of GM products, which can enhance both local and national economic activities (Sujadmiko *et al.*, 2021). Consumer acceptance remains a significant challenge globally, with varying levels of acceptance influenced by awareness of benefits and risks, knowledge, trust, and personal values. This has led to public resistance in some regions, affecting the commercialization of GM crops (Krishna *et al.*, 2023). In the EU, regions play a significant role in the regulation of GM crop cultivation. Some regions have declared themselves GM-free zones, highlighting the importance of subnational governance in shaping GM crop policies. This reflects the complex multi-level governance structure within the EU, where regions can function as significant legal and political actors (Hunt, 2011). While region-specific policies are essential for the successful commercialization of GM crops, they must be balanced with global trade considerations and consumer acceptance. The development of comprehensive regulatory frameworks that address biosafety, economic, and social factors is crucial for integrating GM crops into local agricultural systems. Additionally, fostering public trust and acceptance through transparent communication and education is vital for overcoming resistance and ensuring the sustainable adoption of GM technologies.

3.7 Ethical Implications of Genome Editing of Medicinal Plants

The ethical implications of genome editing in medicinal plants are multifaceted, encompassing concerns about safety, environmental impact, and socio-economic effects. Genome-editing technologies, particularly CRISPR-Cas9, have revolutionized the ability to enhance the production of secondary metabolites in medicinal plants, which are crucial for therapeutic applications. However, these advancements bring ethical challenges that need careful consideration to ensure responsible use. The ethical concerns primarily revolve around potential unintended consequences, the impact on traditional practices, and the need for robust regulatory frameworks.

Genome editing can lead to off-target effects, which may result in unintended genetic changes. These changes could potentially affect the plant's safety for human consumption and its ecological interactions (Wolt, 2017). There is a risk of gene flow from genetically modified plants to wild relatives, which could disrupt local ecosystems and biodiversity (Munawar *et al.*, 2024a). The potential for unintended environmental impacts necessitates comprehensive biosafety assessments to prevent adverse effects on non-target species and ecosystems (Munawar *et al.*, 2024b). The introduction of genome-edited plants may undermine traditional farming and medicinal practices, leading to socio-economic disruptions in communities that rely on these practices (Munawar *et al.*, 2024b). The control of genome-editing technologies by a few corporations could lead to monopolies, affecting the accessibility and affordability of medicinal plants for small-scale farmers and traditional practitioners (Munawar *et al.*, 2024a). Many countries lack clear guidelines for the regulation of genome-edited crops, leading to uncertainties among stakeholders. This highlights the need for international cooperation to establish standardized regulations (Idris *et al.*, 2023). Ethical concerns include the moral implications of altering natural organisms and the potential long-term impacts on human health and the environment. These concerns necessitate transparent and inclusive discussions involving diverse stakeholders (Idris *et al.*, 2023).

The distribution of genome-edited plants involves complex considerations of public perception, ecological risks, and equity. Public perception is shaped by a variety of factors, including ethical concerns, scientific uncertainty, and the perceived benefits of genome editing. Ecological risks pertain to potential unintended consequences for biodiversity and the environment, while equity issues revolve around access and the distribution of benefits from these technologies. These aspects are interlinked and influence the regulatory frameworks and acceptance of genome-edited plants globally.

Public perception of genome-edited plants is influenced by ethical concerns, scientific uncertainty, and media framing. Studies show that people tend to have a higher acceptance of genome editing compared to traditional genetic modification, especially when the benefits are clearly communicated (Bearth *et al.*, 2022; Atimango *et al.*, 2024). In Europe, awareness and concerns about genome editing have increased, influencing regulatory frameworks and public opinion (Ichim, 2024). In the USA, public opinion is divided, with socio-economic and belief-based factors playing significant roles (Cummings and Peters, 2023). Transparency and effective communication are crucial for gaining public trust and acceptance. Tailored information that highlights tangible benefits can improve attitudes toward genome editing (Munawar *et al.*, 2024a).

Ecological risks associated with genome-edited plants include potential biodiversity loss, the emergence of superweeds, and unintended environmental impacts (Choriyeva, 2024). The precision of genome-editing techniques like CRISPR-Cas9 is often highlighted as a safety feature, but it does not eliminate the need for thorough risk assessments. Unintended changes can still occur, necessitating case-specific premarket risk assessments (Eckerstorfer *et al.*, 2019a). Regulatory frameworks need to adapt to address these risks effectively, ensuring that genome-edited plants do not compromise ecological integrity (Eckerstorfer *et al.*, 2019a).

Equity issues in the distribution of genome-edited plants involve access to technology and the distribution of benefits. There is concern about corporate dominance and the potential marginalization of traditional farming practices (Munawar *et al.*, 2024a). Regulatory disparities across regions can affect equity. For instance, countries like the USA, Canada, and Australia have more relaxed regulations, potentially leading to faster market entry and benefits, while stricter regulations in the EU may limit access (Dayé *et al.*, 2023). Ensuring equitable access requires international cooperation and harmonization of regulations to prevent disparities in technology adoption and benefit distribution (Spök *et al.*, 2022). Genome-edited plants offer promising solutions for food security and sustainability; they also present challenges that need to be addressed. Public perception is a critical factor that can influence regulatory decisions and the acceptance of these technologies. Ecological risks must be carefully managed through robust regulatory frameworks, and equity considerations should guide the distribution of benefits to ensure that all stakeholders, including smallholder farmers and developing countries, can access and benefit from genome-editing advancements. Balancing these aspects is essential for the responsible development and deployment of genome-edited plants.

3.8 Challenges and Future Directions

The advancement of CRISPR applications in medicinal plants faces significant scientific and technical hurdles. A primary challenge is the prevalence of genomic knowledge gaps, where many species, particularly understudied ones, lack high-quality reference genomes and annotated gene databases, which are a fundamental prerequisite for targeted editing. Compounding this is the issue of transformation inefficiency, as species-specific recalcitrance to genetic transformation—often due to poor tissue culture response or low regeneration rates—severely limits the scalability of genome editing. Furthermore, despite advancements in precision, the risk of off-target effects remains a persistent concern, especially in polyploid medicinal species with highly complex genomes (Table 3.5 and Fig. 3.5).

Beyond the lab, the regulatory landscape presents a formidable obstacle course. Divergent definitions of what constitutes a GMO across regions, such as the EU's strict process-based regulation versus the USA's more product-focused approach, create international confusion and delay global commercialization. Navigating overlapping national and international biosafety compliance, like the Cartagena Protocol, adds substantial cost and complexity. Moreover, regulatory frameworks universally struggle to keep pace with emerging technologies like CRISPR-Cas9 or epigenetic editing, leading to a state of legal ambiguity that stifles innovation.

The ethical and socio-economic dimensions of this technology are equally complex. Environmental concerns center on potential biodiversity threats, such as gene flow from edited plants to wild relatives, which could disrupt ecosystems or create invasive "superweeds." Socio-economic inequity is a major risk, as patent monopolies on CRISPR technologies may marginalize small-scale farmers and traditional healers who rely on medicinal plants. This is coupled with the danger of cultural erosion, where the rapid adoption of genome-edited varieties could displace heirloom cultivars and the Indigenous knowledge systems tied to traditional medicine.

Finally, public perception and resource limitations create significant barriers. Widespread misinformation, fueled by sensationalized media narratives and a lack of public scientific literacy, often fuels unwarranted fears. This is exacerbated by cultural resistance in regions with strong anti-GMO sentiment and transparency gaps in regulatory decisions that breed distrust. Practically, the high costs of necessary infrastructure—from speed breeding facilities to CRISPR toolkits—and critical skill shortages in genomics and bioinformatics exclude low-income countries. The challenge of data management, requiring robust computational infrastructure to integrate omics data, further impedes adoption in resource-limited settings, creating a stark global divide in who can benefit from these advancements.

To realize the transformative potential of CRISPR in medicinal plants, a multifaceted and collaborative future direction is essential. Scientifically, this entails launching global consortia to sequence and annotate the genomes of high-priority species, with data housed in open-access databases to accelerate discovery for all researchers. Technologically, the focus must be on designing modular CRISPR toolkits compatible with diverse plant species to overcome transformation bottlenecks, while also integrating multi-omics data and leveraging AI-driven breeding to predict gene-trait associations and holistically engineer valuable biosynthetic pathways for novel metabolites. A critical parallel effort involves developing ecosystem-safe designs, such as containment strategies, to prevent unintended gene flow.

On the regulatory front, there is an urgent need for harmonized international policies, advocated through bodies like the WHO and Food and Agriculture Organization (FAO), which clearly distinguish transgenic organisms from transgene-free edited plants. This should be supported by a tiered risk assessment framework that is proportional to the editing technique's precision, potentially exempting edits that mimic natural mutations, and by establishing fast-track approvals for edited crops with demonstrable benefits like climate resilience.

To ensure ethical and equitable development, new models must be implemented. This includes creating benefit-sharing frameworks to

Table 3.5. Future directions and recommendations for genome editing in medicinal plants. Author's own table.

Domain	Key challenges	Proposed future directions	Citations/rationale
Scientific & technical	Genomic knowledge gaps: lack of high-quality reference genomes and annotated gene databases for non-model medicinal plants.	Global sequencing consortia: launch initiatives to sequence high-priority species and create open-access genomic databases.	(Implied from text on "genomic knowledge gaps")
	Transformation inefficiency: species-specific recalcitrance to genetic transformation and regeneration in tissue culture.	Modular toolkits: develop standardized, species-optimized CRISPR toolkits and protocols to overcome transformation bottlenecks.	(Implied from text on "transformation inefficiency")
	Off-target effects: risk of unintended edits, especially in complex, polyploid genomes.	Advanced editing & AI: utilize high-fidelity Cas variants and integrate AI to improve guide RNA design and predict off-target sites.	(Wolt, 2017; Eckerstorfer *et al.*, 2019a)
Regulatory & policy	Regulatory fragmentation: divergent international frameworks (e.g. EU's process-based vs USA's product-based).	Harmonized policies: advocate for internationally harmonized, science-based policies through WHO/FAO that distinguish transgenic from transgene-free edits.	(Dayé *et al.*, 2023; Idris *et al.*, 2023)
	Pace of regulation: legal ambiguity and slow adaptation to emerging technologies (e.g. epigenetic editing).	Tiered risk assessment: implement a proportional risk assessment framework, with fast-track approvals for edits mimicking natural mutations.	(Eckerstorfer *et al.*, 2019a; implied from text)
Ethical & socio-economic	Socio-economic inequity: corporate patent monopolies marginalizing small-scale farmers and traditional healers.	Benefit-sharing & open-source: create fair benefit-sharing models and promote open-source licensing to democratize technology access.	(Spök *et al.*, 2022; Munawar *et al.*, 2024a)
	Cultural erosion: displacement of heirloom cultivars and associated Indigenous knowledge.	Indigenous partnerships: foster collaborative research with Indigenous communities to preserve cultural heritage and ensure equitable development.	(Munawar *et al.*, 2024b)

Continued

Table 3.5. Continued

Domain	Key challenges	Proposed future directions	Citations/rationale
	Environmental risk: gene flow to wild relatives, potentially disrupting ecosystems or creating invasive species.	Ecosystem-safe design: develop and implement genetic containment strategies (e.g. terminator genes, chloroplast engineering) to prevent gene flow.	(Choriyeva, 2024; Munawar *et al.*, 2024a)
Public perception & capacity	Public mistrust: fueled by misinformation, anti-GMO sentiment, and lack of transparency.	Science communication & participatory governance: launch transparent public engagement campaigns and include diverse stakeholders in regulatory dialogues.	(Bearth *et al.*, 2022; Ichim, 2024)
	Resource & capacity gap: high infrastructure costs and skill shortages in genomics/bioinformatics, especially in developing countries.	Capacity building & low-cost tech: invest in affordable infrastructure (e.g. portable kits) and intensive training programs for scientists in developing nations.	(Implied from text on "resource limitations")
Overarching application	Climate vulnerability: medicinal plants are threatened by climate change, pests, and diseases.	Engineering climate resilience: direct CRISPR applications to develop drought-tolerant, pest-resistant, and climate-adaptive medicinal plant varieties.	(Implied from text on "climate-adaptive traits")

ensure fair profit distribution for source communities, fostering Indigenous partnerships to preserve cultural heritage and enhance acceptance, and promoting open-source licensing to democratize access and prevent corporate monopolies. Furthermore, bolstering public trust requires robust science communication campaigns using relatable narratives, participatory governance that includes farmers and non-governmental organizations (NGOs) in regulatory dialogues, and comprehensive educational programs to integrate genome-editing literacy into public discourse.

Finally, bridging the global capacity gap demands targeted investment in low-cost infrastructure, such as affordable growth chambers and portable CRISPR kits, alongside intensive capacity building to train scientists in developing nations. The ultimate application of these concerted efforts should be channeled toward engineering climate-adaptive traits—like drought tolerance and pest resistance—aligning the advancement of medicinal plants with the urgent goals of climate-smart agriculture.

3.9 Conclusion

The advent of CRISPR-Cas9 and synthetic biology has ushered in a transformative era for precision breeding in medicinal plants, offering unparalleled opportunities to address pressing agricultural and pharmaceutical challenges. By enabling precise modifications in biosynthetic pathways, these technologies enhance the production of therapeutic secondary metabolites—such as alkaloids,

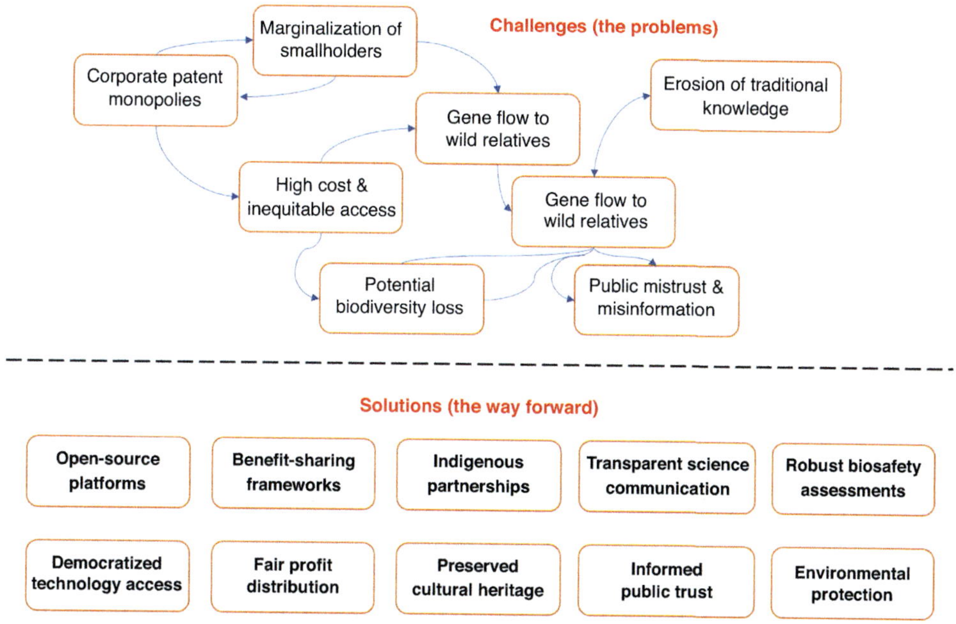

Fig. 3.5. Addressing ethical and socio-economic implications. (Author's own figure.)

terpenoids, and flavonoids—while improving stress resilience and disease resistance. Case studies like *Mentha* spp. exemplify the industrial scalability of CRISPR, where synergistic approaches combining gene editing, growth promoters, and optimized cultivation practices have dramatically increased essential oil yields. Similarly, the integration of MAS and GS bridges the gap between trait-specific precision and polygenic complexity, accelerating the development of high-value medicinal cultivars. Speed breeding further amplifies these gains by compressing breeding cycles, though its reliance on controlled environments underscores the need for cost-effective infrastructure to democratize access.

However, the promise of these innovations is tempered by significant challenges. Regulatory frameworks remain fragmented globally, with the EU's precautionary stance contrasting sharply with the USA's product-oriented approach, creating barriers to international collaboration and market entry. Ethical dilemmas—such as ecological risks from gene flow, corporate monopolization of patented technologies, and inequitable access to genome-editing tools—threaten to undermine public trust and exacerbate socio-economic disparities. Moreover, the lack of genomic data for many medicinal species limits the application of CRISPR, while high infrastructure costs exclude resource-poor regions from leveraging advanced breeding techniques. Public perception, shaped by misinformation and cultural values, further complicates adoption, necessitating transparent dialogue to align technological advancements with societal needs.

Looking ahead, the future of medicinal plant breeding hinges on a balanced approach that marries innovation with responsibility. Expanding genomic databases, refining species-specific transformation protocols, and developing next-generation editing tools (e.g. base/prime editing, Cas9-NG) will enhance precision and reduce off-target risks. Harmonizing global regulations through frameworks like the Cartagena Protocol could streamline commercialization while safeguarding biosafety. Equally critical is fostering equity—ensuring smallholder farmers and developing nations benefit from CRISPR advancements through open-access platforms and affordable technologies. Public engagement initiatives must demystify genome editing, emphasizing its role in sustainable agriculture and health care. By addressing these challenges collaboratively, the scientific community can unlock the full potential of medicinal plants, paving the way for climate-resilient crops, novel therapeutics, and a more equitable bioeconomy. The path forward demands not only technical excellence but also ethical foresight, ensuring that the benefits of genome editing are shared globally and sustainably.

Acknowledgments

The authors express their gratitude for the financial support received through the Distinguished High-Level Talents Research Grant from the Guizhou Science and Technology Corporation Platform Talents Fund (Grant nos.: [2017]5733-001 and CK-1130-002), high-level specialists project fund of Zunyi Medical University (1092/2000006)F-1092, the National Natural Science Foundation of China (82373981), and the support provided by Zunyi Medical University, China. Special appreciation is extended to all laboratory colleagues and research staff members for their valuable insights, constructive guidance, and assistance throughout this study.

References

Abdul Fiyaz, R., Ajay, B.C., Ramya, K.T., Aravind Kumar, J., Sundaram, R.M. *et al.* (2020) Speed breeding: Methods and applications. In: Singh Gosal, S. and Hussain Wani, S. (eds) *Accelerated Plant Breeding, Volume 1*. Springer International Publishing, Cham, Switzerland, pp. 31–49.

Alahmad, S., Rambla, C., Voss-Fels, K.P. and Hickey, L.T. (2022) Accelerating breeding cycles. In: Reynolds, M.P. and Braun, H.-J. (eds) *Wheat Improvement*. Springer International Publishing, Cham, Switzerland, pp. 557–571.

Ali, E. and Zhang, K. (2023) CRISPR-mediated technology for seed oil improvement in rapeseed: Challenges and future perspectives. *Frontiers in Plant Science* 14, 1086847. DOI: 10.3389/fpls.2023.1086847.

Ardalani, H., Hadipanah, A., Pazoki, A. and Zolfaghar, M. (2017) Phytochemical, morphological and yield responses of *Mentha canadensis* to organic and nitrogen fertilizers. *Journal of Essential Oil Bearing Plants* 20(3), 752–757. DOI: 10.1080/0972060X.2017.1322005.

Arun, M., Halka, J. and Kowsalya, K. (2023) CRISPR/Cas9: A novel genetic tool to manipulate plant secondary metabolite pathways. In: Singh, R. and Kumar, N. (eds) *Genetic Manipulation of Secondary Metabolites in Medicinal Plant*. Springer Nature Singapore, Cham, Switzerland, pp. 45–57.

Asghar, G., Saleem, S., Bibi, H., Rozina, R., Bibi, B. *et al.* (2025) Exploring the intersection of plant genetics and biotechnology for global challenges. *Scholars Journal of Agriculture and Veterinary Sciences* 12(1), 64–73. DOI: 10.36347/sjavs.2025.v12i01.006.

Asha, A. and Sunita, J. (2024) CRISPR-Cas-mediated optimization of plant-derived natural products of biomedical significance. In: Mérillon, J.-M. and Gopal Ramawat, K. (eds) *Plant Specialized Metabolites: Phytochemistry, Ecology and Biotechnology*. Springer Nature Switzerland, Cham, Switzerland, pp. 1–40.

Atimango, A.O., Wesana, J., Kalule, S.W., Verbeke, W. and De Steur, H. (2024) Genome editing in food and agriculture: From regulations to consumer perspectives. *Current Opinion in Biotechnology* 87, 103127. DOI: 10.1016/j.copbio.2024.103127.

Bearth, A., Kaptan, G. and Kessler, S.H. (2022) Genome-edited versus genetically-modified tomatoes: An experiment on people's perceptions and acceptance of food biotechnology in the UK and Switzerland. *Agriculture and Human Values* 39(3), 1117–1131. DOI: 10.1007/s10460-022-10311-8.

Belay, G., Kimani, V. and Wafula, D. (2014) Harmonization of regional biosafety and regulatory services to remove future trade barriers in the COMESA region. In: Wambugu, F. and Kamanga, D. (eds) *Biotechnology in Africa: Emergence, Initiatives and Future*. Springer International Publishing, Cham, Switzerland, pp. 207–222.

Bhuvaneswari, R., Saravanan, K.R., Vennila, S. and Suganthi, S. (2020) Precision breeding techniques: CRISPR-Cas9 and beyond in modern plant improvement. *Plant Science Archives* 5(1), 17–21. DOI: 10.51470/PSA.2020.5.1.17.

Boniecka, J. (2024) CRISPR/Cas-based precision breeding of oilseed rape (*Brassica napus* L.) – recent improvements. In: Ricroch, A., Eriksson, D., Miladinović, D., Sweet, J. and Van Laere, K. *et al* (eds) *A Roadmap for Plant Genome Editing*. Springer Nature Switzerland, Cham, Switzerland, pp. 291–307.

Choriyeva, Z. (2024) Genetically modified ecosystems: Innovative approaches in agriculture and their environmental impact. *BIO Web of Conferences* 149, 01009. DOI: 10.1051/bioconf/202414901009.

Chu, Y., Wu, C.L., Holbrook, C.C., Tillman, B.L., Person, G. *et al.* (2011) Marker-assisted selection to pyramid nematode resistance and the high oleic trait in peanut. *The Plant Genome* 4(2), 110–117. DOI: 10.3835/plantgenome2011.01.0001.

Cummings, C.L. and Peters, D. (2023) Gene-edited foods and the public: The first representative survey study of the United States. *Environmental Communication* 17(8), 965–974. DOI: 10.1080/17524032.2022.2086894.

Das, S., Kwon, M. and Kim, J.-Y. (2024) Enhancement of specialized metabolites using CRISPR/Cas gene editing technology in medicinal plants. *Frontiers in Plant Science* 15, 1279738. DOI: 10.3389/fpls.2024.1279738.

Dayé, C., Spök, A., Allan, A.C., Yamaguchi, T. and Sprink, T. (2023) Social acceptability of cisgenic plants: Public perception, consumer preferences, and legal regulation. In: Chaurasia, A. and Kole, C. (eds) *Cisgenic Crops: Safety, Legal and Social Issues*. Springer International Publishing, Cham, Switzerland, pp. 43–75.

Degen, B. and Müller, N. (2023) Advanced marker-assisted selection versus genomic selection in breeding programs. *Genomics*. DOI: 10.1101/2023.02.20.529263.

Eckerstorfer, M.F., Dolezel, M., Heissenberger, A., Miklau, M., Reichenbecher, W. *et al.* (2019a) An EU perspective on biosafety considerations for plants developed by genome editing and other new genetic modification techniques (NGMs). *Frontiers in Bioengineering and Biotechnology* 7, 31. DOI: 10.3389/fbioe.2019.00031.

Eckerstorfer, M.F., Engelhard, M., Heissenberger, A., Simon, S. and Teichmann, H. (2019b) Plants developed by new genetic modification techniques—comparison of existing regulatory frameworks in the EU and non-EU countries. *Frontiers in Bioengineering and Biotechnology* 7, 26. DOI: 10.3389/fbioe.2019.00026.

Endale, G.K., Karim, M., Joseph, G. and Muffy, K. (2022) Commercialization of genetically modified crops in Africa: Opportunities and challenges. *African Journal of Biotechnology* 21(5), 188–197. DOI: 10.5897/AJB2021.17434.

Fang, J. (2024) Breeding 5.0 AI-driven revolution in designed plant breeding. *Molecular Plant Breeding* 15(1). DOI: 10.5376/mpb.2024.15.0004.

Fazili, M.A., Wani, A.H. and Bhat, Z.A. (2017) Effect of oligomeric sodium alginate and chitosan on growth attributes, physiology and essential oil composition in *Mentha arvensis* L. in Northern Himalayas. *Journal of Functional and Environmental Botany* 7(2), 101. DOI: 10.5958/2231-1750.2017.00012.9.

Fuchs, L.K., Holland, A.H., Ludlow, R.A., Coates, R.J., Armstrong, H. *et al.* (2022) Genetic manipulation of biosynthetic pathways in mint. *Frontiers in Plant Science* 13, 928178. DOI: 10.3389/fpls.2022.928178.

Garro Monge, G. (2012) Desarrollo de cultivos y alimentos por técnicas de biotecnología moderna en centroamérica. *Revista Tecnología En Marcha* 25(4), 40. DOI: 10.18845/tm.v25i4.618.

Glover, B., Akinbo, O., Savadogo, M., Timpo, S., Lemgo, G. *et al.* (2018) Strengthening regulatory capacity for gene drives in Africa: Leveraging NEPAD's experience in establishing regulatory systems for medicines and GM crops in Africa. *BMC Proceedings* 12(Suppl. 8), 11. DOI: 10.1186/s12919-018-0108-y.

Guimarães, R.C., Freire, M.D.S. and Almeida, A.E.C.C. (2014) Produção de biológicos em plataformas vegetais e seus impactos na segurança alimentar – análise de lacunas na regulação vigente. *Vigilância Sanitária Em Debate* 2(4), 76–85. DOI: 10.3395/vd.v2i4.399.

Hajinajaf, N., Fayyazbakhsh, A., Kamal Shahsavar, S., Sanjarian, F. and Rahnama, H. (2024) Boosting plant oil yields: The role of genetic engineering in industrial applications. *Biofuel Research Journal* 11(2), 2105–2145. DOI: 10.18331/BRJ2024.11.2.5.

Hu, T.Y., Gao, W. and Huang, L.Q. (2016) Prospecting application of CRISPR/Cas9 genome editing technology in research of medicinal plants. *China Journal of Chinese Materia Medica* 41(16), 2953–2957. DOI: 10.4268/cjcmm20161602.

Huang, T., Men, W., Myanganbayar, A. and Davaasambuu, U. (2025) Transcriptome analysis reveals regulatory mechanism of methyl jasmonate-induced monoterpenoid biosynthesis in *Mentha arvensis* L. *Frontiers in Plant Science* 15, 1517851. DOI: 10.3389/fpls.2024.1517851.

Hunt, J. (2011) Ploughing their own furrow: Subnational regions and the regulation of GM crop cultivation. In: Barnard, C. and Odudu, O. (eds) *Cambridge Yearbook of European Legal Studies 13*. Cambridge University Press, Cambridge, UK, pp. 135–159.

Ichim, M.C. (2024) The citizens' awareness and concerns during the transition from genetically modified to genome edited plants in Europe about their use in agriculture and food production. In: Ricroch, A., Eriksson, D., Miladinović, D., Sweet, J. and Laere, K. *et al* (eds) *A Roadmap for Plant Genome Editing*. Springer Nature Switzerland, Cham, Switzerland, pp. 519–532.

Idris, S.H., Mat Jalaluddin, N.S. and Chang, L.W. (2023) Ethical and legal implications of gene editing in plant breeding: A systematic literature review. *Journal of Zhejiang University–SCIENCE B* 24(12), 1093–1105. DOI: 10.1631/jzus.B2200601.

Insaf, M., Saddique, M.A.B., Sher, M.A., Khan, M.A., Usman, M. *et al.* (2023) Genome editing for biotic stress resistance in medicinal plants. In: Khan, Z., Yasmin, A. and Safnar, N. (eds) *Medicinal Plants: Microbial Interactions, Molecular Techniques and Therapeutic Trends*. Bentham Science Publishers, Singapore, pp. 174–187.

Jamwal, M., Ghora, B., Kumar, S. and Gupta, A.P. (2023) Applications of CRISPR/Cas9 in the synthesis of bioactive compounds from medicinal plants. In: Kumar, N. and Singh, R.S. (eds) *Biosynthesis of Bioactive Compounds in Medicinal and Aromatic Plants: Manipulation by Conventional and Biotechnological Approaches*. Springer Nature Switzerland, Cham, Switzerland, pp. 81–97.

Jarvis, B.A., Romsdahl, T.B., McGinn, M.G., Nazarenus, T.J., Cahoon, E.B. *et al.* (2021) CRISPR/Cas9-induced Fad2 and Rod1 mutations stacked with Fae1 confer high oleic acid seed oil in pennycress (*Thlaspi arvense* L.). *Frontiers in Plant Science* 12, 652319. DOI: 10.3389/fpls.2021.652319.

Kaur, N. and Ahmed, T. (2021) Bioactive secondary metabolites of medicinal and aromatic plants and their disease-fighting properties. In: Kumar, N. and Singh, R.S. (eds) *Medicinal and Aromatic Plants*. Springer International Publishing, Cham, Switzerland, pp. 113–142.

Komen, J. and Wafula, D.K. (2021) Authorizing GM crop varieties: Policy implications for seed systems in Sub-Saharan Africa. *Agronomy* 11(9), 1855. DOI: 10.3390/agronomy11091855.

Kopertekh, L. and Wilhelm, R. (2023) Biosafety, risk analysis, and regulatory framework for molecular farming in Europe. In: Kole, C., Chaurasia, A., Hefferon, K.L. and Panigrahi, J. (eds) *Tools & Techniques of Plant Molecular Farming*. Springer Nature Singapore, Singapore, pp. 359–381.

Krishna, R., Soumia, P.S., Ansari, W.A., Khandagale, K. and Singh, M. (2023) Genetically modified cereal crops regulation policies. In: Nadaf, A., Sonah, H., Singh, K., Deshmukh, R. and Akhter Ansari, W. (eds) *Biofortification in Cereals*. Springer Nature Singapore, Singapore, pp. 347–366.

Kumar, K.P. and Walia, P. (2024) Speed breeding in cereal crops: Accelerating genetic improvement for rapid agricultural advancement. *Journal of Experimental Agriculture International* 46(6), 465–477. DOI: 10.9734/jeai/2024/v46i62498.

Kumar, D., Tiwari, A.K., Suryavanshi, P., Singh Verma, P.P. and Padalia, R.C. (2022) Variation in yield and essential oil composition of *Mentha spicata* cultivars. *The Indian Journal of Agricultural Sciences* 92(2), 245–248. DOI: 10.56093/ijas.v92i2.122230.

Kumar, R., Das, S.P., Choudhury, B.U., Kumar, A., Prakash, N.R. *et al.* (2024) Advances in genomic tools for plant breeding: Harnessing DNA molecular markers, genomic selection, and genome editing. *Biological Research* 57(1), 80. DOI: 10.1186/s40659-024-00562-6.

Lemke, S.L. (2022) Gene editing in plants. *Nutrition Today* 57(2), 57–63. DOI: 10.1097/NT.0000000000000532.

Li, L., Zhang, D., Zhang, Z. and Zhang, B. (2025) CRISPR/Cas: A powerful tool for designing and improving oil crops. *Trends in Biotechnology* 43(4), 773–789. DOI: 10.1016/j.tibtech.2024.09.007.

Moniruzzaman, M., Khatun, R. and Mintoo, A. (2015) Application of marker assisted selection for livestock improvement in Bangladesh. *Bangladesh Veterinarian* 31(1), 1–11. DOI: 10.3329/bvet.v31i1.22837.

Munawar, N., Ahsan, K. and Ahmad, A. (2024a) CRISPR-edited plants' social, ethical, policy, and governance issues. In: Abd-Elsalam, K.A. and Ahmad, A. (eds) *Global Regulatory Outlook for CRISPRized Plants*. Academic Press, Cambridge, Massachusetts, pp. 367–396.

Munawar, N., Faheem, M., Niamat, A., Munir, A., Habibullah Khan, S. *et al.* (2024b) Regulatory, ethical, social, and biosafety concerns in genome-edited horticultural crops. In: Kamel, A., Abd-Elsalam, K.A., Ahmad, A. and Zhang, B. (eds) *CRISPRized Horticulture Crops*. Academic Press, Cambridge, Massachusetts, pp. 421–438.

Naik, B.J., Vasamsetti, B.M.K., Kim, S.-C., Gunti, M., Swamy, M.K. *et al.* (2022) Genome editing in medicinal plants using CRISPR/Cas9 tool. In: Kumara Swamy, M. and Kumar, A. (eds) *Phytochemical Genomics*. Springer Nature Singapore, Singapore, pp. 571–589.

Nang'ayo, F., Simiyu-Wafukho, S. and Oikeh, S.O. (2014) Regulatory challenges for GM crops in developing economies: The African experience. *Transgenic Research* 23(6), 1049–1055. DOI: 10.1007/s11248-014-9805-0.

Nayak, S.N., Singh, V.K. and Varshney, R.K. (2017) Marker-assisted selection. In: Thomas, B., Murray, B.G. and Murphy, D.J. (eds) *Encyclopedia of Applied Plant Sciences*. Academic Press, Cambridge, Massachusetts, pp. 183–197.

Ona, A.D., Muntean, L., Berindean, I., Costin, A.D., Racz, I. *et al.* (2024) Modern insights into breeding of medicinal plants for health and industry. *Hop and Medicinal Plants* 32(1–2), 119–138. DOI: 10.15835/hpm.v32i1-2.15021.

Patel, S. and Gurjar, D. (2024) Speed breeding: A transformative approach to accelerate crop improvement. *PLANT CELL BIOTECHNOLOGY AND MOLECULAR BIOLOGY* 25(7–8), 24–35. DOI: 10.56557/pcbmb/2024/v25i7-88729.

Picking, D. (2017) The global regulatory framework for medicinal plants. In: Badal, S. and Delgoda, R. (eds) *Pharmacognosy*. Academic Press, Cambridge, Massachusetts, pp. 663–675.

Picking, D. (2024) The global regulatory framework for medicinal plants. In: Badal, S. and Delgoda, R. (eds) *Pharmacognosy*. Academic Press, Cambridge, Massachusetts, pp. 769–782.

Ray, S., Sneha, K. and Jangid, C. (2023) CRISPR-Cas9 for sustainable food production: Impacts, recent advancements and future perspectives. *Food and Humanity* 1, 1458–1471. DOI: 10.1016/j.foohum.2023.10.014.

Roy, A., Khan, A., Ahmad, I., Alghamdi, S., Rajab, B.S. *et al.* (2022) Flavonoids a bioactive compound from medicinal plants and its therapeutic applications. *BioMed Research International* 2022(1), 5445291. DOI: 10.1155/2022/5445291.

Safdar, N., Yasmin, A. and Khan, Z. (2023) Medicinal plants: Traditional trends to modern therapeutics. In: Khan, Z., Yasmin, A. and Safnar, N. (eds) *Medicinal Plants: Microbial Interactions, Molecular Techniques and Therapeutic Trends*. Bentham Science Publishers, Singapore, pp. 188–205.

Shah, P., Patel, P., Hariwal, M., Verma, S., Yadav, R. *et al.* (2023) Targeted improvement of medicinal plant under stress condition through CRISPR-Cas-mediated genome engineering. In: Singh, D., Kumar Mishra, A. and Kumar Srivastava, A. (eds) *Stress-Responsive Factors and Molecular Farming in Medicinal Plants*. Springer Nature Singapore, Singapore, pp. 345–357.

Shanmugavel, P., Ramasamy, G., Vellingiri, G., Marimuthu, R. and Thiyagarajan, K. (2023) Speed breeding: A propitious technique for accelerated crop improvement. In: Wang, H. (ed.) *Case Studies of Breeding Strategies in Major Plant Species. IntechOpen*. London. DOI: 10.5772/intechopen.105533.

Sharma, A., Pandey, H., Misra, V., Devadas, V.S., Kesavan, A.K. *et al.* (2024a) *Harnessing speed breeding in controlled environment ecosystem for millets sustainability*. Plant Breeding. DOI: 10.1111/pbr.13258.

Sharma, S., Chauhan, H. and Singh, K. (2024b) Genome-editing advances for disease resistance in plants. In: Singh, K., Kaur, R. and Deshmukh, R. (eds) *Biotechnological Advances for Disease Tolerance in Plants*. Springer Nature Singapore, Singapore, pp. 293–316.

Shinde, C.S., Nishtha, Salunke, A.J. and Deshmukh, V.V. (2023) Speed breeding: One of the rapid generation advancement approaches. *Current Journal of Applied Science and Technology* 42(48), 63–69. DOI: 10.9734/cjast/2023/v42i484331.

Shyamala, S.L. and Sasipriya, S. (2023) Speed breeding–an emerging trend in crop improvement. *International Journal of Plant & Soil Science* 35(22), 697–703. DOI: 10.9734/ijpss/2023/v35i224180.

Singh, S., Uddin, M., Chishti, A.S., Bhat, U.H., Singh, S. *et al.* (2023) Plant-derived smoke water and karrikinolide (KAR1) enhance physiological activities, essential oil yield and bioactive constituents of *Mentha arvensis* L. *Frontiers in Plant Science* 14, 1129130. DOI: 10.3389/fpls.2023.1129130.

Smajlovic, S., Frkatovic, A., Sabit, H., Tombuloglu, H. and Unver, T. (2021) Applications of CRISPR/Cas9 in oil crops to improve oil composition. In: Tombuloglu, H., Unver, T., Tombuloglu, G. and Rehman Hakeem, K. (eds) *Oil Crop Genomics*. Springer International Publishing, Cham, Switzerland, pp. 383–391.

Sparrow, P., Broer, I., Hood, E.E., Eversole, K., Hartung, F. *et al.* (2013) Risk assessment and regulation of molecular farming – a comparison between Europe and US. *Current Pharmaceutical Design* 19(31), 5513–5530. DOI: 10.2174/1381612811319310007.

Spök, A., Sprink, T., Allan, A.C., Yamaguchi, T. and Dayé, C. (2022) Towards social acceptability of genome-edited plants in industrialised countries? Emerging evidence from Europe, United States, Canada, Australia, New Zealand, and Japan. *Frontiers in Genome Editing* 4, 899331. DOI: 10.3389/fgeed.2022.899331.

Sujadmiko, B., Tisnanta, H. and Melati Davey, O. (2021) Local certification: Genetically modified organisms and commercialization. *Kertha Patrika* 43(1), 1. DOI: 10.24843/KP.2021.v43.i01.p01.

Swami, P., Deswal, K., Rana, V., Yadav, D. and Munjal, R. (2023) Speed breeding—a powerful tool to breed more crops in less time accelerating crop research. In: Kamran Khan, M., Pandey, A., Hamurcu, M., Prakash Gupta, O. and Gezgin, S. (eds) *Abiotic Stresses in Wheat*. Academic Press, Cambridge, Massachusetts, pp. 33–49.

Temesgen, B. (2022) Speed breeding to accelerate crop improvement. *International Journal of Agricultural Science and Food Technology* 8(2), 178–186. DOI: 10.17352/2455-815X.000161.

Ugandhar, T. (2025) Advances in plant breeding: Enhancing crop productivity, resilience, and sustainability through modern techniques. *Science Reviews. Biology* 3(4), 1–6. DOI: 10.57098/SciRevs.Biology.3.4.1.

Unêda-Trevisoli, S.H., Silva, F.M. and Mauro, A.O. (2017) Marker-assisted selection and genomic selection. In: Lopes da Silva, F., Borém, A., Sediyama, T. and Hytalo Ludke, W. (eds) *Soybean Breeding*. Springer International Publishing, Cham, Switzerland, pp. 275–291.

Varshney, R.K., Roorkiwal, M. and Sorrells, M.E. (2017) Genomic selection for crop improvement. In: Varshney, R.K., Roorkiwal, M. and Sorrells, M.E. (eds) *Genomic Selection for Crop Improvement*. Springer International Publishing, Cham, Switzerland, pp. 1–6.

Verma, S., Gupta, A., Yalla, S., Shreya, P., Sharma, P.J. *et al.* (2024) Integrating marker-assisted (MAS) and genomic selection (GS) for plant functional trait improvement. In: Kumar, N. and Singh, H. (eds) *Plant Functional Traits for Improving Productivity*. Springer Nature Singapore, Singapore, pp. 203–215.

Watson, A., Ghosh, S., Williams, M.J., Cuddy, W.S., Simmonds, J. *et al.* (2018) Speed breeding is a powerful tool to accelerate crop research and breeding. *Nature Plants* 4(1), 23–29. DOI: 10.1038/s41477-017-0083-8.

Wattoo, F.M., Khalid, T., Rana, R.M. and Ahmad, F. (2024) Speed breeding for rapid crop improvement. In: Fiaz, S. and Prakash, C.S. (eds) *OMICs-based Techniques for Global Food Security*. Wiley, Oxford, UK, pp. 139–158.

Wolt, J.D. (2017) Safety, security, and policy considerations for plant genome editing. *Progress in Molecular Biology and Translational Science* 149, 215–241. DOI: 10.1016/bs.pmbts.2017.03.005.

Yaseen, S., Yasmeen, A., Safdar, N. and Khan, Z. (2023) Genome editing and its applications in plants. In: Khan, Z., Yasmin, A. and Safnar, N. (eds) *Medicinal Plants: Microbial Interactions, Molecular Techniques and Therapeutic Trends*. Bentham Science Publishers, Singapore, pp. 140–158.

4 Metabolic Engineering and Biosynthetic Pathway Optimization

Archana Jain, Surendra Sarsaiya and Qihai Gong*
Key Laboratory of Basic Pharmacology and Joint International Research Laboratory of Ethnomedicine of Ministry of Education, Zunyi Medical University, Zunyi, Guizhou, China

4.1 Introduction

Metabolic engineering and biosynthetic pathway optimization have emerged as cornerstone strategies in modern biotechnology, addressing the critical need for sustainable and scalable production of plant-derived secondary metabolites. These compounds—such as alkaloids, terpenoids, and phenolics—are indispensable to pharmaceuticals, agriculture, and industrial applications due to their diverse bioactivities (Elshafie *et al.*, 2023). For instance, alkaloids like morphine and vinblastine are pivotal in pain management and cancer therapy, while terpenoids such as artemisinin and paclitaxel serve as frontline treatments for malaria and cancer, respectively (Ballout *et al.*, 2019). Despite their immense value, secondary metabolites are typically synthesized in trace amounts by plants as part of their defense mechanisms, rendering traditional extraction methods economically unviable and ecologically unsustainable (de Oliveira *et al.*, 2023). The overharvesting of endangered species like the Pacific yew (for paclitaxel) and the seasonal variability of plant growth further exacerbates these challenges, necessitating innovative biotechnological solutions to meet global demand (Zilberman *et al.*, 2023).

The advent of metabolic engineering has revolutionized this landscape by enabling the reconstruction and optimization of biosynthetic pathways in heterologous hosts such as *Escherichia coli*, *Saccharomyces cerevisiae*, and engineered plant cell cultures (Liu *et al.*, 2022). By transplanting plant-derived genes into microbial systems, researchers can bypass the limitations of native plant metabolism, achieving higher yields under controlled fermentation conditions (Trantas *et al.*, 2015). For example, the heterologous expression of cytochrome P450 enzymes (CYPs) in yeast has facilitated the hydroxylation of triterpenoid precursors, a critical step in synthesizing bioactive molecules like ginsenosides (Zhang *et al.*, 2023). Similarly, CRISPR-Cas9 (clustered regularly interspaced short palindromic-mediated repeats-associated protein 9) genome editing has been employed to silence competitive pathways in *Taxus* cell cultures, redirecting metabolic flux toward taxadiene—the precursor of paclitaxel—thereby boosting yields by over 50% (Halder and Roy, 2023). These advancements underscore the potential of synthetic biology tools to rewire metabolic networks, yet significant hurdles remain, including enzyme promiscuity,

*Corresponding author: gqh@zmu.edu.cn

DOI: 10.1079/9781836994206.0004

metabolic burden, and the complexity of plant pathway regulation.

A key innovation in this field is the integration of multi-omics technologies—genomics, transcriptomics, proteomics, and metabolomics—which provide unprecedented insights into the spatial and temporal dynamics of metabolic pathways (Rai *et al.*, 2017). For instance, metabolic flux analysis (MFA) has been instrumental in mapping carbon allocation in *Artemisia annua*, identifying bottlenecks in the artemisinin biosynthesis pathway that were subsequently alleviated through targeted overexpression of amorpha-4,11-diene synthase (Allen, 2016; Soni *et al.*, 2022). Complementing these experimental approaches, computational tools like flux balance analysis (FBA) and machine learning algorithms enable predictive modeling of metabolic networks (Sampaio *et al.*, 2022). Recent studies have leveraged AlphaFold-predicted enzyme structures to redesign CYP active sites, enhancing substrate specificity and reducing off-target reactions in terpenoid biosynthesis (Apostolidou *et al.*, 2023). These integrative approaches not only accelerate pathway optimization but also pave the way for *de novo* design of novel metabolites with tailored functionalities.

Despite these breakthroughs, the industrial scalability of metabolic engineering remains fraught with challenges. The metabolic burden imposed by heterologous pathway expression often strains host organisms, leading to reduced growth rates and viability (Halder and Roy, 2023). In *Taxus* cell cultures, for example, the high energy demand of paclitaxel biosynthesis necessitates precise optimization of culture conditions, including elicitor treatments (e.g. methyl jasmonate) and two-phase bioreactor systems to separate growth and production phases (Yin *et al.*, 2025). Furthermore, the functional expression of plant-derived enzymes—particularly membrane-bound CYPs—in microbial hosts is hindered by incompatible cellular environments, necessitating advanced protein engineering strategies (Pyne *et al.*, 2018). Recent advances in directed evolution and computational design have yielded CYP variants with improved solubility and activity in *E. coli*, yet achieving industrial-scale production of complex metabolites like vinblastine remains a work in progress (D'Amelia *et al.*, 2021; Sethi *et al.*, 2022).

Looking ahead, the convergence of synthetic biology, machine learning, and sustainable bioprocessing holds immense promise (Goshisht, 2024). Engineered photosynthetic organisms, such as cyanobacteria and microalgae, are being explored as green platforms for terpenoid production, leveraging their ability to harness solar energy and CO_2 (Tanvir *et al.*, 2021). Similarly, plant molecular farming—using fast-growing species like *Nicotiana benthamiana* to produce high-value compounds—offers a scalable alternative to traditional agriculture (Singla, 2020). On the regulatory front, harmonizing biomanufacturing practices with good manufacturing practices (GMP) and ethical guidelines will be critical to ensuring the safety and acceptance of engineered metabolites in global markets (Cauchon *et al.*, 2019).

This chapter delves into the methodologies, challenges, and future directions of metabolic engineering, with a focus on secondary metabolite optimization. Through case studies on paclitaxel biosynthesis, microbial production of alkaloids, and omics-guided pathway design, it highlights the transformative potential of this field while addressing persistent technical and economic barriers. By bridging cutting-edge science with industrial pragmatism, metabolic engineering is poised to redefine the sustainable production of bioactive compounds, offering solutions to some of the most pressing challenges in health care and environmental conservation (Fig. 4.1).

4.2 Engineering Secondary Metabolites: Alkaloids and Terpenoids

The engineering and optimization of secondary metabolites biosynthesis have emerged as pivotal strategies, particularly for producing alkaloids and terpenoids. These compounds, although not essential for the primary metabolism of plants, play significant roles in ecological interactions and have substantial industrial and pharmaceutical applications. The engineering of these pathways involves manipulating the metabolic networks to enhance the yield and productivity of these valuable compounds. This introduction will explore the significance of metabolic

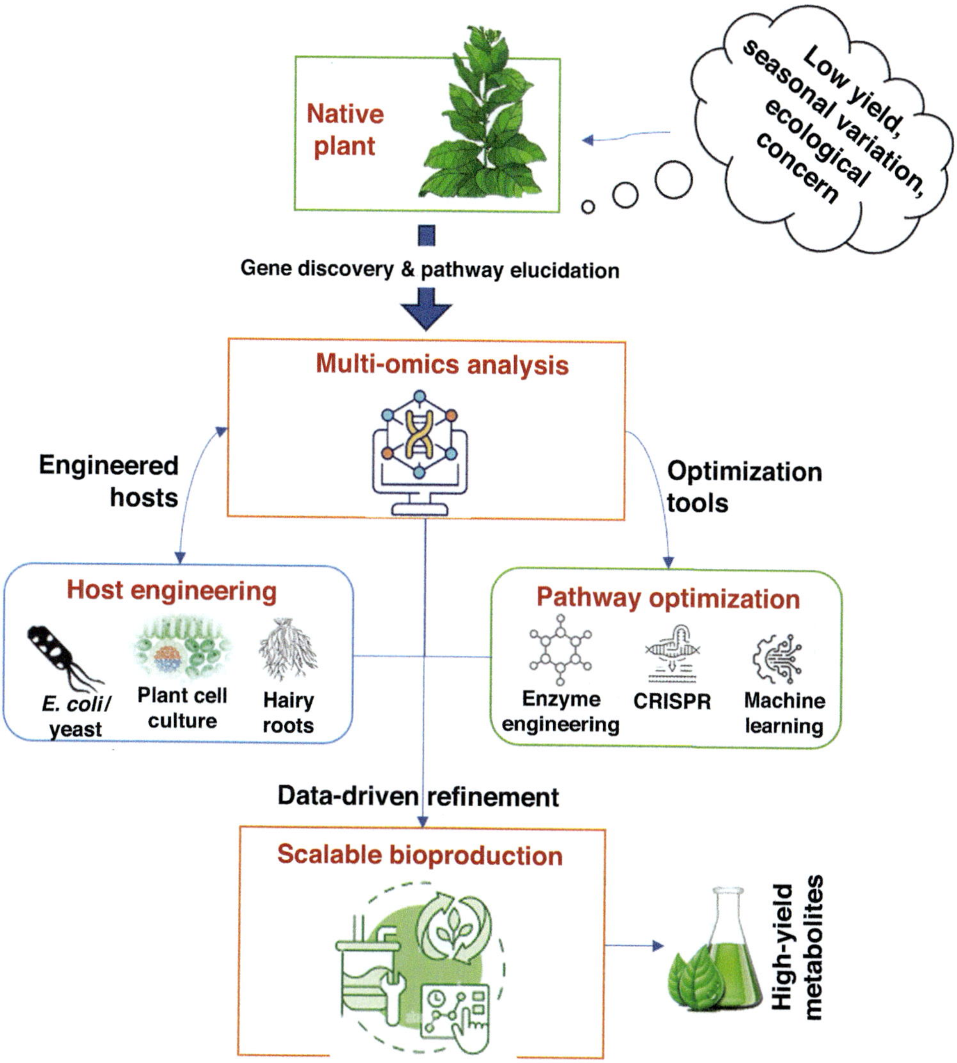

Fig. 4.1. Overall metabolic engineering workflow for plant metabolite production. (Author's own figure.)

engineering in the production of alkaloids and terpenoids, highlighting the methodologies and challenges involved.

Alkaloids, which are nitrogen-containing compounds, are known for their potent biological activities and are widely used in pharmaceuticals. Examples include morphine, nicotine, and vinblastine, which are derived from complex biosynthetic pathways in plants (Sato, 2020). As the largest class of secondary metabolites, terpenoids have diverse applications ranging from pharmaceuticals to biofuels. They are valued for their therapeutic properties and are used in flavors and fragrances (Shukla *et al.*, 2023). Metabolic engineering often involves reconstructing biosynthetic pathways in heterologous hosts, such as microbes, to overcome the limitations of plant-based production. This approach allows for the production of alkaloids and terpenoids in more controlled and scalable environments (Shukla *et al.*, 2023).

Techniques such as overexpression of key biosynthetic genes, silencing of competitive pathways, and the use of synthetic biology tools are employed to enhance the production of target metabolites. These strategies have been successfully applied to increase the yield of compounds like isoquinoline alkaloids and terpenoid indole alkaloids (Kumari *et al.*, 2024). A primary challenge in utilizing secondary metabolites is their low yield in native plants. Metabolic engineering aims to redirect carbon fluxes toward the desired products, thereby improving yield and productivity (Lehari and Kumar, 2024). The right host for metabolic engineering is crucial. Microbial hosts offer advantages such as rapid growth and ease of genetic manipulation, making them suitable for the production of terpenoids and other secondary metabolites (Abdallah and Quax, 2017). While metabolic engineering offers promising solutions for the production of secondary metabolites, it is not without its challenges. The complexity of plant metabolic pathways and the need for precise control over gene expression and metabolic fluxes require advanced techniques and a deep understanding of plant physiology. Additionally, the integration of omics approaches and synthetic biology is essential to overcome these challenges and fully realize the potential of metabolic engineering in the production of alkaloids and terpenoids.

4.2.1 Designing pathways for alkaloids, terpenoids, and phenolics

Designing pathways for the biosynthesis of alkaloids, terpenoids, and phenolics involves understanding the complex biochemical and genetic processes that lead to their formation. These secondary metabolites are crucial for plant defense and have significant pharmacological applications. Recent advances in biotechnology and synthetic biology have opened new avenues for engineering these pathways to enhance the production of these valuable compounds. This analysis will explore the strategies and technologies used in designing these pathways, focusing on the integration of omics approaches, metabolic engineering, and synthetic biology.

Omics technologies, including genomics, transcriptomics, proteomics, and metabolomics, provide comprehensive data that help elucidate the biosynthetic pathways of secondary metabolites in plants. These approaches are particularly useful for non-model medicinal plants where genomic information is limited (Chung *et al.*, 2016). By integrating omics data, researchers can identify key genes and enzymes involved in the biosynthesis of alkaloids, terpenoids, and phenolics, facilitating the reconstruction of these pathways in heterologous systems (Chung *et al.*, 2016). Metabolic engineering involves the modification of metabolic pathways within organisms to increase the production of desired compounds. In the context of secondary metabolites, this can involve overexpressing genes, silencing competing pathways, or introducing novel biosynthetic routes (Mohiuddin, 2019). For alkaloids, metabolic engineering has been successfully applied to produce benzylisoquinoline alkaloids and monoterpenoid indole alkaloids in microbial hosts like *E. coli* and yeast, demonstrating the potential for large-scale production (Schäfer and Wink, 2009). Terpenoids and phenolics have also been targets for metabolic engineering, with efforts focusing on enhancing their biosynthesis in plants and microbial systems (Staniek *et al.*, 2013).

Synthetic biology offers tools to design and construct new biological parts, devices, and systems. This approach is used to reconstruct entire biosynthetic pathways in microbial hosts, allowing for the production of complex secondary metabolites (Huang *et al.*, 2022). Techniques such as CRISPR-Cas9 gene editing and the use of transcription factors are employed to manipulate the expression of genes involved in alkaloid biosynthesis, enhancing their production in plants (Bhushan *et al.*, 2023). Synthetic biology also facilitates the creation of gene cassettes that can be introduced into plants or microbes to produce high-value compounds like terpenoids and phenolics (Schäfer and Wink, 2009). Understanding the regulatory mechanisms of these pathways is essential for achieving efficient production in heterologous systems, yet the complexity of secondary metabolite biosynthesis, involving multiple steps and regulatory layers, requires a detailed understanding of the underlying biology (Staniek *et al.*, 2013). Future research should focus on integrating multiomics data to uncover unknown pathway components and regulatory networks. Additionally, developing more efficient and scalable bioreactor systems for the production of these metabolites will be crucial for industrial applications

(Chung *et al.*, 2016). By engineering pathways for alkaloids, terpenoids, and phenolics, the field continues to evolve with new technologies and insights. The integration of omics data and advanced biotechnological tools holds promise for overcoming current limitations and achieving sustainable production of these valuable compounds. However, the complexity of these pathways and the need for precise regulation present ongoing challenges that require further research and innovation.

4.3 Metabolic Flux Analysis for Pathway Optimization

Metabolic flux analysis (MFA) is a powerful tool for optimizing metabolic pathways in medicinal plants, aiming to enhance the production of valuable natural compounds. By quantifying the flow of metabolites through biochemical pathways, MFA provides insights into the regulation and efficiency of plant metabolism. This approach is particularly relevant for medicinal plants, where the production of secondary metabolites like essential oils and alkaloids is of significant interest. The integration of isotopic labeling and computational modeling in MFA allows for a detailed understanding of metabolic networks, which is crucial for pathway optimization. The following sections delve into the key aspects of MFA in the context of medicinal plants (Table 4.1 and Fig. 4.2).

Isotopically nonstationary MFA (INST-MFA) uses stable isotopic tracers to quantify metabolic reaction rates, offering insights into plant metabolism under various conditions (Koley *et al.*, 2024). ^{13}C-MFA involves supplying isotopically labeled substrates to plants and analyzing the resulting labeling patterns using techniques like NMR (nuclear magnetic resonance) or MS (mass spectrometry), combined with mathematical modeling to calculate fluxes (Calheiros Gomes and Simoes, 2012). These methods help elucidate complex metabolic networks and identify bottlenecks in pathways, which can be targeted for optimization (Allen *et al.*, 2009).

Metabolic engineering strategies aim to enhance the production of high-value compounds in medicinal plants, such as paclitaxel and artemisinin, by optimizing biosynthetic pathways (Halder and Roy, 2023). MFA can identify key enzymes and regulatory mechanisms in the biosynthesis of compounds like triterpenes, which have significant therapeutic effects (Noushahi *et al.*, 2022). The integration of MFA with synthetic biology allows for the reconstitution of biosynthetic pathways in heterologous systems, potentially increasing the yield of desired products (Halder and Roy, 2023). The complexity of plant metabolic networks, with their high degree of compartmentation and diversity of secondary metabolites, poses challenges for MFA (Allen *et al.*, 2009). Advances in analytical techniques and computational algorithms are improving the accuracy and reliability of MFA in plants, enabling better characterization of compartmentalized behaviors (Calheiros Gomes and Simoes, 2012). The transition from a pathway-centered to a network-wide perspective in plant metabolism requires tools for structural network description and model verification (Kruger and Ratcliffe, 2012).

Combining pathway analysis with FBA offers a comprehensive approach to studying metabolic systems, allowing for the prediction of optimal flux distributions under varying conditions. The development of dynamic labeling experiments and enhanced flux calculation algorithms opens new possibilities for investigating specific pathways and improving metabolic engineering efforts (Calheiros Gomes and Simoes, 2012). The application of MFA in conjunction with synthetic biology holds promise for the exploitation of medicinal plants, aiming to increase the production of natural products and improve their therapeutic efficacy (Halder and Roy, 2023). While MFA provides a robust framework for pathway optimization in medicinal plants, it is important to consider the broader context of plant metabolic engineering. The integration of MFA with other approaches, such as synthetic biology and FBA, can enhance our understanding of plant metabolism and facilitate the development of more efficient production systems. However, challenges remain in terms of accurately modeling complex plant metabolic networks and translating these insights into practical applications. Continued advancements in analytical and computational techniques will be crucial for overcoming these challenges and realizing the full potential of MFA

Table 4.1. Key analytical and computational tools for pathway optimization. Author's own table.

Tool/methodology	Principle function	Key applications in metabolic engineering	Limitations/hallenges	Citations
Metabolic flux analysis (MFA)	Quantifies the *in vivo* flow of metabolites through a metabolic network using isotopic tracers (e.g. ^{13}C) and computational modeling.	• Identifying rate-limiting steps (bottlenecks) in pathways (e.g. in artemisinin biosynthesis). • Quantifying carbon allocation under different conditions.	• Requires extensive experimental data. • Complex in plants due to compartmentalization. • Provides a snapshot, not always dynamic.	(Allen *et al.*, 2009; Allen, 2016; Koley *et al.*, 2024)
Flux balance analysis (FBA)	A constraint-based modeling approach that predicts metabolic flux distributions at genome-scale to maximize/minimize an objective (e.g. biomass or product yield).	• Predicting outcomes of gene knockouts or overexpression. • Identifying essential genes for metabolite production. • Guiding strain design *in silico*.	• Does not incorporate enzyme kinetics or regulation by default. • May predict multiple, equally optimal solutions.	(Schilling *et al.*, 2000; Curran *et al.*, 2012; Lopes *et al.*, 2024)
Multi-omics integration (genomics, transcriptomics, proteomics, metabolomics)	Provides a comprehensive, systems-level view of biological systems by integrating data on genes, RNA, proteins, and metabolites.	• Elucidating unknown biosynthetic pathways in non-model plants • Identifying key genes and transcription factors for engineering (e.g. in *Artemisia annua*). • Functional characterization of enzymes like P450s.	• Data integration and interpretation is complex. • Requires sophisticated bioinformatics expertise. • High cost of comprehensive omics studies.	(Chung *et al.*, 2016; Rai *et al.*, 2017; Zheng *et al.*, 2019)

Machine learning (ML) & artificial intelligence (AI)	Uses algorithms to find patterns in large, complex datasets (e.g. omics data) to predict optimal designs and outcomes.	• Predicting enzyme function and stability (e.g. with AlphaFold). • Guiding protein engineering campaigns. • Optimizing biosynthetic pathways and fermentation conditions.	• Dependent on the quality and quantity of training data. • "Black box" nature can make predictions difficult to interpret	(Ramzi *et al.*, 2020; Ao *et al.*, 2024; Goshisht, 2024)

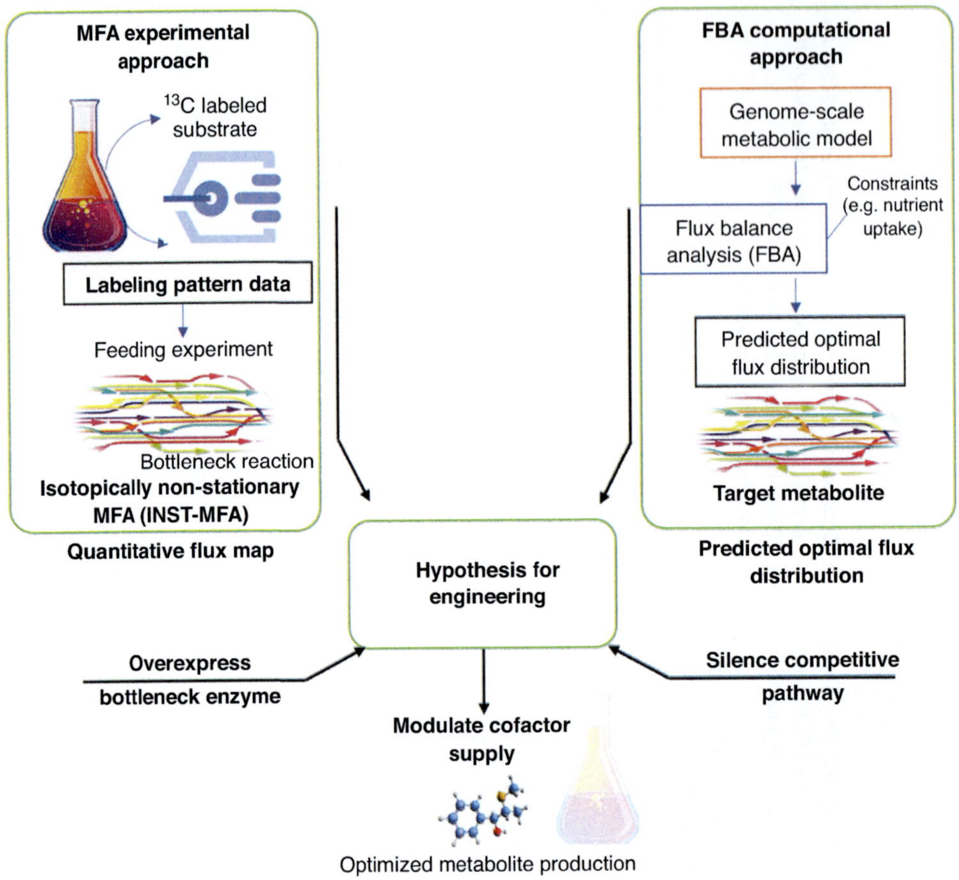

Fig. 4.2. Flux optimization using MFA and FBA. (Author's own figure.)

in optimizing metabolic pathways in medicinal plants.

4.3.1 Using FBA to optimize pathway efficiency

FBA is a powerful computational tool used to optimize metabolic pathways, including those in medicinal plants, by predicting the distribution of metabolic fluxes in a network. This approach is particularly useful in the context of metabolic engineering, where the goal is to enhance the production of specific metabolites. FBA leverages genome-scale metabolic models to simulate and optimize metabolic changes, providing insights into pathway efficiency and potential genetic modifications. The integration of FBA with other analytical methods can further refine these predictions and enhance pathway optimization.

FBA is used to predict and optimize metabolic changes by simulating the effects of gene knockouts and overexpression in microbial systems. This approach can be adapted to medicinal plants to identify key genes and pathways for optimization (Curran *et al.*, 2012). The method involves constructing a stoichiometric model of the metabolic network and applying linear programming to find the optimal flux distribution that maximizes or minimizes a specific objective, such as biomass production or metabolite yield (Lopes *et al.*, 2024).

A challenge of FBA involves the presence of multiple flux distributions that can achieve the same objective function, which can complicate the interpretation of results. The reaction route (RR) approach offers a more rigorous definition of pathways, potentially resolving some of these ambiguities (Fishtik, 2009). Incorporating feedback inhibition into FBA models can improve the accuracy of predictions by accounting for regulatory mechanisms that affect enzyme activity and pathway efficiency (Schilling *et al.*, 2000). Combining FBA with pathway analysis allows for a comprehensive study of metabolic systems, providing insights into the systemic properties of cellular metabolism. This integration can help identify extreme pathways and phenotypic phase planes, which are crucial for understanding pathway utilization and optimization (Schilling *et al.*, 2000). Thermodynamic constraints can be added to FBA models to ensure the selection of feasible pathways, enhancing the prediction of optimal reaction conditions and pathway distributions. FBA has been successfully applied to optimize the production of bioethanol in yeast by analyzing metabolic fluxes under different substrate conditions. This approach can be extended to medicinal plants to enhance the production of valuable metabolites. The identification of essential genes and minimal genomes using FBA can lead to improved metabolite production by focusing on significant pathways and reducing genome complexity (Salleh *et al.*, 2015). While FBA provides a robust framework for optimizing metabolic pathways, it is important to consider its limitations and the need for integration with other methods. The presence of multiple flux distributions and the need for accurate stoichiometric models are challenges that can be addressed through complementary approaches such as the RR framework and thermodynamic analysis. Additionally, the application of FBA in medicinal plants requires careful consideration of plant-specific metabolic networks and regulatory mechanisms. By addressing these challenges, FBA can be a valuable tool for optimizing pathway efficiency in medicinal plants and other biological systems.

4.4 Enzyme Engineering: Cytochrome P450s and Beyond

Enzyme engineering, particularly of CYPs, plays a pivotal role in the biosynthesis of secondary metabolites in medicinal plants. These enzymes are crucial for the production of a wide array of structurally diverse and pharmacologically significant compounds, such as terpenoids, flavonoids, and alkaloids. The engineering of CYPs has emerged as a promising strategy to enhance the production of these high-value metabolites, which are often difficult to obtain in large quantities from natural sources. This approach not only aids in the sustainable production of these compounds but also opens new avenues for drug development and therapeutic applications. The following sections delve into the various aspects of CYP engineering in the context of medicinal plants. CYPs are versatile catalysts involved in the biosynthesis of secondary metabolites, which are essential for plant defense and human health applications (Sethi *et al.*, 2022). They catalyze a variety of reactions, including hydroxylation, which is crucial for the structural diversification of metabolites like alkaloids and terpenoids (Nguyen and Dang, 2021). The structural diversity of these metabolites is often achieved through the modification of chemical groups and rearrangement of skeletal structures, processes heavily reliant on CYP activity (Zhao *et al.*, 2014).

Protein engineering techniques, such as directed evolution and structure-guided design, have been employed to enhance the stability and activity of CYPs (Li *et al.*, 2019). Successful cases include the engineering of CYPs to alter substrate specificity and create novel metabolic pathways, thereby producing new compounds with potential therapeutic benefits (Shang and Huang, 2020). Computational approaches, integrating protein design and evolutionary information, have been used to optimize CYPs for specific reactions, such as the hydroxylation of cucurbitadienol to produce mogrol (Li *et al.*, 2019). A major challenge in CYP engineering is the difficulty in expressing these enzymes in heterologous systems due to their membrane-bound nature (Zhang *et al.*, 2021). Emerging tools like high-throughput sequencing, machine learning, and *de novo* protein design are being leveraged to overcome these challenges and accelerate the discovery and functional characterization of CYPs (Shang and Huang, 2020). The integration of genomics and transcriptomics data has facilitated the identification of CYPs involved in complex biosynthetic pathways, such as those of alkaloids and terpenoids (Nguyen and Dang, 2021). Engineered CYPs have significant potential in the production of phytopharmaceuticals, offering a sustainable approach to drug development (Villa-Ruano *et al.*, 2015). The ongoing discovery of new CYPs and their pathways continues to expand the possibilities for metabolic engineering, enabling the production of novel compounds with enhanced bioactivity (Sethi *et al.*, 2022).

While the engineering of CYPs presents numerous opportunities for enhancing the production of medicinal plant metabolites, it also poses challenges, particularly in terms of enzyme expression and activity optimization. The integration of advanced computational tools and high-throughput techniques is crucial for overcoming these hurdles and fully realizing the potential of CYPs in drug development and other applications. As research progresses, the role of CYPs in the biosynthesis of secondary metabolites will likely expand, offering new insights and opportunities in the field of enzyme engineering (Fig. 4.3).

4.4.1 Protein engineering to enhance catalytic activity and substrate range

Protein engineering has emerged as a pivotal tool in enhancing the catalytic activity and substrate

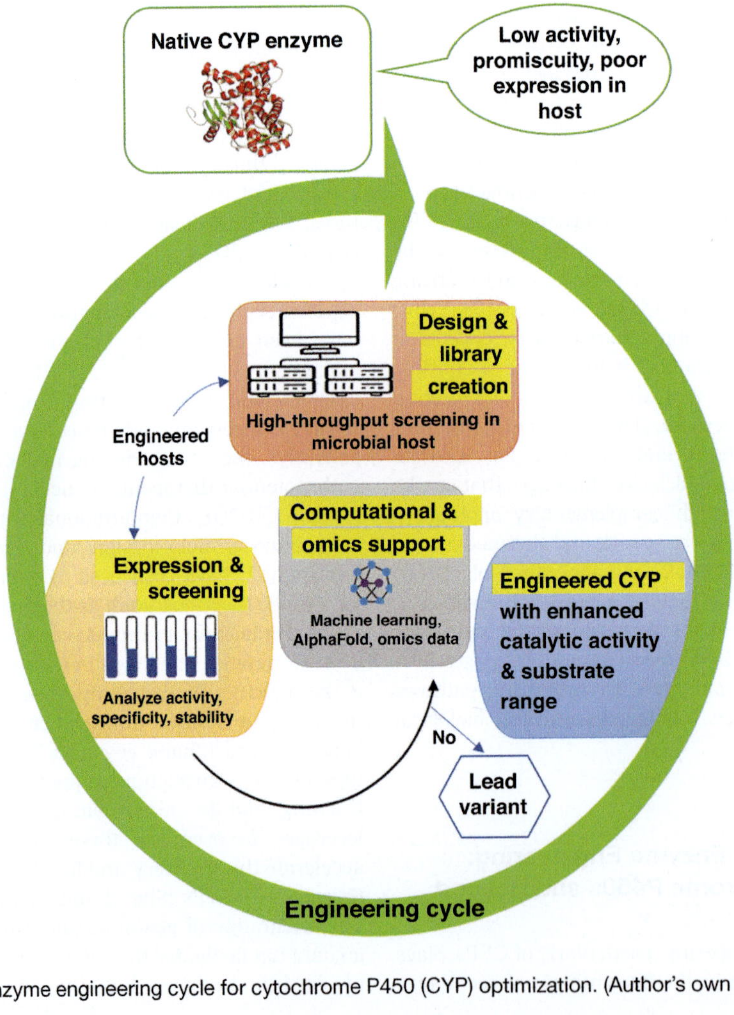

Fig. 4.3. Enzyme engineering cycle for cytochrome P450 (CYP) optimization. (Author's own figure.)

range of enzymes derived from medicinal plants, which are crucial for the biosynthesis of high-value secondary metabolites. These metabolites often have therapeutic applications, and their production can be limited by the natural capabilities of the enzymes involved. By employing various protein engineering strategies, researchers aim to overcome these limitations, thereby expanding the potential of medicinal plants in pharmaceutical applications.

Traditional methods like directed evolution and rational design have been extensively used to modify enzyme activity and substrate specificity. These approaches involve iterative cycles of mutation and selection to identify beneficial variants, although they can be labor-intensive due to the vast protein mutation space that needs to be explored (Antikainen and Martin, 2005). Recent advancements in machine learning have provided new avenues for protein engineering by predicting protein fitness landscapes and identifying catalytic patterns with limited experimental data. These tools help guide protein engineering campaigns more efficiently, reducing the need for extensive experimental screening (Shang and Huang, 2020; Ao *et al.*, 2024). Structural insights into enzyme active sites have facilitated the design of enzymes with enhanced catalytic properties. For instance, the engineering of unspecific peroxygenases (UPOs) has led to significant improvements in oxyfunctionalization activity and selectivity, demonstrating the potential of structure-guided approaches (Knorrscheidt *et al.*, 2021; Ni *et al.*, 2021).

Engineering efforts in CYPs, which are versatile catalysts pivotal for biosynthesizing a diverse array of plant metabolites, have focused on targeted enhancements to both their catalytic activity and their substrate range; these initiatives aim not only to improve inherent enzyme stability and reaction kinetics but also to deliberately alter their functional specificity, thereby enabling the creation of novel, non-natural metabolic pathways for the production of high-value compounds. For example, modifications in the CYP716A subfamily have enhanced triterpenoid production by altering key amino acid residues to improve catalytic activity and substrate specificity (Zhang *et al.*, 2021; Romsuk *et al.*, 2022; Sethi *et al.*, 2022). Protein engineering has also been used to enhance the selectivity of enzymes. For instance, variants of UPOs have been developed with high chemo- and regioselectivities, which are crucial for the synthesis of specific compounds with desired pharmacological properties (Knorrscheidt *et al.*, 2021).

Case studies demonstrate the successful application of engineered enzymes in synthesizing complex plant-derived pharmaceuticals, particularly in the production of alkaloids and other high-value compounds. For instance, enzymatic methods facilitating C–C bond formation have been developed to construct the core scaffolds of these important pharmaceutical agents (Patil *et al.*, 2018). Through strategic protein engineering, the native substrate scope of key enzymes has been expanded, enabling the synthesis of non-natural alkaloids that exhibit novel pharmacological activities. Furthermore, to enhance the efficiency and selectivity of these complex syntheses, researchers have integrated individual engineered enzymes into sophisticated multi-enzymatic cascades. This approach mimics natural biosynthetic pathways and has proven particularly powerful for the streamlined production of intricate natural products that are difficult to obtain through traditional chemical methods (Patil *et al.*, 2018). While protein engineering has significantly advanced the capabilities of enzymes derived from medicinal plants, challenges remain. The complexity of plant metabolic pathways and the difficulty in expressing plant enzymes in heterologous systems are ongoing issues. Additionally, the need for high-throughput screening methods to efficiently evaluate large libraries of enzyme variants is critical. Despite these challenges, the integration of computational tools and machine learning with traditional protein engineering approaches holds promise for further advancements in this field, potentially leading to the discovery and production of new therapeutic compounds.

4.5 Heterologous Production in Microbial Hosts

Heterologous production in microbial hosts is a promising approach to produce medicinal plant-derived compounds, addressing the challenges of low yield and high production costs associated with traditional extraction methods. This technique involves the expression of plant biosynthetic pathways in microorganisms, which can be genetically engineered to produce valuable secondary metabolites. The integration of plant genes into microbial systems not only enhances the production efficiency but also allows for the exploration of novel chemical variants. This approach is increasingly being adopted to meet the growing demand for plant-based pharmaceuticals and nutraceuticals. Table 4.2 summarizes the key aspects of heterologous production in microbial hosts in the context of medicinal plants

The utilization of microbial hosts, such as *E. coli* and *S. cerevisiae*, for the heterologous production of plant secondary metabolites presents significant advantages in terms of sustainability and scalability, offering a viable and efficient alternative to traditional plant extraction methods. By engineering these well-characterized microbial systems to express plant-derived biosynthetic pathways, it is possible to achieve higher yields of valuable compounds—such as specific N-methylpyrrolinium-derived alkaloids and the polyphenol resveratrol—under controlled fermentation conditions, thereby overcoming the limitations of field cultivation, seasonal variability, and low natural abundance (Braga *et al.*, 2018; Ping *et al.*, 2019).

Microbes are genetically tractable, allowing for the refactoring of complex plant biosynthetic pathways. This facilitates the production of medicinal alkaloids and other secondary metabolites, which are otherwise difficult to synthesize chemically (Zhang *et al.*, 2008). By engineering microbial hosts, researchers can explore new chemical variants of plant compounds, potentially leading to the discovery of novel drugs with enhanced pharmacological properties (Bradley *et al.*, 2020). Techniques such as overexpression of key enzymes, gene silencing, and co-localization of enzymes are employed to optimize the production of plant metabolites in microbial hosts (Halder and Roy, 2023). Advances in enzyme engineering and process optimization have significantly improved the yields of complex compounds like macrocyclic sesqui- and diterpenes in microbial systems (Brück *et al.*, 2014). While *E. coli* and *S. cerevisiae* are common hosts, other bacteria, such as Burkholderia, have shown potential for higher yields in the production of specific compounds, such as lasso peptides (Kunakom and Eustáquio, 2020).

The structural complexity of plant-derived compounds poses challenges in fully reconstructing their biosynthetic pathways in microbial hosts. For instance, while partial pathways of complex molecules like paclitaxel have been reconstituted, complete biosynthesis remains a challenge (Zhang *et al.*, 2016). The success of heterologous production often depends on the compatibility of the host with the plant genes. Close phylogenetic relationships between the host and the source plant can improve expression and precursor supply (Kunakom and Eustáquio, 2020). While heterologous production in microbial hosts offers numerous advantages, it is not without its challenges. The complexity of plant biosynthetic pathways and the need for compatible host systems are significant hurdles. However, ongoing research and technological advancements continue to improve the feasibility and efficiency of this approach, making it a viable alternative to traditional plant extraction methods. As the demand for plant-derived pharmaceuticals grows, the role of microbial hosts in their production is likely to expand, offering new opportunities for drug discovery and development.

4.5.1 Microbial systems for scalable metabolite production

Microbial systems have emerged as a promising approach for scalable metabolite production in medicinal plants, addressing the challenges of low yield and environmental impact associated with traditional extraction methods. These systems leverage the capabilities of microorganisms to enhance the biosynthesis of plant secondary metabolites, which are crucial for pharmaceutical applications. By integrating microbial interventions with advanced biotechnological techniques, researchers aim to optimize the production of valuable phytochemicals. The following sections explore various aspects of microbial systems in this context.

A promising strategy for the microbial enhancement of metabolite production is microbial fortification, whereby beneficial

Table 4.2. Comparison of host systems for heterologous production of plant secondary metabolites. (Author's own table.)

Host system	Key advantages	Major limitations & challenges	Notable applications & case studies	Citations
E. coli (bacterium)	• Rapid growth & high-density cultivation. • Well-established genetic tools. • Simple metabolism.	• Lack of eukaryotic post-translational modifications. • Difficulty expressing membrane-bound plant P450s. • Limited native supply of terpenoid precursors.	• Production of resveratrol. • Partial reconstruction of alkaloid pathways.	(Braga *et al.*, 2018; Pyne *et al.*, 2018; Bradley *et al.*, 2020)
S. cerevisiae (yeast)	• Eukaryotic protein processing & compartmentalization. • Robust native mevalonate pathway for terpenoids. • Generally recognized as safe (GRAS) status.	• Metabolic burden can reduce growth. • Competition with endogenous pathways. • Lower transformation efficiency than *E. coli*.	• Heterologous production of terpenoids (e.g. artemisinin precursors). • Expression of plant P450s for ginsenoside synthesis. • Engineering of *N*-methylpyrrolinium-derived alkaloids.	(Andersen-Ranberg *et al.*, 2016; Ping *et al.*, 2019; Zhang *et al.*, 2023)
Plant cell cultures (e.g. *Taxus* spp.)	• Contains native biosynthetic machinery & compartments. • Capable of complex, multi-step modifications. (e.g. paclitaxel biosynthesis). • Avoids harvest of wild plants.	• Very slow growth rates. • Complex and costly culture media. • Production instability over time. • Requires elicitors (e.g. methyl jasmonate).	• Industrial-scale production of Taxol (paclitaxel). • High-yielding cell lines (CL5, CL12) established.	(Osuna *et al.*, 2015; McElroy and Jennewein, 2018; Wang *et al.*, 2018; Pant *et al.*, 2025)
Transgenic hairy roots (e.g. via *A. rhizogenes*)	• Genetically stable. • Fast-growing compared to some cell cultures. • Often exhibit high secondary metabolite production.	• Scaling up in bioreactors is challenging. • Species-specific transformation efficiency.	• Enhanced production of various medicinal alkaloids and phenolics.	(Wawrosch and Zotchev, 2021; Gangopadhyay *et al.*, 2024)

microorganisms—including certain endophytic bacteria and fungi associated with medicinal plants—are harnessed to significantly boost the synthesis of pharmacological metabolites such as terpenoids, polyphenols, and alkaloids. These associated microbes can influence the plant's internal metabolic pathways, often by eliciting defense responses or directly participating in biosynthetic processes, which in turn stimulates the plant's own production machinery and leads to a marked increase in the yield of valuable compounds (Wu *et al.*, 2023). Plant growth-promoting rhizobacteria (PGPR) enhance the production of bioactive compounds by improving nutrient uptake and protecting plants from pathogens. This microbial interaction is particularly beneficial under abiotic stress conditions, which can otherwise hinder plant growth and metabolite production (Vaghela and Gohel, 2023).

Microbial production systems utilizing genetically engineered microbes, such as specialized strains of bacteria and yeast, have emerged as powerful cell factories for the sustainable synthesis of complex plant natural products. These systems are highly advantageous as they are cost-effective, environmentally friendly, and capable of being optimized in controlled fermenters to produce high yields of valuable metabolites, including various flavonoids and carotenoids, thereby providing a reliable and scalable alternative to traditional extraction from plant biomass (*Suzuki et al., 2014; Fidan, 2024*). Microbes from various habitats, including extreme environments, contribute to the production of secondary metabolites. This diversity expands the potential applications of microbial systems in agriculture, medicine, and industry (Yadav *et al.*, 2019). Understanding the regulatory mechanisms by which microorganisms influence plant secondary metabolism is crucial. This includes the role of hormones, nutrient absorption, and gene expression in promoting metabolite accumulation (Lv *et al.*, 2024). While microbial systems offer a sustainable alternative to traditional methods, challenges remain in scaling up production and ensuring consistent quality and yield. Continued research is needed to optimize these systems for commercial applications (Singh *et al.*, 2020). While microbial systems present a promising avenue for scalable metabolite production, it is important to consider the potential limitations and challenges. The complexity of plant–microbe interactions and the need for precise control over metabolic pathways can pose significant hurdles. Additionally, the integration of microbial systems into existing production frameworks requires careful consideration of economic and regulatory factors. Nonetheless, the ongoing advancements in microbial biotechnology hold great potential for transforming the production of medicinal plant metabolites.

4.6 Industrial Production of Taxol in *Taxus* Cell Cultures

The industrial production of Taxol (paclitaxel) using *Taxus* cell cultures represents a significant advancement in the field of medicinal plant biotechnology. Paclitaxel is a highly effective anti-cancer drug, originally derived from the bark of the Pacific yew tree, but its natural extraction is unsustainable due to low yield and environmental concerns. Consequently, biotechnological methods, particularly the use of *Taxus* spp. cell cultures, have emerged as a promising alternative for large-scale production. This approach not only addresses the supply–demand gap but also offers a sustainable and environmentally friendly solution.

The establishment of *Taxus* cell cultures offers a significant advantage in sustainability and conservation, as this biotechnological approach effectively circumvents the need to harvest slow-growing yew trees from their natural habitats. By producing the potent anti-cancer compound paclitaxel in controlled bioreactors, this method directly contributes to conserving natural resources and preventing the ecological damage associated with large-scale tree harvesting, thereby ensuring a stable and reliable supply of this essential pharmaceutical without depleting vulnerable natural yew populations (Pant *et al.*, 2025). *Taxus* cell cultures can be optimized to produce high yields of paclitaxel. For instance, specific cell lines such as CL5, CL12, and CL21 have been developed to consistently produce high levels of paclitaxel, with content reaching up to 0.0477% of dry mass (Wang *et al.*, 2018). Additionally, the use of elicitors like methyl jasmonate has been shown to significantly enhance paclitaxel production in cell cultures (Osuna *et al.*, 2015). Several successful industrial operations in countries like Germany and South Korea have demonstrated the feasibility of using *Taxus* cell cultures for commercial paclitaxel production (Tomilova *et al.*, 2023). Initiating and maintaining *Taxus* cell cultures can be difficult due to their low growth rates and specific media requirements. Overcoming these challenges requires careful optimization of culture conditions (Tomilova *et al.*, 2023).

Advances in metabolic engineering have allowed for the enhancement of paclitaxel biosynthesis pathways. This includes the upregulation of key biosynthetic steps and the use of two-phase cultivation systems to increase yield (McElroy and Jennewein, 2018). Techniques such as immobilization and strain improvement have been employed to further enhance paclitaxel production. For example, immobilized cell cultures have shown increased productivity, and gamma irradiation mutagenesis has been used to develop high-yielding strains (Osuna *et al.*, 2015; El-Sayed *et al.*, 2020). While *Taxus* cell cultures offer a promising solution for paclitaxel production, alternative methods are also being explored. These include the use of endophytic fungi, which can produce paclitaxel in culture, and chemical synthesis, although the latter remains economically unfeasible due to low yields (El-Sayed *et al.*, 2020). Additionally, research into the complete biosynthesis pathway of paclitaxel may eventually allow for its production in non-native platforms, potentially increasing efficiency and reducing costs (McElroy and Jennewein, 2018). These alternative approaches highlight the ongoing efforts to diversify and optimize paclitaxel production methods.

4.6.1 Scaling terpenoid biosynthesis using omics-guided optimization

The scaling of terpenoid biosynthesis in medicinal plants through omics-guided optimization involves leveraging advanced biotechnological tools to enhance the production of these valuable compounds. Terpenoids, a diverse class of plant metabolites, have significant pharmaceutical applications but are often produced in limited quantities in their native plant sources. Omics technologies, including genomics, transcriptomics, proteomics, and metabolomics, provide comprehensive insights into the biosynthetic pathways of terpenoids, enabling targeted metabolic engineering strategies to optimize their production. This approach not only enhances the yield of terpenoids but also facilitates the discovery of novel compounds with potential therapeutic benefits.

4.6.2 Metabolic engineering and synthetic biology

Metabolic engineering is a key strategy for enhancing terpenoid production by manipulating biosynthetic pathways in plants and microorganisms. This involves overexpressing key genes, modulating transcription factors, and downregulating competitive pathways to increase the flux toward desired terpenoids (Lu *et al.*, 2016; Kumari *et al.*, 2024). Synthetic biology enables the construction of microbial cell factories, such as engineered strains of *S. cerevisiae*, to produce terpenoids at a large scale. This includes the heterologous expression of biosynthetic pathways and optimization of host metabolism to enhance production efficiency (Andersen-Ranberg *et al.*, 2016; Wang *et al.*, 2025).

4.6.3 Omics technologies in terpenoid biosynthesis

Omics technologies facilitate the elucidation of complex biosynthetic pathways and the identification of key enzymes involved in terpenoid synthesis. For instance, cytochrome P450 monooxygenases play a crucial role in modifying terpenoid skeletons, and their discovery is essential for pathway optimization (Zheng *et al.*, 2019). Multi-omics approaches have led to the identification of novel triterpenoid biosynthetic pathways and the development of engineered microbes for their production. This includes the use of squalene cyclases, CYPs, and other enzymes to achieve structural diversification (Zhang *et al.*, 2025).

4.6.4 Machine learning and data-driven optimization

Machine learning and high-throughput analytical tools are increasingly used to optimize biosynthetic pathways and enhance microbial production capacity. These technologies enable data-guided optimization, streamlining the biomanufacturing process of natural products, including terpenoids (Ramzi *et al.*, 2020). The integration of omics data with machine learning models allows for the prediction and manipulation of metabolic fluxes, leading to improved yields and the discovery of new terpenoid structures (Ramzi *et al.*, 2020). While omics-guided optimization offers promising avenues for scaling terpenoid biosynthesis, it is important to consider the ecological and economic implications of these technologies. The integration of sustainable practices and the exploration of alternative production systems, such as plant cell cultures and fast-growing host plants, could further enhance the scalability and accessibility of terpenoids for medicinal use. Additionally, the ethical considerations of genetic modifications and their impact on biodiversity should be addressed to ensure responsible development and application of these biotechnological advancements.

4.7 Technical Hurdles in Commercial Biomanufacturing

The commercial biomanufacturing of medicinal plants faces several technical hurdles that hinder its full potential. These challenges are primarily related to the complexity of plant systems, the need for advanced biotechnological methods, and the regulatory landscape. Despite these hurdles, advancements in plant biotechnology and molecular farming are paving the way for more efficient and scalable production of plant-derived pharmaceuticals. A fundamental challenge in harnessing medicinal plants lies in the inherent structural complexity of their secondary metabolites; these compounds often possess intricate molecular architectures with specific stereochemistry that render them difficult and economically unviable to synthesize through conventional chemical means. This structural complexity directly necessitates the extraction and processing of exceedingly large amounts of raw plant material to isolate sufficient quantities of the desired pure compound, an approach that is often neither feasible due to ecological constraints nor cost-effective for large-scale pharmaceutical production (Wawrosch and Zotchev, 2021).

The natural yield of secondary metabolites in plants is often low, requiring significant biomass to produce commercially viable quantities. This low yield is compounded by the fact that production can be tissue-, organ-, or developmental stage-specific (Halder and Roy, 2023). While genetic transformation techniques, such as those using *Rhizobium rhizogenes* (formerly *Agrobacterium rhizogenes*), can enhance the production of secondary metabolites, they require sophisticated technology and expertise. The success of these techniques is variable and often species-specific (Gangopadhyay *et al.*, 2024). *In vitro* technologies, such as cell suspension cultures and transgenic hairy roots, offer a sustainable alternative to traditional cultivation. However, these methods are not yet widely commercialized due to technical challenges in scaling up and maintaining stable production systems (Wawrosch and Zotchev, 2021). Manipulating metabolic pathways to increase the production of specific compounds is a promising approach. However, it involves complex genetic modifications and a deep understanding of plant metabolic networks, which can be technically demanding (Halder and Roy, 2023). Plant-derived pharmaceuticals must meet stringent regulatory standards to ensure safety and efficacy. This requires comprehensive testing and validation, which can be time-consuming and costly (Fischer *et al.*, 2013).

The economic principles of plant-based biomanufacturing must be satisfied to ensure that processes are financially viable. This includes considerations of production costs, scalability, and market demand (Fischer *et al.*, 2013). There are significant opportunities for overcoming technical hurdles in commercial biomanufacturing. Advances in plant molecular farming, such as the use of *N. benthamiana* as an expression host, have shown promise in reducing costs and production timelines (Eidenberger and Steinkellner, 2023). Additionally, the development of large-scale facilities, like the Caliber Biotherapeutics facility in Bryan, Texas, USA, demonstrates the potential for rapid and high-quality production of plant-based pharmaceuticals (Holtz *et al.*, 2015). These advancements suggest that with continued research and development, the technical hurdles in commercial biomanufacturing of medicinal plants can be addressed effectively. While these hurdles are significant, they are not insurmountable. The integration of advanced biotechnological methods and a better understanding of plant systems can lead to more efficient production processes. However, it is crucial to balance these technological advancements with economic and regulatory considerations to ensure the sustainable and safe production of plant-derived pharmaceuticals (Table 4.3 and Fig. 4.4).

4.7.1 Metabolic burden, enzyme promiscuity, and cost-effectiveness

Metabolic burden, enzyme promiscuity, and cost-effectiveness in the context of medicinal plants are significant hurdles in the production and utilization of plant-derived secondary metabolites. These challenges impact the efficiency and feasibility of using medicinal plants for therapeutic purposes. Metabolic engineering and synthetic biology offer promising solutions to these issues, but they also come with their own set of challenges and limitations. Medicinal plants often produce secondary metabolites in low quantities, which is insufficient for commercial use. The structural complexity of these compounds further complicates their extraction and synthesis (Cardoso *et al.*, 2019). Strategies such as overexpression of key enzymes, gene silencing, and metabolic channeling are employed to enhance the production of desired metabolites. However, these approaches can impose a metabolic burden on the host plant, affecting its growth and viability (Fatima *et al.*, 2022). CYPs are crucial in the biosynthesis of secondary metabolites but are known for their enzyme promiscuity, which can lead to the production of unwanted by-products. This promiscuity complicates the optimization of metabolic pathways for specific metabolite production (Thiengsusuk *et al.*, 2020; Sethi *et al.*, 2022).

Identifying and engineering specific CYPs to enhance the production of target metabolites without generating side products is a significant challenge. High-throughput sequencing and mass spectrometry are being used to better understand and manipulate these pathways (Sethi *et al.*, 2022). The cultivation and extraction of medicinal plants are often costly due to factors like geographic limitations, seasonal variations, and the need for specific growth conditions (Halder and Roy, 2023). Synthetic biology and metabolic engineering fields aim to reduce costs by optimizing metabolic pathways and using heterologous systems for metabolite production. However, the initial investment in technology and research can be substantial (Stephanopoulos, 2012).

While metabolic engineering and synthetic biology offer solutions to the challenges faced in the production of plant-derived secondary metabolites, they also introduce new complexities. The integration of multiple genes and the reconstitution of biosynthetic pathways in heterologous hosts require advanced knowledge and technology, which may not be readily available in all research settings (Choudhury *et al.*, 2023). Additionally, the acceptance of natural health products in the medical community hinges on addressing concerns related to active ingredients, mechanisms of action, and potential drug interactions (Choudhury *et al.*, 2023). Despite these challenges, the potential benefits of these technologies in improving the yield and cost-effectiveness of medicinal plant products make them a promising area of research.

4.8 Challenges and Future Prospects

The engineering of biological systems for the production of plant secondary metabolites faces a multitude of interconnected challenges. A primary issue is metabolic burden, where the overproduction of target metabolites in engineered organisms diverts essential resources like ATP and NADPH from primary metabolism, often leading to reduced cellular growth and viability, as observed in *Taxus* cell cultures which exhibit slow proliferation under paclitaxel-producing conditions. This is compounded by enzyme promiscuity, where crucial catalysts like CYPs can catalyze off-target reactions, generating undesirable by-products that complicate purification. Furthermore, the complex

Table 4.3. Major technical hurdles and potential mitigation strategies in commercial biomanufacturing. Author's own table.

Technical hurdle	Description & impact	Potential mitigation strategies	Citations
Metabolic burden	Expression of heterologous pathways diverts resources (adenosine triphosphate—ATP, nicotinamide adenine dinucleotide phosphate—NADPH, precursors) from host growth, leading to reduced viability and productivity.	• Use of inducible promoters to separate growth and production phases. • Engineering host chassis to be more robust. • Two-phase bioreactor systems.	(Stephanopoulos, 2012; Halder and Roy, 2023; Yin *et al.*, 2025)
Enzyme promiscuity (e.g. of CYPs)	Cytochrome P450s and other enzymes may catalyze off-target reactions, leading to undesired by-products, low yields, and complicated purification.	• Protein engineering to enhance substrate specificity (rational design, directed evolution). • Metabolic channeling to sequester intermediates. • Computational redesign of active sites.	(Shang and Huang, 2020; Zhang *et al.*, 2021; Sethi *et al.*, 2022)
Host compatibility	Plant-derived enzymes, especially membrane-bound P450s, often misfold, have low activity, or lack required cofactors in microbial hosts.	• Co-expression of partner proteins (e.g. cytochrome P450 reductases). • Engineering P450s for improved solubility and activity in microbes. • Using plant-based hosts or cell cultures for complex pathways.	(Pyne *et al.*, 2018; Li *et al.*, 2019; Zhang *et al.*, 2021)
Pathway complexity & regulation	Plant secondary metabolite pathways are long, compartmentalized, and subject to complex transcriptional and feedback regulation.	• Use of synthetic biology to refactor pathways into modular "parts." • Multi-omics to unravel regulatory networks. • CRISPR-Cas9 to precisely edit regulatory elements.	(Staniek *et al.*, 2013; Bhushan *et al.*, 2023; Halder and Roy, 2023)
Cost-effectiveness & scalability	High costs of bioreactor operation, nutrient media, and downstream processing make it difficult to compete with plant extraction on price for some compounds.	• Optimization of fermentation media and conditions. • Development of high-yielding strains through iterative engineering. • Exploring low-cost hosts like cyanobacteria or plant molecular farming in *N. benthamiana*.	(Fischer *et al.*, 2013; Tanvir *et al.*, 2021; Eidenberger and Steinkellner, 2023)

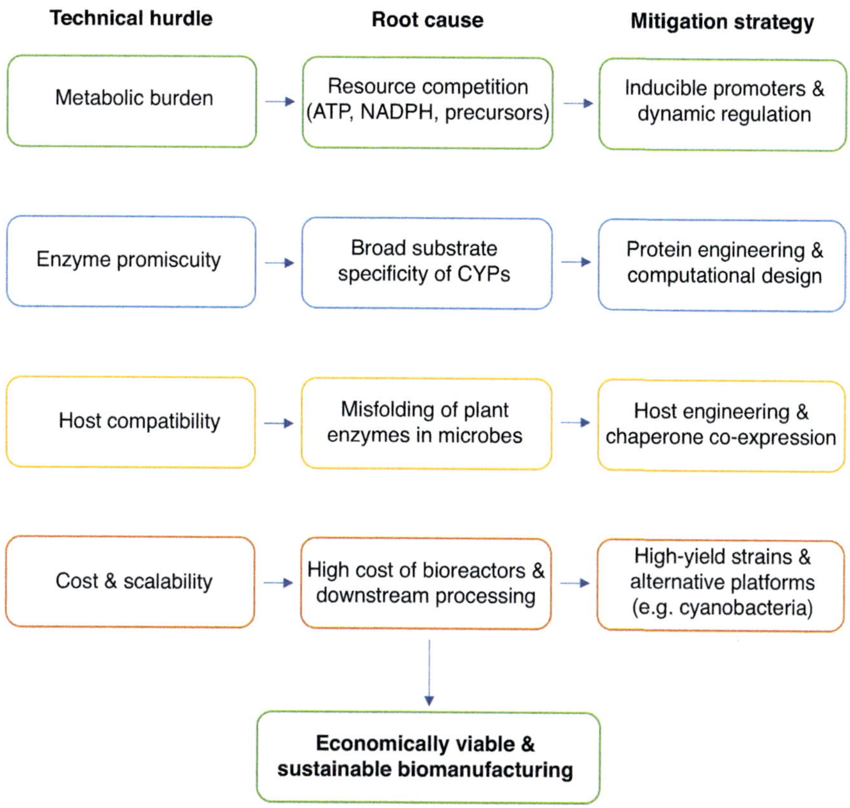

Fig. 4.4. Overcoming technical hurdles in biomanufacturing. (Author's own figure.)

pathway regulation inherent to plants, involving compartmentalization and multi-layered feedback mechanisms, is difficult to reconstruct in heterologous hosts like *E. coli*, where issues with enzyme localization and cofactor availability hinder efficiency. From a practical standpoint, cost and scalability present significant hurdles; while traditional plant extraction is ecologically unsustainable, microbial and cell culture systems incur high expenses from bioreactor operation, nutrient media, and downstream processing. Host compatibility also remains a barrier, as plant enzymes, particularly membrane-bound CYPs, often misfold in microbial hosts due to incompatible cellular environments. Finally, all these technical challenges are capped by stringent regulatory hurdles, requiring adherence to costly and time-consuming protocols such as US Food and Drug Administration (FDA) approval and GMP before these innovative production methods can be commercialized.

The future of engineering plant metabolite production is poised for transformation through the deep integration of multi-omics technologies. By combining genomics, proteomics, and fluxomics, researchers can systematically map elusive pathway components and critical regulatory nodes. For instance, transcriptomic profiling of *A. annua* has already identified key transcription factors that, when overexpressed, significantly enhance artemisinin yields. This systems-level understanding provides a blueprint for targeted genetic interventions.

4.9 Conclusion

This chapter describes a revolutionary approach to addressing the global demand for plant-derived secondary metabolites, such as alkaloids, terpenoids, and phenolics, which are indispensable to pharmaceuticals and industrial applications. By integrating strategies like heterologous pathway reconstruction, MFA, and enzyme engineering, this field has overcome many limitations of traditional plant extraction methods. Innovations such as *Taxus* cell cultures for paclitaxel production and microbial systems engineered with CYPs exemplify the successful translation of laboratory-scale research into scalable biomanufacturing processes. These approaches not only enhance yields but also enable the sustainable and cost-effective synthesis of complex metabolites. However, significant challenges remain. Metabolic burden in engineered hosts, enzyme promiscuity, and the intricate regulation of plant pathways complicate large-scale production. Technical hurdles, including host compatibility issues for plant-specific enzymes and the high costs of bioreactor systems, further impede commercialization. Regulatory compliance and the need for rigorous optimization of growth conditions add layers of complexity to industrial scalability. Future progress hinges on interdisciplinary integration. Multi-omics technologies, machine learning, and AI-driven predictive modeling will refine pathway design by identifying bottlenecks and optimizing metabolic fluxes. Advances in synthetic biology, such as CRISPR-based editing and modular pathway assembly, promise to enhance precision in metabolic engineering. Sustainable platforms, including photosynthetic microbes and plant molecular farming, offer eco-friendly alternatives to conventional methods. Additionally, dynamic metabolic modeling and real-time bioprocess adjustments will bridge the gap between experimental insights and industrial implementation. By addressing these challenges through innovation and collaboration, metabolic engineering is poised to transform the biomanufacturing landscape. The convergence of cutting-edge tools with ecological and economic considerations will drive the sustainable production of bioactive compounds, ensuring their accessibility for therapeutic and industrial applications. This evolution not only safeguards biodiversity but also underscores the critical role of biotechnology in meeting societal needs while fostering environmental stewardship.

Acknowledgments

The authors express their gratitude for the financial support received through the Distinguished High-Level Talents Research Grant from the Guizhou Science and Technology Corporation Platform Talents Fund (Grant nos.: [2017]5733-001 and CK-1130-002), high-level specialists project fund of Zunyi Medical University (1092/2000006)F-1092, the National Natural Science Foundation of China (82373981), and the support provided by Zunyi Medical University, China. Special appreciation is extended to all laboratory colleagues and research staff members for their valuable insights, constructive guidance, and assistance throughout this study.

References

Abdallah, I.I. and Quax, W.J. (2017) A glimpse into the biosynthesis of terpenoids. *KnE Life Sciences* 3(5), 81–98. DOI: 10.18502/kls.v3i5.981.

Allen, D.K. (2016) Quantifying plant phenotypes with isotopic labeling & metabolic flux analysis. *Current Opinion in Biotechnology* 37, 45–52. DOI: 10.1016/j.copbio.2015.10.002.

Allen, D.K., Libourel, I.G.L. and Shachar-Hill, Y. (2009) Metabolic flux analysis in plants: Coping with complexity. *Plant, Cell & Environment* 32(9), 1241–1257. DOI: 10.1111/j.1365-3040.2009.01992.x.

Andersen-Ranberg, J., Kongstad, K.T., Nielsen, M.T., Jensen, N.B., Pateraki, I. *et al.* (2016) Expanding the landscape of diterpene structural diversity through stereochemically controlled combinatorial biosynthesis. *Angewandte Chemie (International Ed. in English)* 55(6), 2142–2146. DOI: 10.1002/anie.201510650.

Antikainen, N.M. and Martin, S.F. (2005) Altering protein specificity: Techniques and applications. *Bioorgic and Medicinal Chemistry* 13(8), 2701–2716. DOI: 10.1016/j.bmc.2005.01.059.

Ao, Y.-F., Dörr, M., Menke, M.J., Born, S., Heuson, E. *et al.* (2024) Data-driven protein engineering for improving catalytic activity and selectivity. *ChemBioChem* 25(3), e202300754. DOI: 10.1002/cbic.202300754.

Apostolidou, D., Ding, Y. and Marszalek, P.E. (2023) Track: High throughput protein science thermal and mechanical stability of highly-luminescent protein NanoLuc in presence and absence of chaperones; award winners/abstracts from the 36th Annual Symposium of the Protein Society, July 7–10, 2022. *Protein Science* 10, 2022. DOI: 10.1002/pro.4494.

Ballout, F., Habli, Z., Monzer, A., Rahal, O.N., Fatfat, M. *et al.* (2019) Anticancer alkaloids: Molecular mechanisms and clinical manifestations. In: Sharma, A.K. (ed.) *Bioactive Natural Products for the Management of Cancer: From Bench to Bedside*. Springer, Singapore, pp. 1–35.

Bhushan, S., Sharma, D., Rakshant, Kaul, S., Dhar, M.K. *et al.* (2023) Recent strategies to engineer alkaloid biosynthesis in medicinal plants. In: Husen, A. and Iqbal, M. (eds) *Medicinal Plants*. Springer Nature, Singapore, pp. 391–416.

Bradley, S.A., Zhang, J. and Jensen, M.K. (2020) Deploying microbial synthesis for halogenating and diversifying medicinal alkaloid scaffolds. *Frontiers in Bioengineering and Biotechnology* 8, 594126. DOI: 10.3389/fbioe.2020.594126.

Braga, A., Ferreira, P., Oliveira, J., Rocha, I. and Faria, N. (2018) Heterologous production of resveratrol in bacterial hosts: Current status and perspectives. *World Journal of Microbiology and Biotechnology* 34(8), 122. DOI: 10.1007/s11274-018-2506-8.

Brück, T., Kourist, R. and Loll, B. (2014) Production of macrocyclic sesqui- and diterpenes in heterologous microbial hosts: A systems approach to harness nature's molecular diversity. *ChemCatChem* 6(5), 1142–1165. DOI: 10.1002/cctc.201300733.

Calheiros Gomes, L. and Simoes, M. (2012) ^{13}C metabolic flux analysis: From the principle to recent applications. *Current Bioinformatics* 7(1), 77–86. DOI: 10.2174/157489312799304404.

Cardoso, J.C., Oliveira, M.E.B. de and Cardoso, F. de C. (2019) Advances and challenges on the *in vitro* production of secondary metabolites from medicinal plants. *Horticultura Brasileira* 37(2), 124–132. DOI: 10.1590/s0102-053620190201.

Cauchon, N.S., Oghamian, S., Hassanpour, S. and Abernathy, M. (2019) Innovation in chemistry, manufacturing, and controls—a regulatory perspective from industry. *Journal of Pharmaceutical Sciences* 108(7), 2207–2237. DOI: 10.1016/j.xphs.2019.02.007.

Choudhury, A., Singh, P.A., Bajwa, N., Dash, S. and Bisht, P. (2023) Pharmacovigilance of herbal medicines: Concerns and future prospects. *Journal of Ethnopharmacology* 309, 116383. DOI: 10.1016/j.jep.2023.116383.

Chung, H.-H., Sung, Y.-C. and Shyur, L.-F. (2016) Deciphering the biosynthetic pathways of bioactive compounds in planta using omics approaches. In: Tsay, H.-S., Shyur, L.-F., Chandra Agrawal, D., Wu, Y.-C. and Wang, S.-Y. (eds) *Medicinal Plants - Recent Advances in Research and Development*. Springer, Singapore, pp. 129–165.

Curran, K.A., Crook, N.C. and Alper, H.S. (2012) Using flux balance analysis to guide microbial metabolic engineering. In: Santos, C.N.S. and Kumaran Ajikumar, P. (eds) *Microbial Metabolic Engineering: Methods and Protocols*. Springer, New York, pp. 197–216.

D'Amelia, V., Docimo, T., Crocoll, C. and Rigano, M.M. (2021) Specialized metabolites and valuable molecules in crop and medicinal plants: The evolution of their use and strategies for their production. *Genes* 12(6), 936. DOI: 10.3390/genes12060936.

de Oliveira, I., Albuquerque, B., Ueda, J.M., Alves, M.J., Ferreira, I.C.F.R. *et al.* (2023) New challenges and opportunities from secondary metabolites. In: Carocho, M., Heleno, S.A. and Barros, L. (eds) *Natural Secondary Metabolites*. Springer International Publishing, Cham, Switzerland, pp. 925–965.

Eidenberger, L. and Steinkellner, H. (2023) Engineering biopharmazeutischer produkte in der pflanzenbiotechnologie. *BIOspektrum* 29(4), 418–420. DOI: 10.1007/s12268-023-1949-0.

El-Sayed, E.-S.R., Zaki, A.G., Ahmed, A.S. and Ismaiel, A.A. (2020) Production of the anticancer drug taxol by the endophytic fungus *Epicoccum nigrum* TXB502: Enhanced production by gamma irradiation mutagenesis and immobilization technique. *Applied Microbiology and Biotechnology* 104(16), 6991–7003. DOI: 10.1007/s00253-020-10712-x.

Elshafie, H.S., Camele, I. and Mohamed, A.A. (2023) A comprehensive review on the biological, agricultural and pharmaceutical properties of secondary metabolites based-plant origin. *International Journal of Molecular Sciences* 24(4), 3266. DOI: 10.3390/ijms24043266.

Fatima, M., Saleem, M.Z., Zafar, S., Askri, S.M.H., Ayub, M.A. *et al.* (2022) Improving the quality of medicinal and aromatic plants through metabolic engineering. In: Aftab, T. and Hakeem, K.R. (eds) *Metabolic Engineering in Plants*. Springer Nature Singapore, Singapore, pp. 321–339.

Fidan, O. (2024) Microbial production of pharmaceutically important plant natural products. In: Kole, C., Chaurasia, A., Hefferon, K.L. and Panigrahi, J. (eds) *Applications of Plant Molecular Farming*. Springer Nature, Singapore, pp. 487–513.

Fischer, R., Schillberg, S., Buyel, J. and Twyman, R. (2013) Commercial aspects of pharmaceutical protein production in plants. *Current Pharmaceutical Design* 19(31), 5471–5477. DOI: 10.2174/1381612811319310002.

Fishtik, I. (2009) A reaction route approach to flux balance analysis. *Nature Precedings*. DOI: 10.1038/npre.2009.2788.1.

Gangopadhyay, M., Sanyamat, S. and Dewanjee, S. (2024) Agrobacterium rhizogenes-mediated genetic transformation: A potential approach to enhance the antiallergic potential of medicinal plants by endorsing the production of responsible phytochemicals. In: Gantait, S., Majumder, J. and Baran Sharangi, A. (eds) *Biotechnology of Medicinal Plants with Antiallergy Properties*. Springer Nature Singapore, Singapore, pp. 521–545.

Goshisht, M.K. (2024) Machine learning and deep learning in synthetic biology: Key architectures, applications, and challenges. *ACS Omega* 9(9), 9921–9945. DOI: 10.1021/acsomega.3c05913.

Halder, M. and Roy, S. (2023) Current status of metabolic engineering of medicinal plants for production of plant-derived secondary metabolites. In: Jha, S. and Halder, M. (eds) *Medicinal Plants: Biodiversity, Biotechnology and Conservation*. Springer Nature, Singapore, pp. 819–869.

Holtz, B.R., Berquist, B.R., Bennett, L.D., Kommineni, V.J.M., Munigunti, R.K. *et al.* (2015) Commercial-scale biotherapeutics manufacturing facility for plant-made pharmaceuticals. *Plant Biotechnology Journal* 13(8), 1180–1190. DOI: 10.1111/pbi.12469.

Huang, W., Wang, Y., Tian, W., Cui, X., Tu, P. *et al.* (2022) Biosynthesis investigations of terpenoid, alkaloid, and flavonoid antimicrobial agents derived from medicinal plants. *Antibiotics* 11(10), 1380. DOI: 10.3390/antibiotics11101380.

Knorrscheidt, A., Soler, J., Hünecke, N., Püllmann, P., Garcia-Borràs, M. *et al.* (2021) Accessing chemo- and regioselective benzylic and aromatic oxidations by protein engineering of an unspecific peroxygenase. *ACS Catalysis* 11(12), 7327–7338. DOI: 10.1021/acscatal.1c00847.

Koley, S., Jyoti, P., Lingwan, M. and Allen, D.K. (2024) Isotopically nonstationary metabolic flux analysis of plants: Recent progress and future opportunities. *The New Phytologist* 242(5), 1911–1918. DOI: 10.1111/nph.19734.

Kruger, N.J. and Ratcliffe, R.G. (2012) Pathways and fluxes: Exploring the plant metabolic network. *Journal of Experimental Botany* 63(6), 2243–2246. DOI: 10.1093/jxb/ers073.

Kumari, M., Checker, V.G., Kathpalia, R., Srivastava, V., Singh, I.K. *et al.* (2024) Metabolic engineering for enhanced terpenoid production: Leveraging new horizons with an old technique. *Plant Physiology and Biochemistry* 210, 108511. DOI: 10.1016/j.plaphy.2024.108511.

Kunakom, S. and Eustáquio, A.S. (2020) Heterologous production of lasso peptide capistruin in a *Burkholderia* host. *ACS Synthetic Biology* 9(2), 241–248. DOI: 10.1021/acssynbio.9b00438.

Lehari, K. and Kumar, D. (2024) Metabolic engineering and production of secondary metabolites. In: Kumar, A. and Kumar, S. (eds) *Secondary Metabolites and Biotherapeutics*. Academic Press, Cambridge, Massachusetts, pp. 215–244.

Li, D., Ma, Y., Zhou, Y., Gou, J., Zhong, Y. *et al.* (2019) A structural and data-driven approach to engineering a plant cytochrome P450 enzyme. *Science China Life Sciences* 62(7), 873–882. DOI: 10.1007/s11427-019-9538-3.

Liu, J., Wang, X., Dai, G., Zhang, Y. and Bian, X. (2022) Microbial chassis engineering drives heterologous production of complex secondary metabolites. *Biotechnology Advances* 59, 107966. DOI: 10.1016/j.biotechadv.2022.107966.

Lopes, J.S., Rodrigues, J.A. and Serpa, J. (2024) Mathematical modeling of metabolic flux: Integrating COBRApy and Escher for enhanced analysis. In: *4th ROME International Conference on Challenges in Engineering, Medical, Economics and Education: Research & Solutions*, Rome (Italy) CEMEERS-24b, RIBES-24 & LLSSH-24, December 4–6, 2024, pp. 26–31.

Lu, X., Tang, K. and Li, P. (2016) Plant metabolic engineering strategies for the production of pharmaceutical terpenoids. *Frontiers in Plant Science* 7, 1647. DOI: 10.3389/fpls.2016.01647.

Lv, J., Yang, S., Zhou, W., Liu, Z., Tan, J. *et al.* (2024) Microbial regulation of plant secondary metabolites: Impact, mechanisms and prospects. *Microbiological Research* 283, 127688. DOI: 10.1016/j.micres.2024.127688.

McElroy, C. and Jennewein, S. (2018) Taxol® biosynthesis and production: From forests to fermenters. In: Schwab, W., Lange, B.M. and Wüst, M. (eds) *Biotechnology of Natural Products*. Springer International Publishing, Cham, Switzerland, pp. 145–185.

Mohiuddin, A.K. (2019) Chemistry of secondary metabolites. *Annals of Clinical Toxicology* 2(1), 1014. DOI: 10.25107/2641-905X-v2-id1014.

Nguyen, T.-D. and Dang, T.-T.T. (2021) Cytochrome P450 enzymes as key drivers of alkaloid chemical diversification in plants. *Frontiers in Plant Science* 12, 682181. DOI: 10.3389/fpls.2021.682181.

Ni, Z.-F., Xu, P., Zong, M.-H. and Lou, W.-Y. (2021) Structure-guided protein engineering of ammonia lyase for efficient synthesis of sterically bulky unnatural amino acids. *Bioresources and Bioprocessing* 8(1), 103. DOI: 10.1186/s40643-021-00456-5.

Noushahi, H.A., Khan, A.H., Noushahi, U.F., Hussain, M., Javed, T. *et al.* (2022) Biosynthetic pathways of triterpenoids and strategies to improve their biosynthetic efficiency. *Plant Growth Regulation* 97(3), 439–454. DOI: 10.1007/s10725-022-00818-9.

Osuna, L., Tapia, N., Cusidó, R., Palazón, J., Bonfill, M. *et al.* (2015) Taxane production induced by methyl jasmonate in free and immobilized cell cultures of Mexican yew (*Taxus globosa* Schltdl). *Acta Physiologiae Plantarum* 37(10), 199. DOI: 10.1007/s11738-015-1947-z.

Pant, M., Sohrab, S.S. and Husen, A. (2025) *Genetic Improvement and Conservation Practices of Medicinal Plants*. CRC Press, Boca Raton, Florida.

Patil, M.D., Grogan, G. and Yun, H. (2018) Biocatalyzed C–C bond formation for the production of alkaloids. *ChemCatChem* 10(21), 4783–4804. DOI: 10.1002/cctc.201801130.

Ping, Y., Li, X., Xu, B., Wei, W., Wei, W. *et al.* (2019) Building microbial hosts for heterologous production of *N*-methylpyrrolinium. *ACS Synthetic Biology* 8(2), 257–263. DOI: 10.1021/acssynbio.8b00483.

Pyne, M.E., Narcross, L., Hansen, C.I.C., Campbell, A., Davies, M.E. *et al.* (2018) Microbial synthesis of plant alkaloids. In: Schwab, W., Lange, B.M. and Wüst, M. (eds) *Biotechnology of Natural Products*. Springer International Publishing, Cham, Switzerland, pp. 99–130.

Rai, A., Saito, K. and Yamazaki, M. (2017) Integrated omics analysis of specialized metabolism in medicinal plants. *The Plant Journal* 90(4), 764–787. DOI: 10.1111/tpj.13485.

Ramzi, A.B., Baharum, S.N., Bunawan, H. and Scrutton, N.S. (2020) Streamlining natural products biomanufacturing with omics and machine learning driven microbial engineering. *Frontiers in Bioengineering and Biotechnology* 8, 608918. DOI: 10.3389/fbioe.2020.608918.

Romsuk, J., Yasumoto, S., Seki, H., Fukushima, E.O. and Muranaka, T. (2022) Identification of key amino acid residues toward improving the catalytic activity and substrate specificity of plant-derived cytochrome P450 monooxygenases CYP716A subfamily enzyme for triterpenoid production in *Saccharomyces cerevisiae*. *Frontiers in Bioengineering and Biotechnology* 10, 955650. DOI: 10.3389/fbioe.2022.955650.

Salleh, A.H.M., Mohamad, M.S., Deris, S. and Illias, R.M. (2015) Metabolites production improvement by identifying minimal genomes and essential genes using flux balance analysis. *International Journal of Data Mining and Bioinformatics* 12(1), 85–99. DOI: 10.1504/IJDMB.2015.068955.

Sampaio, M., Rocha, M. and Dias, O. (2022) Exploring synergies between plant metabolic modelling and machine learning. *Computational and Structural Biotechnology Journal* 20, 1885–1900. DOI: 10.1016/j.csbj.2022.04.016.

Sato, F. (2020) Plant alkaloid engineering. In: Liu, H. -w. and Begley, T. (eds) *Comprehensive Natural Products III*. Elsevier, Amsterdam, pp. 700–755.

Schäfer, H. and Wink, M. (2009) Medicinally important secondary metabolites in recombinant microorganisms or plants: Progress in alkaloid biosynthesis. *Biotechnology Journal* 4(12), 1684–1703. DOI: 10.1002/biot.200900229.

Schilling, C.H., Edwards, J.S., Letscher, D. and Palsson, B.Ø. (2000) Combining pathway analysis with flux balance analysis for the comprehensive study of metabolic systems. *Biotechnology and Bioengineering* 71(4), 286–306. DOI: 10.1002/1097-0290(2000)71:43.0.CO;2-R.

Sethi, A., Bhandawat, A. and Pati, P.K. (2022) Engineering medicinal plant-derived CYPs: A promising strategy for production of high-valued secondary metabolites. *Planta* 256(6), 119. DOI: 10.1007/s00425-022-04024-9.

Shang, Y. and Huang, S. (2020) Engineering plant cytochrome p450s for enhanced synthesis of natural products: Past achievements and future perspectives. *Plant Communications* 1(1), 100012. DOI: 10.1016/j.xplc.2019.100012.

Shukla, V., Gupta, P. and Phulara, S.C. (2023) Metabolic engineering for the biosynthesis of terpenoids from microbial cell factories. In: Soni, R., Chandra Suyal, D. and Morales-Oyervides, L. (eds) *Microbial Bioactive Compounds*. Springer Nature Switzerland, Switzerland, pp. 59–84.

Singh, J., Meshram, V. and Gupta, M. (eds) (2020) *Bioactive Natural Products in Drug Discovery*. Springer Singapore, Singapore.

Singla, P. (2020) Plant synthetic biology: A paradigm shift targeting stress mitigation, reduction of ecological footprints and sustainable transformation in agriculture. In: Giri, B. and Prasad Sharma, M. (eds) *Plant Stress Biology*. Springer Singapore, Singapore, pp. 435–489.

Soni, R., Shankar, G., Mukhopadhyay, P. and Gupta, V. (2022) A concise review on *Artemisia annua* L.: A major source of diverse medicinal compounds. *Industrial Crops and Products* 184, 115072. DOI: 10.1016/j.indcrop.2022.115072.

Staniek, A., Bouwmeester, H., Fraser, P.D., Kayser, O., Martens, S. *et al.* (2013) Natural products - modifying metabolite pathways in plants. *Biotechnology Journal* 8(10), 1159–1171. DOI: 10.1002/biot.201300224.

Stephanopoulos, G. (2012) Synthetic biology and metabolic engineering. *ACS Synthetic Biology* 1(11), 514–525. DOI: 10.1021/sb300094q.

Suzuki, S., Koeduka, T., Sugiyama, A., Yazaki, K. and Umezawa, T. (2014) Microbial production of plant specialized metabolites. *Plant Biotechnology* 31(5), 465–482. DOI: 10.5511/plantbiotechnology.14.1003a.

Tanvir, R.U., Zhang, J., Canter, T., Chen, D., Lu, J. *et al.* (2021) Harnessing solar energy using phototrophic microorganisms: A sustainable pathway to bioenergy, biomaterials, and environmental solutions. *Renewable & Sustainable Energy Reviews* 146, 111181. DOI: 10.1016/j.rser.2021.111181.

Thiengsusuk, A., Boonprasert, K. and Na-Bangchang, K. (2020) A systematic review of drug metabolism studies of plants with anticancer properties: Approaches applied and limitations. *European Journal of Drug Metabolism and Pharmacokinetics* 45(2), 173–225. DOI: 10.1007/s13318-019-00582-8.

Tomilova, S.V., Globa, E.B., Demidova, E.V. and Nosov, A.M. (2023) Secondary metabolism in *Taxus* spp. plant cell culture *in vitro*. *Russian Journal of Plant Physiology* 70(3), 23. DOI: 10.1134/S102144372270008X.

Trantas, E.A., Koffas, M.A.G., Xu, P. and Ververidis, F. (2015) When plants produce not enough or at all: Metabolic engineering of flavonoids in microbial hosts. *Frontiers in Plant Science* 6, 7. DOI: 10.3389/fpls.2015.00007.

Vaghela, N. and Gohel, S. (2023) Medicinal plant-associated rhizobacteria enhance the production of pharmaceutically important bioactive compounds under abiotic stress conditions. *Journal of Basic Microbiology* 63(3–4), 308–325. DOI: 10.1002/jobm.202200361.

Villa-Ruano, N., Pacheco-Hernandez, Y., Lozoya-Gloria, E., Castro-Juarez, C., Mosso-Gonzalez, C. *et al.* (2015) Cytochrome P450 from plants: Platforms for valuable phytopharmaceuticals. *Tropical Journal of Pharmaceutical Research* 14(4), 731–742. DOI: 10.4314/tjpr.v14i4.24.

Wang, J., Ji, X., Yi, R., Li, D., Shen, X. *et al.* (2025) Heterologous biosynthesis of terpenoids in *Saccharomyces cerevisiae*. *Biotechnology Journal* 20(1), e202400712. DOI: 10.1002/biot.202400712.

Wang, S., Wang, H., Li, T., Li, C., Zhou, Y. *et al.* (2018) The selection and stability analysis of stable and high taxol-producing cell lines from *Taxus cuspidata*. *Journal of Forestry Research* 29(1), 65–71. DOI: 10.1007/s11676-017-0406-7.

Wawrosch, C. and Zotchev, S.B. (2021) Production of bioactive plant secondary metabolites through *in vitro* technologies-status and outlook. *Applied Microbiology and Biotechnology* 105(18), 6649–6668. DOI: 10.1007/s00253-021-11539-w.

Wu, X., Yang, Y. and Zhang, H. (2023) Microbial fortification of pharmacological metabolites in medicinal plants. *Computational and Structural Biotechnology Journal* 21, 5066–5072. DOI: 10.1016/j.csbj.2023.10.024.

Yadav, A.N., Kour, D., Rana, K.L., Yadav, N., Singh, B. *et al.* (2019) Metabolic engineering to synthetic biology of secondary metabolites production. In: Bahadur Singh, H., Jogaiah, S. and Kumar Gupta, V.J. (eds) *New and Future Developments in Microbial Biotechnology and Bioengineering*. Elsevier, Amsterdam, pp. 279–320.

Yin, J.-Y., Lai, M., Yu, X.-Y., Su, D.-D., Xiong, X.-Y. *et al.* (2025) Comprehensive strategies for paclitaxel production: Insights from plant cell culture, endophytic microorganisms, and synthetic biology. *Horticulture Research* 12(3). DOI: 10.1093/hr/uhae346.

Zhang, F., Hao, X., Liu, J., Hou, H., Chen, S. *et al.* (2025) Herbal multiomics provide insights into gene discovery and bioproduction of triterpenoids by engineered microbes. *Journal of Agricultural and Food Chemistry* 73(1), 47–65. DOI: 10.1021/acs.jafc.4c08372.

Zhang, H., Wang, Y. and Pfeifer, B.A. (2008) Bacterial hosts for natural product production. *Molecular Pharmaceutics* 5(2), 212–225. DOI: 10.1021/mp7001329.

Zhang, S., Wang, S. and Zhan, J. (2016) Engineered biosynthesis of medicinally important plant natural products in microorganisms. *Current Topics in Medicinal Chemistry* 16(15), 1740–1754. DOI: 10.2174/1568026616666151012112637.

Zhang, X., Luo, W., Yao, Y., Luo, X., Han, C. *et al.* (2021) Enhanced chemoselectivity of a plant cytochrome P450 through protein engineering of surface and catalytic residues. *aBIOTECH* 2(3), 215–225. DOI: 10.1007/s42994-021-00056-z.

Zhang, Y., Ma, L., Su, P., Huang, L. and Gao, W. (2023) Cytochrome p450s in plant terpenoid biosynthesis: Discovery, characterization and metabolic engineering. *Critical Reviews in Biotechnology* 43(1), 1–21. DOI: 10.1080/07388551.2021.2003292.

Zhao, Y.-J., Cheng, Q.-Q., Su, P., Chen, X., Wang, X.-J. *et al.* (2014) Research progress relating to the role of cytochrome P450 in the biosynthesis of terpenoids in medicinal plants. *Applied Microbiology and Biotechnology* 98(6), 2371–2383. DOI: 10.1007/s00253-013-5496-3.

Zheng, X., Li, P. and Lu, X. (2019) Research advances in cytochrome P450-catalysed pharmaceutical terpenoid biosynthesis in plants. *Journal of Experimental Botany* 70(18), 4619–4630. DOI: 10.1093/jxb/erz203.

Zilberman, D., Perloff, J.M. and Spindell Berck, C. (eds) (2023) *Sustainable Resource Development in the 21st Century*, Vol. 57. Springer International Publishing, Cham, Switzerland.

5 Transcriptional Regulation of High-value Metabolites

Archana Jain, Surendra Sarsaiya and Qihai Gong*
Key Laboratory of Basic Pharmacology and Joint International Research Laboratory of Ethnomedicine of Ministry of Education, Zunyi Medical University, Zunyi, Guizhou, China

5.1 Introduction

Medicinal plants have been indispensable to human health for millennia, producing a vast array of high-value metabolites—alkaloids, terpenoids, and phenolics—that serve as the foundation for over 50% of modern pharmaceuticals (Alamgir, 2017). Compounds like vinblastine from *Catharanthus roseus* (used in leukemia treatment) and paclitaxel from *Taxus* species (a potent anti-cancer agent) exemplify the therapeutic and economic significance of these specialized metabolites (Alam *et al.*, 2017). However, their low natural abundance and extraction challenges impede scalable production, driving an urgent need to unravel the molecular mechanisms governing biosynthesis (Wang *et al.*, 2024). Central to this endeavor is transcriptional regulation, a dynamic process where transcription factors (TFs), epigenetic modifications, and environmental cues converge to control metabolic pathways (Meraj *et al.*, 2020).

Transcriptional regulation in medicinal plants operates through hierarchical networks. TFs, such as the MYB (myeloblastosis), bHLH (basic helix-loop-helix), and WRKY (highly conserved sequence of amino acids: tryptophan-arginine-lysine-tyrosine) families, act as molecular switches that bind to *cis*-regulatory elements in gene promoters, activating or repressing biosynthetic enzymes (Li *et al.*, 2024b). For instance, MYB TFs regulate terpenoid indole alkaloid (TIA) pathways in *Catharanthus* by coordinating jasmonate-responsive genes, while in *Taxus*, R2R3-MYB (protein contains two MYB domain repeats) TFs like TcMYB29a (first paralog/variant of the 29th identified MYB TF gene from *Taxus cuspidata*, belonging to the R2R3 structural subfamily) enhance paclitaxel production by upregulating taxadiene synthase (Sui *et al.*, 2018). These TFs rarely act in isolation; they form intricate complexes with co-regulators (e.g. MYC-WRKY interactions) and respond to phytohormones like methyl jasmonate (MeJA), which reprograms metabolism under stress (Bandurska, 2022). Complementing TF networks, epigenetic mechanisms—DNA methylation, histone modifications, and non-coding RNAs—integrate environmental signals into heritable metabolic traits (Hao and Xiao, 2018). Drought or UV stress, for example,

*Corresponding author: gqh@zmu.edu.cn

DOI: 10.1079/9781836994206.0005

can induce methylation changes that silence or activate genes, boosting metabolite synthesis in *Salvia officinalis* or *C. roseus* (Jampílek and Kráľová, 2023).

Despite these advances, the field grapples with spatial and temporal complexity. Secondary metabolite biosynthesis is often restricted to specific cell types (e.g. glandular trichomes for terpenoid storage) and developmental stages, necessitating technologies like single-cell RNA sequencing (scRNA-seq) to resolve tissue-specific regulation. Recent studies in *Gynostemma pentaphyllum* and *Catharanthus* have leveraged scRNA-seq to map triterpenoid and alkaloid pathways to mesophyll and idioblast cells, respectively. Yet, the integration of such high-resolution data with metabolic flux remains challenging.

Biotechnological interventions like promoter engineering and metabolic engineering offer promising solutions. Synthetic promoters, designed using machine learning (ML), can enhance TF activity and pathway flux, while heterologous expression in model systems (e.g. *Nicotiana benthamiana*) aims to reconstitute pathways like paclitaxel biosynthesis. However, these approaches face hurdles in precision, scalability, and biosafety (Szymczyk and Majewska, 2024).

5.2 TF Networks in Secondary Metabolism

TF networks play a crucial role in the regulation of secondary metabolism in medicinal plants, which is essential for the biosynthesis of pharmacologically active compounds. Among these TFs, the MYB family is particularly significant due to its involvement in various biological processes, including the regulation of secondary metabolites. These metabolites are vital for plant defense and have substantial therapeutic value. The regulation of secondary metabolism is a complex process involving multiple layers of genetic and epigenetic control, as well as environmental interactions. This analysis explores the role of TF networks, particularly MYB TFs, in the context of secondary metabolism in medicinal plants (Table 5.1 and Fig. 5.1).

MYB TFs are one of the largest families of TFs in plants and are extensively involved in the regulation of secondary metabolism (Huang *et al.*, 2023). They regulate the biosynthesis of various secondary metabolites by controlling the expression of multiple synthase genes, which are crucial for the production of active ingredients in medicinal plants (Tong *et al.*, 2024). MYB TFs are involved in the regulation of morphological and developmental processes, stress responses, and the biosynthesis of secondary metabolites, making them versatile regulators in plant biology (Cao *et al.*, 2020).

MYB TFs regulate secondary metabolism at multiple levels, including protein, post-transcriptional, and transcriptional levels (Cao *et al.*, 2020). They interact with other TFs such as MYC, WRKY, and AP2/ERF to form complex networks that modulate the expression of genes involved in secondary metabolism (Gupta *et al.*, 2023). The regulation of secondary metabolism is also influenced by signaling molecules like jasmonic acid (JA), which can induce the biosynthesis and accumulation of secondary metabolites through transcriptional regulation (Afrin *et al.*, 2015). The low yield of secondary metabolites in medicinal plants poses a challenge for their extraction and utilization. TF-targeted metabolic engineering offers a promising approach to enhance the production of these valuable compounds (Zhao *et al.*, 2023). Advances in next-generation sequencing and transcriptomic resources are facilitating the study of transcriptional networks and biosynthetic pathways in medicinal plants, providing new insights into the regulation of secondary metabolism (Tripathi *et al.*, 2016).

While MYB TFs are central to the regulation of secondary metabolism, other factors and environmental conditions also play significant roles. For instance, epigenetic modifications such as DNA methylation and histone modification can influence gene expression and secondary metabolite production (Zhao *et al.*, 2023). Additionally, environmental stresses such as UV radiation can trigger the synthesis of secondary metabolites, highlighting the dynamic interplay between genetic regulation and environmental factors (Thakur *et al.*, 2023). Understanding these complex networks and interactions is crucial

Table 5.1. TF networks in regulating secondary metabolism of medicinal plants. Author's own table.

Aspect	Key findings & mechanisms	Examples/functional role	Regulation & interaction	Citations
Central regulator: MYB TFs	One of the largest TF families in plants; master regulators of secondary metabolite biosynthesis.	Regulate multiple synthase genes for active ingredients. In *Taxus*, TcMYB29a upregulates taxadiene synthase to enhance paclitaxel production.	Act at protein, post-transcriptional, and transcriptional levels.	(Cao *et al.*, 2020; Huang *et al.*, 2023; Tong *et al.*, 2024)
Key TF families & complexes	TFs rarely act alone; they form intricate complexes to modulate pathway genes.	MYB, bHLH (MYC), WRKY, AP2/ERF families. In *Catharanthus*, MYB TFs coordinate jasmonate-responsive genes in the terpenoid indole alkaloid (TIA) pathway.	Form MYB-MYC-WRKY interactions to create hierarchical regulatory networks.	(Gupta *et al.*, 2023; Li *et al.*, 2024b)
Environmental & hormonal cues	External and internal signals reprogram metabolic networks, often via TFs.	JA and MeJA are key signaling molecules that induce biosynthesis of defense-related metabolites.	Phytohormones like MeJA activate TF complexes under stress, leading to metabolic reprogramming.	(Afrin *et al.*, 2015; Bandurska, 2022)
Epigenetic layer	Epigenetic mechanisms integrate environmental signals into heritable metabolic traits.	DNA methylation & histone modifications. Drought/UV stress can induce methylation changes that activate/silence genes in *S. officinalis* and *C. roseus*.	Provides a memory of stress, leading to sustained enhancement of metabolite synthesis.	(Hao and Xiao, 2018; Jampílek and Kráľová, 2023; Zhao *et al.*, 2023)
Spatial & temporal complexity	Metabolite biosynthesis is restricted to specific cell types and developmental stages.	In *G. pentaphyllum* (triterpenoids) and *Catharanthus* (alkaloids), pathways are mapped to mesophyll and idioblast cells, respectively.	Necessitates technologies like single-cell RNA-seq (scRNA-seq) to resolve tissue-specific regulation.	(Wang *et al.*, 2024)
Biotech application: metabolic engineering	TF-targeted engineering is a promising strategy to overcome low natural yields.	Promoter engineering using synthetic promoters designed with ML to enhance TF activity and pathway flux.	Aims to reconstitute pathways in heterologous systems (e.g. *N. benthamiana*). Faces hurdles in precision and scalability.	(Zhao *et al.*, 2023; Szymczyk and Majewska, 2024)
Research enablers	Modern tools are facilitating the dissection of complex transcriptional networks.	Next-generation sequencing and rich transcriptomic resources are providing new insights into biosynthetic pathways.	Enables the identification and characterization of novel TFs and their target genes.	(Tripathi *et al.*, 2016)

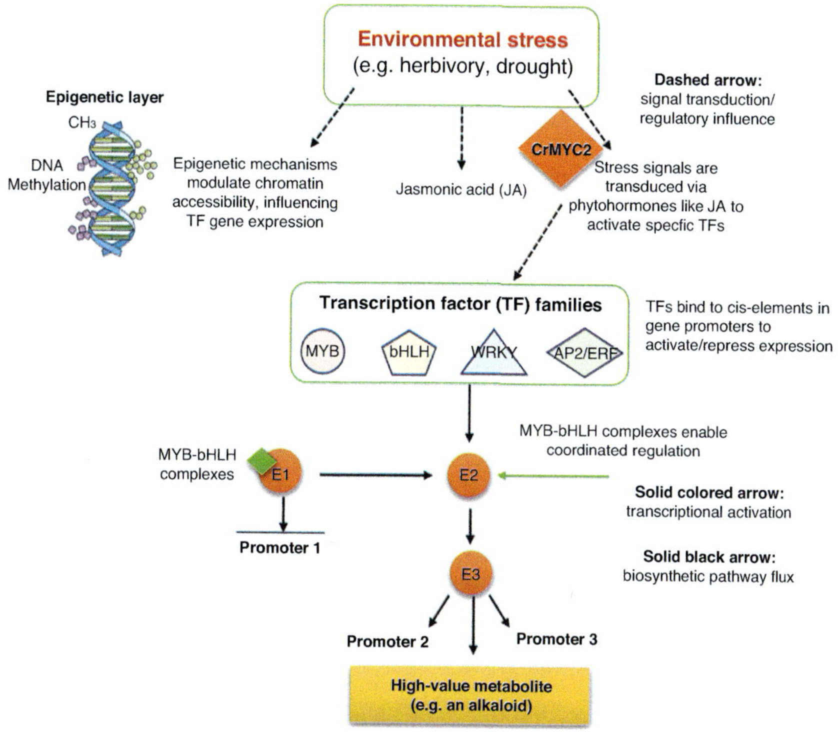

Fig. 5.1. Hierarchical transcriptional regulation of secondary metabolism in medicinal plants. (Author's own figure.)

for optimizing the production of secondary metabolites in medicinal plants, which can lead to enhanced therapeutic applications and economic benefits.

of valuable metabolites. The following sections delve into the specific roles of TFs in each of these pathways (Fig. 5.2).

5.2.1 Role of TFs in orchestrating alkaloid, terpenoid, and phenolic pathways

TFs play a pivotal role in orchestrating the biosynthesis of alkaloids, terpenoids, and phenolics in medicinal plants. These secondary metabolites are crucial for plant defense and have significant pharmaceutical applications. TFs regulate the expression of biosynthetic genes, thereby controlling the production and accumulation of these compounds. Recent advances in molecular biology have enhanced our understanding of the transcriptional regulatory networks involved in these pathways, offering potential strategies for metabolic engineering to increase the yield

5.2.2 Alkaloid, terpenoid, and phenolic pathway regulation

TFs such as bHLH, APETALA2/ethylene-responsive factor (AP2/ERF), and WRKY are crucial in regulating alkaloid biosynthesis. These TFs bind to *cis*-elements in the promoters of biosynthetic genes, modulating their expression (Yamada and Sato, 2021). In *C. roseus*, TFs have been identified using co-expression analysis and *cis*-regulatory motif identification, which are essential for understanding the regulation of TIAs (Singh *et al.*, 2022). Genetic modulation of alkaloid pathways through TFs has been explored to enhance the production of commercially important alkaloids like vincristine and

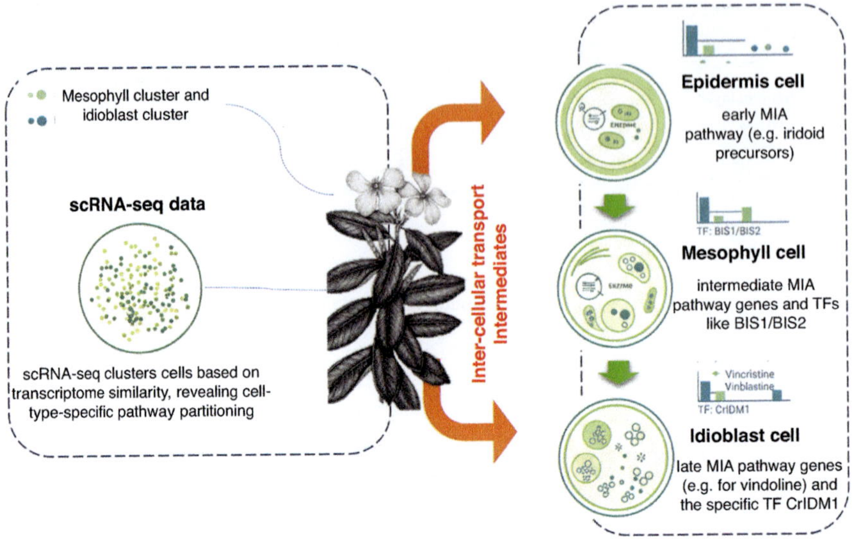

Fig. 5.2. Single-cell omics reveals spatial partitioning of the MIA pathway in *C. roseus*. (Author's own figure.)

vinblastine, which are used in cancer treatment (Bhushan *et al.*, 2023; Prakash *et al.*, 2023).

Terpenoids, the largest class of secondary metabolites, are regulated by TFs that control terpene synthase activities at the transcriptional level. This regulation is often influenced by environmental cues (Kumar *et al.*, 2023). TFs are also involved in the differentiation of glandular trichomes, which are key sites for terpenoid accumulation. This highlights the dual role of TFs in both metabolic and developmental processes (Wang *et al.*, 2023). In addition to their role in biosynthesis, TFs are implicated in the transport and storage of terpenoids, as seen in the spatial-temporal regulation of TIAs in *C. roseus* (Liu et al., 2016).

Although the research articles primarily focus on alkaloids and terpenoids, TFs are known to regulate phenolic pathways by activating or inhibiting the expression of genes involved in phenolic acid biosynthesis (Shi *et al.*, 2024). Jasmonate-responsive TFs, including AP2/ERFs and MYB, are particularly important in the regulation of phenolic compounds, as they respond to stress signals and modulate secondary metabolism (Meraj *et al.*, 2020). While TFs are central to the regulation of secondary metabolite pathways, the complexity of these networks poses challenges. The specificity of TFs to particular pathways and their interaction with other regulatory proteins can vary significantly across plant species and environmental conditions. Moreover, the integration of TFs with other regulatory mechanisms, such as post-transcriptional modifications and epigenetic changes, adds another layer of complexity. Understanding these interactions is crucial for developing effective strategies for metabolic engineering and enhancing the production of valuable plant metabolites (Fig. 5.2).

5.3 Epigenetic Control: Stress-induced DNA Methylation and Heritable Traits

Epigenetic control through stress-induced DNA methylation plays a significant role in the adaptation and heritability of traits in plants, including medicinal plants. DNA methylation, a key epigenetic mechanism, can lead to heritable phenotypic changes without altering the DNA sequence, allowing plants to adapt to environmental stresses. This process is crucial for plant survival and evolution, as it enables plants to respond to both biotic and abiotic stresses by modulating gene expression. The heritability of these methylation changes suggests a potential for transgenerational adaptation, although the stability and persistence of these changes can vary depending on the type of stress and plant lineage.

5.3.1 Mechanisms and heritability of stress-induced DNA methylation in plants

Environmental stresses such as drought, salinity, and pathogen attacks can trigger DNA methylation changes in plants, leading to altered gene expression patterns that help plants cope with these stresses (Akhter *et al.*, 2021; Talarico *et al.*, 2024). In dandelions, various stresses, including chemical induction of defenses, have been shown to cause significant methylation variation, which is often heritable across generations (Preite, 2016). DNA methylation can also affect transposable elements (TEs) in the genome, which can influence gene expression and plant adaptability under stress conditions (Galindo-González *et al.*, 2018).

Stress-induced methylation changes are often heritable, allowing plants to pass on adaptive traits to their offspring. This has been observed in apomictic dandelions, where methylation changes persisted in offspring generations (Galindo-González *et al.*, 2018). The heritability of these changes suggests that epigenetic mechanisms can play an independent role in evolutionary processes, potentially leading to adaptive divergence in plant populations (Khan *et al.*, 2015). However, the stability of these heritable changes can be stress- and lineage-dependent, with some changes fading in the absence of stressors (Preite, 2016; Gupta and Nair, 2022).

DNA methylation involves the addition of methyl groups to cytosine bases, which can lead to gene silencing or activation, depending on the context (Viggiano and de Pinto, 2017). The process is regulated by a complex interplay of enzymes and small RNAs, which can modulate methylation patterns in response to environmental cues (Zang *et al.*, 2023). Non-coding RNAs and histone modifications also play roles in the epigenetic regulation of stress responses, contributing to the dynamic nature of the epigenome (Akhter *et al.*, 2021; Talarico *et al.*, 2024). While DNA methylation offers a mechanism for rapid adaptation, its stability and persistence can be variable, posing challenges for its use in crop improvement and breeding programs (Khan *et al.*, 2015). The potential fitness costs associated with maintaining stress-induced methylation marks suggest that these changes may not always be beneficial in the long term (Gupta and Nair, 2022). Further research is needed to fully understand the ecological and evolutionary consequences of epigenetic variations and their potential applications in agriculture (Viggiano and de Pinto, 2017; Zang *et al.*, 2023). While stress-induced DNA methylation provides a mechanism for plants to adapt to environmental changes, the variability in the stability and heritability of these changes highlights the complexity of epigenetic regulation. The potential for epigenetic mechanisms to contribute to plant evolution and adaptation is significant, but further studies are necessary to explore their long-term implications and applications in crop improvement.

5.3.2 Epigenetic mechanisms linking environmental stress to metabolite production

Epigenetic mechanisms play a crucial role in linking environmental stress to metabolite production in medicinal plants. These mechanisms, including DNA methylation, histone modification, and non-coding RNA activity, regulate gene expression in response to stress, thereby influencing the synthesis and accumulation of secondary metabolites. These metabolites, such as terpenoids, alkaloids, and phenolic compounds, are vital for plant defense and have significant medicinal properties. The interaction between environmental stressors and epigenetic modifications can lead to either an increase or a decrease in metabolite production, depending on the type and intensity of the stress. This complex relationship underscores the adaptability of medicinal plants to changing environments and their potential for enhanced therapeutic efficacy under stress conditions (Table 5.2).

DNA methylation is a key epigenetic modification that affects gene expression by adding methyl groups to DNA, often leading to

Table 5.2. Epigenetic mechanisms linking stress to metabolite production. Author's own table.

Epigenetic mechanism	Role in stress response & gene regulation	Impact on metabolite production & examples	Key insights & complexity	Citations
DNA methylation	Adds methyl groups to DNA, often leading to gene silencing. Stress alters methylation patterns, activating or repressing stress-responsive genes.	Can boost defense metabolite synthesis. In *C. roseus* and *S. officinalis*, drought/UV stress induces methylation changes that activate alkaloid/terpenoid pathways.	Provides a "molecular memory" of stress, potentially leading to heritable traits. Stability of these changes can be variable.	(Jampílek and Kráľová, 2023; Singh *et al*., 2023; Talarico *et al*., 2024)
Histone modifications	Acetylation, methylation, phosphorylation alter chromatin structure, making genes more or less accessible for transcription.	Crucial for regulating biosynthetic gene expression under stress. Can lead to sustained activation of pathways for compounds like terpenoids and phenolics.	Works in concert with DNA methylation; a key part of the complex regulatory network reprogramming metabolism during stress.	(Singh *et al*., 2023)
Non-coding RNAs	Modulate gene expression post-transcriptionally (e.g. via mRNA degradation or translational inhibition). Involved in fine-tuning stress responses.	Affects the production of secondary metabolites by regulating key enzymes in their biosynthetic pathways.	Adds a layer of precise, dynamic control to the epigenetic response, integrating multiple stress signals.	(Talarico *et al*., 2024)
Abiotic stressors (drought, salinity, UV)	Act as signals that trigger the above epigenetic mechanisms, reprogramming the plant's gene expression profile.	Alters metabolite profiles. Salinity in *Catharanthus* affects vincristine; drought in *Salvia* influences terpenes; salinity in *A. vera* increases aloin but decreases phenolics.	Response is metabolite- and species-specific. Stress can be used as an "elicitor" to enhance valuable compounds, but extreme stress is detrimental.	(Pant *et al*., 2021; Kanso *et al*., 2022; Sharma, 2024)
Interaction of multiple stressors	Multiple stresses activate interacting signaling pathways (e.g. JA, abscisic acid—ABA), leading to complex, non-additive epigenetic and transcriptional responses.	Can create unique metabolic outcomes that cannot be predicted from single-stress exposure. May enhance or suppress specific metabolite production.	Highlights the complexity of predicting plant metabolic responses in real-world, multi-stress environments.	(Jan *et al*., 2021)

Continued

Table 5.2. Continued

Epigenetic mechanism	Role in stress response & gene regulation	Impact on metabolite production & examples	Key insights & complexity	Citations
Biotechnological application	Understanding epigenetic regulation allows for the development of strategies to control metabolite yield.	Controlled cultivation (e.g. applying mild stress) and biotech tools can be used to steer epigenetic states and optimize the medicinal value of plants.	A promising approach for sustainable production, but requires fine-tuning to avoid yield penalties from severe stress.	(Punetha *et al.*, 2022; Jampílek and Kráľová, 2023)

gene silencing. Under stress conditions, changes in DNA methylation patterns can activate or repress genes involved in stress responses and secondary metabolite production (Singh *et al.*, 2023; Talarico *et al.*, 2024). Histone modifications such as acetylation, methylation, and phosphorylation of histone proteins alter chromatin structure, influencing gene accessibility and expression. These modifications are crucial for regulating genes involved in the synthesis of secondary metabolites during stress (Singh *et al.*, 2023). Non-coding RNAs molecules can modulate gene expression post-transcriptionally and are involved in stress responses, affecting the production of secondary metabolites (Talarico *et al.*, 2024).

Abiotic stressors such as drought, salinity, and temperature extremes can significantly alter the biochemical pathways in medicinal plants, leading to changes in secondary metabolite profiles. For instance, salinity stress in *C. roseus* affects vincristine levels, while drought stress influences terpene levels in *S. officinalis* (Pant *et al.*, 2021; Kanso *et al.*, 2022). Biochemical changes caused by stress conditions can lead to both upregulation and downregulation of specific metabolites. For example, *Aloe vera* under salinity stress showed decreased phenolic content but increased aloin production, demonstrating the plant's adaptive response to stress (Sharma, 2024). The interaction of multiple stressors can lead to unique responses that are not simply additive. This complexity is due to the interaction of multiple signaling pathways, which can enhance or suppress secondary metabolite production (Jan *et al.*, 2021).

The stress-induced modulation of secondary metabolites can enhance or diminish the therapeutic properties of medicinal plants. Understanding these changes is crucial for optimizing cultivation practices to maximize the medicinal value of these plants (Jampílek and Kráľová, 2023; Sharma, 2024). Conservation and sustainability strategies, such as controlled cultivation environments and the use of biotechnological tools, can help maintain the quality and quantity of secondary metabolites in medicinal plants, ensuring their sustainability and efficacy (Punetha *et al.*, 2022; Jampílek and Kráľová, 2023). While epigenetic mechanisms provide a robust framework for understanding how environmental stress influences metabolite production in medicinal plants, it is important to consider the potential negative impacts of stress. Extreme stress conditions can lead to detrimental effects on plant health and reduce the overall yield of valuable secondary metabolites. Therefore, balancing stress exposure to optimize metabolite production without compromising plant vitality is crucial for sustainable medicinal plant cultivation.

5.4 Single-cell Omics for Tissue-specific Transcriptional Profiling

Single-cell omics technologies have revolutionized the field of plant biology by enabling tissue-specific transcriptional profiling, which is particularly beneficial for studying medicinal plants (Table 5.3). These technologies allow researchers to dissect the complex biosynthetic pathways of specialized metabolites at a cellular level, providing insights into the spatial and temporal distribution of gene expression and

Table 5.3. Single-cell omics applications in medicinal plants. Author's own table.

Plant species	Technology used	Key finding	Metabolite pathway	References
G. pentaphyllum	scRNA-seq	Gypenoside biosynthesis localized to mesophyll cells.	Triterpenoid saponins	(Li *et al.*, 2024a)
C. roseus	Single-cell multi-omics (scRNA-seq + metabolomics)	MIA pathway partitioned across leaf and root cell types.	Monoterpene indole alkaloids (MIAs)	(Li *et al.*, 2023a; Vu *et al.*, 2024)
Taxus mairei	Single-cell ATAC-seq	Cell-type-specific *cis*-regulatory elements in paclitaxel biosynthesis.	Paclitaxel	(Zang *et al.*, 2023)
A. thaliana	scRNA-seq	Identification of glucosinolate-producing cell types.	Glucosinolates	(Maeda *et al.*, 2023)

metabolite production. This approach is crucial for understanding the unique properties of medicinal plants, which often rely on specific cell types for the synthesis of bioactive compounds. The following sections explore the application of single-cell omics in medicinal plants, highlighting key findings and methodologies. Single-cell RNA sequencing (scRNA-seq) was used to profile over 50,000 cells from shoot apexes and leaves of *G. pentaphyllum*, identifying cell-type-specific expression patterns for triterpenoid saponin biosynthesis. This study revealed that gypenoside biosynthesis predominantly occurs in mesophyll cells, with higher activity in shoot apexes compared to leaves (Li *et al.*, 2024a).

A multi-omics approach, including single-cell RNA sequencing, was employed to study the monoterpene indole alkaloid (MIA) biosynthetic pathway. This research identified gene clusters involved in MIA biosynthesis and demonstrated cell-type-specific partitioning of the pathway in leaves and roots (Li *et al.*, 2023a). Single-cell omics provide a high-resolution view of gene expression and metabolite distribution, overcoming the limitations of bulk tissue analyses that average out cellular heterogeneity (Talukdar, 2019; Lv *et al.*, 2024). Gene regulatory networks technologies enable the investigation of gene regulatory networks at a cellular level, facilitating the understanding of how specific cell types contribute to the overall function and productivity of the plant (Hurgobin and Lewsey, 2022). Despite the advancements, challenges remain in detecting and quantifying metabolites at single-cell resolution, particularly in plants with complex tissue structures (Vu *et al.*, 2024).

Combining transcriptomics with other omics data, such as metabolomics and proteomics, is essential for a comprehensive understanding of biosynthetic pathways. This integrative approach can reveal novel insights into the regulation and function of specialized metabolites (Burlat *et al.*, 2023). *Polygonum cuspidatum* tissue-specific transcriptome analyses identified candidate genes for the biosynthesis of stilbenes, flavonoids, and anthraquinones, highlighting the differential expression of these genes across root, stem, and leaf tissues (Wang *et al.*, 2021). *Arabidopsis thaliana* single-cell RNA sequencing identified specialized cell types involved in the glucosinolate-myrosinase defense system, demonstrating the potential of single-cell omics to uncover cell-type-specific metabolic pathways (Maeda *et al.*, 2023).

While single-cell omics technologies have significantly advanced our understanding of tissue-specific transcriptional profiling in medicinal plants, they also present new challenges and opportunities. The integration of multi-omics data and the development of more refined analytical techniques will be crucial for further elucidating the complex biosynthetic pathways of plant metabolites. As these technologies continue to evolve, they hold the promise of unlocking new potential in the biotechnological production of plant-based drugs and enhancing our understanding of plant biology at an unprecedented level.

5.4.1 Resolving cell-type-specific regulatory mechanisms (e.g. glandular trichomes)

Resolving cell-type-specific regulatory mechanisms in glandular trichomes (GTs) of medicinal plants involves understanding the complex interplay of genetic, molecular, and environmental factors that govern their development and function. GTs are specialized structures that synthesize and store a variety of secondary metabolites, which are crucial for the plant's defense and have significant medicinal value. The regulatory mechanisms of GTs are intricate, involving specific TFs, signaling pathways, and environmental influences that determine their formation and metabolic activity.

The development of GTs is regulated by specific TFs, such as MYB, which play a crucial role in the differentiation and metabolic activity of these structures. For instance, the MYB TFs are involved in the regulation of terpene synthesis and hormonal signaling, which are essential for GT development (Kabir *et al.*, 2024; Li *et al.*, 2025). The biosynthesis of secondary metabolites in GTs is controlled by complex gene networks. In *C. roseus*, for example, the monoterpenoid indole alkaloid (MIA) pathway is regulated by cell-type-specific TFs and *cis*-regulatory elements, highlighting the importance of gene expression and chromatin accessibility in GT function (Li *et al.*, 2025).

The formation of GTs involves distinct regulatory pathways, including the GLABRA1 (GL1), GLABRA3 (GL3), and ENHANCER OF GLABRA3 (EGL3) pathways for unicellular trichomes, and MYB TFs for multicellular GTs (Kabir *et al.*, 2024). The growth and development of GTs is influenced by environmental factors such as temperature, UV radiation, and herbivore presence. These conditions can affect the density, morphology, and metabolic activity of GTs, thereby influencing the plant's adaptive responses (Kabir *et al.*, 2024). Hormonal signals, particularly phytohormones, are crucial for the initiation and development of GTs. They integrate with transcriptional networks to modulate trichome formation and secondary metabolite production (Yuan *et al.*, 2023; Kabir *et al.*, 2024).

In thyme, the TF TqHD1 has been identified as a key regulator of GT formation and volatile component changes, demonstrating the role of specific genes in GT development (Zhang *et al.*, 2025). Similarly, in *Nepeta tenuifolia*, the gene NtMIXTA1 is associated with GT development and monoterpene biosynthesis, illustrating the genetic basis of GT function (Zhou *et al.*, 2022). The CmMYC2–CmMYBML1 module in chrysanthemum regulates both trichome development and terpene biosynthesis, providing insights into the coordinated regulation of physical and chemical defenses in plants (Guan *et al.*, 2024). While GT regulatory mechanisms have advanced significantly, challenges remain in fully elucidating the cell-type-specific pathways and their interactions with environmental factors. The integration of omics technologies and advanced genetic tools continues to provide deeper insights into the molecular basis of GT function, offering potential for metabolic engineering and the development of stress-tolerant plant varieties. This ongoing research is crucial for harnessing the full potential of GTs in medicinal plants and improving their pharmaceutical applications.

5.5 Regulatory Networks in Paclitaxel Biosynthesis: From Genes to Metabolites

The biosynthesis of paclitaxel, a potent anti-cancer compound, involves a complex regulatory network that spans from genes to metabolites in medicinal plants, particularly in the genus *Taxus*. This network is orchestrated by a variety of genetic and biochemical factors, including TFs, microRNAs, and small interfering RNAs, which collectively regulate the expression of biosynthetic genes and the production of metabolites. Understanding these regulatory mechanisms is crucial for enhancing paclitaxel production through biotechnological approaches.

A significant number of TFs, including MYB, AP2, and WRKY families, have been identified as key regulators in the paclitaxel biosynthesis pathway. These TFs are involved in the transcriptional regulation of genes responsible for the synthesis of paclitaxel, influencing the expression levels of enzymes involved in the pathway (Wang *et al.*, 2019; Sun *et al.*, 2024). miRNAs such as miR5298b play a crucial role in post-transcriptional regulation by targeting specific mRNAs for degradation or translational repression. miR5298b, for instance, enhances paclitaxel biosynthesis by cleaving the mRNA of TcNPR3, thereby alleviating the inhibition of the TcNPR3-TcTGA6 complex, which negatively regulates paclitaxel biosynthesis genes (Ying *et al.*, 2023).

In the areas of gene expression and pathway elucidation, despite the complexity of the paclitaxel biosynthetic pathway, significant progress has been made in characterizing the genes involved. Out of the 19 steps in the pathway, 14 genes have been well characterized, and recent studies have identified additional genes that contribute to the biosynthesis of intermediates like baccatin III (McElroy and Jennewein, 2018; McClune *et al.*, 2024). The paclitaxel biosynthetic pathway involves multiple enzymatic steps, including cyclization, oxygenation, and acyl/aroyl transfers. These steps are mediated by enzymes such as cytochrome P450s and CoA-dependent acyltransferases, which are crucial for the formation of intermediates like baccatin III (Li *et al.*, 2017; Zhou *et al.*, 2019).

The processes of metabolite production and regulation show that the production of paclitaxel and its precursors is influenced by the expression levels of key biosynthetic genes. For instance, high expression of genes like taxadienol acetyltransferase (TAT) and taxadiene 5α-hydroxylase (T5H) is associated with increased paclitaxel content in certain *Taxus* cultivars (Wang *et al.*, 2019). Advances in synthetic biology have enabled the reconstitution of the paclitaxel biosynthetic pathway in heterologous systems, such as *N. benthamiana*, allowing for the production of paclitaxel precursors at levels comparable to natural sources. This approach involves the combinational expression of multiple biosynthetic genes to optimize production (McClune *et al.*, 2024).

The use of *Taxus* cell suspension cultures, induced by methyl jasmonate, has been a foundational method for studying and enhancing paclitaxel biosynthesis. These cultures provide a controlled environment for manipulating gene expression and enzyme activity to increase paclitaxel yield (Li *et al.*, 2017; Zhou *et al.*, 2019). Despite advances in understanding the regulatory networks of paclitaxel biosynthesis, challenges remain in fully elucidating the pathway and optimizing production. The integration of genetic, biochemical, and biotechnological approaches holds promise for overcoming these challenges and meeting the growing demand for this valuable anti-cancer drug (Table 5.4 and Fig. 5.3).

5.5.1 TF-driven coordination of terpenoid pathways in *Taxus* spp.

The coordination of terpenoid pathways in *Taxus* spp., particularly in the context of medicinal plants, is a complex process driven by TFs that regulate the biosynthesis of paclitaxel, a highly valued anti-cancer drug. The biosynthesis of paclitaxel involves a series of enzymatic reactions and regulatory mechanisms that are influenced by various TFs and environmental factors. Understanding these pathways is crucial for enhancing paclitaxel production and ensuring a sustainable supply of this important compound.

TFs such as MYB, ERF, and RAP (Related to APETALA2) play significant roles in the regulation of paclitaxel biosynthesis. For instance, the R2R3-MYB transcription factor TcMYB29a has been shown to upregulate paclitaxel biosynthesis in *Taxus chinensis* by activating the expression of key biosynthetic genes like taxadiene synthase (TS) and taxane 5α-hydroxylase (T5OH) (Cao *et al.*, 2022). In *Taxus mairei*, single-cell ATAC (assay for transposase-accessible chromatin) sequencing has identified cell-type-specific TF motifs that are involved in paclitaxel biosynthesis, highlighting the importance of chromatin accessibility in regulating gene expression (Zhan *et al.*, 2023).

ABA and MeJA are two hormones that influence paclitaxel biosynthesis. ABA has been shown to upregulate TcMYB29a, thereby enhancing paclitaxel production, while MeJA has a suppressive effect on this TF (Cao *et al.*, 2022). The differential expression of JA-related TF genes in response to exogenous JA applications suggests that environmental factors can significantly impact paclitaxel biosynthesis across different *Taxus* species (Zhou *et al.*, 2019).

Comparative transcriptome analyses have revealed significant differences in the expression patterns of paclitaxel biosynthesis-related genes among various *Taxus* species, which

Table 5.4. Regulatory networks in paclitaxel biosynthesis in *Taxus* spp. Author's own table.

Regulatory layer	Key regulators & mechanisms	Target/function	Impact on paclitaxel biosynthesis	Citations
Transcription factors (TFs)	MYB TFs (e.g. TcMYB29a): master positive regulators. Binds promoters of key biosynthetic genes.	TS (taxadiene synthase), T5H (taxadiene 5α-hydroxylase). Upregulates core pathway enzymes.	Directly enhances flux through the early, rate-limiting steps of the pathway.	(Cao *et al*., 2022; Sun *et al*., 2024)
	AP2/ERF, WRKY, RAP TFs: part of a complex TF network.	Various biosynthetic gene promoters. Modulate pathway in response to stress and hormones.	Fine-tunes the pathway; integrates multiple environmental and developmental signals.	(Wang *et al*., 2019; Zhan *et al*., 2023)
Post-transcriptional regulation	microRNAs (e.g. miR5298b): cleave specific mRNA targets.	TcNPR3 mRNA. Degrades a negative regulator of the pathway.	Indirectly enhances biosynthesis by relieving repression (TcNPR3 inhibits TF TcTGA6).	(Ying *et al*., 2023)
Hormonal & environmental cues	Abscisic acid (ABA): signaling hormone.	Upregulates TcMYB29a expression.	Stimulates paclitaxel biosynthesis by activating a key master TF.	(Cao *et al*., 2022)
	MeJA: defense signaling hormone.	Suppresses TcMYB29a expression.	Suppresses paclitaxel biosynthesis in some contexts; effect can be species-specific.	(Zhou *et al*., 2019; Cao *et al*., 2022)
Epigenetic & cell-type specificity	Chromatin accessibility: revealed by single-cell ATAC-seq.	Cell-type-specific TF binding motifs in regulatory regions.	Explains spatial regulation of the pathway within different plant tissues.	(Zhan *et al*., 2023)

Continued

Table 5.4. Continued

Regulatory layer	Key regulators & mechanisms	Target/function	Impact on paclitaxel biosynthesis	Citations
Pathway elucidation & engineering	Gene characterization: 14 of ~19 enzymatic steps well-characterized.	Enzymes like TAT (tyrosine amino transferase), T5H (taxadiene 5-alpha-hydroxylase), P450s, acyltransferases.	Enables metabolic engineering and heterologous reconstruction.	(McElroy and Jennewein, 2018; McClune *et al.*, 2024)
	Heterologous reconstitution: expression in *N. benthamiana*.	Combinatorial expression of multiple *Taxus* genes.	Produces paclitaxel precursors at levels comparable to natural sources.	(McClune *et al.*, 2024)
Production systems	Cell suspension cultures: induced by elicitors like MeJA.	Provides a controlled system to study and enhance pathway gene expression and enzyme activity.	Foundational method for studying regulation and a bioproduction platform.	(Li *et al.*, 2017; Zhou *et al.*, 2019)
Key challenges	Pathway complexity: not all genes identified; regulatory network is not fully mapped.	–	Hinders complete optimization and engineering of the pathway.	(McElroy and Jennewein, 2018)
	Interspecific variation: differing gene expression and regulatory responses across *Taxus* species.	–	Complicates the development of universal strategies for yield enhancement.	(Zhou *et al.*, 2019)

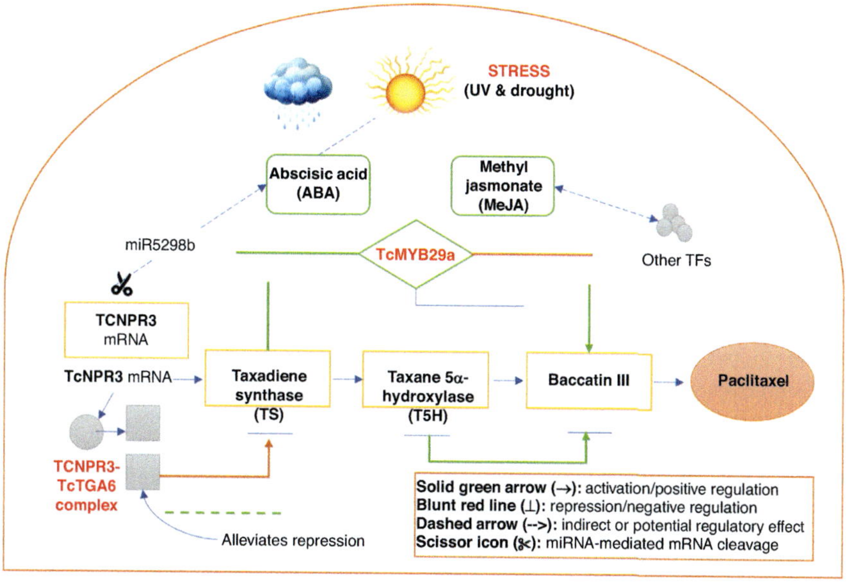

Fig. 5.3. Multi-layered regulatory network of paclitaxel biosynthesis in *Taxus* spp. (Author's own figure.)

could explain the interspecific variations in paclitaxel accumulation (Zhou *et al.*, 2019). Biotechnological methods, such as *in vitro* cultivation and genetic modification, are being explored to enhance paclitaxel production. These approaches aim to optimize the expression of biosynthetic genes and improve the yield of paclitaxel in plant cell cultures (Cusido *et al.*, 2014). Despite the potential of endophytic fungi as alternative sources of paclitaxel, inconsistencies in fungal taxane synthesis have been reported, indicating that the biosynthetic machinery may have been acquired from the plant host (El-Sayed *et al.*, 2020). The elucidation of the paclitaxel biosynthetic pathway remains incomplete, with ongoing research focused on identifying and characterizing the remaining genes and enzymes involved in this complex process. The coordination of terpenoid pathways in *Taxus* spp. is primarily driven by TFs and environmental cues, but challenges such as the low natural yield of paclitaxel and the complexity of its biosynthetic pathway persist. Future research may focus on further unraveling the genetic and regulatory networks involved in paclitaxel biosynthesis, potentially leading to more efficient production methods and a stable supply of this critical anti-cancer drug.

5.6 Promoter Engineering for Transcriptional Amplification

Promoter engineering in medicinal plants is a promising approach to enhance transcriptional amplification, which is crucial for increasing the production of valuable secondary metabolites. This technique involves modifying promoter sequences to regulate gene expression more effectively, thereby improving the yield and quality of bioactive compounds in medicinal plants. The integration of synthetic biology and genetic engineering tools has enabled the development of novel promoters with enhanced strength and specificity, which can be applied to medicinal plants to optimize their metabolic pathways. The following sections explore the key aspects of promoter engineering in this context.

Ubiquitous promoters, such as those derived from *A. thaliana*, have been shown to induce systemic expression in various plants, including medicinal ones like ginseng and *Codonopis lanceolata*. These promoters can be used to express useful genes or improve agricultural traits in transgenic plants (Yang *et al.*, 2016). Synthetic promoters are designed to overcome the limitations of native promoters, such as insufficient transcriptional activity. They are engineered to drive the expression of foreign genes at maximum amplitude, which is crucial for enhancing the production of secondary metabolites in medicinal plants (Khan *et al.*, 2023). The architecture of promoters, including the arrangement of *cis*-regulatory elements, plays a critical role in transcription regulation. Advances in understanding these elements have led to the development of promoters that can tightly control gene expression, achieving desired plant phenotypes and product yields (Pandey *et al.*, 2019).

ML techniques are being employed to model and predict promoter characteristics, enabling the design of highly active promoters with desirable features. This approach facilitates the creation of promoters that can be tailored for specific applications in metabolic engineering and cellular therapeutics (Cazier and Blazeck, 2021). Promoter engineering is a key tool in metabolic engineering, allowing for the manipulation of biosynthetic pathways to enhance the production of secondary metabolites. This is particularly important for medicinal plants, where the demand for bioactive substances is high (Yembaturova and Cheryatova, 2023). Genetic manipulation, including the use of engineered promoters, can lead to higher accumulation of target secondary metabolites in plant-based bioreactors. This approach is essential for meeting the pharmaceutical industry's needs for these compounds (Bharti *et al.*, 2023).

Despite the potential of promoter engineering, challenges remain in achieving precise control over gene expression. The complexity of plant genomes and the need for species-specific promoter designs are significant hurdles (Jores *et al.*, 2021). Future research should focus on refining promoter engineering techniques, exploring new design paradigms, and expanding the use of computational models to predict promoter activity across different plant species (Ji *et al.*, 2024). While promoter engineering offers significant advantages for enhancing transcriptional amplification in medicinal plants, it is important to consider the broader implications of these technologies. The development of transgenic plants raises concerns about biosafety and environmental impact, which must be addressed through rigorous testing and regulation. Additionally, the integration of promoter engineering with other biotechnological tools, such as gene editing and tissue culture, can further optimize the production of secondary metabolites, providing a comprehensive approach to improving medicinal plant yields (Table 5.5).

5.6.1 Designing promoters to boost TF activity and pathway flux

Designing promoters to enhance TF activity and pathway flux in medicinal plants involves a strategic approach to optimize gene expression for increased production of valuable secondary metabolites. This process is crucial for improving the yield of therapeutic compounds in plants like *C. roseus*, which are known for their complex biosynthetic pathways. The integration of synthetic biology and transcriptional regulation offers promising avenues to achieve these goals. The following sections outline key strategies and considerations in designing promoters for this purpose.

Identifying *cis*-regulatory motifs (CRMs) in gene promoters provides insights into potential TFs that bind to these regions, which is essential for understanding pathway regulation in plants like *C. roseus* (Singh *et al.*, 2022). The co-expression analysis approach assumes that TFs regulating a metabolic pathway exhibit expression pattern similar to those of the pathway genes, aiding in the identification of relevant TFs (Le *et al.*, 2015). By isolating known TFs from other species, researchers can identify TFs that regulate similar or diverse metabolites (Singh *et al.*, 2022). Core promoter elements such as the TATA-box (thymine-adenine-thymine-adenine-box), GC (guanine-cytosine) content, and proximal TF binding sites significantly influence promoter strength. Computational models can predict promoter strength, facilitating the design of highly active synthetic promoters (Jores *et al.*, 2021).

Minimal synthetic promoters (MinSyns) are designed using ML to create small, standardized *cis*-regulatory elements that allow for tunable expression in plants, enhancing the production of secondary metabolites (Tagliani, 2020). TFs can be targeted to enhance the biosynthesis of

Table 5.5. Promoter engineering strategies for metabolic engineering. Author's own table.

Strategy/aspect	Core concept & methodology	Application in medicinal plants	Advantages & outcomes	Challenges & limitations	Citations
Ubiquitous promoters	Using strong, constitutive promoters from model plants (e.g. *Arabidopsis*) to drive gene expression systemically.	Expressing useful genes or improving traits in transgenic medicinal plants like ginseng and *C. lanceolata*.	Provides reliable, high-level expression across different tissues and species.	Lack of specificity can lead to metabolic burden or undesirable phenotypes; not tailored for pathway-specific regulation.	(Yang *et al.*, 2016)
Synthetic promoter design	Engineering artificial promoters by combining core elements with specific *cis*-regulatory motifs to enhance strength and control.	Overcoming insufficient activity of native promoters to maximize expression of biosynthetic genes for secondary metabolites.	Enables maximum transcriptional amplitude, tight regulation, and custom-designed expression patterns.	Requires deep knowledge of *cis*-element function; design can be complex and species-specific.	(Pandey *et al.*, 2019; Khan *et al.*, 2023)
Computational design & ML	Using ML models to predict promoter activity and design novel promoters with desired features based on sequence.	Facilitating the creation of highly active, tailored promoters for metabolic engineering without extensive trial and error.	Accelerates the design process and allows for the generation of optimized promoters with predictable performance.	Dependent on large, high-quality training datasets; model accuracy may vary across diverse plant genomes.	(Cazier and Blazeck, 2021; Ji *et al.*, 2024)
Application in metabolic engineering	Using engineered promoters to manipulate and control flux through biosynthetic pathways.	Enhancing the production of high-value secondary metabolites (alkaloids, terpenoids) in plant-based bioreactors.	Directly increases the yield of target bioactive compounds, addressing supply limitations for pharmaceuticals.	Precise balancing of multiple genes in a pathway is challenging; overexpression can cause metabolic imbalance.	(Bharti *et al.*, 2023; Yembaturova and Cheryatova, 2023)

Continued

Table 5.5. Continued

Strategy/aspect	Core concept & methodology	Application in medicinal plants	Advantages & outcomes	Challenges & limitations	Citations
Integration with other biotech tools	Combining promoter engineering with gene editing (CRISPR) and tissue culture systems.	Creating non-transgenic edited plants with enhanced metabolic pathways or optimizing production in cell cultures.	Provides a comprehensive toolkit for rational plant design and bioproduction.	Technical complexity of integration; regulatory hurdles for transgenic plants.	(Jores *et al.*, 2021)
Overarching challenges	–	–	–	Biosafety & environmental impact of transgenic plants necessitates rigorous testing and regulation.	(Implied from text)
	–	–	–	Species-specific complexity requires customized promoter designs for each medicinal plant.	(Jores *et al.*, 2021)

secondary metabolites by binding to specific gene promoters, thus activating or suppressing gene expression (Biswas *et al.*, 2022; Tong *et al.*, 2024). Synthetic dual hormone-responsive promoters respond to signals like salicylic acid and JA, enabling plants to resist a broad spectrum of pathogens and potentially increasing metabolite production (Li *et al.*, 2023b). With quantitative characterization of promoters using systems like the dual-luciferase reporter, researchers can quantitatively assess promoter strength, which is crucial for optimizing gene expression in plant synthetic biology (Tian *et al.*, 2022).

By combining synthetic promoters with metabolic engineering, researchers can enhance the yield and diversity of compounds like flavonoids, which have significant health benefits (Jiang *et al.*, 2023). While the design of synthetic promoters holds great promise for boosting TF activity and pathway flux in medicinal plants, challenges remain. The complexity of plant genomes and the variability in promoter activity across different species and tissues can complicate the design process. Additionally, the integration of synthetic promoters with existing metabolic pathways requires a deep understanding of plant biochemistry and genetics. Despite these challenges, advancements in computational modeling and synthetic biology continue to drive progress in this field, offering new opportunities for the production of valuable plant-based compounds (Table 5.6).

5.7 Transcriptional Switches in *C. roseus* Alkaloid Production

The production of alkaloids in *C. roseus*, a medicinal plant known for its therapeutic TIAs, is intricately regulated by transcriptional switches. These transcriptional mechanisms are crucial for the plant's ability to produce compounds like vinblastine and vincristine, which are used in cancer treatment. The regulation of these pathways involves a complex network of TFs that respond to various environmental and developmental cues. This section explores the transcriptional regulation of alkaloid production in *C. roseus*, focusing on the roles of specific TFs and the influence of external stimuli.

5.7.1 TFs in alkaloid biosynthesis

The herbivore-induced vindoline-gene expression (HIVE) TFs plays a pivotal role in coordinating the production of TIAs in response to herbivory and cold stress. HIVE allows *C. roseus* to allocate resources for chemical defense when herbivory pressure is high and switch to cold tolerance under cooler temperatures (Liu *et al.*, 2025). The bHLH transcription factor CrMYC2 is a major activator of jasmonate-responsive expression of ORCA (octadecanoid-responsive *Catharanthus* AP2-domain) genes, which regulate alkaloid biosynthesis. CrMYC2 binds to jasmonate-responsive elements in the ORCA3 promoter, facilitating the expression of alkaloid biosynthesis genes (Yamada and Sato, 2021). BIS2, another bHLH TF, is essential for MIA production. It transactivates promoters of iridoid biosynthesis genes and is part of an amplification loop with BIS1, crucial for MIA production (Van Moerkercke *et al.*, 2016).

Phytohormone influence in MeJA and ethylene significantly affect TIA biosynthesis. MeJA induces the expression of several pathway genes, while ethylene's role varies depending on the plant variety and developmental stage (Fraser *et al.*, 2020). The vindoline pathway, a part of the TIA biosynthetic process, is highly activated by light and is tissue-specific, mostly expressed in young leaves. This pathway is also influenced by stress conditions such as water and light stress, which affect the transcriptional regulation of biosynthetic genes (Cole, 2022).

MeJA is a well-known elicitor of TIA biosynthesis. It induces the expression of several TFs, including CrMYC2, which in turn activate the expression of biosynthetic genes. This regulation is crucial for the plant's defense response and secondary metabolite production (Yamada and Sato, 2021). The vindoline pathway, a part of the TIA biosynthesis, is highly activated by light and is predominantly expressed in young leaves. This tissue-specific expression is regulated by TFs such as GOLDEN2-LIKE (GLK), which are involved in photomorphogenesis (Cole, 2022). Identification of TFs often involves co-expression analysis and the study of CRMs in gene promoters. These methods help in understanding the spatiotemporal expression patterns of TFs and their target genes (Singh *et al.*, 2022).

Table 5.6. Designing promoters to boost TF activity and pathway flux. Author's own table.

Strategy/ component	Core concept & methodology	Application/ outcome	Key tools & enablers	Citations
Identification of regulatory elements	Identifying CRMs in gene promoters to understand which TFs bind and regulate the pathway.	Provides the foundational "parts list" for building synthetic promoters tailored to specific pathways (e.g. in *C. roseus*).	Bioinformatics, genomic sequencing, footprinting assays.	(Singh *et al.*, 2022)
TF discovery & validation	Using co-expression analysis and homolog isolation to identify candidate TFs that control a metabolic pathway.	Discovers key TFs (e.g. MYB, bHLH) that can be targeted for overexpression or whose binding sites can be engineered into promoters.	RNA-seq, phylogenetic analysis, functional genomics.	(Le *et al.*, 2015; Singh *et al.*, 2022)
Core promoter engineering	Optimizing core elements (TATA-box, GC content, TF binding site number/affinity) to maximize transcription initiation frequency.	Creates a strong foundational structure for any synthetic promoter, ensuring high transcriptional activity.	Computational models, ML to predict strength.	(Jores *et al.*, 2021)
MinSyns	Designing small, synthetic promoters using ML-optimized *cis*-regulatory elements for tunable, strong expression.	Enables precise control over gene expression with minimal genetic footprint, enhancing metabolite production.	ML, synthetic biology.	(Tagliani, 2020)
Inducible & responsive systems	Designing promoters that respond to specific signals (e.g. hormones like JA/SA, stress, chemicals).	Allows external, temporal control over pathway activation. Enables plants to resist pathogens and can be used to induce metabolite synthesis on demand.	Synthetic biology, characterization of hormone-responsive elements.	(Li *et al.*, 2023b)
Quantitative characterization	Using reporter systems (e.g. dual-luciferase) to precisely measure promoter strength and functionality *in planta*.	Provides essential data for selecting and optimizing the best-performing promoters before full pathway integration.	Transient expression assays, protoplast systems.	(Tian *et al.*, 2022)

Continued

Table 5.6. Continued

Strategy/ component	Core concept & methodology	Application/ outcome	Key tools & enablers	Citations
Integration with metabolic engineering	Combining engineered promoters with metabolic engineering to control flux through multi-step biosynthetic pathways.	Enhances the yield and diversity of valuable compounds (e.g. flavonoids, alkaloids) by coordinately overexpressing multiple pathway genes.	Synthetic biology, genetic transformation.	(Biswas *et al.*, 2022; Jiang *et al.*, 2023)
Overarching goal	To create a toolkit of strong, specific, and tunable promoters that recruit and enhance the activity of endogenous or synthetic TFs.	Amplifies pathway flux by ensuring key biosynthetic genes are transcribed at high levels at the right time and place.	Integration of all the above strategies.	(Tong *et al.*, 2024)

bHLH TFs, such as BIS1 and BIS2, play a crucial role in regulating the iridoid pathway, a branch of the TIA biosynthesis pathway. BIS2, in particular, is essential for the jasmonate-induced expression of iridoid pathway genes, and its expression is part of an amplification loop with BIS1 (Van Moerkercke *et al.*, 2016). CrMYC2, another bHLH TF, is a major activator of jasmonate-responsive expression of ORCA genes, which regulate alkaloid biosynthesis. CrMYC2 binds to jasmonate-responsive elements in the ORCA3 promoter, facilitating the expression of downstream biosynthetic genes (Yamada and Sato, 2021). CrMYC2 interacts with bZIP factors CrGBF1 and CrGBF2, which act as repressors. These interactions modulate TIA production by forming complexes that prevent CrMYC2 from binding to target promoters, thus fine-tuning the expression of TIA pathway genes (Sui *et al.*, 2018). The TIA biosynthetic pathway involves over 30 enzymes, and the regulation of these enzymes is not fully understood. The downstream portion of the pathway, particularly the vindoline pathway, requires further investigation to identify additional regulatory TFs (Cole, 2022).

While significant progress has been made in understanding the transcriptional regulation of TIA biosynthesis, the transport and trafficking of TIAs and their intermediates remain less explored. Understanding these processes is essential for optimizing TIA production in *C. roseus* (Liu *et al.*, 2016). While the transcriptional regulation of alkaloid production in *C. roseus* is well studied, challenges remain in fully understanding the subcellular and inter-cellular trafficking of TIA end products and intermediates. The transport processes are crucial for the biosynthesis, storage, and function of these compounds across different plant tissues. Further research is needed to elucidate the regulatory networks controlling tissue-specific biosynthesis and transport of TIAs, which could lead to enhanced production of these valuable compounds (Liu *et al.*, 2016). The production of TIAs in *C. roseus* is also influenced by the plant's genetic background and environmental conditions. Different varieties of *C. roseus* may exhibit distinct patterns of TIA biosynthesis in response to phytohormones like MeJA and ethylene, highlighting the importance of variety selection in research and cultivation (Fraser *et al.*, 2020). This complexity underscores the need for comprehensive "omics" resources and a deeper understanding of the regulatory networks involved in TIA production (Table 5.7).

5.7.2 How TFs regulate vincristine/vinblastine biosynthesis under stress

TFs play a crucial role in regulating the biosynthesis of vincristine and vinblastine, two important anti-cancer alkaloids, in medicinal plants like *C. roseus*. These TFs modulate the expression of genes involved in the complex TIA biosynthetic pathway, which is sensitive to environmental stressors. Understanding how TFs regulate this pathway under stress conditions can help enhance the production of these valuable compounds. The following sections detail the mechanisms by which TFs influence vincristine and vinblastine biosynthesis under stress.

TFs in alkaloid biosynthesis are key regulators of gene expression in plants, controlling various processes including secondary metabolite biosynthesis. They modulate the expression of genes in the alkaloid biosynthesis pathways, which are crucial for the production of vincristine and vinblastine (Prakash *et al.*, 2023). In *C. roseus*, the TIA pathway involves over 30 enzymes, and TFs can activate multiple enzymes within this pathway, potentially increasing the yield of these alkaloids (Cole, 2022). The biosynthesis of MIAs in *C. roseus* is partitioned into specialized cell types. For instance, the idioblast cell type is crucial for the final steps of the MIA pathway (Li *et al.*, 2025). A novel idioblast-specific TF, CrIDM1, has been identified, which activates the expression of late-stage MIA biosynthetic genes, highlighting the importance of cell-type-specific regulation in alkaloid biosynthesis (Li *et al.*, 2025).

Environmental factors such as light and nutrient availability can influence the expression of TFs and, consequently, the biosynthesis of alkaloids. For example, the vindoline pathway, a part of the TIA pathway, is highly activated by light, and TFs like GOLDEN2-LIKE (GLKs) are involved in its regulation (Cole, 2022). Potassium and ascorbic acid treatments have been shown to increase the production of vinblastine and vincristine by enhancing the expression of specific genes in the biosynthetic pathway, indicating that TFs may mediate these stress responses (Sahi *et al.*, 2022). Advances in genomics and transcriptomics have facilitated the identification of TFs involved in the TIA pathway. Techniques such as co-expression analysis and *cis*-regulatory motif identification are used to discover and characterize these TFs (Singh *et al.*, 2022). Genetic engineering approaches, including the overexpression of specific TFs, are being explored to enhance the production of vincristine and vinblastine in *C. roseus* and other host plants (Grzech *et al.*, 2023). While TFs are pivotal in regulating the biosynthesis of vincristine and vinblastine, the complexity of the TIA pathway and its regulation under stress conditions presents challenges. The interplay between different TFs, environmental factors, and cellular contexts needs further exploration to fully harness the potential of TFs in enhancing alkaloid production. Additionally, the development of synthetic biology approaches could provide alternative strategies for producing these valuable compounds in non-native hosts, potentially overcoming the limitations of natural plant sources.

5.8 Challenges and Future Directions

The field of transcriptional regulation for high-value metabolites is confronted by a series of multifaceted challenges, beginning with the primary hurdle of network complexity. TF interactions are not isolated events but are embedded within intricate webs involving epigenetic modifiers, environmental signals, and post-transcriptional mechanisms, creating highly context-dependent regulatory networks that vary significantly across species and tissues. For instance, while MYB TFs are known to regulate terpenoid pathways in *Taxus*, they exhibit divergent functions in other plants, complicating the development of universal applications. Adding to this complexity is the issue of epigenetic instability; although

Table 5.7. Transcriptional switches in *C. roseus* alkaloid biosynthesis. Author's own table.

Transcription factor (TF)	TF family	Key role & mechanism	Regulatory input/stress	Impact on vincristine/ vinblastine pathway	Citations
HIVE	Not specified	Master coordinator; switches resource allocation between chemical defense (TIAs) and cold tolerance.	Herbivory and cold stress	Induces TIA biosynthesis in response to herbivory.	(Liu *et al.*, 2025)
CrMYC2	bHLH	Central JA-signaling integrator. Binds jasmonate-responsive elements in the ORCA3 promoter.	Jasmonates (e.g. MeJA)	Activates the expression of ORCA TFs, which in turn upregulate multiple TIA biosynthetic genes.	(Yamada and Sato, 2021)
BIS1/BIS2	bHLH	Essential regulators of the iridoid (early) branch of the pathway. BIS2 transactivates iridoid gene promoters.	Jasmonates	Forms an amplification loop crucial for the flux of precursors into the MIA pathway.	(Van Moerkercke *et al.*, 2016)
ORCA3	AP2/ERF	Key downstream TF activated by CrMYC2. Binds to promoters of strictosidine synthase (*STR*) and other biosynthetic genes.	Jasmonates (via CrMYC2)	Directly activates core biosynthetic genes, increasing flux through the central TIA pathway.	(Yamada and Sato, 2021)
CrIDM1	Novel	Idioblast-specific TF regulating the late-stage steps of the MIA pathway.	Cell-type specific development	Activates the final steps in the biosynthesis, directly impacting the production of complex alkaloids like vindoline, a precursor to vinblastine/ vincristine.	(Li *et al.*, 2025)
GLK (GOLDEN2-LIKE)	GARP	Regulates the vindoline pathway in response to light.	Light stress/ photomorphogenesis	Activates light-induced expression of vindoline biosynthesis genes, crucial for assembling the final dimers.	(Cole, 2022)

Continued

Table 5.7. Continued

Transcription factor (TF)	TF family	Key role & mechanism	Regulatory input/stress	Impact on vincristine/ vinblastine pathway	Citations
CrGBF1/CrGBF2	bZIP	Act as repressors by interacting with and inhibiting CrMYC2.	Part of fine-tuning mechanism	Modulates the jasmonate response, preventing over-activation and providing a balance in the regulation of the TIA pathway.	(Sui *et al.*, 2018)
Key environmental & hormonal cues					
MeJA	–	Master elicitor hormone that activates the CrMYC2-ORCA regulatory module.	Simulated herbivory, defense signaling	Strongly induces the entire TIA pathway, leading to increased alkaloid accumulation.	(Fraser *et al.*, 2020; Yamada and Sato, 2021)
Light	–	A critical environmental signal that activates specific branches of the pathway.	Light exposure/photoperiod	Specifically activates the vindoline pathway via TFs like GLK, which is essential for the dimerization to form vinblastine/vincristine.	(Cole, 2022)
Nutrient stress (e.g. K+, ascorbic acid)	–	Treatments can enhance gene expression and alkaloid yield, likely mediated through unknown or specific TFs.	Nutrient availability/oxidative stress	Increases the production of vinblastine and vincristine by upregulating pathway genes.	(Sahi *et al.*, 2022)

stress-induced DNA methylation can transiently enhance metabolite production, as seen with terpenes under UV stress, its heritability is often transient and stress-dependent, which limits its utility in stable breeding programs. Technical limitations further impede progress, particularly in the realm of single-cell omics. While scRNA-seq has successfully mapped cell-type-specific alkaloid biosynthesis in *C. roseus*, the detection of metabolites at cellular resolution remains constrained by plant tissue complexity and low metabolite abundance. Similarly, the integration of multi-omics data—spanning transcriptomics, proteomics, and metabolomics—is computationally intensive and hampered by a lack of standardized pipelines. Finally, significant metabolic engineering barriers persist, including the inefficient reconstitution of heterologous pathways, exemplified by the production of only precursor-level yields in attempts to reconstruct paclitaxel biosynthesis in *N. benthamiana*, as well as unintended fitness costs from TF overexpression. Although promoter engineering holds promise, it struggles with species-specific design complexities and biosafety concerns associated with transgenic plants, presenting a substantial obstacle to practical application.

Looking ahead, future directions must prioritize a systems-level integration of these complex regulatory networks. ML offers a powerful tool to model intricate TF–promoter interactions, enabling the rational design of synthetic promoters—such as hormone-responsive MinSyns—for precise and tunable gene expression. Coupling this computational approach with CRISPR-based epigenome editing could stabilize beneficial, stress-induced methylation marks, transforming transient epigenetic changes into heritable trait enhancements for crop improvement. Advanced omics technologies should also focus on spatial transcriptomics and metabolomics to resolve critical gaps in our understanding of subcellular metabolite trafficking, particularly in pathways like alkaloid biosynthesis where the transport mechanisms for valuable compounds such as vincristine remain poorly defined. Concurrently, biotechnological innovation will be key; for instance, endophyte-mediated paclitaxel production requires rigorous validation to address current biosynthetic inconsistencies, while cell culture systems could be significantly optimized through TF-targeted elicitors like methyl jasmonate. For sustainable and scalable production, controlled cultivation environments must be engineered to strategically balance stress exposure—maximizing metabolite yields without compromising plant health, as demonstrated in salinity-stressed *A. vera*.

Lastly, translational research must bridge lab and field applications. Collaborative frameworks between geneticists, ecologists, and industry can accelerate CRISPR-engineered medicinal crops, epigenetic biomarkers for stress resilience, and AI-driven metabolic flux prediction. Addressing ethical and environmental risks of transgenic plants will be crucial for public acceptance. By converging these strategies, the field can overcome existing bottlenecks and unlock scalable production of plant-derived therapeutics (Table 5.8).

5.9 Conclusion

Transcriptional regulation stands as the cornerstone of high-value metabolite biosynthesis in medicinal plants, intricately balancing genetic, epigenetic, and environmental inputs. This chapter has detailed how TF networks—spearheaded by MYB, bHLH, and AP2/ERF families—orchestrate alkaloid, terpenoid, and phenolic pathways, enabling plants to fine-tune defense and therapeutic compounds. The MYB TFs, for instance, emerge as master regulators, modulating terpene synthase genes in *Taxus* and interacting with jasmonate signaling to amplify metabolite yields under stress. Epigenetic mechanisms further potentiate this adaptability, with DNA methylation and histone modifications serving as molecular conduits that convert environmental cues (e.g. drought, UV) into heritable metabolic traits, albeit with variability in stability.

Technological strides in single-cell omics have revolutionized our spatial understanding of regulation, unmasking cell-type-specific biosynthetic hotspots like GTs in *C. roseus* and mesophyll-specific saponin production in *Gynostemma*. These insights, coupled with promoter engineering, offer unprecedented tools for metabolic engineering. Synthetic promoters and TF-targeted designs can amplify

Table 5.8. Challenges and future directions in transcriptional regulation research. Author's own table.

Domain	Key challenges	Proposed future directions & solutions	Citations
Network complexity	Intricate, context-dependent TF networks involving epigenetic modifiers and environmental signals. Functions of TFs (e.g. MYB) are not universal across species.	Systems-level integration & ML: use ML to model TF–promoter interactions and design rational, tunable synthetic systems (e.g. MinSyn promoters).	(Implied from overall context)
Epigenetic instability	Transient heritability of stress-induced epigenetic marks (e.g. DNA methylation for terpene enhancement). Limits utility for stable crop improvement.	CRISPR-based epigenome editing: use tools like dCas9 to write stable, heritable epigenetic marks that lock in beneficial metabolic traits.	(Implied from text on epigenetic instability)
Technical & omics limitations	Single-cell metabolomics is constrained by plant tissue complexity and low metabolite abundance. Multi-omics integration is computationally intensive and lacks standardized pipelines.	Spatial multi-omics: develop and apply spatial transcriptomics and metabolomics to map subcellular metabolite trafficking and resolve pathway localization (e.g. for alkaloid transport).	(Liu *et al.*, 2016; Li *et al.*, 2025)
Metabolic engineering barriers	Inefficient heterologous pathway reconstitution (e.g. low paclitaxel precursor yields in *N. benthamiana*). Fitness costs from TF overexpression. Species-specific promoter design complexity and biosafety concerns.	Advanced chassis & optimization: engineer microbial or plant chassis with optimized cofactors. Use inducible systems and fine-tuning promoters to avoid fitness costs. Develop non-transgenic editing approaches.	(Li *et al.*, 2023b; McClune *et al.*, 2024)
Sustainable production & scaling	Balancing stress to maximize metabolite yield (e.g. aloin in *A. vera*) without harming plant health and overall yield.	Precision controlled cultivation: engineer growth environments to apply precise, mild stress elicitors (e.g. UV, nutrient) to boost metabolites without significant growth penalties.	(Jampílek and Kráľová, 2023; Sharma, 2024)

Continued

Table 5.8. Continued

Domain	Key challenges	Proposed future directions & solutions	Citations
Translational research gap	Disconnect between lab discoveries and field applications. Ethical and environmental concerns over transgenic plants hinder public acceptance and deployment.	Collaborative frameworks: foster partnerships between geneticists, ecologists, and industry. Develop epigenetic biomarkers for breeding. Use AI for flux prediction. Proactively address biosafety and public engagement.	(Implied from text on translational research)
Validation of alternative sources	Inconsistencies in biosynthetic capability of endophytic fungi reported as alternative sources for compounds like paclitaxel.	Rigorous validation: conduct thorough genetic and biochemical validation to confirm the endogenous ability of endophytes to produce complex plant metabolites.	(El-Sayed *et al.*, 2020)

pathway flux, as demonstrated in vincristine biosynthesis, where CrMYC2 and ORCA gene switches optimize alkaloid production under elicitation.

Nevertheless, the journey from mechanistic insight to industrial application demands resolving persistent challenges. The complexity of TF–epigenome cross-talk, limitations in single-cell metabolite detection, and ecological risks of transgenic approaches necessitate interdisciplinary collaboration. Future success hinges on leveraging ML for predictive network modeling, CRISPR for stable epigenetic editing, and multi-omics integration for holistic pathway mapping. As we refine these strategies, the potential to sustainably harness medicinal plants—for cancer therapeutics like paclitaxel or vinblastine—will expand, ultimately bridging traditional knowledge with cutting-edge synthetic biology. In this era of climate change and pharmaceutical demand, mastering transcriptional regulation is not merely academic but a pragmatic imperative for global health and biodiversity conservation.

Acknowledgments

The authors express their gratitude for the financial support received through the Distinguished High-Level Talents Research Grant from the Guizhou Science and Technology Corporation Platform Talents Fund (Grant nos.: [2017]5733-001 and CK-1130-002), high-level specialists project fund of Zunyi Medical University (1092/2000006)F-1092, the National Natural Science Foundation of China (82373981), and the support provided by Zunyi Medical University, China. Special appreciation is extended to all laboratory colleagues and research staff members for their valuable insights, constructive guidance, and assistance throughout this study.

References

Afrin, S., Huang, J.-J. and Luo, Z.-Y. (2015) JA-mediated transcriptional regulation of secondary metabolism in medicinal plants. *Science Bulletin* 60(12), 1062–1072. DOI: 10.1007/s11434-015-0813-0.

Akhter, Z., Bi, Z., Ali, K., Sun, C., Fiaz, S. *et al.* (2021) In response to abiotic stress, DNA methylation confers epigenetic changes in plants. *Plants (Basel, Switzerland)* 10(6), 1096. DOI: 10.3390/plants10061096.

Alamgir, A.N.M. (2017) Cultivation of herbal drugs, biotechnology, and *in vitro* production of secondary metabolites, high-value medicinal plants, herbal wealth, and herbal trade. In: *Therapeutic Use of Medicinal Plants and Their Extracts: Volume 1: Pharmacognosy*. Springer International Publishing, Cham, Switzerland, pp. 379–452.

Alam, M.M., Naeem, M., Khan, M.M.A. and Uddin, M. (2017) Vincristine and vinblastine anticancer *Catharanthus* alkaloids: Pharmacological applications and strategies for yield improvement. In: Naeem, M., Aftab, T., Masroor, M. and Khan, A. (eds) *Catharanthus Roseus*. Springer International Publishing, Cham, Switzerland, pp. 277–307.

Bandurska, H. (2022) Drought stress responses: Coping strategy and resistance. *Plants (Basel, Switzerland)* 11(7), 922. DOI: 10.3390/plants11070922.

Bharti, P.K., Santhosh, B., Singh, S.K. and Kumari, S. (2023) Genetic manipulation in medicinal plants for enhancement of secondary metabolites. In: Singh, R. and Kumar, N. (eds) *Genetic Manipulation of Secondary Metabolites in Medicinal Plant*. Springer Nature, Singapore, pp. 129–145.

Bhushan, S., Sharma, D., Rakshant,Kaul, S., Dhar, M.K. *et al.* (2023) Recent strategies to engineer alkaloid biosynthesis in medicinal plants. In: Husen, A. and Iqbal, M. (eds) *Medicinal Plants*. Springer Nature, Singapore, pp. 391–416.

Biswas, P., Ghorai, M., Pandey, D.K., Singh, J., Al-Tawaha, A.R. *et al.* (2022) Regulation of expression of transcription factors for enhanced secondary metabolites production under challenging conditions. In: Aftab, T. (ed.) *Environmental Challenges and Medicinal Plants: Sustainable Production Solutions under Adverse Conditions*. Springer International Publishing, Cham, Switzerland, pp. 249–280.

Burlat, V., Papon, N. and Courdavault, V. (2023) Medicinal plants enter the single-cell multi-omics era. *Trends in Plant Science* 28(11), 1205–1207. DOI: 10.1016/j.tplants.2023.08.005.

Cao, X., Xu, L., Li, L., Wan, W. and Jiang, J. (2022) TcMYB29a, an ABA-responsive R2R3-MYB transcriptional factor, upregulates Taxol biosynthesis in *Taxus chinensis*. *Frontiers in Plant Science* 13(13), 804593. DOI: 10.3389/fpls.2022.804593.

Cao, Y., Li, K., Li, Y., Zhao, X. and Wang, L. (2020) MYB transcription factors as regulators of secondary metabolism in plants. *Biology* 9(3), 61. DOI: 10.3390/biology9030061.

Cazier, A.P. and Blazeck, J. (2021) Advances in promoter engineering: Novel applications and predefined transcriptional control. *Biotechnology Journal* 16(10), e2100239. DOI: 10.1002/biot.202100239.

Cole, L.F. (2022) Transcription factors regulating vindoline biosynthesis in the medicinal plant, *Catharanthus roseus*. Doctoral dissertation, Northeastern University, Boston, Massachusetts.

Cusido, R.M., Onrubia, M., Sabater-Jara, A.B., Moyano, E., Bonfill, M. *et al.* (2014) A rational approach to improving the biotechnological production of taxanes in plant cell cultures of *Taxus* spp. *Biotechnology Advances* 32(6), 1157–1167. DOI: 10.1016/j.biotechadv.2014.03.002.

El-Sayed, A.S.A., El-Sayed, M.T., Rady, A.M., Zein, N., Enan, G. *et al.* (2020) Exploiting the biosynthetic potency of Taxol from fungal endophytes of conifers plants; genome mining and metabolic manipulation. *Molecules* 25(13), 3000. DOI: 10.3390/molecules25133000.

Fraser, V.N., Philmus, B. and Megraw, M. (2020) Metabolomics analysis reveals both plant variety and choice of hormone treatment modulate vinca alkaloid production in *Catharanthus roseus*. *Plant Direct* 4(9), e00267. DOI: 10.1002/pld3.267.

Galindo-González, L., Sarmiento, F. and Quimbaya, M.A. (2018) Shaping plant adaptability, genome structure and gene expression through transposable element epigenetic control: Focus on methylation. *Agronomy* 8(9), 180. DOI: 10.3390/agronomy8090180.

Grzech, D., Hong, B., Caputi, L., Sonawane, P.D. and O'Connor, S.E. (2023) Engineering the biosynthesis of late-stage vinblastine precursors precondylocarpine acetate, catharanthine, tabersonine in *Nicotiana benthamiana*. *ACS Synthetic Biology* 12(1), 27–34. DOI: 10.1021/acssynbio.2c00434.

Guan, Y., Jiang, L., Wang, Y., Liu, G., Wu, J. *et al.* (2024) CmMYC2 – CmMYBML1 module orchestrates the resistance to herbivory by synchronously regulating the trichome development and constitutive terpene biosynthesis in *Chrysanthemum*. *New Phytologist* 244(3), 914–933. DOI: 10.1111/nph.20081.

Gupta, A. and Nair, S. (2022) Heritable epigenomic modifications influence stress resilience and rapid adaptations in the brown planthopper (*Nilaparvata lugens*). *International Journal of Molecular Sciences* 23(15), 8728. DOI: 10.3390/ijms23158728.

Gupta, A., Gupta, K.K., Gupta, S.K. and Mishra, P.K. (2023) Transcriptional regulation in biosynthesis of phytochemicals in medicinal plants under stress conditions. In: Singh, D., Kumar Mishra, A. and Kumar Srivastava, A. (eds) *Stress-Responsive Factors and Molecular Farming in Medicinal Plants*. Springer Nature, Singapore, pp. 123–140.

Hao, D.-C. and Xiao, P.-G. (2018) Deep in shadows: Epigenetic and epigenomic regulations of medicinal plants. *Chinese Herbal Medicines* 10(3), 239–248. DOI: 10.1016/j.chmed.2018.02.003.

Huang, X., Yang, Q. and Gao, H. (2023) Research progress in the regulation of secondary metabolism in medicinal plants by MYB transcription factors. *Journal of Holistic Integrative Pharmacy* 4(4), 287–292. DOI: 10.1016/j.jhip.2024.01.006.

Hurgobin, B. and Lewsey, M.G. (2022) Applications of cell- and tissue-specific omics to improve plant productivity. *Emerging Topics in Life Sciences* 6(2), 163–173. DOI: 10.1042/ETLS20210286.

Jampílek, J. and Kráľová, K. (2023) Impact of abiotic stresses on production of secondary metabolites in medicinal and aromatic plants. In: Aftab, T. (ed.) *New Frontiers in Plant-Environment Interactions: Innovative Technologies and Developments*. Springer Nature Switzerland, Cham, Switzerland, pp. 169–252.

Jan, R., Asaf, S., Numan, M., Lubna and Kim, K.-M. (2021) Plant secondary metabolite biosynthesis and transcriptional regulation in response to biotic and abiotic stress conditions. *Agronomy* 11(5), 968. DOI: 10.3390/agronomy11050968.

Ji, C.-H., Je, H.-W., Kim, H. and Kang, H.-S. (2024) Promoter engineering of natural product biosynthetic gene clusters in actinomycetes: Concepts and applications. *Natural Product Reports* 41(4), 672–699. DOI: 10.1039/D3NP00049D.

Jiang, L., Gao, Y., Han, L., Zhang, W. and Fan, P. (2023) Designing plant flavonoids: Harnessing transcriptional regulation and enzyme variation to enhance yield and diversity. *Frontiers in Plant Science* 14, 1220062. DOI: 10.3389/fpls.2023.1220062.

Jores, T., Tonnies, J., Wrightsman, T., Buckler, E.S., Cuperus, J.T. *et al.* (2021) Synthetic promoter designs enabled by a comprehensive analysis of plant core promoters. *Nature Plants* 7(6), 842–855. DOI: 10.1101/2021.01.07.425784.

Kabir, N., Wahid, S., Rehman, S.U. and Qanmber, G. (2024) The intricate world of trichome development: From signaling pathways to transcriptional regulation. *Environmental and Experimental Botany* 217, 105549. DOI: 10.1016/j.envexpbot.2023.105549.

Kanso, M., Hijazi, M., Aboul Ela, M. and El-Lakany, A. (2022) MEDICINAL PLANTS' STRESS FACTORS: EFFECTS ON METABOLITES AND NOVEL PERSPECTIVES FOR TOLERANCE. *BAU Journal - Health and Well-Being* 5(1). DOI: 10.54729/XWWG9154.

Khan, A., Nasim, N., Pudhuvai, B., Koul, B., Upadhyay, S.K. *et al.* (2023) Plant synthetic promoters: Advancement and prospective. *Agriculture* 13(2), 298. DOI: 10.3390/agriculture13020298.

Khan, A.R., Shah, S.M. and Irshad, M. (2015) Immediate and transgenerational regulation of plant stress response through DNA methylation. *Journal of Agricultural Science* 7(4), 144–151. DOI: 10.5539/jas.v7n4p144.

Kumar, A., Sharma, P., Srivastava, R. and Verma, P.C. (2023) Demystifying the role of transcription factors in plant terpenoid biosynthesis. In: Srivastava, V., Mishra, S., Mehrotra, S. and Kumar Upadhyay, S. (eds) *Plant Transcription Factors*. Academic Press, Cambridge, Massachusetts, pp. 233–249.

Le, T.D., Liu, L., Zhang, J., Liu, B. and Li, J. (2015) From miRNA regulation to miRNA-TF co-regulation: Computational approaches and challenges. *Briefings in Bioinformatics* 16(3), 475–496. DOI: 10.1093/bib/bbu023.

Li, B.-J., Wang, H., Gong, T., Chen, J.-J., Chen, T.-J. *et al.* (2017) Improving 10-deacetylbaccatin III-10-β-o-acetyltransferase catalytic fitness for Taxol production. *Nature Communications* 8(1), 15544. DOI: 10.1038/ncomms15544.

Li, C., Wood, J.C., Vu, A.H., Hamilton, J.P., Rodriguez Lopez, C.E. *et al.* (2023a) Single-cell multi-omics in the medicinal plant *Catharanthus roseus*. *Nature Chemical Biology* 19(8), 1031–1041. DOI: 10.1038/s41589-023-01327-0.

Li, C., Colinas, M., Wood, J.C., Vaillancourt, B., Hamilton, J.P. *et al.* (2025) Cell-type-aware regulatory landscapes governing monoterpene indole alkaloid biosynthesis in the medicinal plant *Catharanthus roseus*. *The New Phytologist* 245(1), 347–362. DOI: 10.1111/nph.20208.

Li, R., Du, K., Zhang, C., Shen, X., Yun, L. *et al.* (2024a) Single-cell transcriptome profiling reveals the spatiotemporal distribution of triterpenoid saponin biosynthesis and transposable element activity in *Gynostemma pentaphyllum* shoot apexes and leaves. *Frontiers in Plant Science* 15, 1394587. DOI: 10.3389/fpls.2024.1394587.

Li, S., Khoso, M.A., Xu, H., Zhang, C., Liu, Z. *et al.* (2024b) WRKY transcription factors (TFs) as key regulators of plant resilience to environmental stresses: Current perspective. *Agronomy* 14(10), 2421. DOI: 10.3390/agronomy14102421.

Li, X., Niu, G., Fan, Y., Liu, W., Wu, Q. *et al.* (2023b) Synthetic dual hormone-responsive promoters enable engineering of plants with broad-spectrum resistance. *Plant Communications* 4(4), 100596. DOI: 10.1016/j.xplc.2023.100596.

Liu, J., Cai, J., Wang, R. and Yang, S. (2016) Transcriptional regulation and transport of terpenoid indole alkaloid in *Catharanthus roseus*: Exploration of new research directions. *International Journal of Molecular Sciences* 18(1), 53. DOI: 10.3390/ijms18010053.

Liu, Y., Shi, J., Patra, B., Singh, S.K., Wu, X. *et al.* (2025) Transcriptional reprogramming deploys a compartmentalized 'timebomb' in *Catharanthus roseus* to fend off chewing herbivores. *Plant, Cell & Environment* 48(5), 3236–3256. DOI: 10.1111/pce.15324.

Lv, Z., Jiang, S., Kong, S., Zhang, X., Yue, J. *et al.* (2024) Advances in single-cell transcriptome sequencing and spatial transcriptome sequencing in plants. *Plants (Basel, Switzerland)* 13(12), 1679. DOI: 10.3390/plants13121679.

Maeda, T., Sugano, S.S., Shirakawa, M., Sagara, M., Ito, T. *et al.* (2023) Single-cell RNA sequencing of Arabidopsis leaf tissues identifies multiple specialized cell types: Idioblast myrosin cells and potential glucosinolate-producing cells. *Plant & Cell Physiology* 64(2), 234–247. DOI: 10.1093/pcp/pcac167.

McClune, C.J., Liu, J.C.-T., Wick, C., Peña, R., Lange, B.M. *et al.* (2024) Multiplexed perturbation of yew reveals cryptic proteins that enable a total biosynthesis of *baccatin* III and *Taxol precursors*. *BioRxiv*. DOI: 10.1101/2024.11.06.622305.

McElroy, C. and Jennewein, S. (2018) Taxol® biosynthesis and production: From forests to fermenters. In: Schwab, W., Markus Lange, B. and Wüst, M. (eds) *Biotechnology of Natural Products*. Springer International Publishing, Cham, Switzerland, pp. 145–185.

Meraj, T.A., Fu, J., Raza, M.A., Zhu, C., Shen, Q. *et al.* (2020) Transcriptional factors regulate plant stress responses through mediating secondary metabolism. *Genes* 11(4), 346. DOI: 10.3390/genes11040346.

Pandey, B., Prakash, P., Chandra Verma, P. and Srivastava, R. (2019) Regulated gene expression by synthetic modulation of the promoter architecture in plants. In: Pandey, A., Negi, S. and Soccol, C.R. (eds) *Current Developments in Biotechnology and Bioengineering*. Elsevier, Amsterdam, pp. 235–255.

Pant, P., Pandey, S. and Dall'Acqua, S. (2021) The influence of environmental conditions on secondary metabolites in medicinal plants: A literature review. *Chemistry & Biodiversity* 18(11), e2100345. DOI: 10.1002/cbdv.202100345.

Prakash, P., Kumar, R. and Gupta, V. (2023) The regulatory aspects of plant transcription factors in alkaloids biosynthesis and pathway modulation. In: Srivastava, V., Mishra, S., Mehrotra, S. and Kumar Upadhyay, S. (eds) *Plant Transcription Factors*. Academic Press, Cambridge, Massachusetts, pp. 199–217.

Preite, V. (2016) Epigenetic inheritance in apomictic dandelions: Stress-induced and heritable modifications in DNA methylation and small RNA. Doctoral dissertation, Wageningen University, Netherlands.

Punetha, A., Kumar, D., Suryavanshi, P., Padalia, R. and Venkatesha, K.T. (2022) Environmental abiotic stress and secondary metabolites production in medicinal plants: A review. *Tarım Bilimleri Dergisi* 28(3), 351–362. DOI: 10.15832/ankutbd.999117.

Sahi, N., Mostajeran, A. and Ghanadian, M. (2022) Changing in the production of anticancer drugs (vinblastine and vincristine) in *Catharanthus roseus* (L.) G. Don by potassium and ascorbic acid treatments. *Plant, Soil and Environment* 68(1), 18–28. DOI: 10.17221/121/2021-PSE.

Sharma, M.K. (2024) Exploring the biochemical profiles of medicinal plants cultivated under stressful environmental conditions. *Current Agriculture Research Journal* 12(1), 81–103. DOI: 10.12944/CARJ.12.1.07.

Shi, M., Zhang, S., Zheng, Z., Maoz, I., Zhang, L. *et al.* (2024) Molecular regulation of the key specialized metabolism pathways in medicinal plants. *Journal of Integrative Plant Biology* 66(3), 510–531. DOI: 10.1111/jipb.13634.

Singh, C., Saini, R., Upadhyay, R. and Tiwari, K.N. (2023) Revealing the epigenetic mechanisms underlying the stress response in medicinal plants. In: Singh, D., Kumar Mishra, A. and Kumar Srivastava, A. (eds) *Stress-Responsive Factors and Molecular Farming in Medicinal Plants*. Springer Nature, Singapore, pp. 107–122.

Singh, S.K., Patra, B., Singleton, J.J., Liu, Y. and Paul, P. (2022) Identification and characterization of transcription factors regulating terpenoid indole alkaloid biosynthesis in *Catharanthus roseus*. In: Courdavault, V. and Besseau, S. (eds) *Catharanthus Roseus: Methods and Protocols*. Springer US, New York, pp. 203–221.

Sui, X., Singh, S.K., Patra, B., Schluttenhofer, C., Guo, W. *et al*. (2018) Cross-family transcription factor interaction between MYC2 and GBFs modulates terpenoid indole alkaloid biosynthesis. *Journal of Experimental Botany* 69(18), 4267–4281. DOI: 10.1093/jxb/ery229.

Sun, M.-S., Jia, Y., Chen, X.-Y., Chen, J.-S., Guo, Y. *et al*. (2024) Regulatory microRNAs and phasiRNAs of paclitaxel biosynthesis in *Taxus chinensis*. *Frontiers in Plant Science* 15, 1403060. DOI: 10.3389/fpls.2024.1403060.

Szymczyk, P. and Majewska, M. (2024) Plant synthetic promoters. *Applied Sciences* 14(11), 4877. DOI: 10.3390/app14114877.

Tagliani, A. (2020) Synthetic promoters went green: MinSyns bridge the gap between tunable expression and synthetic biology in plants. *Synthetic Biology (Oxford, England)* 5(1), ysaa027. DOI: 10.1093/synbio/ysaa027.

Talarico, E., Zambelli, A., Araniti, F., Greco, E., Chiappetta, A. *et al*. (2024) Unravelling the epigenetic code: DNA methylation in plants and its role in stress response. *Epigenomes* 8(3), 30. DOI: 10.3390/epigenomes8030030.

Talukdar, D. (2019) Bulk to individuality: Specifying plants' cellular functions through single-cell omics. In: Barh, D. and Azevedo, V. (eds) *Single-Cell Omics*. Elsevier, Amsterdam, pp. 277–314.

Thakur, K., Kumari, C., Zadokar, A., Sharma, P. and Sharma, R. (2023) Physiological and omics-based insights for underpinning the molecular regulation of secondary metabolite production in medicinal plants: UV stress resilience. *Plant Physiology and Biochemistry* 204, 108060. DOI: 10.1016/j.plaphy.2023.108060.

Tian, C., Zhang, Y., Li, J. and Wang, Y. (2022) Benchmarking intrinsic promoters and terminators for plant synthetic biology research. *BioDesign Research* 2022, 9834989. DOI: 10.34133/2022/9834989.

Tong, Y., Xue, J., Li, Q. and Zhang, L. (2024) A generalist regulator: MYB transcription factors regulate the biosynthesis of active compounds in medicinal plants. *Journal of Experimental Botany* 75(16), 4729–4744. DOI: 10.1093/jxb/erae225.

Tripathi, S., Jadaun, J.S., Chandra, M. and Sangwan, N.S. (2016) Medicinal plant transcriptomes: The new gateways for accelerated understanding of plant secondary metabolism. *Plant Genetic Resources* 14(4), 256–269. DOI: 10.1017/S1479262116000162.

Van Moerkercke, A., Steensma, P., Gariboldi, I., Espoz, J., Purnama, P.C. *et al*. (2016) The basic helix-loop-helix transcription factor BIS 2 is essential for monoterpenoid indole alkaloid production in the medicinal plant *Catharanthus roseus*. *The Plant Journal* 88(1), 3–12. DOI: 10.1111/tpj.13230.

Viggiano, L. and de Pinto, M.C. (2017) Dynamic DNA methylation patterns in stress response. In: Rajewsky, N., Jurga, S. and Barciszewski, J. (eds) *Plant Epigenetics*. Springer International Publishing, Cham, Switzerland, pp. 281–302.

Vu, A.H., Kang, M., Wurlitzer, J., Heinicke, S., Li, C. *et al*. (2024) Quantitative single-cell mass spectrometry provides a highly resolved analysis of natural product biosynthesis partitioning in plants. *Journal of the American Chemical Society* 146(34), 23891–23900. DOI: 10.1021/jacs.4c06336.

Wang, M., Zhang, S., Li, R. and Zhao, Q. (2024) Unraveling the specialized metabolic pathways in medicinal plant genomes: A review. *Frontiers in Plant Science* 15, 1459533. DOI: 10.3389/fpls.2024.1459533.

Wang, Q., Zhao, X., Jiang, Y., Jin, B. and Wang, L. (2023) Functions of representative terpenoids and their biosynthesis mechanisms in medicinal plants. *Biomolecules* 13(12), 1725. DOI: 10.3390/biom13121725.

Wang, T., Chen, Y., Zhuang, W., Zhang, F., Shu, X. *et al*. (2019) Transcriptome sequencing reveals regulatory mechanisms of Taxol synthesis in *Taxus wallichiana* var. *Mairei*. *International Journal of Genomics*20191596895. DOI: 10.1155/2019/1596895.

Wang, X., Hu, H., Wu, Z., Fan, H., Wang, G. *et al*. (2021) Tissue-specific transcriptome analyses reveal candidate genes for stilbene, flavonoid and anthraquinone biosynthesis in the medicinal plant *Polygonum cuspidatum*. *BMC Genomics* 22(1), 353. DOI: 10.1186/s12864-021-07658-3.

Yamada, Y. and Sato, F. (2021) Transcription factors in alkaloid engineering. *Biomolecules* 11(11), 1719. DOI: 10.3390/biom11111719.

Yang, L., Yang, C., Li, C., Zhao, Q., Liu, L. *et al.* (2016) Recent advances in biosynthesis of bioactive compounds in traditional Chinese medicinal plants. *Science Bulletin* 61(1), 3–17. DOI: 10.1007/s11434-015-0929-2.

Yembaturova, E.Y. and Cheryatova, Y.S. (2023) Transgenic medicinal plants as producers of bioactive substances. *Ecological Genetics* 21(3S), 41–42. DOI: 10.17816/ecogen567947.

Ying, C., Meng, Z., Wenli, Z., Yamin, W., Hua, Z. *et al.* (2023) miR5298b regulated Taxol biosynthesis by acting on TcNPR3, resulting in an alleviation of the strong inhibition of the TcNPR3-TcTGA6 complex in *Taxus chinensis*. *International Journal of Biological Macromolecules* 248, 125909. DOI: 10.1016/j.ijbiomac.2023.125909.

Yuan, S., Li, Q., Shen, H., Wang, W., Wang, T. *et al.* (2023) Advances in the regulatory mechanisms of multicellular trichome formation and its secondary metabolite synthesis in vegetable crops. *Vegetable Research* 3(1). DOI: 10.48130/VR-2023-0024.

Zang, Y., Xie, L., Su, J., Luo, Z., Jia, X. *et al.* (2023) Advances in DNA methylation and demethylation in medicinal plants: A review. *Molecular Biology Reports* 50(9), 7783–7796. DOI: 10.1007/s11033-023-08618-8.

Zhan, X., Zhang, H., Liang, X., Kailin, H., Lin, W. *et al.* (2023) Single-cell ATAC sequencing illuminates the cis-regulatory differentiation of Taxol biosynthesis between leaf *Mesophyll* and leaf epidermal cells in *Taxus mairei*. *Industrial Crops and Products* 205, 117411. DOI: 10.1016/j.indcrop.2023.117411.

Zhang, Y., Zhang, J., Miao, J., Sun, G., Bai, H. *et al.* (2025) Micromorphology and molecular insights into glandular trichomes in two different thymes: Glandular trichomes formation process and the function of the main regulator TqHD1. *Cell & Environment*. DOI: 10.22541/au.173676463.30881523/v1.

Zhao, Y., Liu, G., Yang, F., Liang, Y., Gao, Q. *et al.* (2023) Multilayered regulation of secondary metabolism in medicinal plants. *Molecular Horticulture* 3(1), 11. DOI: 10.1186/s43897-023-00059-y.

Zhou, P., Dang, J., Shi, Z., Shao, Y., Sang, M. *et al.* (2022) Identification and characterization of a novel gene involved in glandular trichome development in *Nepeta tenuifolia*. *Frontiers in Plant Science* 13, 936244. DOI: 10.3389/fpls.2022.936244.

Zhou, T., Luo, X., Yu, C., Zhang, C., Zhang, L. *et al.* (2019) Transcriptome analyses provide insights into the expression pattern and sequence similarity of several Taxol biosynthesis-related genes in three *Taxus* species. *BMC Plant Biology* 19(1), 33. DOI: 10.1186/s12870-019-1645-x.

6 Omics-driven Drug Discovery and Development

Surendra Sarsaiya, Archana Jain and Qihai Gong*
Key Laboratory of Basic Pharmacology and Joint International Research Laboratory of Ethnomedicine of Ministry of Education, Zunyi Medical University, Zunyi, Guizhou, China

6.1 Introduction

Omics-driven drug discovery leverages advanced technologies in genomics, proteomics, and metabolomics to enhance the identification and validation of drug targets (Chauhan *et al.*, 2024). The integration of these omics disciplines within high-throughput screening (HTS) platforms allows for the rapid testing of numerous compounds against specific biological targets, significantly accelerating the drug discovery process (Joshi and Makwana, 2024). Genomics-driven target identification utilizes genomic data to pinpoint relevant drug targets, while metabolomics-based compound profiling assesses the effects of compounds on cellular metabolism, identifying those with therapeutic potential (Valencia Quiroz, 2024). Additionally, proteomics-driven biomarker discovery aids in recognizing disease-specific biomarkers, further refining the therapeutic approach (Wang and Yang, 2024). This comprehensive omics technology not only streamlines the screening process but also fosters the development of more targeted and effective therapies, ultimately enhancing the efficacy of drug discovery pipelines (Chauhan *et al.*, 2024).

Omics-enabled HTS platforms play a pivotal role in the exploration of medicinal plants by integrating various omics disciplines, such as metabolomics and transcriptomics, to identify bioactive compounds and their biosynthetic pathways (Garse *et al.*, 2025). Metabolomics allows for the comprehensive analysis of metabolites, facilitating the discovery of novel therapeutic agents from plant extracts (Dzobo, 2022). Concurrently, transcriptomics provides insights into gene expression profiles, which can reveal the genes involved in the production of these bioactive compounds (Sobti, 2025). High-content screening further enhances this process by employing automated microscopy to quantify cellular responses to these compounds, thereby elucidating their biological effects (Booij *et al.*, 2019). Additionally, machine learning algorithms can analyze the vast data generated from these screenings, identifying patterns that predict bioactivity and improving the efficiency of the discovery process (Gupta *et al.*, 2021). Together, these technologies create a robust framework for accelerating drug discovery from natural product libraries derived from medicinal plants.

Network pharmacology plays a pivotal role in precision target identification within

*Corresponding author: gqh@zmu.edu.cn

DOI: 10.1079/9781836994206.0006

the context of medicinal plants. By integrating systems biology, pharmacology, and bioinformatics, this approach allows researchers to analyze complex interactions between phytochemicals, molecular targets, and diseases (Babar *et al.*, 2017; Noor *et al.*, 2022). Through advanced computational tools, network pharmacology facilitates the identification of specific molecular targets that are crucial for developing novel therapeutic compounds with enhanced efficacy and minimized side effects (Son *et al.*, 2024a). Understanding the mechanisms of action of phytochemicals, which are bioactive compounds responsible for the medicinal properties of plants, is essential for harnessing their full therapeutic potential (Dar *et al.*, 2023). Thus, network pharmacology serves as a bridge between traditional medicinal practices and contemporary scientific methodologies.

Multi-omics approaches are pivotal in mapping plant compounds to disease-specific molecular targets, particularly in medicinal plants. Plant metabolomics specifically allows for the exploration of bioactive compounds and their potential therapeutic effects, which can be aligned with disease-specific molecular targets (Buriani *et al.*, 2017; Lin *et al.*, 2024). Furthermore, network pharmacology enhances this process by elucidating the complex interactions between these compounds and their targets, promoting the development of more effective plant-based therapies (Aloufi *et al.*, 2024). Ultimately, the genomic analysis of medicinal plants aids in optimizing the production of these bioactive compounds, thereby improving their therapeutic potential (Aloufi *et al.*, 2024). This multifaceted approach underscores the importance of systems biology in advancing personalized medicine through medicinal plants.

Lead optimization of medicinal plants involves a multifaceted approach that integrates metabolomics, bioactive compound isolation, HTS, machine learning, and natural product chemistry (Zhang *et al.*, 2023). Bioactive compound isolation is critical for extracting potent compounds that can be optimized for drug development (Kamil Hussain *et al.*, 2019). HTS accelerates the identification of promising lead candidates from large compound libraries, streamlining the drug discovery process (Sadybekov and Katritch, 2023). Additionally, machine learning enhances the analysis of omics data, allowing for the prediction of biological activity and prioritization of lead candidates (Koromina *et al.*, 2019). Finally, a solid understanding of natural product chemistry is vital for modifying and optimizing the structures of these compounds to improve pharmacological activity and reduce toxicity (Wang *et al.*, 2023b). Together, these elements create a robust framework for advancing drug candidates derived from medicinal plants.

Refining bioactive scaffolds in medicinal plants can be effectively achieved through a combination of computational and experimental strategies utilizing omics insights. Network pharmacology serves as a foundational approach, integrating omics data to identify bioactive compounds and their targets, thereby shifting the paradigm from "one target, one drug" to a more holistic view of drug discovery (Mansoor *et al.*, 2024; Parmar *et al.*, 2025). Complementing this, metabolomics-based scaffold refining focuses on identifying key metabolites and their structural features, which is crucial for prioritizing bioactive compounds (Gonzalez-Covarrubias *et al.*, 2022). A systems biology approach further enriches our understanding of complex interactions within biological systems, identifying critical pathways and targets (Kohl *et al.*, 2010). Finally, phytochemical profiling allows for the quantification of phytochemicals, providing essential data for identifying novel bioactive compounds with therapeutic potential. Together, these strategies create a robust framework for scaffold refinement in medicinal plant research.

Turmeric's curcuminoids, particularly curcumin, desmethoxycurcumin, and bisdemethoxycurcumin, are gaining attention in the development of anti-inflammatory drugs derived from medicinal plants (Akaberi *et al.*, 2021). Curcumin is recognized for its potent anti-inflammatory properties, which are mediated through the downregulation of inflammatory transcription factors and cytokines, making it a promising lead compound for various diseases, including arthritis and cancer (Peng *et al.*, 2021). Additionally, desmethoxycurcumin and bisdemethoxycurcumin have demonstrated similar anti-inflammatory effects by inhibiting pro-inflammatory enzymes and reducing oxidative stress, further supporting their potential in drug development (Weinberg Sibony *et al.*, 2024). The use of turmeric extracts, which concentrate these bioactive compounds, enhances their efficacy and bioavailability, positioning them as valuable candidates in medicinal formulations aimed at treating inflammatory conditions (Razavi *et al.*, 2021). Overall, the multifaceted benefits of turmeric's curcuminoids underscore their significance in advancing therapeutic strategies against inflammation-related diseases.

The omics-driven validation of curcuminoid efficacy in medicinal plants involves a multifaceted approach that integrates metabolomics and transcriptomics analyses (Younas *et al.*, 2023). These methods help identify biomarkers of efficacy and elucidate the mechanisms of action of curcuminoids, thereby enhancing our understanding of their therapeutic potential (Roy *et al.*, 2024). Additionally, pharmacokinetics plays a crucial role in determining the optimal dosage and formulation for curcuminoid-based treatments, ensuring effective therapeutic outcomes (Howells *et al.*, 2021). However, formulation challenges persist, as developing stable and bioavailable curcuminoid products is essential for commercial viability (Zheng and McClements, 2020). To address these challenges, strategies for enhancing curcuminoid bioavailability, such as nanoparticle formulations, are being explored through omics-driven approaches (Zheng and McClements, 2020). These insights underscore the importance of integrating various omics disciplines to overcome formulation challenges and validate the efficacy of curcuminoids in medicinal plant research.

Translational omics plays a pivotal role in bridging preclinical and clinical development, particularly in the medicinal plants (Wafi and Mirnezami, 2018). This interdisciplinary approach, known as systems biology, allows for a comprehensive understanding of complex biological interactions, which is essential for developing personalized therapies and optimizing treatments derived from medicinal plants (Buriani *et al.*, 2012). Furthermore, translational research ensures that these findings are effectively communicated and applied in clinical settings, addressing the gap between basic research and practical application (Fernandez-Moure, 2016). Ultimately, this synergy not only enhances the efficacy of phytotherapy but also promotes affordable healthcare solutions, especially in resource-limited environments (Rausch *et al.*, 2024).

Omics technologies play a pivotal role in the assessment of toxicity, pharmacokinetics, and clinical trial design for medicinal plants. Metabolomics and proteomics are instrumental in identifying biomarkers that indicate toxic responses, thereby enhancing the safety of drug development by predicting potential toxicities (Chen *et al.*, 2016; Zhang *et al.*, 2023). Additionally, toxicogenomics contributes to understanding the mechanisms of toxicity, facilitating the development of accurate prediction models (Singh *et al.*, 2023). In pharmacokinetics, pharmacogenomics helps tailor drug dosages based on genetic variations, optimizing therapeutic outcomes and minimizing adverse reactions (Papachristos *et al.*, 2023). Furthermore, transcriptomics aids in clinical trial design by identifying gene expression profiles that predict treatment responses, ensuring that trials are more effective and targeted (Mukherjee *et al.*, 2025). The commercialization of phytopharmaceuticals derived from medicinal plants faces several regulatory challenges. First, adherence to good manufacturing practice (GMP) is essential to ensure product quality and compliance with regulatory standards; failure to meet these guidelines can result in product recalls and legal issues (Ahmed, 2024). Additionally, understanding phytopharmaceutical patent law is crucial, as it governs the intellectual property rights associated with these products, impacting their development and marketability (Kapoor *et al.*, 2025). Furthermore, the concept of phytoequivalence is vital as it ensures that different batches maintain consistent quality and efficacy, which is necessary for regulatory approval (Pace and Martinelli, 2022). Lastly, compliance with the International Conference on Harmonization (ICH) guidelines is imperative for global marketing as non-adherence can lead to significant delays in product approvals. Collectively, these factors highlight the complex landscape of regulatory challenges in the commercialization of phytopharmaceuticals.

The standardization, intellectual property, and global compliance for plant-based drugs face significant hurdles, particularly in medicinal plants. Variations in regulatory frameworks across countries complicate the approval and marketing processes, creating barriers to global compliance (Amaral *et al.*, 2024; Jadhav and Khare, 2024). Additionally, the lack of harmonization in pharmacopeial standards undermines the quality and consistency of these drugs, further complicating their international acceptance (Heyward *et al.*, 2018). Intellectual property issues, particularly regarding patent protection, pose challenges in proving novelty and non-obviousness for natural products, which can deter innovation and investment in research and development (Thomford *et al.*, 2018). Furthermore, the standardization of manufacturing processes is critical to ensuring the efficacy and safety of plant-based drugs, necessitating the implementation of GMPs and robust quality control measures. Addressing these challenges is essential for the successful commercialization of plant-based medicinal products on a global scale.

6.2 Omics-enabled High-Throughput Screening Platforms

The integration of omics-enabled HTS platforms in medicinal plants represents a transformative approach in natural drug discovery (Table 6.1 and Fig. 6.1). These platforms leverage advanced technologies to explore the genetic, transcriptomic, proteomic, and metabolomic landscapes of medicinal plants, facilitating the identification and characterization of bioactive compounds. This approach not only accelerates the discovery process but also enhances the understanding of the complex biosynthetic pathways and regulatory networks involved in the production of these compounds. The application of multi-omics technologies in medicinal plant research is crucial for developing new medicinal resources and improving existing ones. Below, key aspects of this integration are discussed.

6.2.1 Genomics and transcriptomics

Genomics and transcriptomics provide insights into the genetic makeup and gene expression profiles of medicinal plants, respectively. These omics approaches help identify key genes and regulatory elements involved in the biosynthesis of bioactive compounds (Zhang *et al.*, 2023; Cardona *et al.*, 2024). In the case of *Aconitum*, a medicinal plant known for its diterpenoid alkaloids, genomics and

Table 6.1. Core omics technologies in medicinal plant drug discovery. Author's own table.

Omics discipline	Primary role in drug discovery	Key applications	Common technologies	Citations
Genomics	Decodes the genetic blueprint and identifies BGCs for specialized metabolites.	Target identification, understanding phytochemical diversity, evolutionary studies.	High-throughput sequencing, whole-genome sequencing.	(Jabeen *et al.*, 2024; Wang *et al.*, 2024a)
Transcriptomics	Reveals gene expression profiles under specific conditions, identifying genes involved in metabolite production.	Elucidating biosynthetic pathways, understanding plant response to stress.	RNA-seq, microarrays.	(Wang *et al.*, 2024b; Sobti, 2025)
Proteomics	Studies the full set of proteins, including enzymes that catalyze the biosynthesis of bioactive compounds.	Biomarker discovery, understanding protein–protein interactions, identifying key enzymes.	2D-GE, mass spectrometry (liquid chromatography-tandem mass spectrometry—LC-MS/MS).	(Pandita *et al.*, 2021; Vijay *et al.*, 2022)
Metabolomics	Provides a comprehensive profile of small-molecule metabolites, the direct source of bioactivity.	Bioactive compound discovery, quality control, understanding metabolic fluxes.	GC-MS (gas chromatography-mass spectrometry), LC-MS, NMR (nuclear magnetic resonance).	(Rai *et al.*, 2017; Zhang *et al.*, 2023)

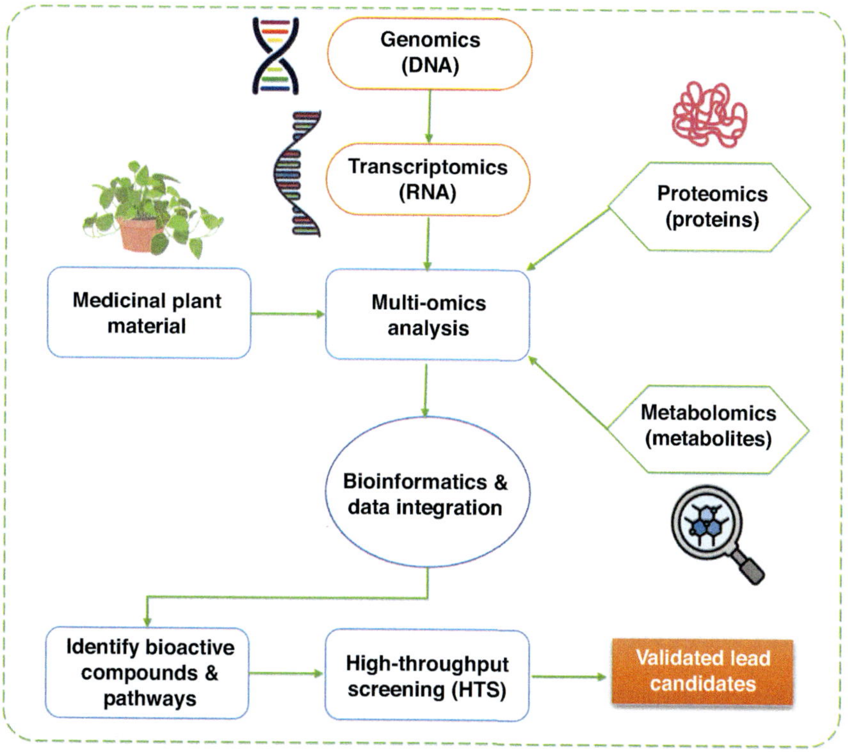

Fig. 6.1. Multi-omics workflow in medicinal plant drug discovery. (Author's own figure.)

transcriptomics have been used to study gene regulation and metabolic pathways, which are crucial for understanding its medicinal properties and toxicity (Wang *et al.*, 2024b).

6.2.2 Proteomics and metabolomics

Proteomics involves the large-scale study of proteins, which are the functional molecules in cells. It helps in understanding the protein interactions and pathways that lead to the production of medicinal compounds (Pandita *et al.*, 2021). Metabolomics focuses on the comprehensive analysis of metabolites, providing a snapshot of the biochemical activity within a plant. This is essential for identifying bioactive compounds and understanding their roles in plant physiology and human health (Rai *et al.*, 2017; Wang *et al.*, 2024b).

6.2.3 High-throughput screening (HTS)

HTS technologies enable the rapid screening of large libraries of natural products, significantly speeding up the drug discovery process. This is particularly useful in identifying potential anti-cancer agents from botanicals, as it allows for the testing of thousands of compounds in a short period (Patel and Patel, 2023). The integration of HTS with omics data allows for a more targeted approach in identifying compounds with specific biological activities, such as anti-inflammatory or analgesic effects (Wang *et al.*, 2024b). Multi-omics data provides a holistic view of the biosynthetic pathways and regulatory networks in medicinal plants. This approach is essential for overcoming the limitations of single-omics studies, which often fail to capture the complexity of biological systems (Ma *et al.*, 2020; Yang *et al.*, 2023). Despite the advancements, challenges remain in managing large datasets, standardizing protocols, and effectively applying synthetic biology for sustainable production of specialized metabolites (Wang *et al.*, 2024b).

6.2.4 Integration of genomics, metabolomics, and proteomics in automated compound screening

Integration of genomics, metabolomics, and proteomics in automated compound screening for medicinal plants represents a transformative approach in natural drug discovery. This integration, often referred to as multi-omics, allows for a comprehensive understanding of the complex biological processes and metabolic pathways involved in the biosynthesis of bioactive compounds. By leveraging high-throughput technologies, researchers can identify and characterize the genetic, proteomic, and metabolic profiles of medicinal plants, facilitating the discovery of novel therapeutic compounds. This approach not only enhances the efficiency of compound screening but also provides insights into the regulatory networks and environmental adaptations of medicinal plants, thereby supporting sustainable drug development.

6.2.4.1 Genomics in medicinal plant research

Genomics provides a foundational understanding of the genetic makeup of medicinal plants, enabling the identification of biosynthetic gene clusters (BGCs) responsible for the production of specialized metabolites (Wang *et al.*, 2024a). High-throughput sequencing technologies have advanced the exploration of genetic variations and evolutionary histories among plant species, which are crucial for understanding the diversity of bioactive compounds (Yang *et al.*, 2023; Jabeen *et al.*, 2024). Genomic data, when integrated with other omics, can reveal the genetic basis of phytochemical diversity and support the development of genetically enhanced plant varieties with improved medicinal properties (Vijay *et al.*, 2022; Kielich *et al.*, 2024).

6.2.4.2 Metabolomics and its role

Metabolomics focuses on the comprehensive profiling of metabolites within a plant, providing insights into the chemical diversity and composition of bioactive compounds (Zhang *et al.*, 2023). Techniques such as GC-MS, LC-MS, and NMR are employed to analyze complex metabolic pathways, identify rate-limiting steps, and elucidate the biosynthesis of phytochemicals (Surendran *et al.*, 2022). Metabolomics data, combined with genomic and proteomic information, can help in understanding the dynamic changes in metabolite production in response to environmental factors and stress conditions (Aborode *et al.*, 2022; Wang *et al.*, 2024a).

6.2.4.3 Proteomics and its integration

Proteomics involves the large-scale study of proteins, including their expression, structure, and functions, which are critical for understanding the enzymatic processes involved in metabolite biosynthesis (Vijay *et al.*, 2022). By integrating proteomic data with genomic and metabolomic analyses, researchers can map out gene–protein–metabolite networks, providing a holistic view of the biosynthetic pathways (Wang *et al.*, 2024a). This integration aids in identifying key enzymes and regulatory proteins that can be targeted for enhancing the production of desired metabolites (Cabral da Silva *et al.*, 2024; Jabeen *et al.*, 2024).

6.2.4.4 Automated compound screening

The integration of multi-omics data facilitates automated compound screening by providing a comprehensive dataset that can be analyzed using bioinformatics tools (Cabral da Silva *et al.*, 2024). Techniques such as molecular docking and molecular modeling are employed to predict interactions between plant-derived compounds and target proteins, streamlining the identification of potential drug candidates (Cabral da Silva *et al.*, 2024). Automated screening processes benefit from the integration of omics data, as it allows for the rapid identification and validation of bioactive compounds with therapeutic potential (Aborode *et al.*, 2022).

While genomics, metabolomics, and proteomics in medicinal plant research offer significant advantages, challenges remain in managing large datasets and standardizing protocols across different omics platforms. Additionally, the dynamic nature of specialized metabolism and the complexity of regulatory networks require continuous advancements in bioinformatics tools and analytical techniques. Despite these challenges, the multi-omics approach holds great promise for the sustainable development of plant-based drugs and the discovery of novel therapeutic compounds.

6.3 Network Pharmacology for Precision Target Identification

Network pharmacology has emerged as a pivotal approach in precision target identification, especially in the context of medicinal plants. This method integrates systems biology and computational tools to understand the complex interactions between bioactive compounds and their biological targets. By mapping these interactions, network pharmacology facilitates the identification of potential therapeutic targets and elucidates the mechanisms of action of medicinal plants. This approach is particularly valuable in drug discovery and development, offering insights into the multi-targeted therapeutic potential of plant-derived compounds. The following sections delve into the key aspects of network pharmacology in the context of medicinal plants.

By providing a systematic framework, network pharmacology allows for a more holistic investigation of medicinal plants. It moves beyond studying single compounds to identify multiple novel bioactive agents and uncover their complex interactions with therapeutic targets in the body (Noor *et al.*, 2022). This approach has been instrumental in understanding the pharmacological properties of traditional medicinal plants, integrating traditional knowledge with modern analytical techniques (Gautam, 2023).

Network pharmacology provides a powerful approach for evaluating the multi-targeted therapeutic potential of medicinal plants, which is especially crucial for treating complex diseases like cancer and diabetes that involve intricate pathophysiology (Gogoi *et al.*, 2022). This methodology was effectively demonstrated in a study on *Sauropus androgynus*, which successfully identified key target genes and pathways involved in cancer and inflammation, thereby highlighting the plant's significant promise for developing treatments for these conditions (Gayathiri *et al.*, 2024). The integration of network pharmacology with metabolomics and molecular docking significantly enhances the understanding of the pharmacological mechanisms of medicinal plants (Sharma and Yadav, 2022). This powerful, combined approach allows researchers to systematically explore the therapeutic claims of traditional medicine, as exemplified by its use in validating the pharmacological potential of *Fibraurea tinctoria* for diabetes management, and in investigating the scientific basis for traditional Indian medicinal plants (Umar *et al.*, 2025).

Network pharmacology serves as a key tool for identifying the bioactive compounds within medicinal plants and elucidating their associated biological pathways, thereby providing crucial insights into the molecular basis of their therapeutic effects (Shawky *et al.*, 2020). This capability was clearly demonstrated in a study on *Galphimia glauca*, where the approach successfully revealed key protein targets and pathways responsible for the plant's sedative and anti-inflammatory effects, effectively underscoring its multi-target therapeutic potential (Acosta and de Anda Jáuregui, 2025). Despite its significant potential, network pharmacology faces considerable challenges, primarily stemming from the inherent complexity of biological networks and the persistent need for more comprehensive, high-quality databases of plant-derived compounds (Fatokun *et al.*, 2024). To overcome these limitations and fully realize the power of this approach, future research must focus on the strategic integration of artificial intelligence (AI). Leveraging AI can significantly enhance the accuracy of predicting drug–target interactions and substantially improve the quality and efficiency of drug repurposing efforts for plant-based therapeutics (Fatokun *et al.*, 2024).

While network pharmacology offers a robust framework for precision target identification in medicinal plants, it is essential to consider the limitations and challenges associated with this approach. The complexity of biological systems and the variability in plant-derived compounds necessitate further research and validation through *in vitro* and *in vivo* studies. Additionally, the integration of advanced computational tools and AI could significantly enhance the predictive power and applicability of network pharmacology in drug discovery and development (Table 6.2 and Fig. 6.2).

6.3.1 Multi-omics-guided mapping of plant compounds to disease-specific molecular targets

The integration of multi-omics approaches in mapping plant compounds to disease-specific molecular targets is a burgeoning field that leverages the comprehensive data from genomics, transcriptomics, proteomics, and metabolomics to elucidate the therapeutic potential of medicinal plants. This approach is particularly valuable in identifying bioactive compounds and understanding their mechanisms of action against specific diseases. By employing multi-omics, researchers can uncover complex interactions and pathways that single-omics studies might miss, thus providing a more holistic view of plant-based therapeutics. The following sections delve into the various aspects of multi-omics applications in medicinal plant research.

6.3.2 Multi-omics approaches in medicinal plant research

Genomics and transcriptomics platforms are crucial for understanding the genetic basis of bioactive compound production in plants. For instance, the study of *Aconitum* species has utilized genomics to explore gene regulation and metabolic pathways involved in the synthesis of diterpenoid alkaloids, which have analgesic and anti-inflammatory properties (Wang *et al.*, 2024b). Similarly, herbgenomics combines genomic data with other omics to unravel the genetic foundations of medicinal activities in plants like those in the Papaveraceae family (Kielich *et al.*, 2024).

Proteomics provides insights into the protein expressions and interactions that underpin plant responses to diseases, while metabolomics focuses on the small-molecule metabolites that are often the active compounds in medicinal plants. Metabolomics has been particularly effective in decoding the bioactive ingredients of herbal medicines and elucidating their molecular targets (Wang *et al.*, 2024b). This approach is consistent with the multi-component and multi-target nature of herbal medicine, allowing for a comprehensive understanding of their pharmacological effects (Wang *et al.*, 2024b).

6.3.3 Applications in disease-specific targeting

In the field of cancer research, multi-omics approaches have become an invaluable tool for systematically studying the anti-cancer potential of medicinal plants, with significant applications in understanding gastrointestinal cancers. By integrating genomic, proteomic, and metabolomic data, these studies have successfully identified key molecular targets and elucidated the complex signaling pathways modulated by plant-derived compounds. This comprehensive analysis provides crucial, multi-layered insights into the mechanisms of action behind their therapeutic effects, moving beyond a single-dimensional view to reveal how these natural products collectively disrupt cancer processes (Dai *et al.*, 2022). Network pharmacology provides a powerful framework for deciphering the complex therapeutic nature of medicinal plants by integrating data from metabolomics to explore the synergistic effects of their numerous compounds. This approach involves systematically mapping the interactions between multiple bioactive constituents and their biological targets within the body. By constructing these intricate networks, researchers can move beyond a single-target, single-drug paradigm to better understand the polypharmacological basis of plant-based treatments, revealing how they collectively modulate entire disease phenotypes (Mukherjee *et al.*, 2021).

Table 6.2. Computational strategies for target identification and lead optimization. Author's own table.

Computational strategy	Primary function	Application in medicinal plants	Key advantages	Citations
Network pharmacology	Models complex interactions between multiple compounds, targets, and diseases.	Identifying polypharmacological mechanisms, validating traditional uses (e.g. *S. androgynus*).	Shifts from "one target, one drug" to a systems-level, holistic view.	(Gogoi *et al.*, 2022; Noor *et al.*, 2022; Gayathiri *et al.*, 2024)
Molecular docking & dynamics	Predicts binding affinity and stability of a compound with a target protein.	Virtual screening of plant compound libraries, lead optimization.	Rapid *in silico* assessment of potential efficacy before experimental validation.	(Cabral da Silva *et al.*, 2024; Mahanayak, 2024)
QSAR & machine learning	Relates chemical structure to biological activity; machine learning models predict bioactivity and ADMET (absorption, distribution, metabolism, excretion, and toxicity) properties.	Fine-tuning molecular designs; prioritizing lead candidates from large datasets.	Handles high-dimensional omics data; improves prediction accuracy for toxicity and efficacy.	(Chihomvu *et al.*, 2024a; Khair *et al.*, 2024)
AI & deep learning	Leverages big data to forecast drug–target interactions and optimize lead compounds.	Predicting multi-target effects, *de novo* drug design, enhancing drug repurposing.	Capable of identifying complex, non-linear patterns missed by traditional methods.	(Fatokun *et al.*, 2024; Sabha, 2025)
ADMET *in silico* profiling	Predicts absorption, distribution, metabolism, excretion, and toxicity.	Early-stage filtering of compounds with poor pharmacokinetic or toxic profiles.	Reduces late-stage attrition by identifying liabilities early.	(Javid *et al.*, 2024)

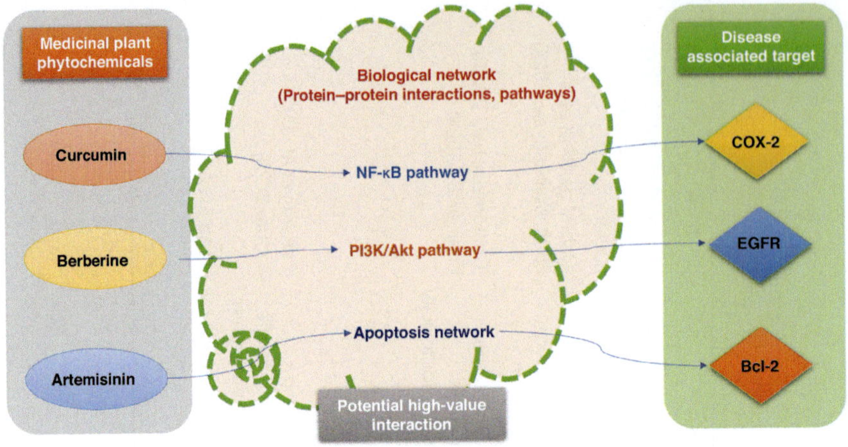

Fig. 6.2. Network pharmacology approach for target identification. (Author's own figure.)

While multi-omics technologies provide an unprecedentedly detailed understanding of plant compounds, the inherent toxicity of certain botanicals, such as the diterpenoid alkaloids found in *Aconitum*, remains a significant challenge. This persistent risk underscores the critical need for careful traditional processing and rigorous, omics-guided validation of safety to de-risk these compounds long before any clinical application can be considered (Wang *et al.*, 2024b). The rich heritage of traditional medicine offers an invaluable and time-tested resource for modern drug discovery. By systematically integrating this empirical knowledge with cutting-edge omics technologies, researchers can create a powerful synergistic pipeline, significantly enhancing both the discovery and the validation of novel plant-based therapeutics (Singh *et al.*, 2019). While multi-omics approaches offer a comprehensive framework for mapping plant compounds to disease-specific targets, challenges such as toxicity and the integration of traditional knowledge must be addressed. The synergy between traditional medicine and modern omics technologies holds promise for advancing the field of plant-based drug discovery, offering new insights into the therapeutic potential of medicinal plants.

6.4 Lead Optimization: From Omics Data to Drug Candidates

Lead optimization from omics data to drug candidates in medicinal plants involves a multifaceted approach that integrates modern technologies and traditional medicinal chemistry principles. This process is crucial for transforming bioactive compounds from plants into viable therapeutic agents. The optimization process focuses not only on enhancing the efficacy of these compounds but also on improving their pharmacokinetic and pharmacodynamic profiles. The integration of omics technologies, computational tools, and AI plays a significant role in this transformation.

6.4.1 Omics technologies in lead optimization

Omics technologies have become integral to the lead optimization phase in drug discovery, providing comprehensive insights into biological systems that enhance the identification and refinement of lead compounds. These technologies, including genomics, proteomics, and metabolomics, offer a multidimensional view of biological interactions, enabling more precise targeting and optimization of drug candidates. The integration of omics data with computational and AI-driven approaches further accelerates the lead optimization process, making it more efficient and effective. Below, the role of omics technologies in lead optimization is explored in detail.

6.4.1.1 *Genomics and proteomics in lead optimization*

Genomic technologies have revolutionized target identification and lead optimization by providing detailed insights into genetic variations and their implications for drug response. Pharmacogenomics, which studies gene expression changes in response to drugs, helps tailor lead compounds to specific genetic profiles, enhancing efficacy and reducing adverse effects (Singh, 2019). Proteomics complements genomics by analyzing the protein expressions and modifications that occur in response to drug candidates. This technology helps identify potential biomarkers and therapeutic targets, facilitating the design of drugs with improved specificity and reduced toxicity (Savino *et al.*, 2012).

6.4.1.2 *Computational and AI-driven approaches*

AI-driven techniques, such as deep neural networks and reinforcement learning, leverage big data to predict the pharmacological properties of lead compounds. These methods enable the rapid screening and optimization of compounds by integrating multi-omics data, thus streamlining the drug discovery process (Sabha, 2025). Computational chemistry and cheminformatics play a crucial role in analyzing structure–activity relationships and predicting ADME (absorption, distribution, metabolism, and excretion) properties. Computer-aided drug design (CADD) techniques like structure-based design and pharmacophore modeling are used to optimize natural leads by modifying their chemical structures to enhance drug-like properties (Xiao *et al.*, 2016). These tools help refine lead compounds by providing insights into molecular interactions and potential toxicological endpoints.

6.4.1.3 *Integration of omics with HTS*

The integration of omics technologies with HTS enables the rapid identification of lead compounds from extensive chemical libraries, allowing researchers to prioritize candidates based on their biological activity and pharmacokinetic profiles for efficient lead optimization (Aborode *et al.*, 2022; Haghir Ebrahim Abadi *et al.*, 2025). Furthermore, the use of expression databases provides critical data on gene and protein expression patterns, which aids in both the identification of novel therapeutic targets and the subsequent optimization of lead compounds to interact with specific biological pathways (Singh *et al.*, 2022).

6.4.1.4 *Medicinal chemistry and structural modifications*

Medicinal chemistry transformations, including scaffold hopping and functional group modifications, are applied to improve the pharmacological properties of lead compounds (Mishra *et al.*, 2024). These modifications are guided by transformation rules derived from historical data and expert knowledge (Lamberth, 2018). Structure–activity relationship (SAR) studies help in understanding the relationship between the chemical structure of a compound and its biological activity, guiding further optimization efforts (Babiker *et al.*, 2025).

6.4.1.5 *Case studies and applications*

The strategic application of these integrated approaches has yielded several successful examples in drug discovery, particularly in the urgent development of anti-cancer and antiviral therapies. For instance, the optimization of plant-derived compounds, such as vinblastine and paclitaxel analogs, has significantly advanced cancer treatment regimens by improving their efficacy and safety profiles. Similarly, in response to the global pandemic, virtual HTS of natural product libraries successfully identified several promising plant-derived compounds as potential anti-COVID-19 drug leads, demonstrating the power of computational methods to rapidly pinpoint candidates that can inhibit key viral proteins (Firouzi and Ashouri, 2022, 2023). The Natural Products Research Laboratories at the University of North Carolina have demonstrated successful lead optimization of bioactive compounds, highlighting the importance of integrating traditional and modern approaches (Xiao *et al.*, 2016).

While omics data and computational tools have significantly advanced the field of drug discovery from medicinal plants, challenges remain. The complexity of plant-derived compounds and the variability in their bioactive constituents due to environmental factors pose significant hurdles. Additionally, the need for large-scale production and standardization of these compounds is critical for their successful development into therapeutic agents. Despite these challenges, the continued advancement in omics technologies and AI holds promise for overcoming these obstacles and accelerating the discovery of novel drug candidates from medicinal plants.

6.4.2 Computational and experimental strategies to refine bioactive scaffolds using omics insights

The refinement of bioactive scaffolds from medicinal plants involves a combination of computational and experimental strategies that leverage the rich chemical diversity of plant-derived compounds. These strategies aim to enhance the drug discovery process by identifying, characterizing, and optimizing bioactive compounds for therapeutic use. Computational methods, such as molecular docking, machine learning, and network pharmacology, play a crucial role in predicting the interactions and efficacy of these compounds, while experimental approaches validate these predictions and ensure the compounds' safety and effectiveness. The integration of these methodologies bridges traditional knowledge with modern scientific techniques, facilitating the development of novel therapeutics.

6.4.2.1 Computational strategies

Molecular docking and dynamics techniques are used to predict the binding affinities and modes of bioactive compounds with target proteins. Molecular dynamics simulations further validate these interactions by assessing their stability and efficacy over time (Laloo *et al.*, 2021; Mahanayak, 2024). Machine learning algorithms, including decision trees and neural networks, are employed to predict the bioactivity of compounds and optimize their therapeutic profiles. QSAR (quantitative structure–activity relationship) modeling helps in fine-tuning molecular designs for improved activity and reduced toxicity (Chihomvu *et al.*, 2024a; Khair *et al.*, 2024). The network pharmacology approach provides insights into the multi-target effects of compounds, elucidating their mechanisms of action within biological systems. It helps in understanding how phytochemicals interact with potential targets in the body (Chihomvu *et al.*, 2024b; Mahanayak, 2024).

In silico screening methods, such as pharmacophore-based screening and 2D similarity searches, are used to identify potentially active compounds from large databases of natural products (Pfisterer *et al.*, 2014). Pharmacokinetic and ADMET profiling analyses predict the absorption, distribution, metabolism, excretion, and toxicity profiles of compounds, ensuring the selection of candidates with favorable pharmacokinetic properties (Javid *et al.*, 2024; Mahanayak, 2024). Experimental validation of computational predictions is crucial for confirming the bioactivity and safety of compounds. This involves standardizing the botanical, chemical, and biological aspects of medicinal plants (Laloo *et al.*, 2021). Structural and computational biology integration aids in lead identification by providing detailed information on how phytochemicals interact with biological targets, thus guiding experimental validation efforts (Chihomvu *et al.*, 2024c).

6.4.2.2 Challenges and perspectives

While computational and experimental strategies offer significant advantages in refining bioactive scaffolds, challenges remain. The integration of diverse data types and the interpretability of complex models are ongoing issues. Additionally, the regulatory demands for new therapeutics require rigorous validation processes. However, the promise of these strategies lies in their ability to streamline drug discovery, reduce costs, and enhance the efficacy of plant-derived therapeutics. Continued interdisciplinary collaborations and investments in advanced technologies are essential to fully harness the potential of these approaches in drug discovery (Pfisterer *et al.*, 2014; Khair *et al.*, 2024).

6.5 Turmeric's Curcuminoids in Anti-inflammatory Drug Development

Turmeric's curcuminoids, particularly curcumin, have garnered significant attention in the development of anti-inflammatory drugs due to their potent biological activities. These compounds have been extensively studied for their ability to modulate inflammation, a key factor in many chronic diseases. Curcumin, the most active curcuminoid, has shown promise as a lead candidate for anti-inflammatory drug development, owing to its ability to interact with multiple molecular targets involved in the inflammatory process. This overview will explore the therapeutic potential of curcuminoids in anti-inflammatory drug development, highlighting their mechanisms of action (Table 6.3 and Fig. 6.3).

6.5.1 Mechanisms of action

Curcuminoids, especially curcumin, exhibit anti-inflammatory properties by modulating various signaling pathways. They inhibit the activity of nuclear factor-kappa B (NF-κB), a transcription factor that plays a crucial role in the expression of pro-inflammatory cytokines and enzymes such as COX-2 and LOX (Fernández-Moriano *et al.*, 2019). Curcumin also affects the production of inflammatory mediators like interleukins and tumor necrosis factor-alpha (TNF-α), reducing inflammation and oxidative stress (Boroumand *et al.*, 2018; da Silva Carneiro *et al.*, 2023). The compound's ability to scavenge reactive oxygen species and enhance antioxidant defenses further contributes to its anti-inflammatory effects (Sharma and Sharma, 2022; Cozmin *et al.*, 2024).

Table 6.3. Turmeric curcuminoids—efficacy, challenges, and advanced solutions. Author's own table.

Aspect	Findings & challenges	Omics-driven insights & advanced solutions	Citations
Mechanism of action	Potent anti-inflammatory activity via inhibition of NF-κB, COX-2, and pro-inflammatory cytokines.	Multi-omics (transcriptomics, proteomics) validate the multi-target mechanisms of action.	(Boroumand *et al.*, 2018; Fernández-Moriano *et al.*, 2019)
Efficacy validation	Demonstrates antiproliferative and pro-apoptotic effects in cancer models (e.g. chemo-resistant colorectal cancer—CRC).	Metabolomics identifies active metabolites (e.g. tetrahydrocurcumin) with enhanced activity.	(Jude *et al.*, 2018; Girst *et al.*, 2021; Wan *et al.*, 2023)
Major challenge: poor bioavailability	Low aqueous solubility, rapid metabolism, and systemic clearance limit therapeutic efficacy.	Omics guides the development of novel formulations to bypass these issues.	(Razavi *et al.*, 2021; El-Saadony *et al.*, 2025)
Solution: advanced formulations	Use of nanocarriers (liposomes, nanoparticles), microemulsions, and solid lipid particles.	These systems significantly enhance solubility, stability, and targeted delivery.	(Zheng and McClements, 2020; Hassan *et al.*, 2024; Jacob *et al.*, 2024)
Solution: structural analogs	Synthesis of derivatives (e.g. thiophene curcuminoids) to improve drug-like properties.	*In silico* ADMET and docking studies guide the rational design of more stable and potent analogs.	(da Cruz *et al.*, 2021; Yusuf and Khan, 2022)

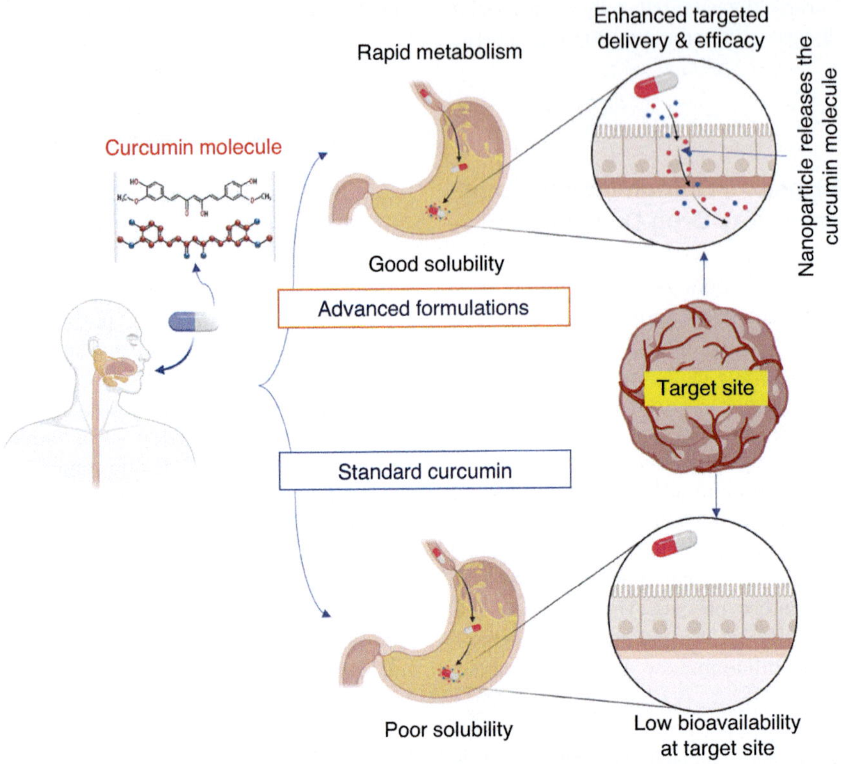

Fig. 6.3. Overcoming curcumin bioavailability challenges. (Created in Biorender.com; author's own figure.)

6.5.2 Therapeutic applications

Curcumin has been investigated for its potential in treating various inflammatory disorders, including arthritis, inflammatory bowel disease, and neurodegenerative diseases (Razavi *et al.*, 2021; da Silva Carneiro *et al.*, 2023). Clinical trials have shown that curcumin can alleviate symptoms of these conditions, although its bioavailability remains a challenge (Razavi *et al.*, 2021). Curcuminoids have also been explored as structural motifs for the synthesis of novel anti-inflammatory agents, such as thiophene curcuminoids and nitric oxide-donating derivatives, which aim to improve efficacy and reduce side effects associated with conventional NSAIDs (non-steroidal anti-inflammatory drugs; da Cruz *et al.*, 2021).

The issue with curcumin as a therapeutic agent is its poor bioavailability, attributed to rapid metabolism and systemic clearance. Strategies such as complexing curcumin with other substances or developing synthetic derivatives are being explored to overcome this limitation (El-Saadony *et al.*, 2025). Further research is needed to elucidate the pharmacokinetics and pharmacodynamics of curcuminoids, as well as to conduct high-quality clinical trials to establish their efficacy and safety in humans (Boroumand *et al.*, 2018; Razavi *et al.*, 2021). The development of curcumin-based drugs also involves evaluating their ADME properties and COX-2 inhibitory potential, which have shown promising results in *in silico* studies (Yusuf and Khan, 2022).

While curcuminoids present a promising avenue for anti-inflammatory drug development, it is important to consider the limitations and challenges associated with their use. The poor bioavailability of curcumin, for instance, necessitates innovative approaches to enhance its therapeutic potential. Additionally, while preclinical and some clinical studies have demonstrated the efficacy of curcuminoids, more rigorous trials are required to confirm these findings and translate them into effective treatments. Ongoing research and development efforts aim to address these challenges, potentially leading to the successful integration of curcuminoids into mainstream therapeutic regimens.

6.5.3 Omics-driven validation of curcuminoid efficacy and formulation challenges

The efficacy of curcuminoids, particularly curcumin, and the challenges associated with their formulation are significant areas of research in medicinal plant studies. Curcumin, derived from turmeric, is known for its diverse pharmacological properties, including anti-cancer, anti-inflammatory, and antioxidant activities. However, its clinical application is hindered by poor bioavailability due to low solubility, rapid metabolism, and instability. Recent advancements in formulation strategies aim to overcome these challenges, enhancing curcumin's therapeutic potential. This response explores the omics-driven validation of curcuminoid efficacy and the formulation challenges of medicinal plants, focusing on curcumin.

6.5.3.1 Omics-driven validation of curcuminoid efficacy

Curcuminoids have shown significant anti-cancer activity, particularly in CRC. A novel surfactant-formulated curcumin, CuminUP60, demonstrated strong antiproliferative effects on chemo-resistant CRC cells, inducing apoptosis and inhibiting Th17 cell differentiation, which is crucial for immune response regulation in gut malignancies (Wan *et al.*, 2023). The pharmacokinetics of curcuminoids and their metabolites, such as tetrahydrocurcumin (THC) and curcumin-O-glucuronide (COG), have been studied using advanced spectrometry techniques. These studies highlight the enhanced concentration of these metabolites in plasma, suggesting improved bioavailability when curcumin is administered in a complete turmeric matrix (Jude *et al.*, 2018). Curcuminoids and their reduced metabolites exhibit antioxidant properties, with hexahydrocurcumin identified as a promising candidate for drug development due to its increased solubility and stability (Girst *et al.*, 2021).

6.5.3.2 Formulation challenges and solutions

Curcumin's poor bioavailability is a major challenge. Various strategies, including structural modifications and novel drug delivery systems, have been explored to improve its pharmacokinetic profile. These include conjugation with biomolecules, use of natural and synthetic polymers, and development of micelles, liposomes, and nanoparticles (Hegde *et al.*, 2023; Chelimela *et al.*, 2024). Nanocarrier systems, such as liposomes, nanoemulsions, and polymeric nanoparticles, have been developed to enhance curcumin's solubility, stability, and therapeutic efficacy. These systems also aim to reduce toxicity and improve targeted delivery (Jacob *et al.*, 2024).

Encapsulation technologies, such as microemulsions and solid lipid particles, have been employed to trap curcumin within small particles, enhancing its solubility and stability. For instance, curcumin encapsulated in geranium oil microemulsion showed elevated antibacterial, antioxidant, and anti-cancer activities (Zheng and McClements, 2020; Hassan *et al.*, 2024). While significant progress has been made in enhancing curcumin's bioavailability and efficacy, challenges remain. The regulatory approval of nanocarrier systems is pending, and there are ongoing safety concerns regarding their use (Jacob *et al.*, 2024). Additionally, the influence of gut microbiota on curcumin delivery systems' metabolism and bioavailability is an area requiring further research (Wang *et al.*, 2024c). Omics technologies and advanced formulation strategies hold promise for the future clinical application of curcumin in treating various diseases.

6.6 Bridging Preclinical and Clinical Development (Translational Omics)

Bridging the gap between preclinical and clinical development of medicinal plants involves a multifaceted approach that integrates scientific research, regulatory frameworks, and traditional knowledge. This process is crucial for transforming promising plant-based compounds into viable therapeutic agents. The transition from preclinical studies, which often involve *in vitro* and *in vivo* models, to clinical trials requires rigorous standardization, safety assessments, and efficacy evaluations. The following sections outline key aspects of this transition, drawing insights from various studies on medicinal plants (Fig. 6.4).

6.6.1 Preclinical studies and standardization

Preclinical development often begins with activity-guided fractionation to identify active constituents in plant extracts. This method helps in isolating compounds with therapeutic potential, ensuring that subsequent clinical trials are based on scientifically validated extracts (Laloo *et al.*, 2021). Preclinical studies utilize various pharmacological models to explore the mechanisms of action of bioactive phytochemicals. These models are essential for understanding how these compounds interact with biological systems and for predicting their therapeutic effects in humans (Al Rashid *et al.*, 2020). Ensuring batch conformity and standardization of extracts is critical. This involves using pharmacologically active markers to maintain consistency in potency and efficacy across different batches of the herbal product (Karthikeyan *et al.*, 2025).

6.6.2 Clinical trials and regulatory considerations

Well-controlled clinical trials are necessary to document the safety and efficacy of plant-based formulations. These trials should be designed with relevant outcome measurements that align with the intended therapeutic use of the herbal product (Maheshwari *et al.*, 2025). The development of botanical drugs must adhere to regulatory standards, such as those set by the US Food and Drug Administration (FDA), which require comprehensive preclinical data and well-designed clinical trials to support investigational new drug (IND) applications (Wang *et al.*, 2023a). Despite promising preclinical results, many botanical products face challenges in clinical translation due to the complexity of their chemical composition and the variability in traditional use practices (Wang *et al.*, 2023a).

6.6.3 Case studies and examples

Research on medicinal plants for Alzheimer's disease highlights the potential of natural

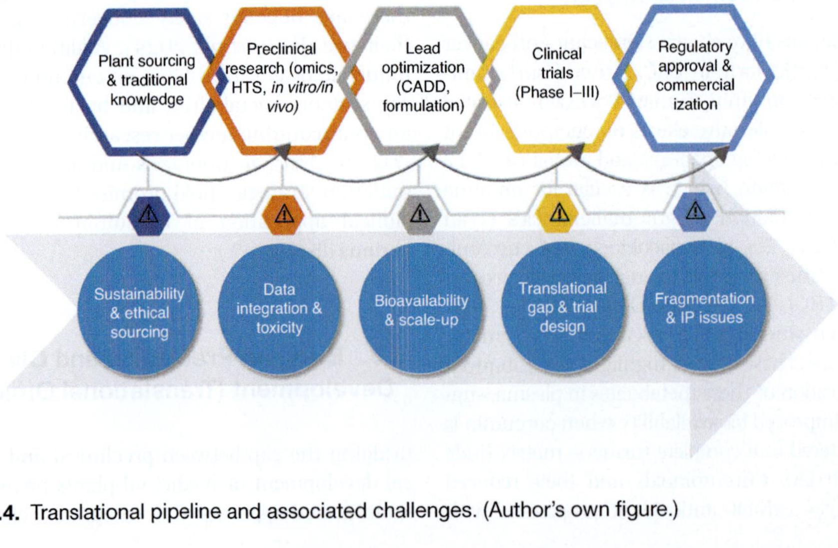

Fig. 6.4. Translational pipeline and associated challenges. (Author's own figure.)

products as neuroprotective agents. Preclinical studies have demonstrated efficacy, and some formulations have progressed to clinical trials, underscoring the need for continued research in this area (Nagori *et al.*, 2023). Several plant-derived compounds, such as paclitaxel and vincristine, have successfully transitioned from preclinical studies to clinical use, demonstrating the potential of medicinal plants in oncology (Desai *et al.*, 2025). The search for novel anti-hepatitis B agents from medicinal plants is ongoing, with preclinical studies focusing on the antiviral properties of various phytoconstituents. These efforts aim to develop safer and more effective treatments (Mondal *et al.*, 2024).

While the integration of medicinal plants into modern therapeutic practices holds great promise, it is not without challenges. The complexity of plant-based compounds and the variability in traditional usage can complicate standardization and regulatory approval processes. Additionally, the lack of comprehensive pharmacological evaluations and robust clinical trials for many herbal medicines impedes their widespread acceptance in contemporary medicine (Aware *et al.*, 2022). Addressing these challenges requires a concerted effort to enhance scientific research, improve regulatory frameworks, and foster collaboration between traditional knowledge holders and modern scientists.

6.6.4 Role of omics in toxicity assessment, pharmacokinetics, and clinical trial design

Omics technologies have revolutionized the fields of toxicity assessment, pharmacokinetics, and clinical trial design for medicinal plants by providing comprehensive insights into the molecular mechanisms underlying these processes. These technologies, which include genomics, transcriptomics, proteomics, and metabolomics, enable researchers to identify biomarkers, understand biosynthetic pathways, and predict potential toxicities and pharmacokinetic behaviors of plant-derived compounds. This integration of omics data with traditional methodologies enhances the safety and efficacy of medicinal plant research and development. The following sections detail the specific roles of omics in these areas.

6.6.4.1 Toxicity assessment

Omics technologies facilitate the early detection of toxic effects by identifying changes in gene expression, protein, and metabolite profiles. This is crucial for removing potentially toxic substances from development pipelines (Patil *et al.*, 2023). Multi-omics approaches enhance the identification of biomarkers that can detect toxicity at early stages, improving the safety profile of therapeutic agents (Son *et al.*, 2024b). Omics data provide insights into the molecular pathways involved in drug-induced toxicity, aiding in the elucidation of mechanisms and the identification of safety biomarkers (Son *et al.*, 2024a).

6.6.4.2 Pharmacokinetics

Computational models, such as physiologically based pharmacokinetic (PBPK) modeling, combined with omics data, offer precise predictions of human responses to medicinal plant compounds (Kurniawati *et al.*, 2023; Son *et al.*, 2024a). Omics technologies help explore the correlation between chemical structures and pharmacokinetic parameters, such as lipophilicity and P-glycoprotein substrate status, enhancing the understanding of compound behavior in the body (Kurniawati *et al.*, 2023).

6.6.4.3 Clinical trial design

The integration of omics data into clinical trial design allows for more informed decision making regarding the selection of candidate compounds and the identification of potential adverse effects (Marx-Stoelting *et al.*, 2015). Omics data can support regulatory toxicology by providing mechanistic information that aids in the classification and labeling of medicinal plant compounds, potentially reducing the need for animal testing (Marx-Stoelting *et al.*, 2015). Sophisticated computational techniques and continued collaboration among researchers, clinicians, and regulatory bodies are essential to fully harness these technologies (Son *et al.*, 2024a, b). Additionally, the application of omics in medicinal plant research is still in its infancy, requiring further development of standardized methodologies and validation processes to ensure reliability and regulatory acceptance (Marx-Stoelting *et al.*, 2015).

Omics technologies play a pivotal role in advancing the fields of toxicity assessment, pharmacokinetics, and clinical trial design for medicinal plants. By providing detailed molecular insights, these technologies enhance the safety and efficacy of plant-derived therapeutics. However, the successful integration of omics data into these fields requires overcoming challenges related to data complexity and ensuring collaboration across disciplines.

6.7 Regulatory Challenges in Phytopharmaceutical Commercialization

The commercialization of phytopharmaceuticals derived from medicinal plants faces numerous regulatory challenges. These challenges stem from the complex nature of herbal products, which are often composed of multiple constituents, and the diverse regulatory frameworks that exist globally. The need for harmonization of these regulations is critical to ensuring the safety, efficacy, and quality of herbal medicines while facilitating their global market expansion. The following sections delve into the key regulatory challenges and considerations in the commercialization of phytopharmaceuticals (Table 6.4).

6.7.1 Diverse regulatory frameworks

Different countries have varying definitions and regulatory approaches for medicinal plants, which complicates international trade and market entry. For instance, the European Union (EU), the USA, China, and India each have distinct regulatory requirements for herbal medicines, affecting classification, licensing, and post-market surveillance (Verma, 2016; Khuspe

Table 6.4. Key regulatory and commercialization hurdles for phytopharmaceuticals. Author's own table.

Hurdle category	Specific challenge	Impact on commercialization	Citations
Standardization & quality control	Complex, multi-component nature of botanicals; lack of universal analytical protocols.	Leads to batch-to-batch variability, difficulties in proving consistent efficacy and safety.	(Umamaheswari *et al.*, 2021; Chaudhary *et al.*, 2024)
Regulatory fragmentation	Divergent definitions and approval processes across regions (e.g. USFDA, European Medicines Agency—EMA, World Health Organization – WHO).	Creates a complex, costly, and time-consuming path for global market access.	(Khuspe *et al.*, 2024; Picking, 2024)
Intellectual property (IP)	Difficulty in patenting natural products; inadequate protection for traditional knowledge leading to biopiracy.	Deters investment in R&D; raises ethical concerns and legal disputes.	(Aware *et al.*, 2022; Wang *et al.*, 2023a)
Clinical translation	Difficulty in isolating a single active pharmaceutical ingredient (API); poorly designed trials; insufficient pharmacokinetic data for complex mixtures.	High failure rate in clinical phases; delays in filing successful IND applications.	(Wijayabandara, 2019; Wang *et al.*, 2024b)
Safety & toxicity	Inherent toxicity of some botanicals (e.g. *Aconitum*); adulteration of raw materials.	Raises safety concerns, requires rigorous omics-guided toxicity profiling, and damages consumer trust.	(Wijayabandara, 2019; Wang *et al.*, 2024b)

et al., 2024). The WHO plays a significant role in attempting to harmonize these regulations, but significant challenges remain. The lack of a unified global regulatory framework hinders the consistent application of safety and efficacy standards (Khuspe *et al.*, 2024; Picking, 2024).

6.7.2 Safety, efficacy, and quality control

Herbal medicines are complex mixtures, and identifying the constituents responsible for therapeutic effects is challenging. This complexity complicates stability testing and quality control, which are crucial for ensuring product safety and efficacy (Verma, 2016). Ensuring the quality of raw materials is essential. The trend toward domestication and biotechnological improvement of medicinal plants aims to provide uniform and high-quality raw materials, which are fundamental to the efficacy and safety of herbal drugs (Halder *et al.*, 2021).

6.7.3 Market and regulatory challenges

The global market for herbal pharmaceuticals is expanding rapidly, driven by consumer perceptions of safety and natural origins. However, regional regulatory differences can affect product quality and consumer trust, necessitating international harmonization of standards (Khuspe *et al.*, 2024). Various regulatory models exist, including those for prescription drugs, over-the-counter substances, traditional medicines, and dietary supplements. The German Commission E-model, which combines scientific studies with traditional knowledge, is often cited as a successful approach (Nayan *et al.*, 2025).

6.7.4 Case studies and regional insights

Brazil has developed specific regulatory frameworks for herbal medicines, improving product profiles in terms of safety, efficacy, and quality. However, challenges remain, such as the limited registration of native species and stringent quality control requirements (Oliveira *et al.*, 2016). In India, botanicals are regulated as dietary supplements, foods, and medicines, with specific requirements for regulatory clearance. The country's rich tradition in herbal medicine, particularly Ayurveda, influences its regulatory approach (Chattopadhyay and Veer, 2016).

While the regulatory landscape for phytopharmaceuticals is complex and varied, efforts toward harmonization and improved quality control are ongoing. The WHO's Traditional Medicine Strategy and other international initiatives aim to standardize herbal medication approval procedures globally (Khuspe *et al.*, 2024). However, challenges such as product adulteration, lack of therapeutic data, and restricted intellectual property rights persist, highlighting the need for calculated regulatory changes to ensure consumer safety and market legitimacy.

6.7.5 Hurdles in standardization, intellectual property, and global compliance for plant-based drugs

The development and commercialization of plant-based drugs face significant hurdles in standardization, intellectual property, and global compliance. These challenges stem from the complex nature of herbal medicines, the variability in plant materials, and the diverse regulatory landscapes across different countries. Standardization is crucial for ensuring the quality, safety, and efficacy of plant-based drugs, but it is often hindered by the lack of consistent methodologies and regulatory frameworks. Intellectual property issues further complicate the landscape, as they involve protecting traditional knowledge and innovations in plant-based drug development. Global compliance is challenged by varying regulations and the need for harmonization to facilitate international trade and ensure consumer safety.

Plant materials in herbal medicines are complex, consisting of numerous compounds, making it difficult to identify active principles and ensure consistent quality. Standardization often relies on chemical markers, but these do not always correlate with therapeutic efficacy (Umamaheswari *et al.*, 2021). A significant challenge in the standardization of plant-based medicines is the current lack of consistent and universally applicable methodologies. There is a pressing need for more reliable analytical techniques to comprehensively profile complex phytochemical compositions and ensure crucial batch-to-batch consistency. While established methods like high-performance liquid chromatography (HPLC) and gas chromatography (GC) are routinely employed, they often require further refinement and development to become robust and standardized enough for global application across the diverse spectrum of medicinal plants and their multi-component extracts (Umamaheswari *et al.*, 2021).

Different countries have varying standards for herbal medicines, complicating efforts to establish a unified standardization process. The WHO provides guidelines, but their adoption is inconsistent globally (Chaudhary *et al.*, 2024; Picking, 2024). Many plant-based drugs are derived from traditional knowledge, which is often not adequately protected by current intellectual property laws. This can lead to biopiracy and exploitation of indigenous resources without fair compensation (Picking, 2024). The development of new plant-based drugs involves significant research and innovation, which need protection through patents. However, the complexity of plant compounds and the traditional nature of many formulations pose challenges in securing patents (Ortega-Berlanga and Pniewski, 2023). The regulatory landscape for plant-based drugs is fragmented, with significant differences between regions such as the USA, EU, and developing countries. This diversity creates barriers to international trade and complicates compliance for manufacturers (Picking, 2024). There is a pressing need for international harmonization of regulations to ensure safety and efficacy while facilitating global market access. The WHO plays a crucial role in this effort, but achieving consensus among countries remains challenging (Picking, 2024). Ensuring the safety and quality of plant-based drugs is paramount, yet the lack of standardized testing and quality control measures poses risks to consumer health (Wijayabandara, 2019).

While the hurdles in standardization, intellectual property, and global compliance for plant-based drugs are significant, they also present opportunities for innovation and collaboration. The development of robust analytical methods and international regulatory frameworks can enhance the quality and safety of herbal medicines. Additionally, protecting traditional knowledge and fostering equitable sharing of benefits can promote sustainable development and innovation in the field. Addressing these challenges requires a concerted effort from governments, industry stakeholders, and international organizations to create a more cohesive and supportive environment for plant-based drug development.

6.8 Challenges and Future Directions

The journey from discovering a plant's bioactive potential to developing a safe and effective therapeutic is fraught with significant scientific and technical challenges. A primary hurdle lies in the immense data complexity and integration required. Modern research relies on multi-omics approaches, generating vast datasets from genomics, proteomics, and metabolomics. However, managing this information requires advanced bioinformatics tools to harmonize the different data types. This effort is further complicated by the inherent variability of plant metabolism, which can shift dramatically with environmental conditions, undermining experimental reproducibility. Additionally, the existence of fragmented and siloed databases often hinders the cross-omics analyses necessary to build a complete picture of a plant's biochemical machinery.

Even after a promising compound is identified, issues of bioavailability and formulation present a major barrier. Many bioactive scaffolds, such as the well-known curcumin, are plagued by poor solubility in water, rapid metabolism in the body, and chemical instability. To overcome these problems, researchers are developing sophisticated delivery systems like nanocarriers and encapsulation techniques. These approaches show great promise for protecting the compound and enhancing its delivery to target sites. Nonetheless, translating these innovations from the lab to the clinic faces its own set of obstacles, including regulatory hurdles for novel formulations and significant challenges in scaling up production in a cost-effective and reliable manner.

Ultimately, ensuring the safety of any plant-derived compound is paramount and non-negotiable. Some plants, such as *Aconitum*, which contains toxic diterpenoid alkaloids, inherently pose serious toxicity risks. Traditional processing methods used in herbal medicine may reduce but do not always fully eliminate these hazards, raising concerns for modern therapeutic applications. Therefore, it is essential to move beyond traditional assessments and employ rigorous, omics-guided toxicity profiling. By using transcriptomics and metabolomics, scientists can gain a comprehensive understanding of a compound's mechanism of toxicity and its effects on a cellular level, which is critical for ensuring patient safety and meeting regulatory standards for new therapeutics.

The inherent multi-component nature of herbal medicines presents a fundamental challenge for standardization and quality control. Unlike single-compound pharmaceuticals, the therapeutic effect of an herbal product is often the result of a complex synergy between numerous active and inactive constituents. This complexity makes it difficult to ensure consistent chemical profiles and biological activity from one batch to another. Compounding this issue is the global lack of universal protocols for identifying and quantifying key marker compounds. Without harmonized analytical standards and validated methods, quality assurance becomes a significant hurdle across international supply chains, leading to variable product potency, efficacy, and safety.

The path to market for plant-based therapeutics is further obstructed by significant regulatory fragmentation across the globe. Divergent frameworks and approval processes from major bodies like the EMA, USFDA, and WHO create a complex and often burdensome compliance landscape for manufacturers. This lack of international harmonization slows down global access and increases development costs. Simultaneously, profound intellectual property conflicts arise, as conventional patent systems often fail to adequately protect collectively held traditional knowledge. This inadequacy can lead to instances of biopiracy, where commercial entities patent compounds derived from traditional remedies without fair benefit-sharing with the source communities, creating ethical and legal disputes.

A critical bottleneck in plant-based drug development is the persistent translational gap between preclinical research and clinical application. Promising results in laboratory models and animal studies rarely translate into clinical efficacy in human trials. This failure is frequently due to insufficient pharmacokinetic profiling—a lack of understanding of how the body absorbs, distributes, metabolizes, and excretes these complex mixtures—and the common use of underpowered or poorly designed clinical trials. Furthermore, the very complexity that defines herbal medicines, specifically the difficulty in isolating a single API from a matrix of synergistic compounds, often delays or prevents the filing of a robust IND application, which is the essential gateway to human clinical testing.

We must integrate AI with multi-omics technologies to comprehensively forecast compound–target interactions, toxicity profiles, and ADMET properties. Deep learning algorithms should be leveraged to prioritize promising lead candidates identified during virtual screening campaigns. Furthermore, expanding multi-omics-guided network models will enable precise mapping of polypharmacological effects, significantly enhancing therapeutic precision for complex multifactorial diseases like cancer and diabetes.

Concurrently, overcoming bioavailability challenges requires developing next-generation nanocarriers—such as engineered lipid nanoparticles and exosomes—specifically designed to improve the delivery of plant metabolites. Strategic investment should also focus on "smart" responsive formulations that react to physiological triggers, exemplified by pH-sensitive curcumin micelles. To address supply limitations, synthetic biology offers transformative potential: we can engineer plant or microbial systems for high-yield metabolite synthesis (e.g. paclitaxel production), with CRISPR-based genome editing serving as a powerful tool to optimize biosynthetic pathways.

Ensuring sustainability demands promoting ethical sourcing through blockchain-tracked cultivation practices, guaranteeing ecological responsibility. Globally, advocacy for WHO-led standards is essential to formally integrate traditional knowledge with rigorous multi-omics validation. Simultaneously, establishing unified intellectual property frameworks is critical—these must protect indigenous resources while effectively incentivizing innovation. On the translational front, implementing PBPK models will bridge gaps between preclinical findings and clinical data. Clinical development should utilize biomarker-driven trial designs incorporating omics-derived safety and efficacy endpoints, such as metabolomic signatures predictive of toxicity.

Success hinges on fostering dedicated consortia that unite ethnobotanists, computational biologists, and regulators to streamline the entire drug development pipeline. Finally, building comprehensive open-access omics repositories—including genomic biosynthetic gene cluster (BGC) databases—will provide the foundational infrastructure to accelerate discovery efforts across the scientific community.

6.9 Conclusion

Omics technologies have indelibly transformed drug discovery from medicinal plants, offering unprecedented insights into bioactive compound biosynthesis, target identification, and therapeutic efficacy. Genomics and transcriptomics unravel genetic blueprints of metabolites like diterpenoid alkaloids, while proteomics and metabolomics decode functional protein networks and metabolic fluxes. HTS, augmented by AI, accelerates lead identification, and network pharmacology elucidates multi-target mechanisms critical for complex diseases. Case studies on turmeric's curcuminoids exemplify how omics validate anti-inflammatory pathways and guide nanoformulations to overcome bioavailability barriers. However, the journey from plant to patient remains fraught with challenges. Translational gaps persist due to pharmacokinetic variability, while regulatory fragmentation and standardization hurdles impede global commercialization. Toxicity concerns, data integration complexities, and ethical sourcing demand innovative solutions. Future success hinges on AI-driven predictive modeling, sustainable bioproduction, and harmonized regulations that honor traditional knowledge. By fostering interdisciplinary collaboration and investing in advanced delivery systems, the field can unlock the full potential of phytopharmaceuticals. In bridging ancient wisdom and cutting-edge science, omics-driven approaches promise a new era of safe, effective plant-based therapeutics. As computational tools evolve and global frameworks mature, medicinal plants will continue to be vital reservoirs for drug discovery, addressing unmet medical needs while upholding ecological and ethical integrity. The integration of omics is not merely transformative—it is essential for the sustainable future of global health care.

Acknowledgments

The authors express their gratitude for the financial support received through the Distinguished High-Level Talents Research Grant from the Guizhou Science and Technology Corporation Platform Talents Fund (Grant nos.: [2017]5733-001 and CK-1130-002), high-level specialists project fund of Zunyi Medical University (1092/2000006)F-1092, the National Natural Science Foundation of China (82373981), and the support provided by Zunyi Medical University, China. Special appreciation is extended to all laboratory colleagues and research staff members for their valuable insights, constructive guidance, and assistance throughout this study.

References

Aborode, A.T., Awuah, W.A., Mikhailova, T., Abdul-Rahman, T., Pavlock, S. *et al.* (2022) OMICs technologies for natural compounds-based drug development. *Current Topics in Medicinal Chemistry* 22(21), 1751–1765. DOI: 10.2174/1568026622666220726092034.

Acosta, M.J.C. and de Anda Jáuregui, G. (2025) Network pharmacognosy of *Galphimia glauca*: Mapping the molecular landscape of a traditional Mexican medicinal plant. *PLoS ONE* 20(7), e0317546. DOI: 10.1101/2025.01.02.631171.

Ahmed, R. (2024) The role of cGMP in the manufacturing unit of a pharmaceutical industry. *Radinka Journal of Health Science* 2(2), 242–253. DOI: 10.56778/rjhs.v2i2.361.

Akaberi, M., Sahebkar, A. and Emami, S.A. (2021) Turmeric and curcumin: From traditional to modern medicine. In: Guest, P.C. (ed.) *Studies on Biomarkers and New Targets in Aging Research in Iran: Focus on Turmeric and Curcumin*. Springer International Publishing, Cham, Switzerland, pp. 15–39.

Al Rashid, H., Kundu, A., Mandal, V., Wangchuk, P. and Mandal, S.C. (2020) Preclinical and clinical trials of Indian medicinal plants in disease control. In: Sen, S. and Chakraborty, R. (eds) *Herbal Medicine in India*. Springer Singapore, pp. 119–142.

Aloufi, B., Alshabrmi, F.M., Sreeharsha, N. and Rehman, A. (2024) Exploring therapeutic targets and drug candidates for obesity: A combined network pharmacology, bioinformatics approach. *Journal of Biomolecular Structure & Dynamics* 42(21), 11879–11900. DOI: 10.1080/07391102.2023.2265491.

Amaral, C., Paiva, M., Rodrigues, A.R., Veiga, F. and Bell, V. (2024) Global regulatory challenges for medical devices: Impact on innovation and market access. *Applied Sciences* 14(20), 9304. DOI: 10.3390/app14209304.

Aware, C.B., Patil, D.N., Suryawanshi, S.S., Mali, P.R., Rane, M.R. *et al.* (2022) Natural bioactive products as promising therapeutics: A review of natural product-based drug development. *South African Journal of Botany* 151, 512–528. DOI: 10.1016/j.sajb.2022.05.028.

Babar, M.M., Zaidi, N.-S.S., Pothineni, V.R., Ali, Z., Faisal, S. *et al.* (2017) Application of bioinformatics and system biology in medicinal plant studies. In: Rehman Hakeem, K., Malik, A., Vardar-Sukan, F. and Ozturk, M. (eds) *Plant Bioinformatics*. Springer International Publishing, Cham, Switzerland, pp. 375–393.

Babiker, N.A., Nadeem, S., Abu Kariem, H., Abdul Hameed, A., Negmeldin, A.T. *et al.* (2025) Medicinal chemistry strategies in targeting TGF-βR1 kinase domain: Unveiling insights into inhibitor structure–activity relationship (SAR). *Pharmaceuticals (Basel, Switzerland)* 18(5), 716. DOI: 10.3390/ph18050716.

Booij, T.H., Price, L.S. and Danen, E.H.J. (2019) 3D cell-based assays for drug screens: Challenges in imaging, image analysis, and high-content analysis. *SLAS Discovery* 24(6), 615–627. DOI: 10.1177/2472555219830087.

Boroumand, N., Samarghandian, S. and Hashemy, S.I. (2018) Immunomodulatory, anti-inflammatory, and antioxidant effects of curcumin. *Journal of Herbmed Pharmacology* 7(4), 211–219. DOI: 10.15171/jhp.2018.33.

Buriani, A., Garcia-Bermejo, M.L., Bosisio, E., Xu, Q., Li, H. *et al.* (2012) Omic techniques in systems biology approaches to traditional Chinese medicine research: Present and future. *Journal of Ethnopharmacology* 140(3), 535–544. DOI: 10.1016/j.jep.2012.01.055.

Buriani, A., Fortinguerra, S. and Carrara, M. (2017) Clinical perspectives in diagnostic-omics and personalized medicine approach to monitor effectiveness and toxicity of phytocomplexes. In: Pelkonen, O., Duez, P., Vuorela, P.M. and Vuorela, H. (eds) *Toxicology of Herbal Products*. Springer International Publishing, Cham, Switzerland, pp. 385–476.

Cabral da Silva, R.C., Silva Alves, M.C., Ferreira Matos, D. and Alves Campos Quaresma, S. (2024) Bioinformatics approaches to explore the discovery of new plant-derived drugs: A brief review. *Concilium* 24(19), 103–125. DOI: 10.53660/CLM-4175-24T09.

Cardona, K.M., Rugeles-Silva, P.A., Tarazona-Pulido, L., Diaz-Ariza, L.A., Muñoz-Florez, J.E. *et al.* (2024) Exploring the potential of omics technologies in medicinal plant research: A review in Colombia. *Boletin Latinoamericano y Del Caribe de Plantas Medicinales y Aromaticas* 23(4), 460–486. DOI: 10.37360/blacpma.24.23.4.32.

Chattopadhyay, P. and Veer, V. (2016) Regulatory approval of botanical products including herbal drugs, food, and insecticides for commercialization. In: Veer, V. and Gopalakrishnan, R. (eds) *Herbal Insecticides, Repellents and Biomedicines: Effectiveness and Commercialization*. Springer India, New Delhi, pp. 243–258.

Chaudhary, P., Gupta, M., Tawar, M.G. and Shrivastava, B. (2024) Need & scope of standardization of herbal medicines – a review. *Journal of Research in Pharmaceutical Science* 7(8), 26–31. DOI: 10.2139/ssrn.4961133.

Chauhan, M., Kumar, S., Biswas, A., Kumar, M., Verma, S.K. *et al*. (2024) A comprehensive review of the advancement in omic technologies in the field of drug discovery and development. *Letters in Drug Design & Discovery* 21(16), 3319–3331. DOI: 10.2174/0115701808287654240126112003.

Chelimela, N., Alavala, R.R. and Satla, S.R. (2024) Curcumin – bioavailability enhancement by prodrug approach and novel formulations. *Chemistry & Biodiversity* 21(5), e202302030. DOI: 10.1002/cbdv.202302030.

Chen, D.-Q., Chen, H., Chen, L., Tang, D.-D., Miao, H. *et al*. (2016) Metabolomic application in toxicity evaluation and toxicological biomarker identification of natural product. *Chemico-Biological Interactions* 252, 114–130. DOI: 10.1016/j.cbi.2016.03.028.

Chihomvu, P., Ganesan, A., Gibbons, S., Johannson, M., Woollard, K. *et al*. (2024a) Phytochemicals in drug discovery - a confluence of tradition and innovation. DOI: 10.20944/preprints202406.0927.v1.

Chihomvu, P., Ganesan, A., Gibbons, S., Johannson, M., Woollard, K. *et al*. (2024b) Phytochemicals in drug discovery - a confluence of tradition and innovation. DOI: 10.20944/preprints202406.0927.v2.

Chihomvu, P., Ganesan, A., Gibbons, S., Woollard, K. and Hayes, M.A. (2024c) Phytochemicals in drug discovery - a confluence of tradition and innovation. *International Journal of Molecular Sciences* 25(16), 8792. DOI: 10.3390/ijms25168792.

Cozmin, M., Lungu, I.I., Gutu, C., Stefanache, A., Duceac, L.D. *et al*. (2024) Turmeric: From spice to cure. A review of the anti-cancer, radioprotective and anti-inflammatory effects of turmeric sourced compounds. *Frontiers in Nutrition* 11, 1399888. DOI: 10.3389/fnut.2024.1399888.

da Cruz, R.M.D., Mendonça-Junior, F.J.B., de Mélo, N.B., Scotti, L., de Araújo, R.S.A. *et al*. (2021) Thiophene-based compounds with potential anti-inflammatory activity. *Pharmaceuticals (Basel, Switzerland)* 14(7), 692. DOI: 10.3390/ph14070692.

da Silva Carneiro, R., Nogueira, T.A., de Barros Sousa, É., da Silva, S.D.C. and Mendes, A.N. (2023) Inflammatory modulation of compounds derived from turmeric (*Curcuma longa*) in neurodegenerative diseases. In: Rai, M. and Feitosa, C.M. (eds) *Curcumin and Neurodegenerative Diseases*. Springer Nature, Singapore, pp. 437–452.

Dai, R., Liu, M., Xiang, X., Li, Y., Xi, Z. *et al*. (2022) OMICS applications for medicinal plants in gastrointestinal cancers: Current advancements and future perspectives. *Frontiers in Pharmacology* 13, 842203. DOI: 10.3389/fphar.2022.842203.

Dar, R.A., Shahnawaz, M., Ahanger, M.A. and Majid, I. ul. (2023) Exploring the diverse bioactive compounds from medicinal plants: A review. *The Journal of Phytopharmacology* 12(3), 189–195. DOI: 10.31254/phyto.2023.12307.

Desai, U., Poojara, L., Goswami, D., Pandya, M., Modi, S. *et al*. (2025) Medicinal plants in combination therapies for cancer. In: Kuca, K., Patocka, J., Kumar, V. and Dhalaria, R. (eds) *Medicinal Plants and Their Bioactives in Human Diseases*. Springer Nature Switzerland, Cham, Switzerland, pp. 259–291.

Dzobo, K. (2022) The role of natural products as sources of therapeutic agents for innovative drug discovery. In: Kenakin, T. (ed.) *Comprehensive Pharmacology*. Elsevier, Amsterdam, pp. 408–422.

El-Saadony, M.T., Saad, A.M., Mohammed, D.M., Alkafaas, S.S., Ghosh, S. *et al*. (2025) Curcumin, an active component of turmeric: Biological activities, nutritional aspects, immunological, bioavailability, and human health benefits - a comprehensive review. *Frontiers in Immunology* 16, 1603018. DOI: 10.3389/fimmu.2025.1603018.

Fatokun, A.A., Nahar, L. and Sarker, S.D. (2024) Network pharmacology in phytochemical research. In: Sarker, S.D. and Nahar, L. (eds) *Computational Phytochemistry*. Elsevier, Amsterdam, pp. 465–493.

Fernández-Moriano, C., González-Burgos, E. and Gómez-Serranillos, M.P. (2019) Curcumin: Current evidence of its therapeutic potential as a lead candidate for anti-inflammatory drugs—an overview. In: Brahmachari, G. (ed.) *Discovery and Development of Anti-Inflammatory Agents from Natural Products*. Elsevier, Amsterdam, pp. 7–59.

Fernandez-Moure, J.S. (2016) Lost in translation: The gap in scientific advancements and clinical application. *Frontiers in Bioengineering and Biotechnology* 4, 43. DOI: 10.3389/fbioe.2016.00043.

Firouzi, R. and Ashouri, M. (2022) Identification of potential anti-COVID-19 drug leads from medicinal plants through virtual high-throughput screening. DOI: 10.26434/chemrxiv-2022-zsk04.

Firouzi, R. and Ashouri, M. (2023) Identification of potential anti-COVID-19 drug leads from medicinal plants through virtual high-throughput screening. *ChemistrySelect* 8(7). DOI: 10.1002/slct.202203865.

Garse, S., Thakur, V., Devarajan, S., Bhori, M., Jobby, R. *et al.* (2025) Systems biology study on bioactive compounds. In: Lahiri, D., Nag, M., Bhattacharya, D., Pati, S. and Sarkar, T. (eds) *Bioactive Ingredients for Healthcare Industry*, Vol. 1. Springer Nature, Singapore, pp. 347–368.

Gautam, G. (2023) Advancement of network pharmacology in multi-targeted therapeutic evaluation of medicinal plants. *Journal of CAM Research Progress* 2(1), 113. DOI: 10.33790/jcrp1100113.

Gayathiri, E., Prakash, P., Ahamed, M., Pandiaraj, S., Venkidasamy, B. *et al.* (2024) Multitargeted pharmacokinetics, molecular docking and network pharmacology-based identification of effective phytocompounds from *Sauropus androgynus* (L.) Merr for inflammation and cancer treatment. *Journal of Biomolecular Structure and Dynamics* 42(15), 7883–7896. DOI: 10.1080/07391102.2023.2243335.

Girst, G., Ötvös, S.B., Fülöp, F., Balogh, G.T. and Hunyadi, A. (2021) Pharmacokinetics-driven evaluation of the antioxidant activity of curcuminoids and their major reduced metabolites—a medicinal chemistry approach. *Molecules (Basel, Switzerland)* 26(12), 3542. DOI: 10.3390/molecules26123542.

Gogoi, B., Gogoi, D., Gogoi, N., Mahanta, S. and Buragohain, A.K. (2022) Network pharmacology based high throughput screening for identification of multi targeted anti-diabetic compound from traditionally used plants. *Journal of Biomolecular Structure and Dynamics* 40(17), 8004–8017. DOI: 10.1080/07391102.2021.1905554.

Gonzalez-Covarrubias, V., Martínez-Martínez, E. and del Bosque-Plata, L. (2022) The potential of metabolomics in biomedical applications. *Metabolites* 12(2), 194. DOI: 10.3390/metabo12020194.

Gupta, R., Srivastava, D., Sahu, M., Tiwari, S., Ambasta, R.K. *et al.* (2021) Artificial intelligence to deep learning: Machine intelligence approach for drug discovery. *Molecular Diversity* 25(3), 1315–1360. DOI: 10.1007/s11030-021-10217-3.

Haghir Ebrahim Abadi, M.H., Ghasemlou, A., Bayani, F., Sefidbakht, Y., Vosough, M. *et al.* (2025) AI-driven covalent drug design strategies targeting main protease (m^{pro}) against SARS-CoV-2: Structural insights and molecular mechanisms. *Journal of Biomolecular Structure and Dynamics* 43(11), 5436–5464. DOI: 10.1080/07391102.2024.2308769.

Halder, M., Majumder, A., Ray, S. and Jha, S. (2021) Medicinal plant research at crossroads: Biotechnological approaches for conservation. In: Ekiert, H.M., Ramawat, K.G. and Arora, J. (eds) *Medicinal Plants: Sustainable Development and Biodiversity*, Vol. 28. Springer, Cham, Switzerland, pp. 459–544.

Hassan, S.F., Asghar, S., Ullah Khan, I., Munir, R. and Khalid, S.H. (2024) Curcumin encapsulation in geranium oil microemulsion elevates its antibacterial, antioxidant, anti-inflammatory, and anticancer activities. *ACS Omega* 9(5), 5624–5636. DOI: 10.1021/acsomega.3c08033.

Hegde, M., Girisa, S., BharathwajChetty, B., Vishwa, R. and Kunnumakkara, A.B. (2023) Curcumin formulations for better bioavailability: What we learned from clinical trials thus far? *ACS Omega* 8(12), 10713–10746. DOI: 10.1021/acsomega.2c07326.

Heyward, J., Padula, W., Tierce, J.C. and Alexander, G.C. (2018) The value of U.S. pharmacopeial standards: A review of the literature. *Journal of Pharmaceutical Sciences* 107(10), 2611–2617. DOI: 10.1016/j.xphs.2018.06.004.

Howells, L., Malhotra Mukhtyar, R., Theofanous, D., Pepper, C., Thomas, A. *et al.* (2021) A systematic review assessing clinical utility of curcumin with a focus on cancer prevention. *Molecular Nutrition & Food Research* 65(13), e2000977. DOI: 10.1002/mnfr.202000977.

Jabeen, F., Khursheed, N., Ahmad, R., Rasool, S., Ali, A. *et al.* (2024) Crosstalk of multiomics approaches with medicinal plants of therapeutic importance. In: Ali, A., Rehman, M.U., Ahmad, S.B. and Arafah, A. (eds) *Biological Insights of Multi-Omics Technologies in Human Diseases*. Academic Press, Cambridge, Massachusetts, pp. 21–51.

Jacob, S., Kather, F.S., Morsy, M.A., Boddu, S.H.S., Attimarad, M. *et al.* (2024) Advances in nanocarrier systems for overcoming formulation challenges of curcumin: Current insights. *Nanomaterials (Basel, Switzerland)* 14(8), 672. DOI: 10.3390/nano14080672.

Jadhav, R.R. and Khare, D. (2024) Green biotherapeutics: Overcoming challenges in plant-based expression platforms. *Plant Biotechnology Reports* 18(4), 465–486. DOI: 10.1007/s11816-024-00910-8.

Javid, A., Fatima, A., Hamad, M. and Ahmed, M. (2024) From roots to codes: Applications of computer-aided drug discovery from medicinal plants. *South African Journal of Botany* 173, 159–174. DOI: 10.1016/j.sajb.2024.08.033.

Joshi, C. and Makwana, D. (2024) Advanced high throughput screening platforms in drug discovery. In: Singh, A., Parikesit, A.A. and Naderi, M. (eds) *Genomics-Driven Drug Discovery Through Pharmacogenomics. IGI Global*. Hershey, Pennsylvania, pp. 35–74.

Jude, S., Amalraj, A., Kunnumakkara, A.B., Divya, C., Löffler, B.-M. *et al.* (2018) Development of validated methods and quantification of curcuminoids and curcumin metabolites and their pharmacokinetic study of oral administration of complete natural turmeric formulation (Cureit™) in human plasma via UPLC/ESI-Q-TOF-MS spectrometry. *Molecules (Basel, Switzerland)* 23(10), 2415. DOI: 10.3390/molecules23102415.

Kamil Hussain, M., Saquib, M. and Faheem Khan, M. (2019) Techniques for extraction, isolation, and standardization of bio-active compounds from medicinal plants. In: Akhtar, M.S., Kumara Swamy, M. and Sinniah, U.R. (eds) *Natural Bio-Active Compounds*. Springer Singapore, pp. 179–200.

Kapoor, R.K., Hafeez, A., Kumar, S., Mishra, D. and Gupta, M.K. (2025) Intellectual property rights of medicinal products. In: Bukhsh Singh, D. and Kumar Upadhyay, S. (eds) *Medicinal Biotechnology*. Academic Press, Cambridge, Massachusetts, pp. 327–341.

Karthikeyan, M., Mehta, A., Kumar, S.S., Mohan, H. and Usha, B. (2025) Safety, dosage, and regulatory aspects in the use of phytochemicals. In: Kuca, K., Patocka, J., Kumar, V. and Dhalaria, R. (eds) *Medicinal Plants and Their Bioactives in Human Diseases*. Springer Nature Switzerland, Cham, Switzerland, pp. 207–233.

Khair, F.B., Bhuiyan, M.M.R., Manik, M.M.T.G., Hossain, S., Islam, M.S. *et al.* (2024) Machine learning approaches to identify and optimize plant-based bioactive compounds for targeted cancer treatments. *British Journal of Pharmacy and Pharmaceutical Sciences* 1(1), 60–67. DOI: 10.32996/bjpps.2024.1.1.7.

Khuspe, P.R., Swapnil, P., Prajakta, I., Sakshi, K., Vaishnavi, I. *et al.* (2024) Overview of herbal drugs: Regulatory perspective with special emphasis on global market. *Yemen Journal of Medicine* 03(3), 207–217. DOI: 10.63475/j.yjom.2024.022.

Kielich, N., Mazur, O., Musidlak, O., Gracz-Bernaciak, J. and Nawrot, R. (2024) Herbgenomics meets Papaveraceae: A promising-omics perspective on medicinal plant research. *Briefings in Functional Genomics* 23(5), 579–594. DOI: 10.1093/bfgp/elad050.

Kohl, P., Crampin, E.J., Quinn, T.A. and Noble, D. (2010) Systems biology: An approach. *Clinical Pharmacology & Therapeutics* 88(1), 25–33. DOI: 10.1038/clpt.2010.92.

Koromina, M., Pandi, M.-T. and Patrinos, G.P. (2019) Rethinking drug repositioning and development with artificial intelligence, machine learning, and omics. *OMICS: A Journal of Integrative Biology* 23(11), 539–548. DOI: 10.1089/omi.2019.0151.

Kurniawati, E.Y., Pramono, N., Hidayat, S.T. and Mahati, E. (2023) *In silico* pharmacokinetic and toxicity analysis on *Clitoria ternatea* flower. *Jurnal Farmasi Indonesia* 20(2), 124–135. DOI: 10.31001/jfi.v20i2.2190.

Laloo, D., Kalita, J.M. and Prasad, S.K. (2021) Molecular docking studies of plant-derived bioactive compounds: A comprehensive *in silico* standardization approach. In: Mandal, S.C., Chakraborty, R. and Sen, S. (eds) *Evidence Based Validation of Traditional Medicines*. Springer Singapore, pp. 371–404.

Lamberth, C. (2018) Agrochemical lead optimization by scaffold hopping. *Pest Management Science* 74(2), 282–292. DOI: 10.1002/ps.4755.

Lin, C., Tian, Q., Guo, S., Xie, D., Cai, Y. *et al.* (2024) Metabolomics for clinical biomarker discovery and therapeutic target identification. *Molecules* 29(10), 2198. DOI: 10.3390/molecules29102198.

Ma, X., Meng, Y., Wang, P., Tang, Z., Wang, H. *et al.* (2020) Bioinformatics-assisted, integrated omics studies on medicinal plants. *Briefings in Bioinformatics* 21(6), 1857–1874. DOI: 10.1093/bib/bbz132.

Mahanayak, B. (2024) Computational strategies for drug discovery: Harnessing Indian medicinal plants. *World Journal of Biology Pharmacy and Health Sciences* 19(1), 165–170. DOI: 10.30574/wjbphs.2024.19.1.0407.

Maheshwari, N.V., Tripathi, M.K., Kumar, S. and Kumar, A. (2025) Translational research in Ayurveda: An introduction to evidence-based approaches. In: Kumar, A., Maheshwari, N.V., Kumar Tripathi, M. and Kumar, S. (eds) *Innovations in Ayurvedic Nutrition*. Springer Nature, Singapore, pp. 1–14.

Mansoor, S., Hamid, S., Tuan, T.T., Park, J.-E. and Chung, Y.S. (2024) Advance computational tools for multi-omics data learning. *Biotechnology Advances* 77, 108447. DOI: 10.1016/j.biotechadv.2024.108447.

Marx-Stoelting, P., Braeuning, A., Buhrke, T., Lampen, A., Niemann, L. *et al.* (2015) Application of omics data in regulatory toxicology: Report of an international BfR expert workshop. *Archives of Toxicology* 89(11), 2177–2184. DOI: 10.1007/s00204-015-1602-x.

Mishra, A., Thakur, A., Sharma, R., Onuku, R., Kaur, C. *et al.* (2024) Scaffold hopping approaches for dual-target antitumor drug discovery: Opportunities and challenges. *Expert Opinion on Drug Discovery* 19(11), 1355–1381. DOI: 10.1080/17460441.2024.2409674.

Mondal, A., Banerjee, S. and Chakraborty, U. (2024) Medicinal plants against hepatitis B infections: Ethnopharmacology, chemistry, and clinical and preclinical studies. In: Pal, D. (ed.) *Anti-Viral Metabolites from Medicinal Plants*. Springer Nature, Cham, Switzerland, pp. 307–340.

Mukherjee, A., Abraham, S., Singh, A., Balaji, S. and Mukunthan, K.S. (2025) From data to cure: A comprehensive exploration of multi-omics data analysis for targeted therapies. *Molecular Biotechnology* 67(4), 1269–1289. DOI: 10.1007/s12033-024-01133-6.

Mukherjee, P.K., Banerjee, S. and Kar, A. (2021) Molecular combination networks in medicinal plants: Understanding synergy by network pharmacology in Indian traditional medicine. *Phytochemistry Reviews* 20(4), 693–703. DOI: 10.1007/s11101-020-09730-4.

Nagori, K., Nakhate, K.T., Yadav, K., Ajazuddinand Pradhan, M. (2023) Unlocking the therapeutic potential of medicinal plants for Alzheimer's disease: Preclinical to clinical trial insights. *Future Pharmacology* 3(4), 877–907. DOI: 10.3390/futurepharmacol3040053.

Nayan, S., Kolay, A., Chauhan, S.B. and Singh, I. (2025) Navigating global regulatory divergence in over-the-counter drugs: Addressing emerging risks and harmonization challenges. *Current Drug Research Reviews* 17. DOI: 10.2174/0125899775370402250622155744.

Noor, F., Tahir ul Qamar, M., Ashfaq, U.A., Albutti, A., Alwashmi, A.S.S. *et al.* (2022) Network pharmacology approach for medicinal plants: Review and assessment. *Pharmaceuticals* 15(5), 572. DOI: 10.3390/ph15050572.

Oliveira, D.R. de, Oliveira, A.C.D. de and Marques, L.C. (2016) O estado regulatório dos fitoterápicos no Brasil: Um paralelo entre a legislação e o mercado farmacêutico (1995–2015). *Vigilância Sanitária Em Debate* 4(4). DOI: 10.22239/2317-269x.00806.

Ortega-Berlanga, B. and Pniewski, T. (2023) Plant molecular pharming: Opportunities, challenges, and future perspectives. In: Kole, C., Chaurasia, A., Hefferon, K.L. and Panigrahi, J. (eds) *Tools & Techniques of Plant Molecular Farming*. Springer Singapore, pp. 35–61.

Pace, R. and Martinelli, E.M. (2022) The phytoequivalence of herbal extracts: A critical evaluation. *Fitoterapia* 162, 105262. DOI: 10.1016/j.fitote.2022.105262.

Pandita, D., Pandita, A., Wani, S.H., Abdelmohsen, S.A.M., Alyousef, H.A. *et al.* (2021) Crosstalk of multi-omics platforms with plants of therapeutic importance. *Cells* 10(6), 1296. DOI: 10.3390/cells10061296.

Papachristos, A., Patel, J., Vasileiou, M. and Patrinos, G.P. (2023) Dose optimization in oncology drug development: The emerging role of pharmacogenomics, pharmacokinetics, and pharmacodynamics. *Cancers* 15(12), 3233. DOI: 10.3390/cancers15123233.

Parmar, G., Shah, A., Chudasama, J.M., Kashav, P. and James, V. (2025) Network-based methods in drug discovery. In: Rudrapal, M. (ed.) *Computational Methods for Rational Drug Design*. Wiley, Oxford, UK, pp. 255–284.

Patel, M. and Patel, V. (2023) High throughput screening techniques in anticancer drug discovery and development from botanicals. In: Chaitanya, M.V.N.L., Pereira, G. and Ali, H. (eds) *Alternative Remedies and Natural Products for Cancer Therapy: An Integrative Approach*. Bentham Science Publishers, Singapore, pp. 97–121.

Patil, R., Satpute, R. and Nalage, D. (2023) The application of omics technologies to toxicology. *Toxicology Advances* 5(2), 6. DOI: 10.53388/TA202305006.

Peng, Y., Ao, M., Dong, B., Jiang, Y., Yu, L. *et al.* (2021) Anti-inflammatory effects of curcumin in the inflammatory diseases: Status, limitations and countermeasures. *Drug Design, Development and Therapy* 15, 4503–4525. DOI: 10.2147/DDDT.S327378.

Pfisterer, P.H., Schuster, D., Rollinger, J.M. and Stuppner, H. (2014) *In silico*-guided strategies for the discovery and rationalization of bioactive natural products. In: Meyers, R.A. (ed.) *Encyclopedia of Analytical Chemistry*. Wiley, Oxford, UK, pp. 1–18.

Picking, D. (2024) The global regulatory framework for medicinal plants. In: Badal Mccreath, S. and Clement, Y.N. (eds) *Pharmacognosy*. Academic Press, Cambridge, Massachusetts, pp. 769–782.

Rai, A., Saito, K. and Yamazaki, M. (2017) Integrated omics analysis of specialized metabolism in medicinal plants. *The Plant Journal* 90(4), 764–787. DOI: 10.1111/tpj.13485.

Rausch, H., Schröder, S., Friedemann, T., Cameron, S., Kuchta, K. *et al.* (2024) History and present of European traditional herbal medicine (phytotherapy). In: Shen, Y. (ed.) *History, Present and Prospect of World Traditional Medicine*. World Scientific, Singapore, pp. 131–234.

Razavi, B.M., Ghasemzadeh Rahbardar, M. and Hosseinzadeh, H. (2021) A review of therapeutic potentials of turmeric (*Curcuma longa*) and its active constituent, curcumin, on inflammatory disorders, pain, and their related patents. *Phytotherapy Research* 35(12), 6489–6513. DOI: 10.1002/ptr.7224.

Roy, D., Ghosh, M. and Rangra, N.K. (2024) Herbal approaches to diabetes management: Pharmacological mechanisms and omics-driven discoveries. *Phytotherapy Research* 39(12), 5464–5490. DOI: 10.1002/ptr.8410.

Sabha, S.U. (2025) Artificial intelligence in human health and diseases. In: Macha, M.A., Assad, A. and Bhat, M.R. (eds) *Artificial Intelligence in Human Health and Diseases*. Springer Nature, Singapore, pp. 75–92.

Sadybekov, A.V. and Katritch, V. (2023) Computational approaches streamlining drug discovery. *Nature* 616(7958), 673–685. DOI: 10.1038/s41586-023-05905-z.

Savino, R., Paduano, S., Preianò, M. and Terracciano, R. (2012) The proteomics big challenge for biomarkers and new drug-targets discovery. *International Journal of Molecular Sciences* 13(11), 13926–13948. DOI: 10.3390/ijms131113926.

Sharma, B. and Yadav, D.K. (2022) Metabolomics and network pharmacology in the exploration of the multi-targeted therapeutic approach of traditional medicinal plants. *Plants* 11(23), 3243. DOI: 10.3390/plants11233243.

Sharma, V. and Sharma, H.D. (2022) Bioactive compounds in turmeric. In: Wani, S.A., Singh, A., and Kumar, P. (eds) *Spice Bioactive Compounds*. CRC Press, Boca Raton, Florida, pp. 71–92.

Shawky, E., Nada, A.A. and Ibrahim, R.S. (2020) Potential role of medicinal plants and their constituents in the mitigation of SARS-CoV-2: Identifying related therapeutic targets using network pharmacology and molecular docking analyses. *RSC Advances* 10(47), 27961–27983. DOI: 10.1039/D0RA05126H.

Singh, A.V., Chandrasekar, V., Paudel, N., Laux, P., Luch, A. *et al.* (2023) Integrative toxicogenomics: Advancing precision medicine and toxicology through artificial intelligence and OMICS technology. *Biomedicine & Pharmacotherapy* 163, 114784. DOI: 10.1016/j.biopha.2023.114784.

Singh, D.B. (2019) The impact of pharmacogenomics in personalized medicine. In: Silva, A.C., Nuno Moreira, J., Sousa Lobo, J.M. and Almeida, H. (eds) *Current Applications of Pharmaceutical Biotechnology*. Springer International Publishing, Cham, Switzerland, pp. 369–394.

Singh, S., Singh, D.B., Singh, S., Shukla, R., Ramteke, P.W. *et al.* (2019) Exploring medicinal plant legacy for drug discovery in post-genomic era. *Proceedings of the National Academy of Sciences, India Section B: Biological Sciences* 89(4), 1141–1151. DOI: 10.1007/s40011-018-1013-x.

Singh, S.K., Kumar, A., Singh, R.B., Ghosh, P. and Bajad, N.G. (2022) Recent applications of bioinformatics in target identification and drug discovery for Alzheimer's disease. *Current Topics in Medicinal Chemistry* 22(26), 2153–2175. DOI: 10.2174/1568026623666221026091010.

Sobti, R.C. (2025) *Biotechnological Innovations for Sustainable Biodiversity and Development*. CRC Press, Boca Raton, Florida.

Son, A., Park, J., Kim, W., Yoon, Y., Lee, S. *et al.* (2024a) Recent advances in omics, computational models, and advanced screening methods for drug safety and efficacy. *Toxics* 12(11), 822. DOI: 10.3390/toxics12110822.

Son, A., Park, J., Kim, W., Yoon, Y., Lee, S. *et al.* (2024b) Utilizing omics technologies and machine learning to improve predictive toxicology. DOI: 10.20944/preprints202408.1827.v1.

Surendran, K., Ranjisha, K.R., Aswati Nair, R. and Pillai, P.P. (2022) Genomics and metabolomics: A strategy for elucidation of metabolic pathways in medicinal plants. In: Kumara Swamy, M. and Kumar, A. (eds) *Phytochemical Genomics*. Springer Nature, Singapore, pp. 343–360.

Thomford, N.E., Senthebane, D.A., Rowe, A., Munro, D., Seele, P. *et al.* (2018) Natural products for drug discovery in the 21st century: Innovations for novel drug discovery. *International Journal of Molecular Sciences* 19(6), 1578. DOI: 10.3390/ijms19061578.

Umamaheswari, D., Muthuraja, R., Kumar, M. and Venkateswarlu, B.S. (2021) Standardization of herbal drugs – an overview. *International Journal of Pharmaceutical Sciences Review and Research* 68(1). DOI: 10.47583/ijpsrr.2021.v68i01.033.

Umar, A.H., Widuri, S.A., Caecilia Sulistyaningsih, Y. and Ratnadewi, D. (2025) Integrating metabolomic analysis, network pharmacology, and molecular docking to underlying pharmacological mechanism and ethnobotanical rationalization for diabetes mellitus: Study on medicinal plant *Fibraurea tinctoria* Lour. *Phytochemical Analysis* 36(4), 912–933. DOI: 10.1002/pca.3477.

Valencia Quiroz, I. (ed.) (2024) *Biotechnology and Drug Development for Targeting Human Diseases*, Vol. 9. Bentham Science Publishers, Singapore.

Verma, N. (2016) Current regulatory challenges and approaches in the registration of herbal drugs in Europe. *Clinical Research and Regulatory Affairs* 33(1), 9–24. DOI: 10.3109/10601333.2016.1130717.

Vijay, A., Kumar, A., Radhakrishnan, A.M., Kumar, S., Singh, K. *et al.* (2022) Emergence of phytochemical genomics: Integration of multi-omics approaches for understanding genomic basis of phytochemicals. In: Kumara Swamy, M. and Kumar, A. (eds) *Phytochemical Genomics*. Springer Nature, Singapore, pp. 219–261.

Wafi, A. and Mirnezami, R. (2018) Translational–omics: Future potential and current challenges in precision medicine. *Methods* 151, 3–11. DOI: 10.1016/j.ymeth.2018.05.009.

Wan, C., Ma, Q., Anderson, S., Zhang, Q.-H., Zhang, C.-F. *et al.* (2023) Effects of curcuminoids and surfactant-formulated curcumin on chemo-resistant colorectal cancer. *The American Journal of Chinese Medicine* 51(6), 1577–1594. DOI: 10.1142/S0192415X23500714.

Wang, M., Zhang, S., Li, R. and Zhao, Q. (2024a) Unraveling the specialized metabolic pathways in medicinal plant genomes: A review. *Frontiers in Plant Science* 15, 1459533. DOI: 10.3389/fpls.2024.1459533.

Wang, T. and Yang, J. (2024) Pharmacoproteomics and precision medicine. In: Amponsah, S.K., Opuni, K.F.M. and Pathak, Y.V. (eds) *Pharmacoproteomics*. Springer Nature, Switzerland, pp. 7–34.

Wang, T., Rangji, C., Liu, W., Ma, J., Zhou, R. *et al.* (2024b) Multi-omics on traditional medicinal plant of the genus aconitum: Current progress and prospect. *Molecules* 30(1), 118. DOI: 10.3390/molecules30010118.

Wang, X., Chan, Y.S., Wong, K., Yoshitake, R., Sadava, D. *et al.* (2023a) Mechanism-driven and clinically focused development of botanical foods as multitarget anticancer medicine: Collective perspectives and insights from preclinical studies, IND applications and early-phase clinical trials. *Cancers* 15(3), 701. DOI: 10.3390/cancers15030701.

Wang, Y., Liang, L. and Zhao, Y. (2024c) Curcumin delivery systems: How far from clinical application in tumor therapy? *Natural Product Communications* 19(2). DOI: 10.1177/1934578X231222102.

Wang, Z., Xiong, Y., Peng, Y., Zhang, X., Li, S. *et al.* (2023b) Natural product evodiamine-inspired medicinal chemistry: Anticancer activity, structural optimization and structure-activity relationship. *European Journal of Medicinal Chemistry* 247, 115031. DOI: 10.1016/j.ejmech.2022.115031.

Weinberg Sibony, R., Segev, O., Dor, S. and Raz, I. (2024) Overview of oxidative stress and inflammation in diabetes. *Journal of Diabetes* 16(10), e70014. DOI: 10.1111/1753-0407.70014.

Wijayabandara, J. (2019) Standardisation and quality control of herbal medicines. In: Ghosh, D. and Mukherjee, P.K. (eds) *Natural Medicines*. CRC Press, Boca Raton, Florida, pp. 13–24.

Xiao, Z., Morris-Natschke, S.L. and Lee, K. (2016) Strategies for the optimization of natural leads to anticancer drugs or drug candidates. *Medicinal Research Reviews* 36(1), 32–91. DOI: 10.1002/med.21377.

Yang, L., Yang, Y., Huang, L., Cui, X. and Liu, Y. (2023) From single- to multi-omics: Future research trends in medicinal plants. *Briefings in Bioinformatics* 24(1), bbac485. DOI: 10.1093/bib/bbac485.

Younas, A., Riaz, N. and Rashid, M. (2023) Multi-omics approaches for breeding in medicinal plants. In: Qayyum, A., Prakash, C.S., Shehzad Baloch, F., Azhar Nadeem, M. and Fiaz, S. (eds) *Sustainable Agriculture in the Era of the OMICs Revolution*. Springer International Publishing, Cham, Switzerland, pp. 165–191.

Yusuf, M. and Khan, S.A. (2022) Assessment of ADME and *in silico* characteristics of natural-drugs from turmeric to evaluate significant COX2 inhibition. *Biointerface Research in Applied Chemistry* 13(1), 5. DOI: 10.33263/BRIAC131.005.

Zhang, W., Zeng, Y., Jiao, M., Ye, C., Li, Y. *et al.* (2023) Integration of high-throughput omics technologies in medicinal plant research: The new era of natural drug discovery. *Frontiers in Plant Science* 14, 1073848. DOI: 10.3389/fpls.2023.1073848.

Zheng, B. and McClements, D.J. (2020) Formulation of more efficacious curcumin delivery systems using colloid science: Enhanced solubility, stability, and bioavailability. *Molecules (Basel, Switzerland)* 25(12), 2791. DOI: 10.3390/molecules25122791.

7 Stress Adaptation and Climate-resilient Medicinal Plants

Surendra Sarsaiya, Archana Jain and Qihai Gong*
Key Laboratory of Basic Pharmacology and Joint International Research Laboratory of Ethnomedicine of Ministry of Education, Zunyi Medical University, Zunyi, Guizhou, China

7.1 Introduction

Medicinal plants face a unique challenge in balancing their survival under environmental stressors with the production of secondary metabolites that confer their medicinal properties. These stressors, such as drought, salinity, and heavy metal exposure, can significantly alter the biochemical pathways responsible for synthesizing these compounds, leading to variations in their therapeutic efficacy. The intricate relationship between environmental stress and secondary metabolite production is crucial for understanding how to conserve and sustainably utilize medicinal flora. This chapter explores the impact of environmental stress on medicinal plants, focusing on the synthesis and accumulation of secondary metabolites. Drought is a major stressor that affects the growth and secondary metabolism of medicinal plants. It often leads to a reduction in plant size and biomass, altering the concentration of secondary metabolites. For instance, drought stress generally increases the concentration of alkaloids and terpenoids, while reducing phenols and flavonoids (Shil and Dewanjee, 2022). This stress-induced variation can enhance the quality of certain secondary metabolites, although not uniformly across all types. Salinity affects the ion balance within plants, influencing the synthesis of secondary metabolites. In *Aloe vera*, elevated soil salinity decreased total phenolic content and antioxidant activity, yet increased the production of aloin, a bioactive compound with anti-inflammatory properties (Sharma, 2024). This indicates that while some metabolites may decrease, others can be upregulated under stress conditions. Exposure to heavy metals like cadmium can lead to a decline in certain secondary metabolites, such as hypericin in *Hypericum perforatum*. However, other compounds, like hyperforin, which has antidepressant properties, may increase under such conditions (Sharma, 2024). Medicinal plants employ various biochemical pathways to adapt to environmental stress. These include the synthesis of osmoprotectants like proline and the activation of detoxification pathways in response to heavy metals (Punetha *et al.*, 2022). Such mechanisms help plants maintain their metabolic functions under adverse conditions.

The impact of environmental stress on secondary metabolites is highly species-specific. For example, *Catharanthus roseus* shows altered levels of vincristine under salinity and drought stress, while *Salvia officinalis* experiences fluctuations

*Corresponding author: gqh@zmu.edu.cn

DOI: 10.1079/9781836994206.0007

in terpene levels (Kanso *et al.*, 2022). This variability underscores the need for species-specific strategies in conservation and cultivation. The effect of climate change on medicinal plants is a multifaceted issue that threatens not only biodiversity but also global health security. Medicinal plants, which have been integral to traditional and modern medicine, are increasingly vulnerable to the effects of climate change, including rising temperatures, altered precipitation patterns, and extreme weather events. These changes affect the growth, distribution, and biochemical properties of medicinal plants, potentially compromising their therapeutic efficacy and availability. This situation necessitates urgent conservation and adaptation strategies to safeguard these vital resources. Climate change is a significant driver of biodiversity loss among medicinal plants, with many species facing extinction due to habitat destruction, overexploitation, and invasive species introduction (Halder and Jha, 2023; Borcă *et al.*, 2024). The geographic distribution of medicinal plants is shifting, with species like *Nepeta glomerulosa* expected to move to higher elevations as their current habitats become unsuitable under future climate scenarios (Karami *et al.*, 2022). The International Union for Conservation of Nature (IUCN) Red List highlights numerous medicinal plant species threatened by climate change, emphasizing the need for targeted conservation efforts (Robiansyah *et al.*, 2023). Changes in temperature and precipitation patterns directly impact the yield and quality of medicinal plants by affecting their growth and secondary metabolite production (Gupta *et al.*, 2019; Sudhakaran, 2025).

Elevated CO_2 levels, ozone, drought, and cold conditions alter the synthesis of secondary metabolites, which are crucial for the medicinal properties of these plants (Gupta *et al.*, 2019). The quality and safety of medicinal products are at risk due to changes in phytochemical constituents, which can affect their therapeutic efficacy (Patni *et al.*, 2022). Conservation strategies include *in situ* and *ex situ* conservation, sustainable harvesting practices, and the cultivation of climate-resilient plant varieties (Applequist *et al.*, 2020; Halder and Jha, 2023). Local cultivation and sustainability training for harvesters are recommended to preserve traditional knowledge and ensure the sustainable use of medicinal plants (Applequist *et al.*, 2020; Patni *et al.*, 2022). Collaborative research and monitoring programs are essential to understand the impacts of climate change on medicinal plants and to develop effective adaptation strategies (Hounsou *et al.*, 2024; Sudhakaran, 2025). The integration of multi-omics technologies has become crucial in understanding the complex mechanisms of stress adaptation in medicinal plants. These technologies, including genomics, transcriptomics, proteomics, metabolomics, and epigenomics, provide a comprehensive view of the molecular responses plants employ to cope with environmental stresses. By examining these responses at various biological levels, researchers can gain insights into the genetic, biochemical, and physiological adaptations that enable plants to survive and thrive under adverse conditions. This holistic approach is essential for developing strategies to enhance plant resilience, which is increasingly important in the context of climate change and global food security.

Genomics provides the foundational genetic blueprint of plants, revealing the potential for adaptation encoded within their DNA. It identifies stress-responsive genes and genetic variations that contribute to stress tolerance (Kumar *et al.*, 2021b; Dar *et al.*, 2023). Advances in high-throughput sequencing have facilitated the discovery of biosynthetic gene clusters and regulatory networks involved in stress responses (Wang *et al.*, 2024). Transcriptomics captures dynamic changes in gene expression under stress conditions, offering insights into the initial molecular defense mechanisms activated in plants (Sharma, 2023; Waseem and Aslam, 2024). This approach helps identify key regulatory genes and pathways that are upregulated or downregulated in response to specific stressors (Ko and Brandizzi, 2024). Proteomics focuses on the functional proteins that execute stress responses, including enzymes, signaling molecules, and structural components. It also highlights post-translational modifications crucial for stress adaptation (Kumar *et al.*, 2021b). By integrating proteomic data with other omics, researchers can better understand the relationship between gene expression and protein abundance under stress conditions (Waseem and Aslam, 2024). Metabolomics profiles the biochemical end products of stress responses, tracking changes in primary and secondary metabolites that are critical for stress tolerance and medicinal value (Jan *et al.*, 2022; Wang *et al.*, 2024). This approach provides insights into metabolic pathways and fluxes that are altered during stress, contributing to the plant's adaptive capacity (Waseem and Aslam, 2024). Epigenomics uncovers heritable, non-genetic regulatory mechanisms, such as DNA methylation and histone modifications, that allow plants to fine-tune their stress responses based on environmental experiences (Singh *et al.*, 2023). These modifications can lead to rapid adaptation without permanent genetic changes, offering a flexible response to fluctuating environmental conditions (Singh *et al.*, 2023).

Multi-omics approaches have elucidated the signaling cascades initiated by stress perception, involving hormones like abscisic acid (ABA), salicylic acid (SA), jasmonate (JA), and ethylene. These hormones play pivotal roles in stress signaling and adaptation (Singh *et al.*, 2015; Morsy and El-Sherif, 2024). Kinases such as MAPKs (mitogen-activated protein kinases) and CDPKs (calcium-dependent protein kinases) are integral to these pathways, mediating the transduction of stress signals and activating downstream responses (de Vasconcelos *et al.*, 2016). Plants employ osmoprotectants like proline and glycine betaine to maintain cellular integrity during drought and salinity stress. These compounds help in osmotic adjustment, protecting cells from dehydration and ionic imbalance (Bagati *et al.*, 2018; Kanojia *et al.*, 2023). Omics studies have identified key genes and enzymes involved in the synthesis of these osmoprotectants, providing targets for genetic engineering to enhance stress tolerance (Debnath *et al.*, 2011).

Reactive oxygen species (ROS) are a common by-product of abiotic stress, and plants have evolved sophisticated antioxidant defense systems to mitigate their damaging effects. Enzymes such as superoxide dismutase (SOD) and catalase (CAT) are crucial in this defense (de Vasconcelos *et al.*, 2016). Secondary metabolites also play a role in ROS scavenging, and their biosynthesis pathways are often upregulated in response to stress, as revealed by metabolomics studies (de Vasconcelos *et al.*, 2016). Maintaining ion homeostasis under salt stress is critical for plant survival. The SOS (salt overly sensitive) signaling pathway is a key mechanism identified through omics approaches that helps regulate ion transport and homeostasis (Khatri and Mudgil, 2017). For heavy metal detoxification, plants utilize chelation and vacuolar sequestration strategies. Phytochelatins are important chelators that bind heavy metals like cadmium and lead, facilitating their sequestration and reducing toxicity (Kanojia *et al.*, 2023). Omics studies have highlighted that stress responses are often tissue-specific, with different tissues exhibiting unique metabolic and gene expression profiles under stress (de Vasconcelos *et al.*, 2016).

The biosynthesis of secondary metabolites, which contribute to the plant's medicinal properties, is closely linked to stress responses. These metabolites not only aid in stress adaptation but also enhance the plant's value for medicinal and industrial applications (Wu *et al.*, 2023). Multi-omics studies into agricultural practices are revolutionizing the development of climate-resilient cultivars. By leveraging genomics, transcriptomics, proteomics, and metabolomics, researchers can identify key genetic markers, transcription factors, and metabolic pathways that contribute to stress tolerance in plants. This knowledge is crucial for breeding strategies aimed at enhancing crop resilience to environmental stressors such as salinity, drought, and heat. These omics-guided approaches not only facilitate the development of robust plant varieties but also unlock new avenues for pathogen defense and phytoremediation. The following sections delve into the specific contributions of multi-omics to these areas. Omics data is instrumental in revolutionizing climate-smart agriculture by informing soil microbiome management, optimizing irrigation schedules, and enabling precision farming approaches (Naqvi *et al.*, 2022; Chaturvedi *et al.*, 2024). These strategies aim to maximize yield and quality while minimizing environmental impact, ensuring sustainable agricultural practices in the face of climate change (Guru *et al.*, 2021; Chaturvedi *et al.*, 2024). Conservation genomics, bolstered by omics technologies, plays a crucial role in safeguarding high-value, critically endangered species like the Pacific yew and goldthread. These species face extinction threats due to climate change and overexploitation, and genomics offers a lifeline by enabling the assessment of genetic diversity, identification of unique populations, and understanding of adaptive potential. This genomic knowledge is vital for prioritizing conservation efforts both *in situ* and *ex situ*, ensuring that strategies are efficient and well directed. The integration of genomics into conservation practices allows for a more comprehensive understanding of species' genetic makeup, which is essential for their long-term survival (Table 7.1 and Fig. 7.1).

7.2 Omics-driven Insights into Abiotic Stress Adaptation in Medicinal Plants

Omics-driven insights into the physiological and biochemical mechanisms of stress adaptation in medicinal plants reveal a complex interplay of molecular responses that enable these plants to withstand various abiotic stresses. These stresses, including UV radiation, temperature extremes, drought, salinity, and heavy metals, significantly impact the growth and productivity of medicinal plants. Omics technologies, such as genomics, proteomics, and metabolomics, provide a comprehensive understanding of the molecular mechanisms underlying stress tolerance, offering potential strategies for enhancing the resilience of medicinal plants. The following sections delve into the specific roles of these omics approaches in elucidating stress adaptation mechanisms in medicinal plants (Table 7.2 and Fig. 7.2).

7.2.1 Proteomics and stress response

Proteomics bridges the gap between genomics and transcriptomics, offering insights into the protein-level changes in response to abiotic stress. Mass-spectroscopy-based proteomics has been instrumental in characterizing stress effects at the protein level, identifying key biomarker proteins and their post-translational modifications. These insights can inform breeding strategies to enhance stress tolerance in medicinal plants (Alugoju and Tencomnao, 2023). Proteomic studies have highlighted the role of specific proteins and enzymes in stress adaptation, such as those involved in the synthesis of osmoprotectants and other protective compounds (Debnath *et al.*, 2011).

7.2.2 Secondary metabolites and UV stress

Medicinal plants synthesize secondary metabolites like alkaloids, terpenoids, flavonoids, and phenolics in response to UV stress. These compounds play a crucial role in counteracting the adverse effects of UV radiation by regulating cellular processes at the molecular level (Thakur *et al.*, 2023). The production of secondary metabolites is modulated by the expression of stress-related genes, enzymes, and transcription factors, which are part of a flexible signaling cascade that enhances the plant's defensive response (Thakur *et al.*, 2023).

7.2.3 Molecular and biochemical adaptation strategies

Medicinal plants employ various molecular and biochemical strategies to adapt to abiotic stresses, including the accumulation of plant hormones (e.g. ABA, JA, ethylene, and SA), changes in redox status, and inorganic ion fluxes (Lamalakshmi Devi *et al.*, 2017; Jan *et al.*, 2022). ROS signaling and systemic acquired resistance (SAR) are also critical components of the plant's defense mechanisms against environmental stressors (Lamalakshmi Devi *et al.*, 2017).

7.2.4 Omics technologies and genetic insights

Advanced omics technologies, including genomics, transcriptomics, and metabolomics, have facilitated the identification of stress-related genes and pathways. These technologies provide a detailed understanding of the genetic and molecular basis of stress tolerance, enabling the development of genetically engineered plants with enhanced resilience (Sachdev *et al.*, 2021; Morsy and El-Sherif, 2024). Omics approaches have also been used to study the effects of combined abiotic stresses, revealing unique responses that cannot be inferred from individual stress responses alone.

While omics technologies have significantly advanced our understanding of stress adaptation

Table 7.1. Environmental stress, climate change, and multi-omics in medicinal plants. Author's own table.

Topic/stressor	Specific impact on medicinal plants	Mechanism/omics insight	Critical analysis/implications	Citations
Drought stress	• Reduction in plant size and biomass. • Increases in alkaloids and terpenoids. • Decreases in phenols and flavonoids.	Alters biochemical pathways and resource allocation; stress signals may upregulate specific biosynthetic gene clusters for certain compound classes while repressing others.	The response is not uniform; "quality" is metabolite-dependent. Increased concentration of some compounds may be due to biomass reduction (a concentrating effect) rather than *de novo* synthesis. This complicates standardization for medicinal use.	(Shil and Dewanjee, 2022)

Salinity stress	• Ion imbalance and osmotic stress.
	• Species-specific metabolite shifts
	  • *A. vera*: ↓ total phenolics, ↑ aloin
	  • *C. roseus*: altered vincristine levels.
		Disruption of ion homeostasis (e.g. Na^+/K^+ ratio) affects enzyme activity. SOS signaling pathway and synthesis of osmoprotectants (proline) are key. Transcriptomics reveals differential expression of biosynthetic genes.	Highlights the critical concept of trade-offs; a gain in one desired metabolite (aloin) can come at the expense of others (antioxidant phenolics). Salinity can be a tool for targeted metabolite enhancement but requires precise control.	(Kanso *et al.*, 2022; Sharma, 2024)

Heavy metal stress	• Decline in certain metabolites (e.g. hypericin in *H. perforatum*). • Increase in others (e.g. hyperforin).	Activation of detoxification pathways: chelation by phytochelatins (PCs) and vacuolar sequestration. Stress signaling (e.g. via ROS, MAPKs) differentially regulates competing metabolic pathways.	Demonstrates that stress responses are highly compound-specific, even within the same plant. The increase in hyperforin suggests a potential re-routing of biochemical precursors, which could be exploited pharmacologically but raises concerns about consistent product composition.	(Kanojia *et al.*, 2023; Sharma, 2024)
General stress adaptation	• Synthesis of osmoprotectants (proline, glycine betaine). • Activation of antioxidant defense systems (SOD, CAT). • Hormonal signaling (ABA, JA, SA, ethylene).	Multi-omics integrates the components: genomics identifies genes; transcriptomics shows expression changes; proteomics confirms enzyme presence; metabolomics profiles end products.	The plant's response is a coordinated, multi-layered network. Understanding this network is key to engineering resilience. The link between secondary metabolites (e.g. phenolics) and ROS scavenging directly connects medicinal value to survival mechanisms.	(de Vasconcelos *et al.*, 2016; Punetha *et al.*, 2022; Morsy and El-Sherif, 2024)
Climate change: biodiversity & distribution	• Habitat loss and species extinction. • Range shifts to higher latitudes/elevations (e.g. *N. glomerulosa*). • Threats documented on IUCN Red List.	Combined effect of temperature rise, altered precipitation, extreme weather, habitat fragmentation, and overexploitation. Genomics helps assess genetic diversity and adaptive potential for conservation prioritization.	Represents an existential threat to medicinal plant resources and associated traditional knowledge. Conservation efforts (*in situ*/*ex situ*) must be informed by predictive modeling and genetic data to be effective.	(Karami *et al.*, 2022; Robiansyah *et al.*, 2023; Borcă *et al.*, 2024)

Climate change: yield & quality	• Altered growth patterns and biomass yield. • Changes in phytochemical constituents (due to elevated CO_2, O_3, drought). • Compromised therapeutic efficacy and safety.	Direct impact of environmental factors on the enzymatic activity and flux of secondary metabolic pathways. Metabolomics is essential for profiling these qualitative changes.	Threatens the reliability and standardization of herbal medicine. The same species grown in different environments may have vastly different medicinal properties, challenging quality control and regulatory standards.	(Gupta *et al.*, 2019; Patni *et al.*, 2022; Sudhakaran, 2025)
Multi-omics integration	• Holistic understanding of stress response from gene to metabolite. • Identification of key regulatory genes, transcription factors, and enzymes. • Discovery of biosynthetic gene clusters.	Genomics: blueprint. Transcriptomics: dynamic gene expression. Proteomics: functional proteins/PTMs. Metabolomics: end-product biochemical profile. Epigenomics: heritable non-genetic regulation.	Provides unprecedented power to deconvolute complex traits. A major challenge is data integration and functional validation. The promise is in moving from correlation to causation in understanding stress adaptation.	(Dar *et al.*, 2023; Sharma, 2023; Singh *et al.*, 2023; Wang *et al.*, 2024)
Applications: climate-smart agriculture	• Development of climate-resilient cultivars. • Precision farming (optimized irrigation, soil management). • Pathogen defense and phytoremediation strategies.	Omics data identifies molecular markers for marker-assisted breeding. Guides management practices by revealing plant status at a molecular level. Identifies pathways for enhanced tolerance or detoxification.	Bridges the gap between basic research and applied agriculture. Enables a proactive rather than reactive approach to crop management under climate change, crucial for ensuring a stable supply of medicinal plants.	(Naqvi *et al.*, 2022; Chaturvedi *et al.*, 2024)
Applications: conservation genomics	• Safeguarding critically endangered species (e.g. Pacific yew, goldthread). • Assessment of genetic diversity and unique populations. • Informing *in situ* and *ex situ* conservation strategies.	Genomic analysis allows for the prioritization of conservation efforts by identifying genetically distinct or resilient populations, ensuring the preservation of adaptive potential.	A vital tool in the race against extinction. For high-value species threatened by overharvesting and climate change, genomics provides a scientific basis for efficient and effective conservation resource allocation.	(Implied from context; Applequist *et al.*, 2020; Halder and Jha, 2023)
Epigenetic regulation	• Rapid, heritable adaptation without DNA sequence change. • Fine-tuning of stress responses based on environmental experience.	Mechanisms include DNA methylation and histone modifications, which alter gene expression patterns. Provides a "molecular memory" of stress exposure.	Offers an explanation for transgenerational plasticity and rapid adaptation to fluctuating environments. This layer of regulation is crucial for understanding the full scope of a plant's adaptive capacity and could be harnessed in priming strategies.	(Singh *et al.*, 2023)

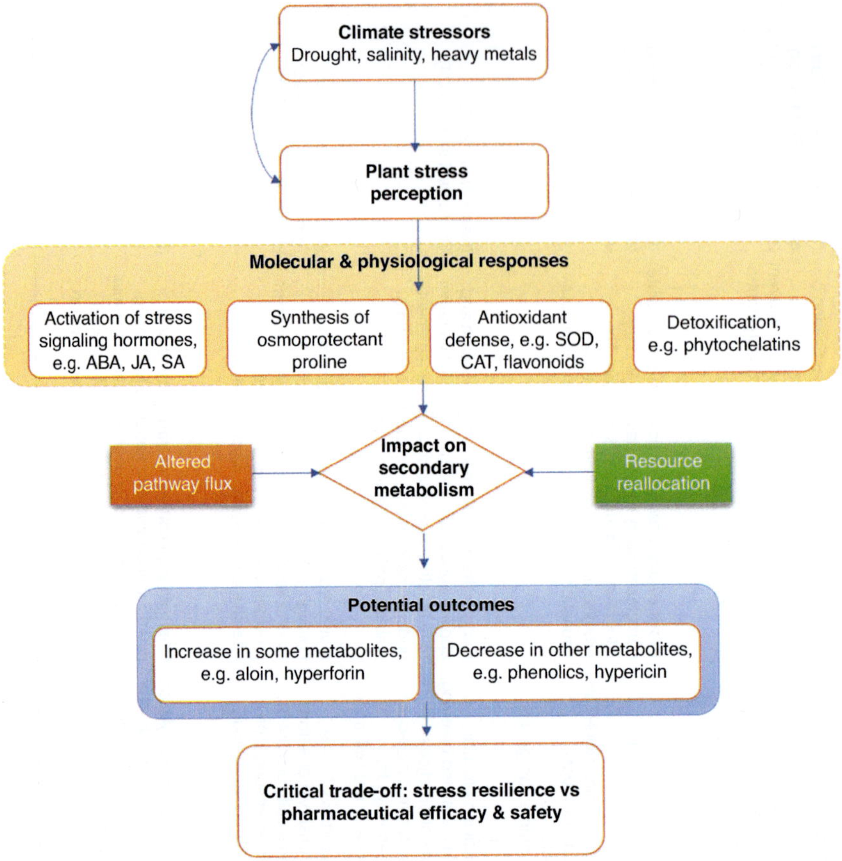

Fig. 7.1. Interplay between climate stress, plant physiology, and medicinal value. (Author's own figure.)

in medicinal plants, challenges remain in translating these insights into practical applications. The complexity of stress responses, involving multiple interacting pathways and genetic factors, poses a challenge for developing stress-tolerant plant varieties. Additionally, the integration of omics data with traditional breeding methods and biotechnological approaches is essential for achieving tangible improvements in plant resilience. Despite these challenges, the continued exploration of omics-driven insights holds promise for enhancing the adaptability and productivity of medicinal plants under abiotic stress conditions.

7.2.5 Unique molecular pathways in medicinal plants under climate resilience drought, salinity, and temperature extremes

Medicinal plants exhibit unique molecular pathways that enable them to withstand climate-induced stresses such as drought, salinity, and temperature extremes. These pathways involve complex physiological, biochemical, and genetic mechanisms that enhance plant resilience and maintain their medicinal properties under adverse conditions. Understanding these pathways is

Table 7.2. Omics-driven insights into abiotic stress adaptation in medicinal plants. Author's own table.

Stress factor/omics approach	Key molecular & physiological mechanisms	Omics-driven insights & identified components	Critical analysis & implications	Citations
General stress signaling & perception	• Stress perception via specific receptors (e.g. kinase receptors). • Activation of complex signaling cascades (MAPKs, CDPKs). • Hormonal cross-talk (ABA, SA, JA, ethylene) for coordinated defense.	Genomics/transcriptomics: identification of genes encoding receptors, kinases (MAPKs, CDPKs), and transcription factors. Proteomics: confirmation of kinase activation and phosphorylation events. Metabolomics: profiling of hormone levels and downstream metabolic changes.	The "cross-talk" between hormones creates a highly flexible but complex network. A stress-specific "hormonal signature" is likely, making it difficult to engineer a universal resilience solution. The fine-tuning of this network is species- and stress-specific.	(Chawla *et al.*, 2022; Elsisi *et al.*, 2024; Lin, 2024)
Drought & salinity (osmotic/ionic stress)	• Osmotic adjustment: accumulation of osmoprotectants (proline, glycine betaine). • Ion homeostasis: regulation of Na^+/K^+ transporters (e.g. SOS pathway). • Physiological adjustments: stomatal closure (via ABA), root architecture changes.	Proteomics: identification of key enzymes in proline biosynthesis (e.g. P5CS). Metabolomics: quantification of osmoprotectant levels and ion fluxes. Genomics/transcriptomics: discovery of genes for ion transporters (NHX, SOS1) and dehydrins.	A major trade-off exists: energy and resources diverted to osmo-protection and ion homeostasis can lead to reduced growth and biomass. Enhancing salinity tolerance might inadvertently reduce yield.	(Debnath *et al.*, 2011; Agrawal *et al.*, 2023; Nazari *et al.*, 2023)
Oxidative stress (a common consequence)	• ROS production: increased reactive oxygen species (e.g. H_2O_2, O_2^-) under most stresses. • Antioxidant system: upregulation of enzymatic (SOD, CAT, ascorbate peroxidase—APX) and non-enzymatic (flavonoids, phenolics) scavengers.	Proteomics: profiling of antioxidant enzyme abundance and activity. Metabolomics: reveals increased synthesis of antioxidant secondary metabolites. Transcriptomics: shows coordinated upregulation of antioxidant gene families.	The link between stress and secondary metabolite production is clear here. However, the dual role of ROS as both damaging agent and signaling molecule creates a delicate balance. Engineering for high antioxidant capacity may disrupt essential ROS signaling.	(Lamalakshmi Devi *et al.*, 2017; Agrawal *et al.*, 2023; Nazari *et al.*, 2023)
UV radiation stress	• Direct damage: to DNA, proteins, and membranes. • Induced defense: synthesis of UV-absorbing and protective secondary metabolites (flavonoids, phenolics, alkaloids).	Transcriptomics: identifies upregulation of key biosynthetic genes (e.g. phenylalanine ammonia-lyase (PAL), chalcone synthase (CHS) for flavonoids). Metabolomics: directly profiles the increase in protective compounds, linking them to UV tolerance.	UV stress is a prime example of where stress adaptation directly enhances medicinal value. The compounds produced for UV protection are often the same ones with therapeutic antioxidant, anti-inflammatory, and anti-cancer properties.	(Thakur *et al.*, 2023)

Continued

Table 7.2. Continued

Stress factor/omics approach	Key molecular & physiological mechanisms	Omics-driven insights & identified components	Critical analysis & implications	Citations
Temperature extremes	• Protein protection: induction of heat-shock proteins (HSPs) or cold-shock proteins. • Membrane integrity: alteration of membrane lipid composition for fluidity.	Proteomics: central role in identifying and characterizing the suite of induced HSPs. Transcriptomics: reveals the rapid transcriptional activation of heat-shock factors (HSFs) and other chaperones. Metabolomics: shows changes in lipid metabolism.	The response is highly transient and dynamic. Proteomics is crucial here as the rapid turnover and re-folding of proteins is a primary line of defense. The effectiveness of this response is tightly linked to the severity and duration of the stress.	(Karthik *et al.*, 2024)
Proteomics as a central hub	• Functional readout: bridges the gap between gene expression and actual physiological response. • Post-translational modifications (PTMs): reveals critical regulation (e.g. phosphorylation of signaling proteins) not visible at the transcript level.	Mass-spectrometry: identifies biomarker proteins indicative of specific stress responses. PTM analysis: uncovers how activities of enzymes and transcription factors are modulated via phosphorylation, ubiquitination, etc.	Proteomics provides the most direct insight into the functional proteome driving adaptation. However, it is technically challenging. The identification of a stress-induced protein does not always confirm its functional necessity without further validation.	(Debnath *et al.*, 2011; Alugoju and Tencomnao, 2023)
Epigenetic regulation	• Phenotypic plasticity: rapid adaptation without changes to DNA sequence. • Molecular memory: heritable modifications (DNA methylation, histone marks) that prime offspring for stress.	Epigenomics: maps changes in DNA methylation patterns and histone modifications in response to stress. Integration with transcriptomics: links epigenetic marks to changes in gene expression of stress-responsive genes.	This explains transgenerational acclimation and how a single plant can exhibit plasticity. It offers a potential mechanism for "priming" plants to be more resilient. The field is young, and applying this knowledge to breeding programs is complex.	(Chawla *et al.*, 2022)
Tissue-specific adaptations	• Spatial specialization: different tissues (roots vs leaves) employ distinct strategies. • Localized defense: roots may focus on suberin/lignin deposition and ion exclusion, while leaves focus on antioxidant production.	Spatially resolved omics: (e.g. single-cell or tissue-specific transcriptomics/proteomics) reveals this compartmentalization. Examples: mangrove root suberization; date palm root amino acid biosynthesis.	Averages from whole-plant omics studies can mask critical tissue-specific responses. Understanding resilience requires a spatially explicit view of molecular events, as the root is often the first point of stress perception.	(Oyarce *et al.*, 2025)
Overarching challenges & trade-offs	• Complexity & redundancy: stress responses involve multi-genic, interconnected pathways. • Resource allocation: energy directed to defense often reduces growth and yield (growth-defense trade-off). • Unintended consequences: engineering for one stress may increase sensitivity to another.	Multi-omics integration: the only way to grasp the full complexity is by combining genomics, transcriptomics, proteomics, and metabolomics datasets. Systems biology modeling: required to predict the outcome of genetic modifications and understand trade-offs.	The major hurdle is translating omics data into practical breeding outcomes. The growth-defense trade-off is a fundamental constraint. Altering secondary metabolite profiles for stress resilience may directly impact the very medicinal efficacy the plant is valued for, creating a critical dilemma for cultivators.	(Sachdev *et al.*, 2021; Morsy and El-Sherif, 2024)

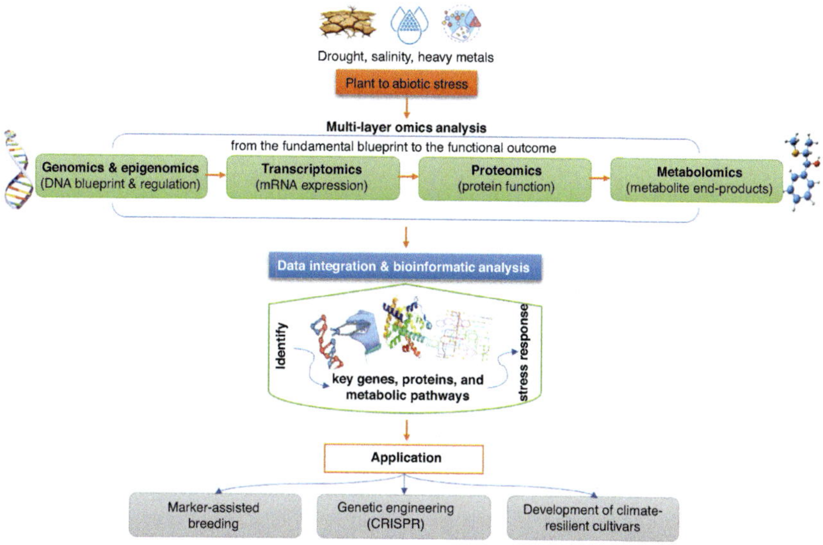

Fig. 7.2. Multi-omics approach to decipher stress adaptation. (Author's own figure.)

crucial for developing strategies to improve the stress tolerance of medicinal plants, which are vital for both ecological balance and pharmaceutical applications. The following sections detail the key molecular pathways and mechanisms involved in the stress resilience of medicinal plants.

Medicinal plants perceive environmental stress through specific receptors that trigger signaling cascades, leading to stress responses. These signaling pathways often involve MAPKs and CDPKs, which modulate gene expression and activate stress-responsive genes (Chawla *et al.*, 2022; Lin, 2024). SA plays a crucial role in stress signaling, particularly in response to drought and salinity. It modulates redox signaling and interacts with other hormones to enhance stress tolerance (Elsisi *et al.*, 2024). Under drought and salinity stress, medicinal plants accumulate osmoprotectants such as proline and glycine betaine to maintain cellular turgor and protect cellular structures (Agrawal *et al.*, 2023; Nazari *et al.*, 2023). Ion homeostasis is critical for salinity tolerance. Plants regulate ion transporters to maintain a balance of sodium and potassium ions, preventing toxic ion accumulation (Agrawal *et al.*, 2023). Oxidative stress is a common consequence of abiotic stress, leading to the production of ROS. Medicinal plants enhance their antioxidant defense systems by upregulating enzymes like SOD and CAT to scavenge ROS (Agrawal *et al.*, 2023; Nazari *et al.*, 2023). The production of secondary metabolites with antioxidant properties is also increased, contributing to the plant's overall stress resilience (Nazari *et al.*, 2023).

Stress tolerance in medicinal plants is often associated with genetic modifications, including the expression of stress-responsive genes such as dehydrins and heat-shock proteins (Karthik *et al.*, 2024). Epigenetic changes, such as DNA methylation and histone modifications, play a role in regulating gene expression in response to stress, allowing plants to adapt to changing environmental conditions (Chawla *et al.*, 2022). Tissue-specific metabolic adaptations are crucial for stress tolerance. For instance, mangrove roots accumulate suberin and lignin in meristematic tissues, enhancing their resilience to salinity (Oyarce *et al.*, 2025). Date palm roots show enrichment in amino acid biosynthesis pathways, which are vital for osmotic adjustment and stress recovery (Oyarce *et al.*, 2025). Hormones such as ABA and ethylene are key regulators of stress responses. They modulate stomatal closure, root growth, and other physiological processes to optimize water use and reduce stress damage (Chawla *et al.*, 2022; Elsisi *et al.*, 2024). Cross-talk between different hormonal pathways fine-tunes the plant's response to multiple stressors, ensuring a coordinated defense strategy (Elsisi *et al.*, 2024). While the molecular pathways discussed provide a robust framework for understanding stress resilience in medicinal plants, it is important to consider the potential trade-offs associated with these adaptations. For instance, while stress tolerance mechanisms can enhance survival under adverse conditions, they may also lead to reduced growth or altered secondary metabolite profiles, impacting the plant's medicinal value. Additionally, the complexity of these pathways poses challenges for breeding and biotechnological interventions aimed at enhancing stress tolerance. Therefore, a balanced approach that integrates traditional breeding, genetic engineering, and sustainable agricultural practices is essential for optimizing the resilience and productivity of medicinal plants under climate change-induced stresses (Fig. 7.2).

7.3 Multi-omics Approaches for Salinity-resilient Cultivars

The application of multi-omics approaches to develop salinity-resilient cultivars of medicinal plants involves integrating various omics technologies to understand and enhance the salt tolerance mechanisms in these plants. This approach is crucial for improving the resilience of medicinal plants to salinity stress, which is a significant challenge in agriculture due to increasing soil salinization. By leveraging genomics, transcriptomics, proteomics, and metabolomics, researchers can identify key genes, proteins, and metabolic pathways that contribute to salinity tolerance, thereby facilitating the development of more resilient cultivars.

Multi-omics approaches help identify genes and transcription factors involved in salinity tolerance. For instance, in rice, genome-wide analyses have been used to identify candidate genes and transcription factors that respond to salinity stress, which can be targeted for breeding salt-tolerant varieties (Das *et al.*, 2015). Similarly, in *Panicum antidotale*, transcriptome analysis revealed differentially expressed genes involved in proline biosynthesis and antioxidant defense, which are crucial for salt resistance (Hussain *et al.*, 2021). Proteomic analysis provides insights into the protein-level changes under salinity stress. In *Sesuvium portulacastrum*, proteomic studies identified differentially expressed proteins related to metabolic pathways and signal transduction, which are essential for salt stress adaptation (Cao *et al.*, 2023). This information can be used to understand the protein networks that confer salinity tolerance. This approach focuses on the metabolic changes in response to salinity stress. In *Salicornia europaea*, metabolomic analysis identified key metabolites such as sucrose, glucose, and proline that are involved in osmotic tolerance and antioxidant activities under salt stress (Duan *et al.*, 2023). These metabolites can be targeted to enhance the salinity tolerance of medicinal plants.

Fennel, ajwain, and anise, medicinal plants from the Apiaceous family, have shown varying degrees of salt tolerance. Fennel and ajwain exhibited better resilience compared to anise, as they maintained higher chlorophyll and carotenoid concentrations and had a more balanced Na^+/K^+ ratio under high salinity conditions (Nouripour-Sisakht *et al.*, 2025). This suggests that specific physiological traits can be targeted to improve salinity tolerance in medicinal plants. As a halophyte, *Portulaca oleracea* has been studied for its ability to tolerate high salinity levels. Multi-omics analyses revealed that it employs salt exclusion mechanisms and has a robust antioxidant defense system, which are critical for its survival under extreme salinity (Silva *et al.*, 2023).

Despite this understanding of the mechanisms of salinity tolerance, challenges remain in translating these findings into practical applications for breeding salinity-resilient cultivars. The integration of omics data with traditional breeding techniques and biotechnological tools, such as CRISPR-Cas9, can accelerate the development of salt-tolerant medicinal plants. Additionally, understanding the cross-talk between different stress response pathways, such as those involving plant hormones, can further enhance the resilience of these plants to salinity stress (Muthuramalingam *et al.*, 2022; Zhou *et al.*, 2024). Future research should focus on the application of these integrated approaches to a wider range of medicinal plants to ensure sustainable agricultural practices in saline environments (Table 7.3).

Table 7.3. Multi-omics approaches for breeding salinity-resilient medicinal plants. Author's own table.

Omics approach/ plant model	Key salinity tolerance mechanisms identified	Specific genes, proteins, & metabolites involved	Critical analysis & implications for breeding	Citations
Genomics & transcriptomics	• Regulatory control: activation of transcription factors (TFs) that orchestrate stress-responsive genes. • Antioxidant defense: upregulation of genes for enzymatic ROS scavenging. • Hormonal signaling: reprogramming of hormone signal transduction pathways (e.g. ABA, JA).	• Candidate genes & TFs identified in rice and *P. antidotale*. • Enzyme genes: SOD, POD. • Pathways: plant hormone signal transduction, phenylpropanoid biosynthesis.	Provides a genetic blueprint for tolerance. However, salt tolerance is a quantitative trait governed by many genes, making single-gene solutions often insufficient. Marker-assisted selection (MAS) is promising but requires validation across genetic backgrounds.	(Das *et al.*, 2015; Hussain *et al.*, 2021; Sui *et al.*, 2024)
Proteomics	• Post-translational modulation: changes in protein abundance and activity beyond transcript levels. • Signal transduction: identification of key proteins in stress perception and signaling cascades. • Metabolic shifts: reprogramming of central and secondary metabolic pathways.	• Differentially expressed proteins in *S. portulacastrum* involved in metabolism and signaling. • Identification of enzymes for osmoprotectant synthesis.	Bridges the gap between gene expression and functional physiology. Reveals critical post-translational modifications (e.g. phosphorylation) that are invisible to transcriptomics. However, protein extraction and analysis remain technically challenging for non-model species.	(Cao *et al.*, 2023)
Metabolomics	• Osmotic adjustment: accumulation of compatible solutes to maintain cell turgor. • ROS scavenging: production of non-enzymatic antioxidants. • Energy & carbon skeleton provision: shifts in primary carbon metabolism.	• Osmoprotectants: proline, glycine betaine, polyamines, sucrose, glucose. • Amino acids: glycine, serine, threonine. • Secondary metabolites: flavonoids, triterpene saponins (e.g. Platycodin D).	Provides the most direct readout of the plant's functional physiological state. The accumulation of specific metabolites like Platycodin D shows a direct link between stress and enhanced medicinal value. This allows for selective breeding for both resilience and quality.	(Ahmad *et al.*, 2021; Duan *et al.*, 2023; Zhang *et al.*, 2023a)
Integrated multi-omics (case studies)	• Coordinated defense: reveals how gene expression changes lead to metabolic shifts for a cohesive stress response. • Pathway elucidation: connects transcriptional upregulation of a pathway to the actual accumulation of its protective end-products.	• *M. tricuspidata*: integration linked hormone signaling and flavonoid biosynthesis genes to metabolic profiles. • *Platycodon grandiflorus*: transcriptomics and metabolomics confirmed the co-activation of lignin and saponin biosynthesis pathways.	This is the most powerful approach, moving from correlation to causality. It allows for the construction of regulatory networks. The major challenge is data integration and interpretation, requiring sophisticated bioinformatics and systems biology models.	(Zhang *et al.*, 2023a; Sui *et al.*, 2024; Zhang *et al.*, 2024a)

Continued

Table 7.3. Continued

Omics approach/ plant model	Key salinity tolerance mechanisms identified	Specific genes, proteins, & metabolites involved	Critical analysis & implications for breeding	Citations
Comparative physiology (Apiaceous family)	• Ion homeostasis: tolerant species (fennel, ajwain) maintain a favorable Na+/K+ratio. • Photosynthetic integrity: protection of photosynthetic pigments (chlorophyll, carotenoids) under stress.	• Physiological traits: chlorophyll concentration, Na+/K+ratio, carotenoid levels.	Provides a phenotypic screening framework to identify resilient germplasm. Confirms that tolerance mechanisms are conserved (ion exclusion, pigment protection) but their efficiency varies significantly even within related species. Useful for traditional selection.	(Nouripour-Sisakht *et al.*, 2025)
Halophyte models for gene discovery	• Specialized adaptations: salt exclusion/secretion mechanisms, superior osmotic adjustment, and robust antioxidant systems. • Constitutive vs induced defenses: some tolerance mechanisms are "always on."	• *S. portulacastrum*, *P. oleracea*, *S. europaea*. • Mechanisms: salt exclusion, vacuolar sequestration, high basal levels of proline and antioxidants.	Halophytes are a goldmine of pre-evolved tolerance genes. The key challenge is the functional transfer of these genes into glycophytic medicinal plants without causing pleiotropic effects or growth penalties.	(Kumari *et al.*, 2015; Duan *et al.*, 2023; Silva *et al.*, 2023)
Translational challenges & future directions	• Trait complexity: salinity tolerance is a multigenic trait with intricate network interactions. • Breeder-practitioner gap: difficulty in translating omics data into field-ready cultivars. • Trade-offs: engineering for salinity tolerance may impact yield, growth, or sensitivity to other stresses.	• Tools: CRISPR-Cas9 for precise gene editing, MAS, system biology models. • Strategies: pyramiding multiple tolerance QTLs (quantitative trait loci)/genes, using halophyte knowledge for engineering glycophytes.	The promise of multi-omics is immense, but its success hinges on interdisciplinary collaboration between molecular biologists, bioinformaticians, and plant breeders. The growth–defense trade-off is a fundamental barrier; the goal is to engineer "resilience without penalty." Prioritizing medicinal plants where stress *enhances* active compounds (e.g. saponins, flavonoids) offers a strategic advantage.	(Daldoul *et al.*, 2014; Ismail and Horie, 2017; Muthuramalingam *et al.*, 2022; Zhou *et al.*, 2024)

7.3.1 Genomics and metabolomics integration to breed salt-tolerant medicinal plants

Integrating genomics and metabolomics to breed salt-tolerant medicinal plants involves understanding the complex interactions between genetic expression and metabolic pathways that confer salt tolerance. This integration is crucial for developing plant varieties that can thrive in saline environments, which is increasingly important due to global soil salinization. By examining various studies, it becomes evident that a multi-omics approach provides comprehensive insights into the mechanisms of salt tolerance, which can be harnessed for breeding purposes. In *Maclura tricuspidata*, the antioxidant enzyme system, including SOD and peroxidase (POD), plays a significant role in salt tolerance. The integration of transcriptomic and metabolomic data revealed that pathways such as plant hormone signal transduction and flavonoid biosynthesis are crucial for adaptation to salt stress (Sui *et al.*, 2024). In *Platycodon grandifloras*, salt stress activates genes involved in lignin and triterpene saponin biosynthesis, which are part of the phenylpropanoid pathway. This pathway is essential for structural reinforcement and stress response, with compounds like Platycodon D increasing significantly under stress (Zhang *et al.*, 2023a). Amino acids such as glycine, serine, and threonine are vital in enhancing salt tolerance, as seen in both *M. tricuspidata* and *Solenstemma argel*. These amino acids contribute to osmotic adjustment and antioxidant defense (Ahmad *et al.*, 2021; Sui *et al.*, 2024). *Sophora alopecuroides* and *Cynomorium songaricum* highlight the importance of integrating transcriptomic and metabolomic data to understand the genetic and metabolic responses to salt stress. This integration helps identify key genes and metabolites involved in pathways like phenylpropanoid biosynthesis, which are crucial for salt tolerance (Ahmad *et al.*, 2021; Zhang *et al.*, 2024a).

In halophytes, metabolites such as proline, glycine betaine, and polyamines are critical for maintaining ion balance and scavenging ROS under salt stress. These metabolites are often identified through metabolomic analyses and are targets for breeding programs (Kumari *et al.*, 2015). The integration of omics data facilitates the identification of candidate genes and pathways that can be targeted in breeding programs. For instance, in grapevines, system biology approaches have been suggested to develop salt-tolerant varieties through marker-assisted selection (Daldoul *et al.*, 2014). Insights from halophytes and other salt-tolerant species can be applied to glycophytes through genetic engineering. This involves transferring genes responsible for salt tolerance traits, such as those involved in ion transport and osmotic regulation, to improve crop resilience (Kumari *et al.*, 2015; Ismail and Horie, 2017). While genomics and metabolomics offer promising avenues for breeding salt-tolerant medicinal plants, challenges remain. The complexity of salt tolerance as a multigenic trait requires comprehensive datasets and advanced analytical tools to unravel the intricate networks involved. Additionally, translating these findings into practical breeding strategies necessitates collaboration between researchers and breeders to ensure the development of viable, salt-tolerant plant varieties.

7.4 Biotic Stress Resistance: Omics-guided Pathogen Defense

The potential applications of biotic stress resistance omics-guided pathogen defense in developing novel medicinal plant-based therapeutics are vast and promising. Omics technologies, which include genomics, transcriptomics, proteomics, and metabolomics, provide comprehensive insights into plant–pathogen interactions and the defense mechanisms plants employ. These insights can be harnessed to develop new plant-based therapeutics by identifying and utilizing plant secondary metabolites and antimicrobial peptides that have therapeutic potential. The integration of omics data can lead to the discovery of novel compounds and enhance the efficacy of existing plant-based medicines. Below are the key aspects of how omics-guided pathogen defense can be applied in medicinal plant-based therapeutics.

Omics technologies facilitate the discovery of plant anti-pathogenic agents (PAAs), which include chemical compounds, proteins, and peptides with antimicrobial properties. Identification of PAAs is crucial in developing new plant-based therapeutics, especially in the face of rising antimicrobial resistance (Ferreira-Neto *et al.*, 2022). High-throughput omics data can be used to mine and develop biotechnological applications of PAAs, providing a rich source of new molecules for medicinal studies (Ferreira-Neto *et al.*, 2022). Secondary metabolites such as phenolic compounds, alkaloids, glycosides, and terpenoids play significant roles in plant defense against pathogens. These compounds can be exploited for their therapeutic properties in developing novel medicines (Al-Khayri *et al.*, 2023). Metabolomics approaches help elucidate the structural and functional properties of these metabolites, aiding in their application in pest management and potential therapeutic uses (Al-Khayri *et al.*, 2023; Muñoz-Hoyos and Stam, 2023). Omics approaches enable the engineering of pathogen-resistant plants by modifying gene expression and metabolic pathways. This can lead to the accumulation of metabolites with potential therapeutic benefits (Gomez-Casati *et al.*, 2016). The integration of genomics, transcriptomics, and proteomics helps identify novel virulence factors and host targets, which can be leveraged to develop disease-resistant plants with enhanced medicinal properties (Gomez-Casati *et al.*, 2016).

Omics integration data provides a comprehensive understanding of plant responses to biotic stress, facilitating the development of biotic stress-resistant cultivars. This approach can be applied to enhance the medicinal properties of plants by altering gene regulation and other regulatory networks (Bisht *et al.*, 2023; Kumar *et al.*, 2024). Multi-omics approaches also aid in understanding the interactions between host plants, pathogens, and biocontrol agents, which can be crucial for sustainable disease management and therapeutic development (Harish *et al.*, 2022). Plants produce a variety of antimicrobial peptides (AMPs) that prevent pathogen establishment. AMPs such as defensins and cyclotides are potential candidates for developing new antimicrobial therapies (Santos-Silva *et al.*, 2020). *In silico* prediction and bioinformatics approaches can identify and characterize these AMPs, providing a valuable resource for developing plant-based antimicrobial agents (Santos-Silva *et al.*, 2020).

While the potential applications of omics-guided pathogen defense in medicinal plant-based therapeutics are promising, challenges remain. The complexity of plant–pathogen interactions and the vast diversity of secondary metabolites require sophisticated analytical tools and methodologies. Additionally, translating omics data into practical applications necessitates interdisciplinary collaboration and advances in bioinformatics and biotechnology. Despite these challenges, the integration of omics technologies holds significant promise for the future of plant-based therapeutics (Table 7.4).

7.4.1 Identifying resistance genes and antimicrobial metabolites via transcriptomics and proteomics

The identification of resistance genes and antimicrobial metabolites in medicinal plants through transcriptomics and proteomics is a promising area of research, offering potential solutions to the growing problem of antibiotic resistance. This approach leverages high-throughput technologies to explore the genetic and protein expressions associated with the biosynthesis of antimicrobial compounds in plants. By integrating genomic, transcriptomic, and proteomic data, researchers can uncover the complex regulatory networks and pathways that govern the production of these bioactive compounds. This synthesis of information not only aids in understanding plant defense mechanisms but also facilitates the discovery of novel antimicrobial agents that could be developed into new therapeutic drugs. The following sections delve into the methodologies and findings from recent studies in this field.

7.4.2 Transcriptomics in identifying resistance genes

Transcriptomics involves the study of RNA transcripts to understand gene expression patterns. In the context of medicinal plants, it

Table 7.4. Omics-guided biotic stress resistance for novel therapeutics. Author's own table.

Omics approach/strategic focus	Key discoveries & identified components	Mechanisms of action & therapeutic potential	Critical analysis & translational challenges	Citations
Discovery of PAAs	• Chemical PAAs: phenolic compounds, alkaloids, terpenoids, glycosides. • Protein/peptide PAAs: AMPs like defensins, cyclotides.	• Direct antimicrobial activity: disruption of microbial membranes, inhibition of essential enzymes. • Biofilm inhibition: preventing pathogen establishment and persistence. • Synergy with antibiotics: potential to rejuvenate existing drugs against resistant strains.	Provides a vast, untapped library for novel drug leads, crucial in the post-antibiotic era. A major hurdle is the isolation, purification, and structural elucidation of these often complex and low-yield compounds. High-throughput screening is needed to prioritize candidates.	(Ferreira-Neto *et al.*, 2022; Al-Khayri *et al.*, 2023)
Transcriptomics for resistance gene discovery	• Biosynthetic pathway genes: identification of unigenes in pathways for phenylpropanoids, terpenoids (e.g. in *O. turpethum*). • Regulatory networks: transcriptional coordination with phytohormone signaling (e.g. JA, SA) in *M. dodecandrum*.	Reveals the genetic blueprint for the production of antimicrobial secondary metabolites. Allows for the identification of key rate-limiting enzymes in bioactive compound synthesis, which can be targets for metabolic engineering.	Shows *potential* for compound production, but gene expression does not always correlate with metabolite accumulation (post-transcriptional regulation). Data is often correlative; functional validation (e.g. via gene knockout) is essential to confirm gene function.	(Biswal *et al.*, 2021; Poh *et al.*, 2023)
Proteomics for functional insight	• Antimicrobial proteins/enzymes: identification of pathogenesis-related (PR) proteins and key enzymes in secondary metabolite pathways. • Signaling proteins: components of defense signal transduction.	Confirms the functional presence of defense-related proteins and provides insight into post-translational modifications (e.g. phosphorylation) that rapidly activate defense responses, a layer invisible to transcriptomics.	Technically challenging for non-model medicinal plants due to lack of annotated genomes. While it identifies effector proteins, it may miss the final, active small molecule metabolites which are often the direct therapeutic agents.	(Radulović *et al.*, 2013; Ferreira-Neto *et al.*, 2022)
Multi-omics integration for systems biology	• Network analysis: constructing interconnected models of genes, proteins, and metabolites in response to pathogen challenge (e.g. in Papaveraceae family). • Host–pathogen–biocontrol interactions: understanding tri-trophic interactions for sustainable disease management.	Moves beyond single components to understand the emergent properties of the defense system. Enables the identification of master regulators and key hubs in defense networks that can be targeted for engineering broad-spectrum resistance.	The "omics" data deluge requires advanced bioinformatics and computational modeling capacity, which is a bottleneck. Integrating disparate data types (gene vs metabolite levels) into a coherent model is non-trivial and often hypothesis-generating rather than conclusive.	(Harish *et al.*, 2022; Bisht *et al.*, 2023; Kielich *et al.*, 2024)

Continued

Table 7.4. Continued

Omics approach/strategic focus	Key discoveries & identified components	Mechanisms of action & therapeutic potential	Critical analysis & translational challenges	Citations
In silico prediction & bioinformatics	• AMP mining: using algorithms to predict novel AMPs from genomic/transcriptomic data based on sequence properties. • Structure–activity relationship modeling.	Dramatically accelerates the discovery pipeline by virtually screening thousands of candidates before lab testing. Lowers cost and time for initial candidate identification.	*In silico* predictions are only as good as the algorithms and training data; they require empirical validation of antimicrobial activity and cytotoxicity. May overlook novel AMPs that do not fit known sequence patterns.	(Santos-Silva *et al.*, 2020)
Engineering pathogen-resistant plants	• Metabolic engineering: overexpressing or silencing genes to enhance the production of therapeutic antimicrobial compounds. • Stacking resistance genes: pyramiding multiple R-genes or AMP genes for durable resistance.	Creates resilient cultivars that consistently produce high levels of desired therapeutics, ensuring a stable and high-quality supply. Also reduces crop losses, enhancing sustainability.	Public perception and regulatory hurdles for genetically modified (GM) medicinal plants are significant. There is a risk of unintended metabolic consequences or yield penalties when re-routing plant metabolic resources.	(Gomez-Casati *et al.*, 2016; Kumar *et al.*, 2024)
Overarching challenges & future directions	• Compound complexity: structural diversity and low abundance of many PAAs. • From lab to clinic: the long, expensive drug development pathway. • Sustainable sourcing: balancing discovery with the conservation of often rare or slow-growing medicinal species.	The promise is a new generation of plant-derived antimicrobials with novel modes of action to combat multidrug-resistant pathogens.	The field is strong on discovery but weak on translation. Interdisciplinary collaboration between plant scientists, pharmacologists, and clinicians is critical. A major ethical and practical consideration is bioprospecting and ensuring equitable benefit-sharing with source countries and communities.	(Caioni *et al.*, 2024; Fik-Jaskółka *et al.*, 2024)

helps identify genes involved in the biosynthesis of antimicrobial compounds. For instance, the transcriptomic analysis of *Melastoma dodecandrum* revealed specialized metabolic pathways and transcriptional coordination with phytohormones, which are crucial for its antibacterial activity against pathogens like *Pseudomonas aeruginosa* and *Staphylococcus aureus* (Poh *et al.*, 2023). Similarly, the *de novo* transcriptome analysis of *Operculina turpethum* identified numerous unigenes associated with secondary metabolite biosynthesis, highlighting pathways like phenylpropanoid and terpenoid biosynthesis (Biswal *et al.*, 2021).

7.4.3 Proteomics in understanding antimicrobial metabolites

Proteomics complements transcriptomics by analyzing the protein expressions that result from gene activity, providing insights into the functional aspects of antimicrobial metabolite production. The integration of proteomic data can reveal the presence of specific proteins and peptides with antimicrobial properties, which are part of the plant's defense system against pathogens (Ferreira-Neto *et al.*, 2022). This approach is particularly useful in identifying the structural diversity and mechanisms of action of plant secondary metabolites (PSMs), which include a wide range of compounds, such as alkaloids, phenolics, and terpenoids, with proven antimicrobial properties (Radulović *et al.*, 2013).

7.4.4 Case studies and applications

The Papaveraceae family, known for its medicinal properties, has been extensively studied using herb genomics, which combines genomic and other -omics data to explore the genetic foundations of its bioactive compounds (Kielich *et al.*, 2024). Caucasian medicinal plants have been investigated for their antimicrobial metabolites, such as phenolic compounds and flavonoids, which exhibit efficacy against various pathogens (Fik-Jaskółka *et al.*, 2024). These studies underscore the potential of plant-derived compounds as alternatives to conventional antibiotics, especially in the face of rising antibiotic resistance (Caioni *et al.*, 2024). While the application of transcriptomics and proteomics in identifying resistance genes and antimicrobial metabolites is significant, challenges remain. The complexity of plant genomes and the diversity of secondary metabolites pose difficulties in data interpretation and application. Moreover, the transition from laboratory findings to clinical applications requires extensive validation and regulatory approval. Despite these challenges, the integration of modern biotechnologies and bio-omics approaches continues to offer promising avenues for the discovery of novel antimicrobial agents from medicinal plants.

7.5 *H. perforatum*'s Omics of Heavy Metal Detoxification

H. perforatum, commonly known as St John's Wort, is a plant with notable capabilities in heavy metal detoxification, which is crucial given the increasing environmental contamination by heavy metals. The plant's ability to accumulate and tolerate heavy metals like cadmium (Cd), lead (Pb), and mercury (Hg), which are often present in polluted environments, reveals its potential in phytoremediation. The detoxification process involves various biochemical and physiological mechanisms that allow the plant to tolerate and mitigate the toxic effects of these metals. Omics technologies, including genomics, transcriptomics, proteomics, and metabolomics, provide insights into the molecular mechanisms underlying this detoxification process. These technologies help identify stress-responsive genes and pathways that enable *H. perforatum* to manage heavy metal stress effectively. The following sections delve into the specific omics approaches and their contributions to understanding heavy metal detoxification in *H. perforatum* (Table 7.5 and Fig. 7.3).

7.5.1 Accumulation of heavy metals

H. perforatum has been shown to accumulate significant levels of heavy metals, particularly cadmium, in its tissues. Studies have reported that the concentration of cadmium in the aerial parts of the plant can exceed the

Table 7.5. *H. perforatum's* heavy metal detoxification and molecular tolerance mechanisms. Author's own table.

Aspect/mechanism	Key findings & components in *H. perforatum*	Broader molecular machinery (genes/ proteins)	Critical analysis & implications	Citations
Heavy metal accumulation capacity	• High accumulation of Cd, Pb, Hg, often exceeding maximum permissible concentrations (MPCs) for medicinal use. • Accumulation levels are region-specific, influenced by soil pollution (e.g. Western Siberia vs Western Serbia).	• Uptake is mediated by broad-spectrum metal transporters (e.g. ZIP (ZRT/IRT-like protein), NRAMP (natural resistance-associated macrophage protein) families) often mistaken for essential nutrient transporters.	Major paradox for medicinal use: the very trait that makes it good for phytoremediation (metal accumulation) makes it a potential health hazard if used in medicine. Sourcing and rigorous quality control are paramount.	(Zagurskaya and Siromlya, 2020)
Chelation & immobilization	• Production of PCs and organic acids to bind free metal ions (Cd^{2+}). • Rich in native flavonoids which can act as chelators and antioxidants.	• Phytochelatin synthase (PCS) genes: key enzyme activated by metal presence to synthesize PCs. • MTs: Cysteine-rich metal-binding proteins encoded by a gene family (e.g. MT1, MT2).	Chelation is the first line of defense, neutralizing ionic toxicity. However, the PC-metal complexes must be sequestered to be truly detoxified. Reliance on flavonoids could alter the medicinal profile intended for use.	(Dubey *et al.*, 2018; Marques *et al.*, 2021; Ejaz *et al.*, 2023)

Vacuolar sequestration (compartmentalization)	• Central mechanism for isolating toxic metals away from cytosolic metabolism.	• Tonoplast transporters:
		  • HMA3 (P-type ATPase): Shoots Cd into the vacuole.
		  • MTP/CDF family: vacuolar transporters for Zn, Cd, Co.
		  • ABC transporters: can transport PC-Cd complexes into the vacuole.
			This is the defining step for hyperaccumulation. The efficiency of these transporters determines the plant's tolerance threshold. Engineering these transporters is a key target for enhancing phytoremediation.	(González-Mendoza and Zapata-Pérez, 2008; Sharma *et al.*, 2016; Keeran *et al.*, 2021)

Antioxidant defense system	• Heavy metals induce ROS. • *H. perforatum* upregulates its inherent antioxidant compounds (e.g. hypericin, flavonoids) and enzymatic defenses.	• Enzymes: SOD, CAT, APX. • Genes: a large family of genes encoding these antioxidants are stress-responsive.	A secondary, crucial defense. The plant must manage the *oxidative* damage caused by metals. The upregulation of antioxidant metabolites may inadvertently increase or decrease the concentration of key medicinal compounds like hypericin.	(Pandey, 2018; Vicidomini *et al.*, 2024)

Omics insights (transcriptomics/ metabolomics)	• Machine learning with metabolomics identifies biomarkers for metal stress and detoxification efficacy. • Reveals rewiring of central metabolic pathways and phenylpropanoid/flavonoid biosynthesis.	• Omics identifies the entire network: from metal perception sensors, signaling cascades (MAPKs), to the full suite of induced genes (PCS, heavy metal ATPases (HMAs), MTs, antioxidants).	Omics provides a systems-level view, moving beyond single genes. The challenge is data integration and functional validation. The use of ML is promising for predicting plant fitness and detoxification potential in the field.	(Rizzo *et al.*, 2020; Kumar *et al.*, 2021a; Zhang *et al.*, 2024b)
Phytoremediation potential	• Demonstrated ability to accumulate metals from soil, making it suitable for phytostabilization or phytoextraction.	• Genetic engineering of model plants (e.g. *Arabidopsis*, tobacco) by expressing *H. perforatum* genes like PCS or HMA3 to create superior phytoremediators.	Trade-off: using a medicinal plant for cleanup poses a cross-contamination risk if the biomass enters the medicinal supply chain. It should be dedicated solely to remediation. Its non-food status reduces ecological risk.	(Goldsbrough, 2020; Pandita and Pamuru, 2022; Ejaz *et al.*, 2023)
Safety & regulatory implications	• Metal content, especially Cd, can exceed safety limits for herbal medicines, posing a direct toxic risk to consumers.	• Not applicable.	This is the most critical constraint. Cannot assume wild-harvested *H. perforatum* is safe. Mandatory heavy metal testing for all medicinal plant raw materials is non-negotiable. Cultivation must be on certified clean soils.	(Zagurskaya and Siromlya, 2020; Ibrahim *et al.*, 2024)
Genetic engineering for enhanced tolerance	• Potential to overexpress its own *HpPCS* and *HpHMA3* genes to create "super-detoxifying" varieties for remediation.	• Successful examples in other species: overexpression of *PCS* and *MT* genes enhances Cd and Cu tolerance, respectively. • CRISPR-Cas9 can be used to fine-tune promoter regions of key transporter genes.	While powerful, public and regulatory acceptance of GM medicinal plants is extremely low. The focus may need to be on using *H. perforatum* genes to engineer non-medicinal, fast-growing plants like poplar for remediation.	(Sebastian *et al.*, 2019; Mathew, 2022)

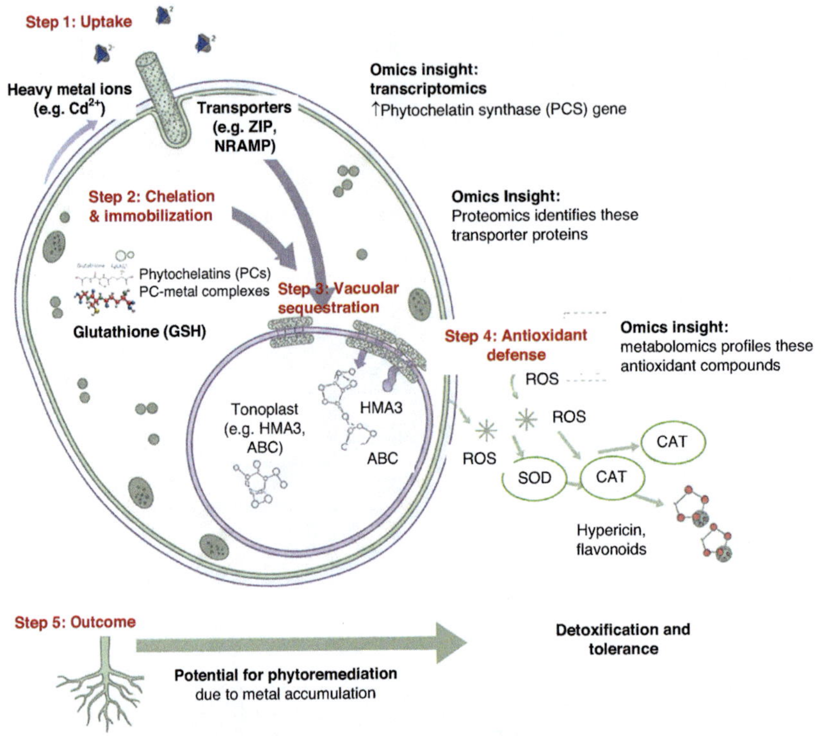

Fig. 7.3. Molecular detoxification pathway of heavy metals in a root cell. (Author's own figure.)

maximum permissible concentration (MPC) for medicinal plant raw materials, indicating its strong accumulation capacity (Zagurskaya and Siromlya, 2020). The plant's ability to accumulate heavy metals varies depending on the region and environmental conditions. For instance, plants from Western Siberia and Western Serbia have shown different levels of heavy metal accumulation, influenced by soil composition and local pollution levels (Zagurskaya and Siromlya, 2020).

7.5.2 Detoxification mechanisms

H. perforatum employs several detoxification strategies to cope with heavy metal stress. These include the chelation of metals by PCs and other organic acids, which bind free metal ions and reduce their toxicity (Dubey *et al.*, 2018; Ejaz *et al.*, 2023). The plant also utilizes compartmentalization, sequestering heavy metals in vacuoles to prevent interference with cellular processes. This is facilitated by tonoplast-located transporters that actively transport metals into vacuoles (Sharma *et al.*, 2016). Additionally, the plant's antioxidant compounds, such as flavonoids, play a crucial role in mitigating oxidative stress induced by heavy metal exposure, protecting cellular structures from damage (Pandey, 2018; Vicidomini *et al.*, 2024).

7.5.3 Omics approaches in understanding detoxification

Recent advances in omics technologies, such as transcriptomics, metabolomics, and proteomics, have provided insights into the molecular basis of heavy metal detoxification in *H. perforatum*. These approaches have identified key genes and metabolic pathways involved in the plant's response to heavy metal stress (Dubey *et al.*, 2018; Rizzo *et al.*, 2020). Machine learning methods combined with metabolomics have been used to identify quality indicators and potential biomarkers for assessing the plant's detoxification capacity. This holistic approach aids in understanding the complex interactions between different metabolic pathways and their role in heavy metal tolerance (Zhang *et al.*, 2024b).

7.5.4 Phytoremediation potential and applications

The ability of *H. perforatum* (St John's Wort) to tolerate, accumulate, and detoxify heavy metals establishes it as a promising candidate for phytoremediation—a sustainable method for removing pollutants from contaminated environments (Ejaz *et al.*, 2023). Planting this species in polluted areas can reduce soil heavy metal concentrations, thereby mitigating associated environmental and health risks. This phytoremediation potential is enhanced by specific genetic and biochemical characteristics, such as the expression of metal-chelating proteins (Pandita and Pamuru, 2022; Parida *et al.*, 2024). Research into these underlying mechanisms is crucial for improving efficacy. Here, integrated multi-omics approaches (genomics, transcriptomics, proteomics) are pivotal, as they provide a comprehensive understanding of the molecular pathways involved in heavy metal detoxification and tolerance (Rizzo *et al.*, 2020; Mathew, 2022; Pandey *et al.*, 2022). This knowledge paves the way for targeted genetic and biotechnological strategies to enhance the plant's remediation efficiency. However, the use of a well-known medicinal plant for heavy metal cleanup raises significant ecological and health considerations. The primary concern is the safety of its products for human consumption if grown on contaminated land, necessitating stringent monitoring and the development of safety standards and guidelines (Zagurskaya and Siromlya, 2020; Ibrahim *et al.*, 2024). Therefore, while *H. perforatum* shows significant promise, further research is needed to optimize its use under diverse field conditions, fully explore the capabilities of this and other hyperaccumulators, and establish robust safety protocols to ensure its safe application in both environmental management and sustainable agriculture.

7.5.5 Molecular mechanisms (e.g. PCs, transporter genes) for metal tolerance

Understanding the molecular mechanisms of metal tolerance in medicinal plants is crucial for developing strategies to mitigate heavy metal contamination in agriculture and enhance phytoremediation efforts. Medicinal plants, like other plant species, have evolved complex biochemical and genetic systems to tolerate and detoxify heavy metals. These mechanisms include the synthesis of metal-binding peptides such as PCs and metallothionein, the expression of transporter genes, and the activation of various stress response pathways. This section explores these molecular mechanisms, focusing on the roles of PCs, metallothioneins, and transporter genes in metal tolerance. PCs are small, cysteine-rich peptides synthesized in response to heavy metal stress. They bind to metal ions through thiolate bonds, forming complexes that are sequestered in vacuoles, thus preventing metal toxicity in the cytosol. The synthesis of PCs is induced by various metal ions, and they play a significant role in cadmium tolerance by modulating its accumulation and detoxification in plants (Marques *et al.*, 2021; Mathew, 2022). Metallothioneins (MTs) are low-molecular-weight, cysteine-rich proteins that bind heavy metals, aiding in their detoxification. MTs are involved in the sequestration of metals within cells, reducing their availability to interfere with cellular processes. The expression of MT genes is often upregulated in response to metal stress, contributing to enhanced metal tolerance (Fig. 7.3; Chatterjee *et al.*, 2020; Liu *et al.*, 2022).

7.5.5.1 Transporter genes

Metal transporter proteins are crucial for the uptake, translocation, and sequestration of metal ions within plant tissues. They help maintain metal homeostasis by regulating the movement of metals across cellular membranes (Ishida and Martinez Corcino, 2022). Transporters such as ATP-binding cassette (ABC) transporters and cation diffusion facilitators (CDFs) are involved in the efflux of metals from the cytosol to vacuoles or out of the cell, thereby reducing cytotoxicity (Kumar *et al.*, 2016; Keeran *et al.*, 2021). Tonoplast transporters located in the vacuolar membrane facilitate the compartmentalization of metal ions into vacuoles, a key detoxification strategy. This compartmentalization prevents metals from interfering with essential cellular functions (González-Mendoza and Zapata-Pérez, 2008; Mathur and Chauhan, 2020).

7.5.5.2 Genetic and biotechnological approaches

Genetic engineering has been employed to enhance metal tolerance in plants by overexpressing genes involved in metal detoxification pathways, such as those encoding phytochelatin synthase and metallothionein. These transgenic plants exhibit improved metal accumulation and tolerance, making them suitable for phytoremediation applications (Sebastian *et al.*, 2019; Goldsbrough, 2020). Advances in genomics, transcriptomics, and proteomics have identified numerous genes and pathways involved in metal tolerance. These approaches provide insights into the complex regulatory networks that govern plant responses to metal stress, offering potential targets for genetic manipulation (Chandrakar *et al.*, 2020; Kumar *et al.*, 2021a). While the molecular mechanisms of metal tolerance in medicinal plants are well documented, it is important to consider the ecological and evolutionary contexts in which these mechanisms have developed. The natural variation in metal tolerance among plant species suggests that these mechanisms are shaped by both genetic factors and environmental pressures. Additionally, the integration of traditional breeding techniques with modern biotechnological approaches could further enhance the development of metal-tolerant plant varieties, contributing to sustainable agriculture and environmental remediation efforts (Fig. 7.3).

7.6 Climate-smart Agriculture: Omics-informed Cultivation Practices

Climate-smart agriculture (CSA) is an approach that integrates sustainable agricultural practices to enhance productivity and resilience and reduce greenhouse gas emissions. In medicinal plants, CSA can be informed by omics technologies to optimize cultivation practices, ensuring the sustainable production of these valuable resources. Omics technologies, which include genomics, transcriptomics, proteomics, and metabolomics, provide insights into the genetic and biochemical pathways of plants, enabling the development of climate-resilient varieties and improved cultivation methods. This integration is crucial for maintaining the pharmaceutical properties of medicinal plants under changing climatic conditions.

CSA emphasizes sustainable cultivation practices that are highly applicable and beneficial for the production of medicinal plants. Key techniques, such as crop diversification, agroforestry, and precision farming, are central to this approach. These practices are essential for maintaining long-term soil health, improving water use efficiency, and enhancing on-farm biodiversity, all of which are critical factors for the sustainable and high-quality production of medicinal species (Bhanuwanti *et al.*, 2024; Safdar *et al.*, 2024). A core objective of CSA is to enhance climate resilience, which is particularly vital for medicinal plants that are often sensitive to environmental changes. The approach aims to develop climate-resilient crop varieties capable of withstanding extreme weather events like drought and heat stress. This is increasingly achieved through advanced omics technologies, which help identify and breed medicinal plants with desirable adaptive traits, thereby securing their supply in the face of a changing climate (Raza *et al.*, 2024; Safdar *et al.*, 2024). Furthermore, CSA contributes to the mitigation of climate change impacts by promoting the adoption of low-emission cultivation practices. Methods such as organic agriculture and precision farming help to reduce the carbon footprint associated with medicinal plant cultivation. This reduction in greenhouse gas emissions is not only important for global climate mitigation efforts but also helps in minimizing the very environmental changes that threaten the cultivation of these valuable plants (Bhanuwanti *et al.*, 2024).

The application of omics technologies is revolutionizing the cultivation of medicinal plants by providing unprecedented insights into their biology. Genomics and transcriptomics allow researchers to identify key genes involved in the biosynthesis of secondary metabolites, which are the primary sources of medicinal properties. This understanding of genetic pathways enables the targeted development of plants with enhanced pharmaceutical qualities (Zhang *et al.*, 2023b). Furthermore, proteomics and metabolomics complement this by revealing the dynamic changes in protein expression and metabolite production in response to environmental stressors. This information is vital for optimizing cultivation practices to consistently maintain the quality and efficacy of medicinal plants under varying conditions (Raza *et al.*, 2024).

Collectively, these technologies provide a theoretical basis for understanding environmental adaptation, enabling the development of strategies to enhance the resilience and productivity of these valuable species (Zhang *et al.*, 2023b). However, the sustainable cultivation of medicinal plants faces significant challenges from climate change, which directly threatens their growth, distribution, and most critically, their production of bioactive compounds. Factors such as elevated CO_2 levels, ozone pollution, and temperature fluctuations can alter the plants' secondary metabolism, thereby impacting their pharmaceutical properties and necessitating the development of adaptive cultivation practices (Gupta *et al.*, 2019; Hounsou *et al.*, 2024). Compounding this challenge is the increasing global demand for medicinal plants, which requires an urgent shift toward sustainable resource management to prevent habitat degradation and ensure long-term availability. In this context, CSA, informed by the deep biological insights from omics technologies, is poised to play a crucial role in achieving a balance between meeting demand and conserving these vital natural resources (Patil *et al.*, 2024).

While CSA and omics technologies offer promising solutions for the sustainable cultivation of medicinal plants, there are challenges that need to be addressed. The complexity of climate change impacts on medicinal plants requires a comprehensive understanding of the interactions between environmental factors and plant physiology. Additionally, the integration of CSA practices with omics technologies necessitates interdisciplinary collaboration and policy support to ensure widespread adoption and implementation. Future research should focus on developing climate-resilient medicinal plant varieties and optimizing cultivation practices to maintain their pharmaceutical properties in the face of climate change (Table 7.6 and Fig. 7.4).

7.6.1 Soil health monitoring, irrigation optimization, and microbiome management using omics data

Integrating omics data into soil health monitoring, irrigation optimization, and microbiome management presents a transformative approach to enhancing the cultivation of medicinal plants. Omics technologies, encompassing genomics, transcriptomics, proteomics, and metabolomics, offer comprehensive insights into the soil microbiome, which is crucial for maintaining soil health and optimizing plant growth. These technologies enable the detailed analysis of microbial communities and their interactions with plants, providing a foundation for sustainable agricultural practices. This approach is particularly beneficial for medicinal plants, which require specific soil conditions and microbiome interactions to maximize their therapeutic properties. The following sections explore the role of omics data in soil health monitoring, irrigation optimization, and microbiome management.

Omics technologies facilitate the assessment of soil microbial diversity and function, which are critical indicators of soil health. These technologies allow for the rapid and accurate evaluation of microbial populations, aiding in the development of sustainable agricultural systems (Bertola *et al.*, 2021; Manda *et al.*, 2023). The soil microbiome plays a pivotal role in biogeochemical cycling, pollutant degradation, and rhizosphere immunity, all of which contribute to soil fertility and crop productivity (Zhang *et al.*, 2021). Advanced molecular tools, such as metagenomics and proteomics, help establish the link between the structure and function of soil microbial communities, providing insights into ecological processes and plant–microbe interactions (Biswas and Sarkar, 2018).

Understanding soil–plant–microbiome interactions through omics data can inform irrigation

Table 7.6. CSA via omics-informed cultivation practices for medicinal plants. Author's own table.

CSA pillar/omics application	Omics-informed strategies & practices	Specific insights & technological tools	Critical analysis & implementation challenges	Citations
Sustainable cultivation & soil health monitoring	• Soil microbiome metagenomics: assessing microbial diversity and functional capacity as a key indicator of soil fertility and disease suppression. • Metabolomics: profiling root exudates and soil metabolites to understand plant–microbe communication and nutrient cycling.	• Metagenomics/ metatranscriptomics: identify which microbes are present and active in the rhizosphere. • Proteomics: analyze microbial enzymes involved in nutrient solubilization (e.g. phosphatases) and organic matter decomposition.	Provides a deep, causal understanding of soil health beyond simple chemical assays. However, the sheer complexity of soil ecosystems makes it difficult to establish universal "healthy microbiome" benchmarks. Data interpretation requires sophisticated bioinformatics.	(Biswas and Sarkar, 2018; Bertola *et al.*, 2021; Manda *et al.*, 2023)
Precision irrigation optimization	• Integrated "digital twin" models: combining omics data (plant transcriptomic stress markers, soil metabolomic profiles) with real-time soil moisture sensors to create a dynamic model of the soil–plant system.	• Plant hydraulics & stress markers: using transcriptomic signatures of drought stress (e.g. ABA-responsive genes) to trigger irrigation *before* visible wilting. • Soil metabolomics: monitoring microbial metabolites that indicate water stress.	Moves irrigation from a schedule-based to a plant-need-based system. The "digital twin" concept is powerful but is currently data-intensive and computationally expensive, limiting its use to high-value medicinal plant cultivation.	(Zeng *et al.*, 2024, 2025)
Microbiome management & biofertilizer development	• Design of synthetic microbial consortia: using omics to identify key beneficial microbes (e.g. plant growth promoting rhizobacteria (PGPR), mycorrhizal fungi) that work synergistically to enhance nutrient uptake and stress resilience. • Microbiome engineering: inoculating soils with tailored consortia to improve plant health and secondary metabolite production.	• Multi-omics network analysis: mapping interaction networks between plants and microbes to identify keystone species. • Genomics: identifying microbial genes for nitrogen fixation, phosphate solubilization, and siderophore production.	A paradigm shift from generic biofertilizers to customized, plant-specific probiotic blends. A major challenge is ensuring the survival and establishment of introduced microbes in a complex, competitive soil environment.	(Niranjan *et al.*, 2023; Bhimani *et al.*, 2024; Dhiman *et al.*, 2024)

Developing climate-resilient varieties	• MAS: using genomic markers linked to drought, heat, or salinity tolerance to speed up conventional breeding. • Gene discovery: omics identifies master regulators (transcription factors) and key biosynthetic genes for resilience traits.	• Genomics/transcriptomics: identify alleles of stress-responsive genes (e.g. for osmoprotectant synthesis, heat-shock proteins). • Metabolomics: select for plants that maintain or enhance secondary metabolite production under stress.	Directly links omics discovery to crop improvement. The growth–defense trade-off is a key concern; selecting for resilience must not come at the cost of reduced biomass or altered medicinal compound profiles.	(Zhang *et al.*, 2023b; Raza *et al.*, 2024)
Mitigating climate impact & ensuring quality	• Optimizing practices to reduce footprint: using omics to refine precision agriculture, minimizing fertilizer and water use, thereby reducing greenhouse gas (GHG) emissions. • Quality assurance: using metabolomics as a quality control tool to ensure that climate-stressed plants still produce the required profile of bioactive compounds.	• Metabolomic fingerprinting: creating chemical profiles to certify the authenticity and potency of medicinal plant material, regardless of environmental fluctuations.	Addresses the core challenge: maintaining therapeutic efficacy under climate change. Metabolomics is essential for standardizing product quality in a changing environment. However, this requires establishing new, dynamic quality standards.	(Gupta *et al.*, 2019; Bhanuwanti *et al.*, 2024; Patil *et al.*, 2024)
Overarching challenges & future directions	• Data integration & interpretation: fusing genomic, transcriptomic, proteomic, metabolomic, and environmental data into actionable insights. • Bridging the lab-to-field gap: translating molecular findings into practical, scalable farming protocols. • Cost & accessibility: making omics technologies affordable and usable for farmers and small-scale growers of medicinal plants.	• Systems biology & AI: using machine learning to model complex interactions and predict optimal management practices. • Interdisciplinary collaboration: essential between plant scientists, data scientists, agronomists, and farmers.	The potential is revolutionary, but the path is littered with technical and practical hurdles. The "omics" approach risks being too top-down and technology-driven. Success depends on co-developing solutions with end-users (farmers) to ensure they are practical and economically viable.	(Hounsou *et al.*, 2024; Safdar *et al.*, 2024)

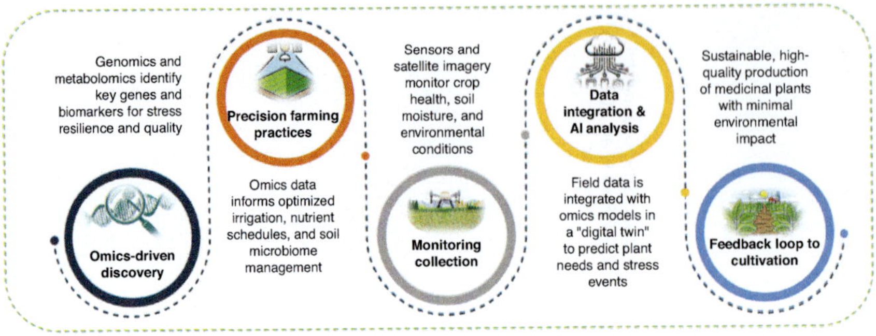

Fig. 7.4. Omics-informed CSA cycle. (Author's own figure.)

practices by identifying the specific water and nutrient needs of medicinal plants (Zeng *et al.*, 2025). The soil–plant digital twin approach, which integrates process-based models and data assimilation, offers a nuanced understanding of soil hydrological processes and plant hydraulics, enabling precise irrigation management (Zeng *et al.*, 2024). By monitoring soil moisture levels and microbial activity, omics data can guide the timing and amount of irrigation, reduce water waste, and enhance plant growth (Zeng *et al.*, 2025).

The manipulation of soil microbial communities, informed by omics data, can enhance soil health and plant productivity. This includes the development of microbial consortia that work synergistically to boost plant growth and maintain soil health (Niranjan *et al.*, 2023). Omics-based characterization of microbial communities allows for the identification of beneficial microbes that can be targeted for bioremediation and regenerative agriculture (Dhiman *et al.*, 2024). Multi-omics approaches enable the creation of detailed interaction networks and regulatory pathways, providing insights into the complex dynamics between plants and microbes, which are essential for optimizing microbiome management (Bhimani *et al.*, 2024).

While omics technologies offer significant potential for improving soil health and plant productivity, challenges remain in fully understanding the complex interactions within the soil microbiome. The integration of omics data with traditional agricultural practices requires careful consideration of environmental factors and the specific needs of medicinal plants. Additionally, the development of user-friendly tools and frameworks for farmers and researchers is essential to maximize the benefits of omics technologies in agriculture. As research progresses, the continued exploration of soil–plant–microbiome interactions will be crucial for advancing sustainable agricultural practices and ensuring the health and productivity of medicinal plants (Fig. 7.4).

7.7 Conservation Genomics for Stress-threatened Medicinal Plant Species

Conservation genomics is a critical tool for addressing the challenges faced by stress-threatened medicinal plant species. These plants are under increasing pressure due to overharvesting, habitat destruction, and climate change, which threaten their survival the sustainable supply of bioactive compounds essential for the pharmaceutical industry. Genomic technologies offer innovative solutions for the conservation and management of these species by providing insights into their genetic diversity, population structure, and adaptive potential. This approach not only aids in the preservation of genetic resources but also supports the development of effective conservation strategies tailored to the specific needs of each species (Table 7.7 and Fig. 7.5).

7.7.1 Genomic tools in conservation

Molecular markers such as DNA barcoding and DNA fingerprinting are pivotal in authenticating

Table 7.7. Conservation genomics for stress-threatened medicinal plants. (Author's own table.)

Conservation strategy/ genomic tool	Application & key insights	Specific technologies & case studies	Critical analysis & implementation challenges	Citations
Genetic diversity & population structure assessment	• Quantifying vulnerability: measures genetic diversity (heterozygosity, allelic richness) and inbreeding levels to assess extinction risk and adaptive potential.	• Simple sequence repeat (SSR) markers and single nucleotide polymorphisms (SNPs) were used in *T. govanianum* to reveal low diversity and high inbreeding, guiding urgent conservation. • High-throughput sequencing: enables genome-wide scans to identify unique and diverse populations for prioritization.	Foundational for setting conservation priorities. Low genetic diversity signals high risk, but even species with moderate diversity can be threatened by habitat fragmentation. This approach identifies *which* populations to save, but not always *how*.	(Rohini, 2020; Ma *et al*., 2023; Ul Islam *et al*., 2023)
DNA barcoding & fingerprinting for authentication	• Combating adulteration: provides a genetic "ID card" to ensure the authenticity of raw materials in trade, protecting both consumers and wild populations from overharvesting of substitute species.	• Standard barcode regions (e.g. *rbcL*, *matK*, *ITS*). • DNA fingerprinting: creates unique profiles for elite germplasms or specific populations.	A critical tool for forensic botany and regulating the herbal trade. However, it does not directly conserve species in the wild; it mitigates a downstream threat. Requires robust reference databases.	(Venkatasubramanian *et al*., 2018; Rohini, 2020)
Epigenomics for adaptive potential	• Assessing phenotypic plasticity: identifies non-genetic (epigenetic) variations (e.g. DNA methylation) that allow plants to adapt to stress without changing DNA sequence.	• Bisulfite sequencing: to map genome-wide DNA methylation patterns in response to environmental stressors.	Extremely valuable for genetically depauperate species, offering a mechanism for rapid adaptation. However, the heritability and stability of epigenetic marks across generations in natural populations are complex and poorly understood.	(Williams *et al*., 2025)
In situ conservation (protected areas)	• Preserving evolutionary processes: protects species in their natural habitat, maintaining ecological interactions and allowing continued adaptation.	• Establishing protected areas: e.g. Sinharaja rainforest for habitat-level conservation. • Community-based conservation: engaging local knowledge and stewardship, as in the north-eastern Himalayas.	The ideal strategy for maintaining evolutionary trajectories. However, it is ineffective alone for small, fragmented populations or when the threat (e.g. climate change) transcends reserve boundaries. Requires long-term political and community commitment.	(Parmin *et al*., 2022; Jayaweera *et al*., 2024)

Continued

Table 7.7. Continued

Conservation strategy/ genomic tool	Application & key insights	Specific technologies & case studies	Critical analysis & implementation challenges	Citations
Ex situ conservation (botanical gardens, seed banks)	• Insurance policy: safeguards genetic material outside the natural habitat as a backup against extinction.	• Seed banks: for species with orthodox seeds. • Field gene banks & living collections: e.g. Government Medicinal Plant Gardens.	Essential for critically endangered species. Major limitations: does not conserve co-adapted microbial communities; high cost of maintenance; and may lead to inadvertent selection for garden-adapted traits, reducing fitness for reintroduction.	(Majumder *et al*., 2023; Jayaweera *et al*., 2024)
Biotechnology-enabled *ex situ* (*in vitro* & cryopreservation)	• Last resort for recalcitrant species: preserves species that cannot be stored in conventional seed banks. • Rapid cloning: micropropagation can quickly build numbers from a few remaining individuals.	• Micropropagation: for species like *Podophyllum* and *Aconitum*. • Cryopreservation: storing tissues in liquid nitrogen for long-term, stable conservation.	A powerful safety net. The critical risk is bottlenecking – a whole restored population may be clones with low genetic diversity, making it vulnerable to disease. Must be integrated with strategies to capture and maintain genetic diversity.	(Sharma *et al*., 2020; Unal *et al*., 2023; Shankar *et al*., 2024)
Identifying & cultivating elite germplasm	• Sustainable sourcing: using biochemical and molecular markers to identify high-yielding, potent genotypes for cultivation, reducing pressure on wild populations.	• Metabolomic profiling: to select for high content of bioactive compounds (e.g. paclitaxel in *Taxus*). • MAS.	Directly addresses the driver of overharvesting. However, cultivation may fail to replicate the complex environmental conditions that induce certain secondary metabolites in the wild, potentially leading to inferior medicinal quality.	(Venkatasubramanian *et al*., 2018; Alum, 2025)
Overarching challenges & synthesis	• Integrated conservation: no single method is sufficient. A meta-population management approach, combining *in situ* reserves with genetically managed *ex situ* populations, is needed. • Socio-economic factors: illicit trade and lack of economic alternatives undermine conservation. • Climate change: renders static protected areas less effective; requires "climate-smart" conservation.	• Genomic-informed action plans: using data on genetic diversity, adaptive loci, and gene flow to design wildlife corridors and assisted migration strategies.	The greatest challenge is translating genomic data into effective policy and on-the-ground action. Conservation is not just a biological problem but a socio-economic one. Prioritization is key: omics helps decide where limited conservation funds can have the greatest impact by identifying evolutionarily significant units (ESUs).	(Vollmer *et al*., 2021; Sudhakaran, 2025)

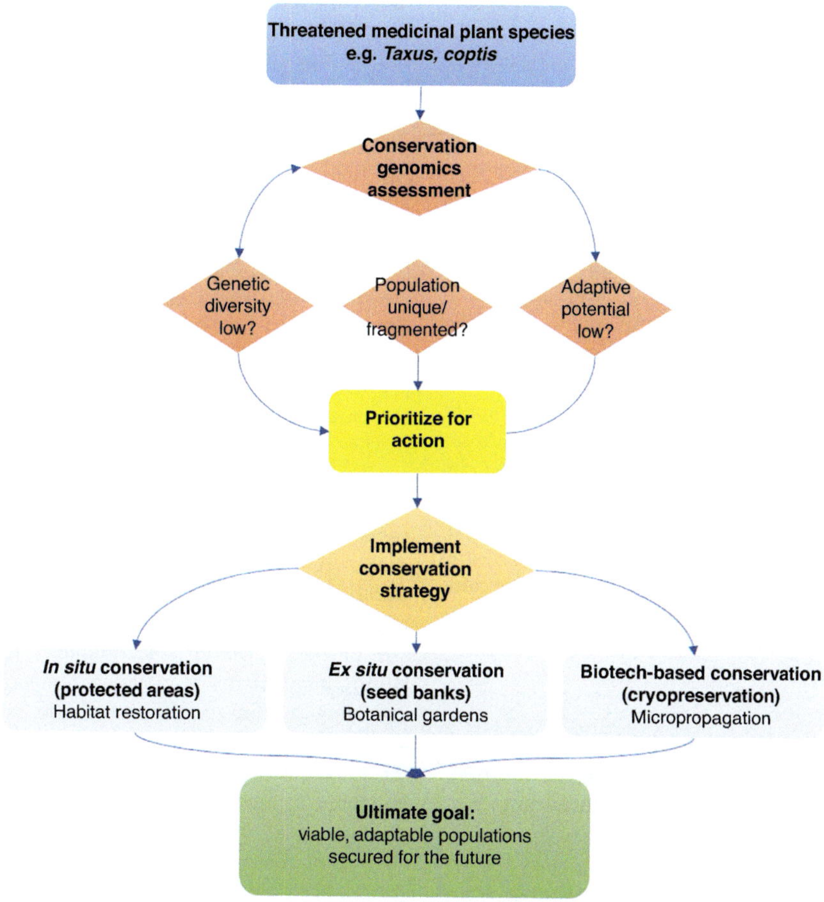

Fig. 7.5. Integrated conservation strategy for endangered medicinal plants. (Author's own figure.)

species identification and preventing adulteration in herbal formulations. These tools help in assessing genetic variability and population structure, which are crucial for understanding species evolution and adaptation to environmental changes (Rohini, 2020). Non-genetic variations, such as DNA methylation, play a significant role in the adaptability of genetically depauperate species. These epigenomic modifications can enhance phenotypic plasticity, allowing plants to better cope with environmental stress, thus serving as a novel target for conservation efforts (Williams *et al.*, 2025). While *in situ* conservation is essential, it may not suffice for species with declining populations. *Ex situ* strategies, including tissue culture and cryopreservation, are vital for maintaining genetic diversity and providing a backup for field gene banks. These methods have been successfully applied to conserve various medicinal plant species (Sharma *et al.*, 2020).

Identifying and cultivating elite germplasms using biochemical markers can help restore threatened species. This approach has been effective in reviving species like *Aconitum* and *Podophyllum*, which were on the brink of extinction (Venkatasubramanian *et al.*, 2018). Understanding the genetic diversity and population structure of endangered species is crucial for their conservation. For instance, studies on *Trillium govanianum* have revealed low genetic diversity and high inbreeding, necessitating targeted conservation strategies to mitigate these issues (Ul Islam *et al.*, 2023). High-throughput sequencing technologies have advanced the field of conservation genomics, enabling detailed analyses of genetic diversity and population dynamics. These insights are essential for developing population-based conservation strategies that ensure the long-term survival of endangered species (Ma *et al.*, 2023).

While genomic tools and conservation strategies offer promising solutions, challenges remain in their implementation. The integration of traditional knowledge with modern biotechnological approaches can enhance conservation efforts. Additionally, addressing socio-economic factors, such as illegal trade and habitat destruction, is crucial for the success of conservation programs. Collaborative efforts involving local communities, governments, and international organizations are necessary to create sustainable conservation frameworks that balance ecological and economic interests.

7.7.2 Using omics to prioritize endangered species (e.g. *Taxus*, *Coptis*) for *ex situ*/*in situ* conservation

The conservation of endangered medicinal plant species, such as *Taxus* and *Coptis*, is crucial due to their significant medicinal value and the threats they face from overexploitation and habitat loss. Both *in situ* and *ex situ* conservation strategies are essential to ensure the survival and sustainable use of these species. *In situ* conservation involves protecting plants in their natural habitats, while *ex situ* conservation involves preserving them outside their natural environments, such as in botanical gardens or through seed banks. Each method has its advantages and challenges, and a combination of both is often necessary for effective conservation.

In situ conservation focuses on protecting the natural habitats of medicinal plants. This approach allows plants to continue evolving in their natural environment, maintaining ecological interactions and genetic diversity. For instance, the establishment of protected areas like the Sinharaja rainforest in Sri Lanka is crucial for conserving species-rich habitats that house numerous threatened medicinal plants (Jayaweera *et al.*, 2024). Engaging local communities in conservation efforts is vital. In the north-eastern Himalayan region, local inhabitants use medicinal plants for traditional medicine, and their involvement in conservation can help protect these resources from overexploitation (Parmin *et al.*, 2022).

Ex situ conservation involves the cultivation and preservation of plants in controlled environments. Government Medicinal Plant Gardens in Sri Lanka and field gene banks in India are examples where endangered species are conserved outside their natural habitats (Majumder *et al.*, 2023; Jayaweera *et al.*, 2024). *In vitro* techniques, such as micropropagation, are increasingly used for the *ex situ* conservation of medicinal plants. These methods allow for the rapid multiplication and preservation of plant species that are difficult to propagate through traditional means (Sharma *et al.*, 2020; Unal *et al.*, 2023).

Cryopreservation technique involves storing plant tissues at ultra-low temperatures, providing a long-term conservation solution for genetic material. It is particularly useful for species with recalcitrant seeds that cannot be stored in seed banks (Vollmer *et al.*, 2021; Shankar *et al.*, 2024). Climate change poses a significant threat to the habitats of medicinal plants, affecting their growth and distribution. Conservation strategies must consider these environmental changes in order to be effective (Sudhakaran, 2025). Overharvesting for commercial purposes is a major threat to medicinal plants. Developing sustainable harvesting practices and promoting the cultivation of medicinal plants can help reduce pressure on wild populations (Alum, 2025).

While *in situ* conservation allows for the natural evolution and adaptation of species, it may not be sufficient for species with limited distribution or those facing immediate threats. *Ex situ* conservation provides a backup for preserving genetic material and can facilitate research and restoration efforts. However, it requires significant resources and expertise to maintain the genetic diversity and viability of the conserved species. A balanced approach that integrates both *in situ* and *ex situ* methods, supported by community involvement and modern biotechnological tools, is essential for the effective conservation of endangered medicinal plants like *Taxus* and *Coptis* (Fig. 7.5).

7.8 Challenges and Future Directions

Bridging the gap between lab-generated omics datasets (genomics, proteomics, metabolomics) and practical agricultural or conservation strategies remains difficult. Complex molecular interactions and environmental variability hinder the implementation of omics insights in real-world cultivation, such as optimizing irrigation or microbiome management under diverse field conditions. Enhancing abiotic stress resilience (e.g. salinity/drought tolerance) often alters secondary metabolite profiles critical for therapeutic value. For example, osmotic protectants like proline may suppress alkaloid biosynthesis, risking diminished pharmaceutical efficacy in prioritized species like *Foeniculum vulgare* (fennel) or *Taxus*. Stress adaptation involves multigenic traits, epigenetic regulation, and tissue-specific pathways, complicating breeding efforts. Integrating omics across these layers—while accounting for species-specific responses (e.g. *H. perforatum*'s heavy metal detoxification)—demands advanced computational tools and large-scale validation. *Ex situ* conservation (e.g. seed banks for *Coptis*) faces genetic diversity erosion, while phytoremediation species (e.g. *Hypericum*) risk heavy metal accumulation, necessitating stringent safety protocols for medicinal use. Scaling genomic tools for endangered species prioritization also requires resource-intensive efforts.

Combining genomics, metabolomics, and epigenomics with machine learning/AI to model stress-response networks could predict trade-offs (e.g. metabolite shifts under drought) and accelerate breeding of "designer" cultivars balancing resilience and compound yield. Integrated multi-omics frameworks represent a powerful frontier, combining genomics, metabolomics, and epigenomics with advanced machine learning/AI to model complex stress-response networks in plants. This approach enables the prediction of critical trade-offs, such as shifts in valuable medicinal metabolites under drought stress, and accelerates the breeding of "designer" cultivars that optimally balance environmental resilience with high yields of therapeutic compounds. Embedding these omics insights into CSA practices creates vital synergies for scalability. Within this framework, microbiome engineering utilizes metagenomics to design and deploy tailored probiotic consortia that enhance root nutrient uptake and bolster plant stress resilience. Concurrently, precision farming leverages dynamic transcriptomic data to inform real-time irrigation and nutrient scheduling through sophisticated soil-plant "digital twins," optimizing resource use and stress mitigation. Biotechnological innovations further enhance resilience: CRISPR-Cas9 can be precisely applied to edit stress-related genes (e.g. enhancing phytochelatin synthases for heavy metal detoxification) while meticulously avoiding disruption to medicinal biosynthetic pathways. Similarly, metabolic engineering pipelines can be developed to overexpress key protective metabolites, like UV-shielding flavonoids, ensuring maintenance of the plant's essential therapeutic profile. Expanding conservation genomics is crucial, utilizing population genomics to identify adaptive genetic variants within threatened medicinal species (e.g. Pacific yew, *Taxus brevifolia*), guiding targeted interventions like *in situ* habitat restoration and strategic *ex situ* cryopreservation. Collaborative global seed banks, underpinned by omics-driven diversity monitoring, are vital for preserving these genetic resources. Finally, realizing the full potential of this integrated approach requires robust policy and interdisciplinary integration, fostering cross-sector frameworks that link cutting-edge omics research with traditional ecological knowledge, agroecological principles, and supportive policy. Prioritizing funding is essential for validating omics-guided field practices in marginal environments, standardizing safety thresholds for plants used in phytoremediation of contaminated land, and implementing strong measures to protect the genetic sovereignty of biodiverse regions housing endangered medicinal plant species (Table 7.8).

Table 7.8. Challenges and future directions in medicinal plant omics. Author's own table.

Domain	Core challenges	Specific examples & implications	Proposed future directions & integrated solutions	Critical analysis of proposed solutions
Translation from lab to field	The "lab-to-field gap": complex omics data fails to translate into robust, scalable agricultural practices due to environmental variability and complex gene–environment interactions.	• An irrigation schedule optimized via transcriptomics in a controlled greenhouse may fail in a field with variable soil and weather. • A beneficial microbiome consortium may not establish in diverse soil types.	Integrated multi-omics + AI: combine genomics, metabolomics, and epigenomics with machine learning to build predictive models of plant responses. Embed these insights into CSA platforms for dynamic management.	Promise: moves from reactive to predictive agriculture. Hurdle: requires massive, high-quality, context-specific datasets. AI models can be "black boxes," and their predictions may lack transparency for farmers.
Resilience–quality trade-off	Enhancing abiotic stress resilience often alters secondary metabolite profiles, risking the very medicinal value of the plant.	• Drought stress may increase proline (osmoprotectant) but suppress alkaloid biosynthesis in species like fennel (*F. vulgare*). • Prioritizing biomass for phytoremediation can dilute compound concentration.	Predictive modeling & "designer cultivars": use AI to model the metabolic trade-off networks and guide the breeding/engineering of cultivars that balance resilience with high, consistent yields of target compounds.	Promise: enables precision breeding for multi-trait optimization. Hurdle: immense biological complexity. A "designer cultivar" for one environment may not be optimal for another, limiting scalability.
Biological complexity	Stress adaptation is a multigenic, multi-layered trait involving complex genetics, epigenetics, and tissue-specific pathways.	• Heavy metal detoxification in *H. perforatum* involves phytochelatin synthesis, transporter genes, and antioxidant systems simultaneously. • Species-specific responses prevent universal solutions.	Multi-omics integration & CRISPR-Cas9: decipher complex networks by integrating data layers. Use CRISPR for precise gene editing (e.g. enhancing phytochelatin synthases) without disrupting non-target medicinal pathways.	Promise: offers causal understanding and precise genetic tools. Hurdle: high risk of unintended pleiotropic effects. Public and regulatory acceptance of GM medicinal plants is a significant barrier.
Conservation & safety	*Ex situ* conservation (seed banks) can lead to genetic erosion. Plants used for phytoremediation (e.g. *H. perforatum*) accumulate heavy metals, creating unsafe medicine.	• Cryopreserved *Coptis* populations may lack the genetic diversity to adapt upon reintroduction. • *H. perforatum* from contaminated sites must be excluded from the medicinal supply chain.	Advanced conservation genomics & safety protocols: use population genomics to prioritize diverse populations for conservation. Establish omics-driven monitoring (metabolomic fingerprinting) to enforce safety standards and segregate phytoremediation biomass.	Promise: ensures long-term genetic health and consumer safety. Hurdle: expensive and requires international regulatory coordination. The economic incentive to sell contaminated biomass may undermine safety protocols.

Microbiome management	Introducing beneficial microbes is unreliable; they often fail to compete with established soil communities.	• A designed probiotic consortium may not colonize the rhizosphere effectively in field conditions.	Microbiome engineering via metagenomics: use metagenomics to identify keystone species and design robust, synergistic microbial consortia that enhance nutrient uptake and stress resilience.	Promise: could reduce fertilizer and pesticide dependency. Hurdle: soil microbiome is a complex, competitive ecosystem. Engineered consortia may not be stable or persistent over time.
Precision farming implementation	Real-time management based on omics data is computationally expensive and inaccessible to most growers.	• Building and running a soil-plant "digital twin" for real-time irrigation is not feasible for smallholder farmers.	Precision farming & digital twins: develop user-friendly decision-support tools that translate complex omics and sensor data into simple, actionable alerts for farmers (e.g. simplified irrigation schedules).	Promise: optimizes resource use and boosts sustainability. Hurdle: high initial cost and digital literacy requirements risk excluding the very communities who often cultivate medicinal plants.
Policy & global equity	Research is often siloed, and there is a risk of biopiracy or exploitation of genetic resources from biodiverse regions.	• Omics data from a rare medicinal plant could be used to synthesize compounds without benefiting the country of origin.	Policy & interdisciplinary integration: create frameworks that link omics research with traditional knowledge, ensure ethical data sharing, and implement access and benefit-sharing (ABS) protocols under the Nagoya Protocol.	Promise: promotes ethical, sustainable, and collaborative research. Hurdle: political and commercial interests often override ethical considerations. Enforcement is weak in many regions.

7.9 Conclusion

The escalating climate crisis poses multifaceted threats to medicinal plants, jeopardizing not only biodiversity but also global health security. This chapter synthesizes how abiotic stressors—drought, salinity, heavy metals—reconfigure physiological and biochemical pathways in medicinal species, often enhancing stress-resilient compounds (e.g. aloin in *A. vera* under salinity) while suppressing others (e.g. phenolics). Multi-omics approaches have revolutionized our understanding of these adaptations, revealing tissue-specific metabolic shifts, hormone signaling (ABA, JA), and genetic regulators like PCs for metal detoxification in *H. perforatum*. Such insights enable targeted breeding of salinity-tolerant cultivars (e.g. fennel, ajwain) and pathogen-resistant lines via omics-guided identification of antimicrobial metabolites. Despite these advances, significant challenges remain. Bridging lab-based omics discoveries with real-world agriculture requires overcoming species-specific complexity; for instance, osmotic protectants like proline may inadvertently dampen alkaloid biosynthesis in *F. vulgare*, undermining therapeutic value. Conservation efforts face similar hurdles: *ex situ* strategies (e.g. seed banking for *Coptis*) risk genetic erosion, while phytoremediation species accumulate toxins, necessitating stringent safety protocols. Moreover, the multigenic nature of stress adaptation demands integrated computational models to predict trade-offs between resilience and metabolite integrity. Looking ahead, synergistic frameworks offer promise. Machine learning can unify genomics, metabolomics, and epigenomics to design "smart" cultivars that balance environmental robustness with pharmaceutical yield. CSA, enhanced by microbiome engineering and precision irrigation informed by soil-plant "digital twins," could optimize resource use in vulnerable ecosystems. Biotechnological tools like CRISPR-Cas9 may edit stress-response genes without disrupting medicinal pathways, while metabolic engineering could amplify protective metabolites (e.g. UV-shielding flavonoids). Concurrently, conservation genomics must prioritize adaptive genetic variants in endangered species (e.g. *T. brevifolia*), coupling *in situ* habitat management with global cryobanking networks. Realizing this vision requires urgent policy action and interdisciplinary collaboration. Funding must prioritize field validation of omics-guided practices, standardization of safety thresholds for phytoremediation-derived medicines, and protection of genetic sovereignty in biodiverse regions. By uniting cutting-edge science with traditional ecological knowledge, we can forge sustainable pathways to conserve medicinal flora—ensuring their survival as both ecological keystones and invaluable sources of future therapeutics in an era of climate uncertainty.

Acknowledgments

The authors express their gratitude for the financial support received through the Distinguished High-Level Talents Research Grant from the Guizhou Science and Technology Corporation Platform Talents Fund (Grant nos.: [2017]5733-001 and CK-1130-002), high-level specialists project fund of Zunyi Medical University (1092/2000006)F-1092, the National Natural Science Foundation of China (82373981), and the support provided by Zunyi Medical University, China. Special appreciation is extended to all laboratory colleagues and research staff members for their valuable insights, constructive guidance, and assistance throughout this study.

References

Agrawal, S., Singh, A., Gawande, V., Bhanuvally, M., Gourkhede, P.H. *et al.* (2023) Salt stress tolerance in medicinal and aromatic plants: A review. *International Journal of Environment and Climate Change* 13(9), 3382–3391. DOI: 10.9734/ijecc/2023/v13i92590.

Ahmad, H., Maher, M., Abdel-Salam, E.M., Li, Y., Yang, C. *et al.* (2021) Integrated *de novo* analysis of transcriptional and metabolic variations in salt-treated solenostemma argel desert plants. *Frontiers in Plant Science* 12, 744699. DOI: 10.3389/fpls.2021.744699.

Al-Khayri, J.M., Rashmi, R., Toppo, V., Chole, P.B., Banadka, A. *et al.* (2023) Plant secondary metabolites: The weapons for biotic stress management. *Metabolites* 13(6), 716. DOI: 10.3390/metabo13060716.

Alugoju, P. and Tencomnao, T. (2023) Medicinal plants proteomics in response to abiotic stresses. In: Husen, A. and Iqbal, M. (eds) *Medicinal Plants*. Springer Nature, Singapore, pp. 35–107.

Alum, E.U. (2025) Sustainable harvesting of medicinal plants: Balancing therapeutic benefits with environmental conservation. *Agroecology and Sustainable Food Systems* 49(3), 380–385. DOI: 10.1080/21683565.2024.2421948.

Applequist, W.L., Brinckmann, J.A., Cunningham, A.B., Hart, R.E., Heinrich, M. *et al.* (2020) Scientists' warning on climate change and medicinal plants. *Planta Medica* 86(1), 10–18. DOI: 10.1055/a-1041-3406.

Bagati, S., Mahajan, R., Nazir, M., Dar, A.A. and Zargar, S.M. (2018) Omics: A gateway towards abiotic stress tolerance. In: Zargar, S.M. and Zargar, M.Y. (eds) *Abiotic Stress-Mediated Sensing and Signaling in Plants: An Omics Perspective*. Springer Singapore, pp. 1–45.

Bertola, M., Ferrarini, A. and Visioli, G. (2021) Improvement of soil microbial diversity through sustainable agricultural practices and its evaluation by -omics approaches: A perspective for the environment, food quality and human safety. *Microorganisms* 9(7), 1400. DOI: 10.3390/microorganisms9071400.

Bhanuwanti,Dar, K.A., Singh, L., Saad, A.A., Rai, U. *et al.* (2024) Climate smart agriculture: Innovating sustainable practices for a changing climate. *Journal of Scientific Research and Reports* 30(11), 568–576. DOI: 10.9734/jsrr/2024/v30i112585.

Bhimani, P., Mahavar, P., Rajguru, B., Bhatt, V.D., Nathani, N.M. *et al.* (2024) Unveiling the green dialogue: Advancements in omics technologies for deciphering plant–microbe interactions in soil. *Discover Plants* 1(1), 4. DOI: 10.1007/s44372-024-00004-3.

Bisht, A., Saini, D.K., Kaur, B., Batra, R., Kaur, S. *et al.* (2023) Multi-omics assisted breeding for biotic stress resistance in soybean. *Molecular Biology Reports* 50(4), 3787–3814. DOI: 10.1007/s11033-023-08260-4.

Biswal, B., Jena, B., Giri, A.K. and Acharya, L. (2021) *De novo* transcriptome and tissue specific expression analysis of gene associated with biosynthesis of medicinally active metabolites in a high valued medicinal plant *Operculina turpethum* (L.). *Scientific Reports* 11, 22539. DOI: 10.21203/rs.3.rs-312726/v1.

Biswas, R. and Sarkar, A. (2018) Omics' tools in soil microbiology: The state of the art. In: Kumar Adhya, T., Lal, B., Mohapatra, B., Paul, D. and Das, D. (eds) *Advances in Soil Microbiology: Recent Trends and Future Prospects*, Vol. 3. Soil-Microbe Interaction. Springer, Singapore, pp. 35–64.

Borcă, M.A., Ona, A.D., Borcă, A.M. and Muntean, L. (2024) Preserving medicinal plant biodiversity: Climate change impacts, IUCN red list insights and conservation strategies. *Hop and Medicinal Plants* 32(1–2), 7–12. DOI: 10.15835/hpm.v32i1-2.15015.

Caioni, G., Reyes, C.P., Laurenti, D., Chiaradia, C., Dainese, E. *et al.* (2024) Biochemistry and future perspectives of antibiotic resistance: An eye on active natural products. *Antibiotics* 13(11), 1071. DOI: 10.3390/antibiotics13111071.

Cao, D., Zhang, W., Yang, N., Li, Z., Zhang, C. *et al.* (2023) Proteomic and metabolomic analyses uncover integrative mechanisms in *Sesuvium portulacastrum* tolerance to salt stress. *Frontiers in Plant Science* 14, 1277762. DOI: 10.3389/fpls.2023.1277762.

Chandrakar, V., Yadu, B., Xalxo, R., Kumar, M. and Keshavkant, S. (2020) Mechanisms of plant adaptation and tolerance to metal/metalloid toxicity. In: Hasanuzzaman, M. (ed.) *Plant Ecophysiology and Adaptation under Climate Change: Mechanisms and Perspectives II*. Springer, Singapore, pp. 107–135.

Chatterjee, S., Kumari, S., Rath, S., Priyadarshanee, M. and Das, S. (2020) Diversity, structure and regulation of microbial metallothionein: Metal resistance and possible applications in sequestration of toxic metals. *Metallomics* 12(11), 1637–1655. DOI: 10.1039/d0mt00140f.

Chaturvedi, P., Pierides, I., Zhang, S., Schwarzerova, J., Ghatak, A. *et al.* (2024) Multiomics for crop improvement. In: Pandey, M.K., Bentley, A., Desmae, H., Roorkiwal, M. and Varshney, R.K. (eds) *Frontier Technologies for Crop Improvement*. Springer Nature Singapore, Singapore, pp. 107–142.

Chawla, R., Mondal, K. and Pankaj, M.S. (2022) Mechanisms of plant stress tolerance: Drought, salinity, and temperature extremes. *Plant Science Archives* 7(2), 4–8. DOI: 10.51470/PSA.2022.7.2.04.

Daldoul, S., Ben Amar, A., Guillaumie, S. and Mliki, A. (2014) Integration of omics and system biology approaches to study grapevine (*Vitis vinifera* L.) response to salt stress: A perspective for functional genomics - A review. *OENO One* 48(3), 189. DOI: 10.20870/oeno-one.2014.48.3.1573.

Dar, A.M., Touseef, H., Nawaz, K., Khan, Y. and Sahu, P.P. (2023) Editorial: Genomics in plant sciences: Understanding and development of stress-tolerant plants. *Frontiers in Plant Science* 14, 1222818. DOI: 10.3389/fpls.2023.1222818.

Das, P., Nutan, K.K., Singla-Pareek, S.L. and Pareek, A. (2015) Understanding salinity responses and adopting "omics-based" approaches to generate salinity tolerant cultivars of rice. *Frontiers in Plant Science* 6, 712. DOI: 10.3389/fpls.2015.00712.

de Vasconcelos, M.W.P.L., Menguer, P.K., Hu, Y., Revers, L.F. and Sperotto, R.A. (2016) Stress signaling responses in plants. *BioMed Research International* 4720280. DOI: 10.1155/2016/4720280.

Debnath, M., Pandey, M. and Bisen, P.S. (2011) An omics approach to understand the plant abiotic stress. *OMICS* 15(11), 739–762. DOI: 10.1089/omi.2010.0146.

Dhiman, N., Uthoff, J., Scharf, B. and Kumar, V. (2024) Plant-microbe interaction to improve soil health. In: Bhatia, R.K. and Walia, A. (eds) *Advancements in Microbial Biotechnology for Soil Health*. Springer Nature Singapore, Singapore, pp. 189–226.

Duan, H., Tiika, R.J., Tian, F., Lu, Y., Zhang, Q. *et al.* (2023) Metabolomics analysis unveils important changes involved in the salt tolerance of *Salicornia europaea*. *Frontiers in Plant Science* 13, 1097076. DOI: 10.3389/fpls.2022.1097076.

Dubey, S., Shri, M., Gupta, A., Rani, V. and Chakrabarty, D. (2018) Toxicity and detoxification of heavy metals during plant growth and metabolism. *Environmental Chemistry Letters* 16(4), 1169–1192. DOI: 10.1007/s10311-018-0741-8.

Ejaz, U., Khan, S.M., Khalid, N., Ahmad, Z., Jehangir, S. *et al.* (2023) Detoxifying the heavy metals: A multipronged study of tolerance strategies against heavy metals toxicity in plants. *Frontiers in Plant Science* 14, 1154571. DOI: 10.3389/fpls.2023.1154571.

Elsisi, M., Elshiekh, M., Sabry, N., Aziz, M., Attia, K. *et al.* (2024) The genetic orchestra of salicylic acid in plant resilience to climate change induced abiotic stress: Critical review. *Stress Biology* 4(1), 31. DOI: 10.1007/s44154-024-00160-2.

Ferreira-Neto, J.R.C., Kido, É.A., Aburjaile, F.F., Silva, M.D., Souza, M.C.P. *et al.* (2022) The journey through the gene: A focus on plant anti-pathogenic agents mining in the omics era. In: Rai, M. and Feitosa, C.M. (eds) *Eco-Friendly Biobased Products Used in Microbial Diseases*. CRC Press, Boca Raton, Florida, pp. 3–22.

Fik-Jaskółka, M., Mittova, V., Motsonelidze, C., Vakhania, M., Vicidomini, C. *et al.* (2024) Antimicrobial metabolites of Caucasian medicinal plants as alternatives to antibiotics. *Antibiotics* 13(6), 487. DOI: 10.3390/antibiotics13060487.

Goldsbrough, P. (2020) Metal tolerance in plants: The role of phytochelatins and metallothioneins. In: Terry, N. and Banuelos, G.S. (eds) *Phytoremediation of Contaminated Soil and Water*. CRC Press, Boca Raton, Florida, pp. 221–233.

Gomez-Casati, D.F., Pagani, M.A., Busi, M.V. and Bhadauria, V. (2016) Omics approaches for the engineering of pathogen resistant plants. *Current Issues in Molecular Biology* 19, 89–98. DOI: 10.21775/9781910190357.09.

González-Mendoza, D. and Zapata-Pérez, O. (2008) Mecanismos de tolerancia a elementos potencialmente tóxicos en plantas. *Botanical Sciences* 82(82), 53–61. DOI: 10.17129/botsci.1781.

Gupta, A., Singh, P.P., Singh, P., Singh, K., Singh, A.V. *et al.* (2019) Medicinal plants under climate change: Impacts on pharmaceutical properties of plants. In: Kumar Choudhary, K. and Kumar, A. (eds) *Climate Change and Agricultural Ecosystems*. Elsevier, Amsterdam, pp. 181–209.

Guru, A., Sahoo, S.K., Dash, G.K., Jena, J. and Dwivedi, P. (2021) Recent advances in multi-omics and breeding approaches towards drought tolerance in crops. In: Kumar Gupta, M. and Behera, L. (eds) *Applications of Bioinformatics in Rice Research*. Springer, Singapore, pp. 333–359.

Halder, M. and Jha, S. (2023) The current status of population extinction and biodiversity crisis of medicinal plants. In: Jha, S. and Halder, M. (eds) *Medicinal Plants: Biodiversity, Biotechnology and Conservation*. Springer Nature Singapore, Singapore, pp. 3–38.

Harish, S., Parthasarathy, S., Prabhukarthikeyan, S.R., Mahanty, A., Teli, B. *et al.* (2022) Omics: A potential tool to delineate the mechanism of biocontrol agents against plant pathogens. In: Sahu, J., Vaishnav, A. and Singh, H.B. (eds) *Plant-Microbe Interactions*. CRC Press, Boca Raton, Florida, pp. 79–100.

Hounsou, E.K., Sonibare, M.A. and Elufioye, T.O. (2024) Climate change and the future of medicinal plants research. *Bioactive Compounds in Health and Disease - Online ISSN* 7(3), 152–169. DOI: 10.31989/bchd.v7i3.1310.

Hussain, T., Asrar, H., Zhang, W., Gul, B. and Liu, X. (2021) Combined transcriptome and proteome analysis to elucidate salt tolerance strategies of the halophyte *Panicum antidotale* Retz. *Frontiers in Plant Science* 12, 760589. DOI: 10.3389/fpls.2021.760589.

Ibrahim, M., Detroja, A., Bhimani, A., Bhatt, T.C., Koradiya, J. *et al.* (2024) Ensuring the safety of Indian herbal drugs: A comprehensive assessment of heavy metal contaminations with emphasis on lead, arsenic and cadmium. DOI: 10.21203/rs.3.rs-3950502/v1.

Ishida, J.K. and Martinez Corcino, D.S. (2022) Role of natural resistance-associated macrophage proteins in metal ion transport in plants. In: Kumar Upadhyay, S. (ed.) *Cation Transporters in Plants*. Elsevier, Amsterdam, pp. 337–356.

Ismail, A.M. and Horie, T. (2017) Genomics, physiology, and molecular breeding approaches for improving salt tolerance. *Annual Review of Plant Biology* 68(1), 405–434. DOI: 10.1146/annurev-arplant-042916-040936.

Jan, M., Mir, T.A., Khare, R.K. and Saini, N. (2022) Adaptation strategies of medicinal plants in response to environmental stresses. In: Aftab, T. (ed.) *Environmental Challenges and Medicinal Plants: Sustainable Production Solutions under Adverse Conditions*. Springer International Publishing, Cham, Switzerland, pp. 133–151.

Jayaweera, C.L., Rajakaruna, R.W.M.T.N., Jayasuriya, K.M.G.G. and Yakan, D.M.D. (2024) Gaps in *in-situ* and *ex-situ* conservation of threatened medicinal plant species of Sri Lanka; towards their effective conservation. DOI: 10.21203/rs.3.rs-4409027/v1.

Kanojia, A., Jaiswal, A. and Mudgil, Y. (2023) Understanding molecular mechanisms of plant physiological responses under drought and salt stresses. In: Chen, J.T. (ed.) *Molecular and Physiological Insights into Plant Stress Tolerance and Applications in Agriculture*. Bentham Science Publishers, Singapore, pp. 156–184.

Kanso, M., Hijazi, M., Aboul Ela, M. and El-Lakany, A. (2022) Medicinal plants' stress factors: Effects on metabolites and novel perspectives for tolerance. *BAU Journal - Health and Well-Being* 5(1). DOI: 10.54729/XWWG9154.

Karami, S., Ejtehadi, H., Moazzeni, H., Vaezi, J. and Behroozian, M. (2022) Climate change impact on the geographic distribution of *Nepeta glomerulosa*, medicinal species endemic to Southwestern and Central Asia. DOI: 10.21203/rs.3.rs-1417775/v1.

Karthik, G., Mehandi, S. and Kumar, T.V. (2024) Enhancing crop resilience to abiotic stress through molecular markers: A review. *Journal of Advances in Biology & Biotechnology* 27(7), 521–535. DOI: 10.9734/jabb/2024/v27i71013.

Keeran, N.S., Usha, B. and Ganesan, G. (2021) P-type atpases and their role in metal homeostasis in plants. In: Roychoudhury, A., Kumar Tripathi, D. and Deshmukh, R. (eds) *Metal and Nutrient Transporters in Abiotic Stress*. Elsevier, Amsterdam, pp. 33–54.

Khatri, N. and Mudgil, Y. (2017) Salinity stress: "Omics" approaches. In: Majeed Zargar, S. and Rai, V. (eds) *Plant Omics and Crop Breeding*. Apple Academic Press, Palm Bay, Florida, pp. 295–310.

Kielich, N., Mazur, O., Musidlak, O., Gracz-Bernaciak, J. and Nawrot, R. (2024) Herbgenomics meets papaveraceae: A promising -omics perspective on medicinal plant research. *Briefings in Functional Genomics* 23(5), 579–594. DOI: 10.1093/bfgp/elad050.

Ko, D.K. and Brandizzi, F. (2024) Multi-omics resources for understanding gene regulation in response to ER stress in plants. In: Kriechbaumer, V. (ed.) *The Plant Endoplasmic Reticulum: Methods and Protocols*. Springer US, New York, pp. 261–272.

Kumar, A., Chaturvedi, A.K., Surendran, U., Shabnam, A.A., Singh, A. *et al.* (2021a) Mechanistic overview of metal tolerance in edible plants: A physiological and molecular perspective. In: Hasanuzzaman, M. and Prasad, N.M.V. (eds) *Handbook of Bioremediation*. Academic Press, Cambridge, Massachusetts, pp. 23–47.

Kumar, D., Jha, S.S., Kumar, A. and Singh, S.K. (2021b) Omics and approaches in plant stress management. In: Kumar, A. and Droby, S. (eds) *Microbial Management of Plant Stresses*. Woodhead Publishing, Cambridge, UK, pp. 107–117.

Kumar, V., Prasad, A., Kashyap, P., Joshi, R. and Kumar, A. (2024) Integrated omics approaches for plant disease resistance. In: Singh, K., Kaur, R. and Deshmukh, R. (eds) *Biotechnological Advances for Disease Tolerance in Plants*. Springer Nature Singapore, pp. 211–237.

Lamalakshmi Devi, E., Kumar, S., Basanta Singh, T., Sharma, S.K., Beemrote, A. *et al.* (2017) Adaptation strategies and defence mechanisms of plants during environmental stress. In: Ghorbanpour, M. and Varma, A. (eds) *Medicinal Plants and Environmental Challenges*. Springer International Publishing, Cham, Switzerland, pp. 359–413.

Kumari, A., Das, P., Parida, A.K. and Agarwal, P.K. (2015) Proteomics, metabolomics, and ionomics perspectives of salinity tolerance in halophytes. *Frontiers in Plant Science* 6, 537. DOI: 10.3389/fpls.2015.00537.

Kumar, R., Mishra, R.K., Mishra, V., Qidwai, A., Pandey, A. *et al.* (2016) Detoxification and tolerance of heavy metals in plants. In: Ahmad, P. (ed.) *Plant Metal Interaction*. Elsevier, Amsterdam, pp. 335–359.

Lin, J. (2024) Molecular mechanisms of plant response to abiotic stresses and breeding for stress tolerance. *Theoretical and Natural Science* 63(1), 123–127. DOI: 10.54254/2753-8818/2024.17986.

Majumder, A., Roychowdhury, D. and Ray, S. (2023) Biotechnological approaches for *ex situ* conservation of medicinal plants. In: Jha, S. and Halder, M. (eds) *Medicinal Plants: Biodiversity, Biotechnology and Conservation*. Springer Nature Singapore, Singapore, pp. 729–800.

Manda, R.R., Addanki, V.A., Giabardo, A., Benjamin, J., Hossain, M.J. *et al.* (2023) Soil health management and microorganisms: Recent development. In: Singh, U.B., Kumar, R. and Singh, H.B. (eds) *Detection, Diagnosis and Management of Soil-Borne Phytopathogens*. Springer Nature, Singapore, pp. 437–493.

Liu, B., Dong, P., Zhang, X., Feng, Z., Wen, Z. *et al.* (2022) Identification and characterization of eight metallothionein genes involved in heavy metal tolerance from the ectomycorrhizal fungus laccaria bicolor. *Environmental Science and Pollution Research* 29(10), 14430–14442. DOI: 10.1007/s11356-021-16776-0.

Ma, Q., Wu, G., Li, W., Yuzuak, S., Guan, F. *et al.* (2023) Research advances and perspectives of conservation genomics of endangered plants. DOI: 10.5772/intechopen.112281.

Marques, D.N., Gaziola, S.A. and Azevedo, R.A. (2021) Phytochelatins and their relationship with modulation of cadmium tolerance in plants. In: Hasanuzzaman, M. and Prasad, N.M.V. (eds) *Handbook of Bioremediation*. Academic Press, Cambridge, Massachusetts, pp. 91–113.

Mathew, S. (2022) Mechanisms of heavy metal tolerance in plants: A molecular perspective. *Plant Science Archives* 7(2), 17–19. DOI: 10.51470/PSA.2022.7.2.17.

Mathur, J. and Chauhan, P. (2020) Mechanism of toxic metal uptake and transport in plants. In: Mishra, K., Kumar Tandon, P. and Srivastava, S. (eds) *Sustainable Solutions for Elemental Deficiency and Excess in Crop Plants*. Springer, Singapore, pp. 335–349.

Morsy, A. and El-Sherif, N. (2024) Integrating omics approaches for abiotic stress tolerance in plants. In: Hasanuzzaman, M. and Nahar, K. (eds) *Abiotic Stress in Crop Plants - Ecophysiological Responses and Molecular Approaches*. IntechOpen, London, pp. 1–13.

Muñoz-Hoyos, L. and Stam, R. (2023) Metabolomics in plant pathogen defense: From single molecules to large-scale analysis. *Phytopathology* 113(5), 760–770. DOI: 10.1094/PHYTO-11-22-0415-FI.

Muthuramalingam, P., Jeyasri, R., Rakkammal, K., Satish, L., Shamili, S. *et al.* (2022) Multi-omics and integrative approach towards understanding salinity tolerance in rice: A review. *Biology* 11(7), 1022. DOI: 10.3390/biology11071022.

Naqvi, R.Z., Siddiqui, H.A., Mahmood, M.A., Najeebullah, S., Ehsan, A. *et al.* (2022) Smart breeding approaches in post-genomics era for developing climate-resilient food crops. *Frontiers in Plant Science* 13, 972164. DOI: 10.3389/fpls.2022.972164.

Nazari, M., Ghasemi-Soloklui, A.A., Kordrostami, M. and Abdel Latef, A.A.H. (2023) Deciphering the response of medicinal plants to abiotic stressors: A focus on drought and salinity. *Plant Stress* 10, 100255. DOI: 10.1016/j.stress.2023.100255.

Niranjan, V., Setlur, A.S., Skariyachan, S. and Chandrashekar, K. (2023) Applications of microbial consortia and microbiome interactions for augmenting sustainable agrobiology. In: Kumar Maheshwari, D. and Dheeman, S. (eds) *Sustainable Agrobiology: Design and Development of Microbial Consortia*. Springer Nature Singapore, Singapore, pp. 275–316.

Nouripour-Sisakht, J., Ehsanzadeh, P. and Ehtemam, M.H. (2025) Comparative response of fennel, ajwain, and anise in terms of osmolytes accumulation, ion imbalance, photosynthetic and growth functions under salinity. *Scientific Reports* 15(1), 2520. DOI: 10.1038/s41598-025-86256-9.

Oyarce, P., Xiao, T.T., Henkel, C., Frederiksen, S.F., Gonzalez-Kise, J.K. *et al.* (2025) Microscopy and spatial-metabolomics identify tissue-specific metabolic pathways uncovering salinity and drought tolerance mechanisms in *Avicennia marina* and *Phoenix dactylifera* roots. *Scientific Reports* 15(1), 1076. DOI: 10.1038/s41598-025-85416-1.

Pandey, A., Prasad, S.M. and Kumar, J. (2022) Genomics (characterization of genes) and molecular aspects of metal tolerance and hyperaccumulation. In: Kumar, J., Gaur, S., Kumar Srivastava, P., Kumar Mishra, R. and Prasad, S.M. (eds) *Heavy Metals in Plants Physiological to Molecular Approach*. CRC Press, Boca Raton, Florida, pp. 310–322.

Pandey, N. (2018) Antioxidant defense system in plants exposed to metal toxicity. In: Hasanuzzaman, M., Nahar, K. and Fujita, M. (eds) *Plants Under Metal and Metalloid Stress*. Springer, Singapore, pp. 107–148.

Pandita, D. and Pamuru, R.R. (2022) Genomic approaches for phytoremediation of trace and hazardous metals. In: Naeem, M., Aftab, T., Ansari, A.A., Singh Gill, S. and Macovei, A. (eds) *Hazardous and Trace Materials in Soil and Plants*. Academic Press, Cambridge, Massachusetts, pp. 341–357.

Parida, T., Riyazuddin, S., Kolli, S.K., Chakraborty, A., Srinivas, N. *et al.* (2024) Phytoremediation of heavy metal(loid)s with integral involvement of the endogenous metal-chelators: Present state-of-the-art and future prospect. *Discover Environment* 2(1), 139. DOI: 10.1007/s44274-024-00170-x.

Parmin, P., Rusilowati, A. and Rahayu, E.F. (2022) Pemberdayaan masyarakat melalui konservasi tanaman obat untuk menunjang penyediaan bahan baku produksi jamu tradisional. *Jurnal Pemberdayaan Masyarakat* 1(1), 10–16. DOI: 10.46843/jmp.v1i1.263.

Patil, P., Sivakumar, B., Bora, N.R., Vas, M., Rajaram, R. *et al.* (2024) A comprehensive review on sustainable cultivation practices of medicinal plants. *International Journal of Plant & Soil Science* 36(9), 536–544. DOI: 10.9734/ijpss/2024/v36i95002.

Patni, B., Bhattacharyya, M., Kumari, A. and Purohit, V.K. (2022) Alarming influence of climate change and compromising quality of medicinal plants. *Plant Physiology Reports* 27(1), 1–10. DOI: 10.1007/s40502-021-00616-x.

Poh, W.H., Ruhazat, N.S., Yang, L.K., Shivhare, D., Lim, P.K. *et al.* (2023) Transcriptomic and metabolomic characterization of antibacterial activity of *Melastoma dodecandrum*. *Plant Biology*. DOI: 10.1101/2023.04.10.536307.

Punetha, A., Kumar, D., Suryavanshi, P., Padalia, R. and Kt, V. (2022) Environmental abiotic stress and secondary metabolites production in medicinal plants: A review. *Tarım Bilimleri Dergisi* 28(3), 351–362. DOI: 10.15832/ankutbd.999117.

Radulović, N.S., Blagojević, P.D., Stojanović-Radić, Z.Z. and Stojanović, N.M. (2013) Antimicrobial plant metabolites: Structural diversity and mechanism of action. *Current Medicinal Chemistry* 20(7), 932–952. DOI: 10.2174/0929867113805219136.

Raza, A., Bashir, S., Khare, T., Karikari, B., Copeland, R.G.R. *et al.* (2024) Temperature-smart plants: A new horizon with omics-driven plant breeding. *Physiologia Plantarum* 176(1), e14188. DOI: 10.1111/ppl.14188.

Rizzo, P., Altschmied, L., Ravindran, B.M., Rutten, T. and D'Auria, J.C. (2020) The biochemical and genetic basis for the biosynthesis of bioactive compounds in *Hypericum perforatum* L., one of the largest medicinal crops in Europe. *Genes* 11(10), 1210. DOI: 10.3390/genes11101210.

Robiansyah, I., Primananda, E., Zulkarnaen, R.N., Helmanto, H., Kusuma, Y.W.C. *et al.* (2023) Climate change impact on medicinal plants: An insight from the IUCN red list of threatened species. In: Jha, S. and Halder, M. (eds) *Medicinal Plants: Biodiversity, Biotechnology and Conservation*. Springer Nature, Singapore, pp. 115–131.

Rohini, M.R. (2020) Genomics and molecular characterization of threatened medicinal plants. In: Rajasekharan, P.E. and Wani, S.H. (eds) *Conservation and Utilization of Threatened Medicinal Plants*. Springer International Publishing, Cham, Switzerland, pp. 317–346.

Sachdev, S., Ansari, S.A., Ansari, M.I., Fujita, M. and Hasanuzzaman, M. (2021) Abiotic stress and reactive oxygen species: Generation, signaling, and defense mechanisms. *Antioxidants (Basel, Switzerland)* 10(2), 277. DOI: 10.3390/antiox10020277.

Safdar, M., Shahid, M.A., Yang, C., Rasul, F., Tahir, M. *et al.* (2024) Climate smart agriculture and resilience. In: Garwi, J., Masengu, R. and Chiwaridzo, O.T. (eds) *Emerging Technologies and Marketing Strategies for Sustainable Agriculture*. IGI Global Scientific Publishing, Hershey, Pennsylvania, pp. 28–52.

Santos-Silva, C.A. dos, Zupin, L., Oliveira-Lima, M., Vilela, L.M.B., Bezerra-Neto, J.P. *et al.* (2020) Plant antimicrobial peptides: State of the art, *in silico* prediction and perspectives in the omics era. *Bioinformatics and Biology Insights* 14, 117793222095273. DOI: 10.1177/1177932220952739.

Sebastian, A., Shukla, P., Nangia, A.K. and Prasad, M.N.V. (2019) Transgenics in phytoremediation of metals and metalloids. In: Prasad, M.N.V. (ed.) *Transgenic Plant Technology for Remediation of Toxic Metals and Metalloids*. Academic Press, Cambridge, Massachusetts, pp. 3–22.

Sharma, M.K. (2024) Exploring the biochemical profiles of medicinal plants cultivated under stressful environmental conditions. *Current Agriculture Research Journal* 12(1), 81–103. DOI: 10.12944/CARJ.12.1.07.

Sharma, N., Pandey, R. and Gowthami, R. (2020) *In vitro* conservation and cryopreservation of threatened medicinal plants of India. In: Rajasekharan, P.E. and Wani, S.H. (eds) *Conservation and Utilization of Threatened Medicinal Plants*. Springer International Publishing, Cham, Switzerland, pp. 181–228.

Shankar, M., Kumar, K.C., Puneeth, G., Deepak, D., Jeevan, N. *et al*. (2024) Synergizing cryopreservation, thermal analysis, and *in vitro* regeneration: Strategies for long-term plant germplasm preservation. *International Journal of Advanced Biochemistry Research* 8(8S), 1224–1235. DOI: 10.33545/26174693.2024.v8.i8Sp.2014.

Sharma, D. (2023) Omics' approaches to analysis of stress response genes in medicinal plants. In: Singh, D., Kumar Mishra, A. and Kumar Srivastava, A. (eds) *Stress-Responsive Factors and Molecular Farming in Medicinal Plants*. Springer Nature, Singapore, pp. 303–321.

Sharma, S.S., Dietz, K. and Mimura, T. (2016) Vacuolar compartmentalization as indispensable component of heavy metal detoxification in plants. *Plant, Cell & Environment* 39(5), 1112–1126. DOI: 10.1111/pce.12706.

Shil, S. and Dewanjee, S. (2022) Impact of drought stress signals on growth and secondary metabolites (SMS) in medicinal plants. *The Journal of Phytopharmacology* 11(5), 371–376. DOI: 10.31254/phyto.2022.11511.

Silva, V.N.B., Silva, T.L.C., Ferreira, T.M.M., Neto, J.C.R., Leão, A.P. *et al*. (2023) Multi-omics analysis of young *Portulaca oleracea* L. plants' responses to high NaCL doses reveals insights into pathways and genes responsive to salinity stress in this halophyte specie. *Phenomics (Cham, Switzerland)* (1), 1–21. DOI: 10.1007/s43657-022-00061-2.

Singh, C., Saini, R., Upadhyay, R. and Tiwari, K.N. (2023) Revealing the epigenetic mechanisms underlying the stress response in medicinal plants. In: Singh, D., Kumar Mishra, A. and Kumar Srivastava, A. (eds) *Stress-Responsive Factors and Molecular Farming in Medicinal Plants*. Springer Nature Singapore, pp. 107–122.

Singh, S., Khatri, N., Katiyar, A. and Mudgil, Y. (2015) Molecular approaches in deciphering abiotic stress signaling mechanisms in plants. In: Pandey, G.K. (ed.) *Elucidation of Abiotic Stress Signaling in Plants*. Springer, New York, pp. 41–73.

Sudhakaran, G. (2025) Impact of climate change on the yield of medicinal plants in recent years. *Natural Product Research* 39(12), 3614–3615. DOI: 10.1080/14786419.2024.2372845.

Sui, D., Wang, B., El-Kassaby, Y.A. and Wang, L. (2024) Integration of physiological, transcriptomic, and metabolomic analyses reveal molecular mechanisms of salt stress in *Maclura tricuspidata* 13(3), 397. DOI: 10.3390/plants13030397.

Thakur, K., Kumari, C., Zadokar, A., Sharma, P. and Sharma, R. (2023) Physiological and omics-based insights for underpinning the molecular regulation of secondary metabolite production in medicinal plants: UV stress resilience. *Plant Physiology and Biochemistry* 204, 108060. DOI: 10.1016/j.plaphy.2023.108060.

Ul Islam, S., Ahmed Mangral, Z., Tariq, L., Ahmad Bhat, B., Waseem Tantray, W. *et al*. (2023) Conservation genetics of endangered *Trillium govanianum* Wall ex D. Don - A pharmaceutically prized medicinal plant from the Himalaya and implications for species recovery. *Gene* 888, 147748. DOI: 10.1016/j.gene.2023.147748.

Unal, B.T., Turker, H. and Ozturk, M. (2023) Ex-situ conservation of medicinal and aromatic plants using in vitro techniques. In: Ansari, M.K.A., Ahmad, M. and Owens, G. (eds) *Plants as Medicine and Aromatics*. CRC Press, Boca Raton, Florida, pp. 13–22.

Venkatasubramanian, P., Balasubramani, S.P., Nandi, S.K. and Tariq, M. (2018) Bioactive metabolite profiling for identification of elite germplasms: A conservation strategy for threatened medicinal plants. *Current Science* 114(3), 554–561. DOI: 10.18520/cs/v114/i03/554-561.

Vicidomini, C., Palumbo, R., Moccia, M. and Roviello, G.N. (2024) Oxidative processes and xenobiotic metabolism in plants: Mechanisms of defense and potential therapeutic implications. *Journal of Xenobiotics* 14(4), 1541–1569. DOI: 10.3390/jox14040084.

Vollmer, R., Espirilla, J., Villagaray, R., Cárdenas, J., Castro, M. *et al.* (2021) Cryopreservation of potato shoot tips for long-term storage. In: Dobnik, D., Gruden, K., Ramšak, Ž. and Coll, A. (eds) *Solanum Tuberosum: Methods and Protocols*. Springer US, New York, pp. 21–54.

Wang, M., Zhang, S., Li, R. and Zhao, Q. (2024) Unraveling the specialized metabolic pathways in medicinal plant genomes: A review. *Frontiers in Plant Science* 15, 1459533. DOI: 10.3389/fpls.2024.1459533.

Waseem, M. and Aslam, M.M. (2024) Integrated approaches of "omics" for the improvement of stress tolerance in plants. In: Ahanger, M.A., Bhat, J.A., Ahmad, P. and John, R. (eds) *Improving Stress Resilience in Plants*. Academic Press, Cambridge, Massachusetts, pp. 321–329.

Williams, B.R., Miller, A.J. and Edwards, C.E. (2025) How do threatened plant species with low genetic diversity respond to environmental stress? Insights from comparative conservation epigenomics and phenotypic plasticity. *Molecular Ecology Resources* 25(5), e13897. DOI: 10.1111/1755-0998.13897.

Wu, M., Northen, T.R. and Ding, Y. (2023) Stressing the importance of plant specialized metabolites: Omics-based approaches for discovering specialized metabolism in plant stress responses. *Frontiers in Plant Science* 14, 1272363. DOI: 10.3389/fpls.2023.1272363.

Zagurskaya, Y.V. and Siromlya, T.I. (2020) Chemical elemental composition of *Hypericum perforatum* – standardized chemical elements (As, Cd, Hg, Pb). *Chemistry of Plant Raw Material* 3(3), 163–170. DOI: 10.14258/jcprm.2020035660.

Zeng, Y., Verhoef, A., Vereecken, H., Ben-Dor, E., Veldkamp, T. *et al.* (2024) Tracking soil health: Monitoring and modeling the soil-plant system. *Authorea Preprints*. DOI: 10.22541/essoar.171804479.91646868/v1.

Zeng, Y., Verhoef, A., Vereecken, H., Ben-Dor, E., Veldkamp, T. *et al.* (2025) Monitoring and modeling the soil-plant system toward understanding soil health. *Reviews of Geophysics* 63(1), e2024RG000836. DOI: 10.1029/2024RG000836.

Zhang, F., Zhu, Y., Peng, J., Shen, Q. and Wei, Z. (2021) Linking the soil microbiome to soil health. *SCIENTIA SINICA Vitae* 51(1), 1–11. DOI: 10.1360/SSV-2020-0320.

Zhang, M., Xing, Y., Ma, J., Zhang, Y., Yu, J. *et al.* (2023a) Investigation of the response of *Platycodon grandiflorus* (Jacq.) A. DC to salt stress using combined transcriptomics and metabolomics. *BMC Plant Biology* 23(1), 589. DOI: 10.1186/s12870-023-04536-w.

Zhang, W., Zeng, Y., Jiao, M., Ye, C., Li, Y. *et al.* (2023b) Integration of high-throughput omics technologies in medicinal plant research: The new era of natural drug discovery. *Frontiers in Plant Science* 14, 1073848. DOI: 10.3389/fpls.2023.1073848.

Zhang, X., Zhang, G., Yan, Q., Ahmad, B., Pei, J. *et al.* (2024a) Quality variation and salt-alkali-tolerance mechanism of *Cynomorium songaricum*: Interacting from microbiome-transcriptome-metabolome. *Science of The Total Environment* 919, 170801. DOI: 10.1016/j.scitotenv.2024.170801.

Zhang, Z., Ying, Z., He, M., Zhang, Y., Nie, W. *et al.* (2024b) UPLC-Q-TOF-MS/MS combined with machine learning methods for screening quality indicators of *Hypericum perforatum* L. *Journal of Pharmaceutical and Biomedical Analysis* 248, 116313. DOI: 10.1016/j.jpba.2024.116313.

Zhou, Y., Feng, C., Wang, Y., Yun, C., Zou, X. *et al.* (2024) Understanding of plant salt tolerance mechanisms and application to molecular breeding. *International Journal of Molecular Sciences* 25(20), 10940. DOI: 10.3390/ijms252010940.

8 Functional Foods and Nutraceuticals from Medicinal Plants

Surendra Sarsaiya, Archana Jain and Qihai Gong*
Key Laboratory of Basic Pharmacology and Joint International Research Laboratory of Ethnomedicine of Ministry of Education, Zunyi Medical University, Zunyi, Guizhou, China

8.1 Introduction

The development of functional foods and nutraceuticals from medicinal plants involves enhancing micronutrient content, engineering metabolic pathways for health-boosting phytochemicals, and employing omics-driven strategies to improve bioavailability. These approaches aim to address nutrient deficiencies and improve health outcomes through dietary means. The integration of biofortification, metabolic engineering, and advanced biotechnological strategies is crucial for the effective development of these functional foods. Biofortification is a sustainable approach to increase the micronutrient content in plants, addressing deficiencies in essential nutrients like iron and zinc. This is particularly important in developing countries where micronutrient deficiencies are prevalent (Białowąs *et al.*, 2024; Naik *et al.*, 2024). Biofortification can be achieved through agronomic practices, conventional breeding, and genetic engineering. These methods have successfully increased the levels of vital micronutrients in staple crops such as rice, wheat, and maize (Sharma and Verma, 2019; Naik *et al.*, 2024). While biofortification holds promise, challenges such as consumer acceptance, regulatory issues, and the need for further research on bioavailability and safety remain (Sharma and Verma, 2019; Naik *et al.*, 2024).

Metabolic engineering involves modifying biosynthetic pathways to enhance the production of nutraceutical compounds like polyphenols and carotenoids. This can improve the nutritional profile of food crops (Kumar *et al.*, 2018). Transgenic plants have been developed with increased levels of vitamins and antioxidants, which are crucial for preventing oxidative stress and related diseases (Zhu *et al.*, 2013).

Recent advancements in genetic engineering, such as CRISPR-Cas9 (clustered regularly interspaced short palindromic repeats-associated protein 9), have facilitated precise modifications in plant genomes, enhancing the nutritional quality of crops (Naik *et al.*, 2024). The use of genomics, proteomics, and metabolomics can optimize the absorption and efficacy of nutraceuticals. These approaches help in understanding the complex interactions within metabolic pathways (Kumar *et al.*, 2018).

*Corresponding author: gqh@zmu.edu.cn

DOI: 10.1079/9781836994206.0008

This technology can improve the bioavailability of nutraceuticals by protecting them from degradation and enhancing their absorption in the human body (Singh *et al.*, 2017). Ongoing research focuses on integrating omics data to develop more effective functional foods and nutraceuticals, ensuring they deliver the intended health benefits (Singh *et al.*, 2017; Misra and Sharma, 2024).

Ginsenoside-enriched ginseng has shown promising potential in targeting neurodegenerative diseases, particularly through omics-guided optimization of ginsenosides. These compounds, primarily derived from ginseng, have been extensively studied for their neuroprotective effects, which are mediated through various molecular pathways. The regulatory frameworks for nutraceutical safety and efficacy, such as those of the US Food and Drug Administration (FDA) and European Food Safety Authority (EFSA), ensure that such products meet stringent quality control standards. Additionally, emerging consumer trends indicate a growing demand for plant-based functional foods, including adaptogen-infused products, personalized nutrition, and clean-label formulations. This chapter will delve into these aspects, highlighting the role of ginsenosides in cognitive health, regulatory compliance, and consumer preferences. Ginsenosides, the active components of ginseng, have been shown to exert neuroprotective effects in neurodegenerative diseases like Alzheimer's and Parkinson's disease by modulating various signaling pathways. They enhance the brain-derived neurotrophic factor (BDNF)/TrkB signaling pathway, which is crucial for neuron growth and survival, thereby potentially improving conditions like Parkinson's disease (Ding *et al.*, 2023). Ginsenosides also decrease β-amyloid production, a hallmark of Alzheimer's disease, by potentiating capacitative calcium entry (CCE; Cho *et al.*, 2024). Specific ginsenosides, such as RK1, have demonstrated efficacy in improving cognitive impairments and mitigating Alzheimer's disease pathology through the AMP-activated protein kinase (AMPK)/nuclear factor erythroid 2–related factor 2 (Nrf2) signaling pathway (She *et al.*, 2024).

Compliance with FDA and EFSA guidelines is critical for the safety and efficacy of nutraceuticals, including ginsenoside-enriched products. These frameworks ensure that health claims are substantiated by scientific evidence and that products meet quality control standards. The pharmacokinetics and delivery modes of ginsenosides are crucial for their clinical application, as highlighted by studies on their neuroprotective effects (Shi *et al.*, 2022). There is a growing consumer interest in plant-based functional foods, driven by the demand for adaptogen-infused products, personalized nutrition, and clean-label formulations. Ginseng, as a natural adaptogen, fits well into this trend, offering potential cognitive health benefits without significant side effects (Lu *et al.*, 2022). The personalization of nutrition, supported by omics technologies, allows for the optimization of ginsenoside formulations to target specific health needs (Liu *et al.*, 2023).

While the development of functional foods and nutraceuticals from medicinal plants offers significant potential for improving public health, it is essential to consider the broader background of food security and nutrition strategies. The success of these initiatives depends on overcoming challenges related to consumer acceptance, regulatory frameworks, and the scalability of production and distribution. Collaboration among various stakeholders, including governments, researchers, and the private sector, is crucial to promoting the widespread adoption of these innovative approaches (Sharma and Verma, 2019; Naik *et al.*, 2024). While ginsenosides show significant promise in neuroprotection and cognitive health, it is essential to consider the broader framework of nutraceutical development. Regulatory compliance ensures that these products are safe and effective, while consumer trends drive innovation in product formulation and marketing. However, the complexity of neurodegenerative diseases means that single-target therapies may not be sufficient, and multi-target approaches, such as those offered by ginsenosides, could be more effective. Additionally, the integration of personalized nutrition and clean-label trends reflects a shift toward more holistic and consumer-driven approaches in the nutraceutical industry (Table 8.1 and Fig. 8.1).

Table 8.1. Development of functional foods and nutraceuticals from medicinal plants. Author's own table.

Aspect	Strategy/finding	Key challenges & critical considerations	Citations
Biofortification	Increasing micronutrient content (e.g. Fe, Zn) in staple crops via agronomic practices, conventional breeding, or genetic engineering to combat deficiencies.	Consumer acceptance of biofortified/GMO (genetically modified organism) crops. Regulatory hurdles for genetically modified products. Uncertainty regarding the bioavailability of enhanced nutrients in the human body and long-term safety.	(Sharma and Verma, 2019; Naik *et al*., 2024)
Metabolic engineering	Modifying plant biosynthetic pathways to boost production of beneficial phytochemicals like polyphenols, carotenoids, vitamins, and antioxidants.	The complexity of metabolic pathways can lead to unintended consequences. Public and regulatory skepticism toward transgenic plants. Ensuring the stability of the engineered trait across generations.	(Zhu *et al*., 2013; Kumar *et al*., 2018)
Advanced biotechnologies (e.g. CRISPR-Cas9, omics)	Using precision gene editing for targeted nutritional improvements. Applying genomics, proteomics, and metabolomics to understand and optimize metabolic pathways and nutraceutical efficacy.	Technical complexity and high cost of omics technologies. Ethical and regulatory debates surrounding gene editing. Translating vast omics data into practical, effective product formulations remains a challenge.	(Kumar *et al*., 2018; Misra and Sharma, 2024; Naik *et al*., 2024)
Bioavailability enhancement	Employing delivery systems (e.g. nanoencapsulation) to protect nutraceuticals from degradation and enhance their absorption in the gastrointestinal tract.	The safety and toxicity of novel delivery materials (e.g. nanoparticles) require extensive evaluation. Scalability and cost-effectiveness of advanced delivery systems for mass-market products.	(Singh *et al*., 2017)
Case study: ginsenosides for neuroprotection	Ginsenosides from ginseng show multi-target neuroprotective effects: enhancing BDNF/TrkB signaling, reducing β-amyloid, and activating AMPK/Nrf2 pathway to improve cognitive impairment.	The pharmacokinetics (absorption, distribution) of ginsenosides is complex and can limit their efficacy. Single-target therapies may be insufficient for complex neurodegenerative diseases. Requires robust clinical trials to substantiate health claims.	(Shi *et al*., 2022; Ding *et al*., 2023; She *et al*., 2024)

Regulatory compliance	Adherence to FDA and EFSA guidelines is mandatory to ensure product safety, efficacy, and quality control. Health claims must be backed by solid scientific evidence.	The regulatory process can be lengthy and expensive, potentially stifling innovation. The burden of proof for health claims is high, requiring significant investment in clinical research.	Implied by text (FDA, EFSA)
Consumer trends & market dynamics	Growing demand for plant-based, adaptogen-infused products (e.g. ginseng), personalized nutrition, and clean-label trends (minimal processing, natural ingredients).	Consumer education is needed to bridge the gap between trend-driven demand and evidence-based benefits. Personalized nutrition relies on expensive omics data and raises privacy concerns. "Clean-label" can be at odds with technologies needed for bioavailability enhancement.	(Lu *et al.*, 2022; Liu *et al.*, 2023)
Overarching critical success factors	Success depends on multi-stakeholder collaboration (government, research, private sector) to address production, distribution, and public awareness.	Scalability from lab to market is a major hurdle. Ensuring equitable access and affordability, especially in developing countries where deficiencies are most prevalent. Balancing innovation with traditional knowledge and public perception.	(Sharma and Verma, 2019; Naik *et al.*, 2024)

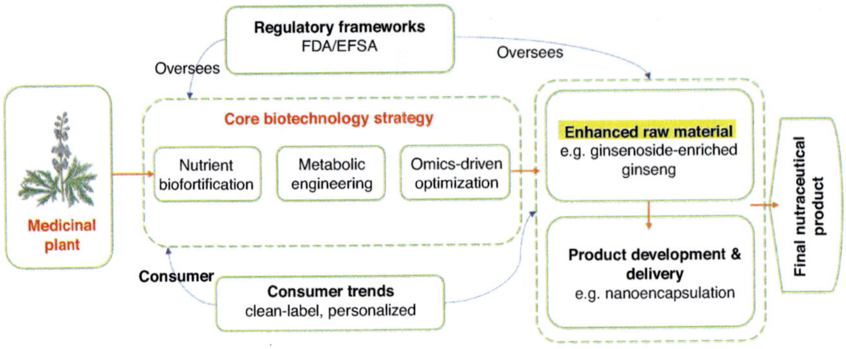

Fig. 8.1. Integrated development pipeline for plant-based nutraceuticals. (Author's own figure.)

8.2 Nutrient Biofortification for Functional Food Development

Biofortification of medicinal plants to enhance micronutrients such as iron and zinc is a promising strategy for developing functional foods that address micronutrient deficiencies. This approach involves increasing the nutrient content of plants through various methods, including conventional breeding, genetic engineering, and agronomic practices. The goal is to improve the nutritional quality of plants, making them more beneficial for dietary applications, particularly in regions where micronutrient deficiencies are prevalent. Biofortification not only aims to improve public health by increasing the intake of essential nutrients but also contributes to food security and crop resilience. Nutrient biofortification in medicinal plants for functional food development is a promising approach to enhance the nutritional quality and therapeutic potential of foods. This strategy involves increasing the concentration of essential nutrients and bioactive compounds in plants, which can then be used to develop functional foods with health benefits beyond basic nutrition. Medicinal plants, known for their therapeutic properties, are particularly suitable for biofortification due to their rich phytochemical content. This approach not only addresses micronutrient deficiencies but also contributes to the prevention and management of chronic diseases. The following sections explore various aspects of nutrient biofortification in medicinal plants for functional food development.

Biofortification can be achieved through genetic engineering and conventional breeding methods. These techniques aim to enhance the levels of micronutrients such as iron, zinc, and vitamins in plants, thereby improving their nutritional profile (Sharma and Verma, 2019; Jabeen, 2023). Agronomic biofortification involves the application of fertilizers and soil amendments to increase the nutrient content of plants. This method is cost-effective and sustainable, making it suitable for large-scale implementation (Raut *et al.*, 2024). Functional foods developed from biofortified medicinal plants can help prevent chronic diseases such as diabetes and cardiovascular diseases. For instance, bioactive compounds in these plants can improve insulin sensitivity and reduce oxidative stress (Devaraj and Mahalingam, 2021; Martinez-Alvarado *et al.*, 2022). Medicinal plants used in functional foods can also support neurological health. They contain bioactive phytochemicals that have been shown to have neuroprotective effects, potentially aiding in the management of neurological disorders (Ahmed, 2020). One of the challenges in biofortification is modulating plant metabolic pathways to increase the production of specific nutrients without affecting plant growth and development. This requires a deep understanding of plant biochemistry and genetics (Zhu *et al.*, 2013). The development and commercialization of biofortified crops face regulatory hurdles and biosafety concerns. Ensuring the stability and safety of these crops is crucial for their acceptance and widespread use (Sharma and Verma, 2019). Flaxseed has been biofortified to enhance its content of omega-3 fatty acids and lignans, which have cancer-preventive properties. This makes flaxseed a valuable ingredient in functional foods aimed at improving nutritional security (Mogali *et al.*, 2020).

Successful examples of biofortified crops include iron-fortified beans and zinc-enhanced rice, which have been developed to combat micronutrient deficiencies in developing countries (Kurdekar *et al.*, 2023). While nutrient biofortification of medicinal plants holds great promise for functional food development, it is essential to consider the broader implications of this approach. The integration of biofortification with nutrition education and agronomic practices can maximize its impact on public health. Additionally, collaboration among governments, research institutions, and the private sector is crucial to scaling up biofortification efforts and building healthier communities worldwide. Despite the challenges, continued research and innovation in this field can lead to significant advancements in nutrition security and disease prevention (Table 8.2).

8.2.1 Enhancing micronutrients (e.g. iron, zinc) in medicinal plants for dietary applications

Enhancing micronutrients such as iron and zinc in medicinal plants for dietary applications is a multifaceted approach that involves improving both the concentration and the bioavailability of these essential nutrients. This can be achieved through various strategies, including the use of organic acids, biofortification techniques, and fermentation processes. Each method offers unique benefits and challenges, contributing to the overall goal of addressing micronutrient deficiencies in plant-based diets.

Organic acids like ascorbic acid and citric acid have been shown to enhance the absorption of iron and zinc in plant-based foods. These acids form soluble chelates with the minerals, improving their bioavailability and counteracting the inhibitory effects of compounds like phytic acid and polyphenols present in the food matrix (Adeyanju *et al.*, 2024). Incorporating fruits rich in these organic acids into the diet can significantly enhance mineral absorption, providing a practical dietary strategy to improve micronutrient intake (Adeyanju *et al.*, 2024). Biofortification, which involves increasing the concentration of micronutrients in crops, is a promising approach to enhance the nutritional quality of plant-based foods. This can be achieved through agronomic practices, conventional breeding, and genetic engineering (Behera *et al.*, 2024; Białowąs *et al.*, 2024).

Agronomic biofortification involves the application of mineral fertilizers to increase the nutrient content in crops. For instance, foliar application of zinc and iron in lentils has been shown to significantly enhance their concentration and uptake, improving both yield and economic returns (Dhaliwal *et al.*, 2021). Genetic engineering offers a targeted approach to biofortification by modifying specific genes involved in nutrient uptake and storage. This method can overcome limitations in genetic variability and enhance the bioavailability of micronutrients in crops (Saini *et al.*, 2022). Fermentation is another effective method to enhance the bioavailability of micronutrients in plant-based foods. It reduces anti-nutritional factors such as phytates and tannins, which can inhibit mineral absorption (Samtiya *et al.*, 2021). Fermented foods have been shown to improve the bioavailability of iron and zinc, making them a valuable addition to diets aimed at addressing micronutrient deficiencies (Samtiya *et al.*, 2021).

Understanding the molecular mechanisms underlying micronutrient uptake and storage in plants can inform biofortification strategies. Research has identified key genes and pathways involved in the accumulation of iron and zinc, which can be targeted to enhance nutrient levels in crops (Manzoor *et al.*, 2024). Plant–microbe interactions also play a crucial role in enhancing soil nutrient status and facilitating the uptake of micronutrients by plants. These interactions can be leveraged to improve the nutritional quality of crops through biofortification (Nazma *et al.*, 2023). While these strategies offer promising solutions to enhance micronutrient levels in medicinal plants, challenges remain in ensuring their effectiveness and sustainability. The complexity of plant nutrient interactions and the variability in soil and environmental conditions can impact the success of biofortification efforts. Additionally,

Table 8.2. Nutrient biofortification in medicinal plants for functional foods. Author's own table.

Aspect	Strategy/application	Key benefits & rationale	Critical challenges & limitations	Citations
Biofortification methods	Agronomic practices: application of mineral fertilizers (e.g. foliar application of Zn/Fe).	Cost-effective and immediately implementable. Improves both crop yield and nutrient content. Suitable for large-scale use.	Effectiveness is highly dependent on soil type, climate, and farming practices. Not a permanent genetic solution; requires repeated application.	(Dhaliwal *et al.*, 2021; Behera *et al.*, 2024)
	Conventional breeding & genetic engineering: cross-breeding or gene editing to enhance nutrient uptake/storage.	Can create sustainable, heritable traits. Overcomes limitations of natural genetic variation. Allows for precise targeting of pathways.	Public and regulatory skepticism, especially toward GMOs. Can be time-consuming (breeding) or technically complex/expensive (engineering). Risk of unintended effects on plant physiology.	(Sharma and Verma, 2019; Saini *et al.*, 2022; Jabeen, 2023)
Bioavailability enhancement	Use of organic acids: incorporating ascorbic/citric acid (e.g. via fruit blends) to chelate minerals.	Simple, dietary-based strategy. Counters anti-nutrients like phytic acid, directly improving human absorption of Fe/Zn.	Requires dietary composition planning, which may not be controlled in final products. Efficacy can vary based on the overall food matrix.	(Adeyanju *et al.*, 2024)
	Fermentation Processes: Using microbial activity to ferment plant materials.	Reduces anti-nutritional factors (phytates, tannins) effectively. Can improve flavor and digestibility.	Alters the taste and nature of the raw material, which may not be desirable for all medicinal plant applications. Requires controlled processes.	(Samtiya *et al.*, 2021)
Therapeutic applications	Chronic disease prevention: developing functional foods for diabetes and cardiovascular health.	Leverages inherent bioactive compounds (e.g. antioxidants) in medicinal plants to improve insulin sensitivity and reduce oxidative stress.	Health claims require robust clinical validation. Synergistic effects of biofortified nutrients with native phytochemicals are not fully understood.	(Devaraj and Mahalingam, 2021; Martinez-Alvarado *et al.*, 2022)
	Neurological health support: utilizing biofortified plants with neuroprotective phytochemicals.	Offers a multi-target approach to managing neurological disorders, capitalizing on plants' natural therapeutic properties.	The blood–brain barrier and complex etiology of neurological diseases make efficacy difficult to ensure and prove.	(Ahmed, 2020)

Scientific & technical foundations	Molecular mechanisms: research into genes and pathways for micronutrient uptake and storage.	Informs targeted biofortification strategies. Identifies specific genetic levers to pull for enhanced nutrient content.	Extreme complexity of plant metabolic networks. Modifying one pathway can have unintended consequences on plant growth, yield, or other compounds.	(Zhu *et al*., 2013; Manzoor *et al*., 2024)
	Plant–microbe interactions: leveraging soil microbes to enhance nutrient availability to plants.	A sustainable, biological approach to improve soil health and plant nutrient status.	Outcomes are highly dependent on specific plant-microbe-soil combinations, making standardization difficult.	(Nazma *et al*., 2023)
Implementation & commercialization	Regulatory hurdles & biosafety: gaining approval for biofortified crops, especially genetically engineered ones.	Ensures public safety and product efficacy, building consumer trust.	Lengthy, expensive, and uncertain regulatory processes can stifle innovation and delay public health benefits.	(Sharma and Verma, 2019)
	Scalability & impact: moving from successful research to widespread adoption.	Potential for significant public health impact, especially in nutrient-deficient regions.	Requires strong collaboration across governments, research institutions, and the private sector. Challenges in supply chain, farmer adoption, and consumer education.	(Sharma and Verma, 2019; Kurdekar *et al*., 2023)

further research is needed to fully understand the physiological links between dietary ligands and mineral bioavailability, as well as the long-term impacts of biofortification on human health (Table 8.2; Zhang *et al.*, 2021; Białowąs *et al.*, 2024).

8.3 Metabolic Engineering of Health-boosting Phytochemicals

Metabolic engineering of medicinal plants is a promising approach to enhance the production of health-boosting phytochemicals, which are secondary metabolites with significant therapeutic potential. These compounds, such as terpenoids, alkaloids, and polyphenols, are often produced in limited quantities in their native plant species, posing challenges for their commercial use. Metabolic engineering offers innovative strategies to overcome these limitations by manipulating plant metabolic pathways to increase the yield and diversity of these valuable compounds. This approach not only addresses the supply issues but also opens new avenues for the development of novel bioactive compounds. Metabolic engineering involves the manipulation of existing metabolic pathways or the introduction of new pathways in plants. This can be achieved by overexpressing key enzymes, silencing competitive pathways, or introducing new genes to enhance the biosynthesis of target compounds (Sharma *et al.*, 2019a; Halder and Roy, 2023).

Genetic modification techniques, such as the use of transgenic plants, allow for the introduction of genes that can enhance the production of specific phytochemicals. This includes the use of RNA interference (RNAi) to silence genes that compete with the desired biosynthetic pathways (Fatima *et al.*, 2022; Kumari *et al.*, 2024). Plant cell and tissue culture techniques provide an alternative to whole plant cultivation, allowing for controlled production of phytochemicals. This method can be optimized to increase the yield of desired compounds regardless of environmental conditions (Nielsen *et al.*, 2019).

The integration of genomics and metabolomics tools has significantly advanced the field of metabolic engineering. These technologies enable the identification and manipulation of genes involved in the biosynthesis of secondary metabolites, facilitating the development of plants with enhanced phytochemical profiles (Nielsen *et al.*, 2019; Sharma *et al.*, 2019a). Synthetic biology approaches, including the use of CRISPR-Cas9 and other genome editing tools, have been employed to precisely edit plant genomes. This allows for the fine-tuning of metabolic pathways to optimize the production of bioactive compounds (Kayser, 2010; Arya *et al.*, 2022).

Enhanced production of phytochemicals through metabolic engineering can lead to the development of new pharmaceuticals and nutraceuticals. These compounds have applications in treating various diseases and improving human health (Saad *et al.*, 2021; Sethi *et al.*, 2022). By increasing the yield of valuable phytochemicals, metabolic engineering can reduce the reliance on wild plant harvesting, thus conserving biodiversity and reducing environmental impact. It also offers economic benefits by lowering production costs and increasing the availability of these compounds for commercial use (Halder and Roy, 2023).

While metabolic engineering holds great promise for enhancing the production of health-boosting phytochemicals, it is not without challenges. The complexity of plant metabolic pathways and the potential for unintended effects on plant physiology and ecology must be carefully considered. Additionally, regulatory hurdles and public perception of genetically modified organisms can impact the adoption of these technologies. Nonetheless, continued advancements in genetic and biotechnological tools are likely to address these challenges, paving the way for more sustainable and efficient production of medicinal plant-derived compounds (Table 8.3).

8.3.1 Tailoring pathways for nutraceutical compounds (e.g. polyphenols, carotenoids)

The tailoring of pathways for nutraceutical compounds such as polyphenols and carotenoids in medicinal plants involves optimizing the production and bioavailability of these

Table 8.3. Metabolic engineering of phytochemicals in medicinal plants. Author's own table.

Aspect	Strategy/technique	Key applications & benefits	Critical challenges & limitations	Citations
Core engineering strategies	Pathway manipulation: overexpressing key enzymes or silencing competitive pathways to redirect metabolic flux.	Increases the yield of low-abundance, high-value compounds (e.g. specific terpenoids, alkaloids). Overcomes natural production bottlenecks.	Complexity of metabolic networks can lead to unintended consequences, such as resource drain, growth inhibition, or accumulation of toxic intermediates.	(Sharma *et al.*, 2019a; (Halder and Roy, 2023)
	Genetic modification: using transgenic approaches and RNAi to introduce new genes or silence existing ones.	Enables the production of novel compounds or completely blocks competing pathways for more efficient target compound synthesis.	Public and regulatory skepticism toward GMOs is a major barrier to commercialization. The stability of transgene expression across generations can be variable.	(Fatima *et al.*, 2022; Kumari *et al.*, 2024)
Enabling technologies	Synthetic biology & genome editing: using CRISPR-Cas9 for precise gene editing and pathway fine-tuning.	Allows for precise, targeted modifications without introducing foreign DNA (in some applications), potentially easing regulatory concerns.	Off-target effects remain a technical risk. The application in non-model medicinal plants is often limited by a lack of genomic information.	(Kayser, 2010; Arya *et al.*, 2022)
	Omics integration: using genomics and metabolomics to map biosynthetic pathways and identify key genes.	Provides a systems-level understanding, enabling data-driven engineering rather than trial-and-error. Identifies all regulatory nodes and potential bottlenecks.	Generates vast, complex datasets that require sophisticated bioinformatics and expertise to interpret and apply effectively.	(Nielsen *et al.*, 2019; Sharma *et al.*, 2019a)
	Plant cell/tissue culture: production of phytochemicals in controlled bioreactor systems.	Independent of environmental factors (climate, soil, pests). Enables year-round production and can protect endangered species from over-harvesting.	Can be extremely expensive to scale. The yields in bioreactors may not match those in whole plants, and cell lines can lose productivity over time.	(Nielsen *et al.*, 2019)

Continued

Table 8.3. Continued

Aspect	Strategy/technique	Key applications & benefits	Critical challenges & limitations	Citations
Target compounds & health applications	Polyphenols: engineered for enhanced antioxidant, anti-inflammatory, and neuroprotective properties.	Potential for managing chronic diseases like cancer, cardiovascular, and neurodegenerative disorders (e.g. Parkinson's).	Bioavailability is often low; enhanced production in the plant does not guarantee efficacy in the human body. Synergistic effects are complex and poorly understood.	(Ananga *et al*., 2013; Kumar *et al*., 2022; Singh *et al*., 2024)
	Carotenoids: engineered as antioxidants and vitamin A precursors to prevent eye diseases and cancer.	Improves the nutritional profile of staple crops (biofortification). Derivatives have industrial applications in food and cosmetics.	Stability during storage and processing is a major issue (sensitivity to light and oxygen). Genetic changes can sometimes alter plant phenotype (color) undesirably.	(Sakure, 2022)
Broader implications & hurdles	Sustainability & conservation: reduced reliance on wild harvesting of endangered medicinal plants.	Promotes biodiversity conservation and offers a more reliable, scalable supply chain for rare compounds.	The high cost of engineered production may not justify compounds that can be cheaply extracted from cultivated plants, limiting its economic viability to high-value products.	(Halder and Roy, 2023)
	Standardization & regulation: ensuring consistent quality, safety, and efficacy of engineered nutraceuticals.	Critical for building consumer trust and meeting the stringent requirements of regulatory bodies (FDA, EFSA).	Lack of universal standards for complex plant extracts. Demonstrating "substantial equivalence" or entirely new safety profiles for engineered products is a significant regulatory hurdle.	(Sharma *et al*., 2019b; Singh *et al*., 2024)

bioactive compounds to enhance their health benefits. Medicinal plants are rich sources of these compounds, which exhibit a range of biological activities including antioxidant, anti-inflammatory, and anti-cancer effects. The potential of these compounds in disease prevention and health promotion is significant, but challenges such as standardization and quality control need to be addressed to fully harness their benefits. The following sections explore the pathways and strategies for enhancing the production of polyphenols and carotenoids in medicinal plants.

Polyphenols are known for their antioxidant, anti-inflammatory, and antimicrobial properties, which contribute to their therapeutic potential in preventing and managing diseases such as cancer and cardiovascular disorders (Ananga *et al.*, 2013; Singh *et al.*, 2024). Advances in metabolic engineering have enabled the manipulation of biosynthetic pathways to increase the yield and diversity of polyphenols in plants. This includes altering structural regulatory genes to optimize levels and composition of polyphenols (Ananga *et al.*, 2013). Polyphenols from medicinal plants have been used in traditional medicine systems like Ayurveda for their therapeutic benefits. They are being explored for their potential in treating neurodegenerative diseases such as Parkinson's disease (Kumar *et al.*, 2022; Zahra *et al.*, 2022). Carotenoids are potent antioxidants and precursors of vitamin A, playing a crucial role in preventing eye diseases, cancer, and cardiovascular diseases (Sakure, 2022).

Genetic engineering and breeding techniques are employed to enhance carotenoid content in crop plants, making them more accessible as nutraceuticals. This includes the introduction of carotenoid pathway genes to increase their concentration in edible plants (Sakure, 2022). Beyond health benefits, carotenoids and their derivatives are used in the food, pharmaceutical, and cosmetic industries due to their colorant properties and bioactivities (Sakure, 2022). The consistent quality and efficacy of medicinal plants present significant challenges for their use in nutraceuticals, necessitating standardized extraction and quantification techniques to ensure reliability (Sharma *et al.*, 2019b; Singh *et al.*, 2024). Developing comprehensive regulatory frameworks is essential to ensure the safety and efficacy of nutraceuticals derived from medicinal plants (Singh *et al.*, 2024). Continued research is needed to explore the full potential of polyphenols and carotenoids, including their synergistic effects and bioavailability in human health applications (Ananga *et al.*, 2013; Singh *et al.*, 2024).

While the potential of polyphenols and carotenoids in medicinal plants is promising, it is important to consider the complexity of their interactions within the human body. The bioavailability and efficacy of these compounds can be influenced by various factors, including their chemical structure and the presence of other dietary components. Therefore, a holistic approach that considers these factors is crucial in the development of effective nutraceuticals. Additionally, the exploration of other bioactive compounds in medicinal plants, such as flavonoids and terpenoids, could further enhance the therapeutic potential of these natural sources (Table 8.3).

8.4 Omics-driven Strategies to Enhance Nutraceutical Bioavailability

Omics-driven strategies are pivotal in enhancing the bioavailability of nutraceuticals derived from medicinal plants. These strategies leverage advanced technologies to understand and manipulate the molecular and biochemical pathways involved in the biosynthesis and absorption of bioactive compounds. By integrating genomics, transcriptomics, proteomics, and metabolomics, researchers can identify key genes and pathways that influence the production and bioavailability of nutraceuticals. This comprehensive approach aids not only in improving the efficacy of these compounds but also in developing novel nutraceutical formulations with enhanced health benefits. The following sections delve into specific omics-driven strategies and their applications in enhancing nutraceutical bioavailability. These approaches help in identifying genetic variations and expression patterns that affect the biosynthesis of bioactive compounds in medicinal plants. By understanding these genetic factors, researchers can manipulate plant genetics to enhance the production of desired compounds (Sahu *et al.*, 2014; Zhang *et al.*, 2023). These techniques provide insights into the protein interactions and metabolic pathways that govern the synthesis and transformation of nutraceuticals. This knowledge is crucial for optimizing the conditions that enhance the bioavailability of these compounds (Dai *et al.*, 2022; Zhang *et al.*, 2023). The utilization of nanoparticle formulations can significantly improve the solubility and stability of nutraceuticals, thereby enhancing their absorption and bioavailability. This approach is particularly effective for compounds with poor aqueous solubility (Singab, 2024). Lipid-based delivery systems improve the permeability and absorption of nutraceuticals by facilitating their transport across biological membranes. Lipid-based carriers are especially useful for delivering hydrophobic compounds (Singab, 2024). Co-administration with certain natural compounds from medicinal plants, such as quercetin and piperine, can enhance the bioavailability of co-administered drugs by inhibiting efflux pumps and oxidative metabolism.

Biofortification techniques such as foliar and soil application of nutrients, as well as genetic modification, are employed to increase the concentration of nutraceuticals in plants. These techniques ensure that the plants produce higher levels of bioactive compounds, which can be more readily absorbed by the human body (Martinez-Alvarado *et al.*, 2022).

The molecular breeding approach, utilizing markers such as microsatellites, allows for the selection and development of plant varieties with enhanced nutraceutical profiles and improved bioavailability of specific compounds (Sahu *et al.*, 2014). While omics-driven strategies offer promising avenues for enhancing nutraceutical bioavailability, challenges remain in standardizing these approaches and ensuring their efficacy across different plant species and environmental conditions. Additionally, regulatory frameworks need to be established to ensure the safety and quality of nutraceutical products derived from genetically modified plants (Singh *et al.*, 2024). Future research should focus on integrating omics data with advanced bioinformatics tools to develop predictive models for nutraceutical bioavailability, thereby facilitating the design of more effective nutraceutical formulations (Table 8.4; Dai *et al.*, 2022).

8.4.1 Using nanotechnology to optimize absorption and efficacy (e.g. nanoencapsulation)

Nanotechnology has emerged as a transformative approach in enhancing the absorption and efficacy of medicinal plants, primarily through nanoencapsulation techniques. This method addresses the inherent limitations of traditional herbal medicine, such as poor solubility, low bioavailability, and inefficient delivery to target sites. By incorporating active plant compounds into nanoscale carriers, nanotechnology significantly improves the therapeutic potential of these compounds.

Nanoencapsulation improves the solubility and bioavailability of poorly soluble phytochemicals, such as flavonoids, tannins, and terpenoids, which are common in medicinal plants. These compounds often struggle to cross lipid membranes due to their hydrophilic nature and large molecular size, leading to poor absorption and efficacy (Bhavsar, 2018; Ratheesh *et al.*, 2018). Nanocarriers, such as nanoparticles, nanocapsules, and nanogels, reduce the particle size of active compounds to the nanoscale, enhancing their absorption and distribution within the body (Chen *et al.*, 2024; Erangika and Kuruppu, 2024). Techniques like nano liposomes, nano emulsions, and lipid nanocarriers have been shown to increase the rate of absorption and bioavailability, resulting in improved therapeutic effects of herbal medicines (Bhavsar, 2018; Chowdhury *et al.*, 2024).

Nanotechnology enables targeted drug delivery, ensuring that active compounds are delivered precisely to the site of action, thereby minimizing side effects on healthy tissues (Chen *et al.*, 2024; Fatima *et al.*, 2024). Nanoformulations can regulate drug release rates, extend the duration of therapeutic action, and enhance the stability of treatment effects (Chen *et al.*, 2024). Nanosystems can maintain a sufficient concentration of active constituents throughout the treatment period, improving the overall efficacy of the treatment (Barik *et al.*, 2022; Fatima *et al.*, 2024).

Table 8.4. Omics and nano-strategies for enhancing nutraceutical bioavailability. Author's own table.

Aspect	Strategy/technology	Mechanism of action	Key benefits	Critical challenges & limitations	Citations
Omics-driven discovery	Genomics & transcriptomics: analysis of genetic variations and gene expression.	Identifies key genes and enzymes in the biosynthetic pathways of bioactive compounds. Enables targeted genetic manipulation to boost production.	Provides a rational foundation for metabolic engineering. Can discover novel genes for high-yield compound production.	Data is complex and requires sophisticated bioinformatics. A genetic change does not always guarantee increased bioavailability in humans.	(Sahu *et al*., 2014; Zhang *et al*., 2023)
	Proteomics & metabolomics: analysis of proteins and metabolic fluxes.	Maps the complete metabolic network and protein interactions governing compound synthesis and transformation.	Offers a systems-level view, revealing regulatory nodes and potential bottlenecks in both plants and human absorption.	Integration of multi-omics data is technically challenging. The cost of high-throughput analyses can be prohibitive.	(Dai *et al*., 2022; Zhang *et al*., 2023)
Nanoencapsulation delivery	Nanoparticle formulations: (e.g. nanocapsules, nanogels).	Encapsulates compounds to improve solubility, protect from degradation, and enhance permeability through biological barriers.	Dramatically improves solubility and stability of poorly soluble phytochemicals (e.g. flavonoids, terpenoids).	Safety and toxicity profiles of nanomaterials are not fully understood, especially with long-term use. Scalability of production is a major hurdle.	(Bhavsar, 2018; Chen *et al*., 2024; Singab, 2024)
	Lipid-based delivery systems (e.g. nano liposomes, emulsions).	Utilizes lipid carriers to solubilize hydrophobic compounds and facilitate transport across lipid-rich biological membranes.	Highly effective for hydrophobic compounds. Can enhance absorption in the gastrointestinal tract.	Stability issues (e.g. oxidation of lipids). Formulation can be complex and expensive.	(Chowdhury *et al*., 2024; Singab, 2024)
Enhanced targeting & release	Targeted drug delivery: functionalized nanocarriers.	Uses surface ligands to direct carriers to specific cells or tissues (e.g. tumor sites), minimizing systemic exposure.	Increases efficacy and reduces side effects by concentrating the active compound at the site of action.	Targeting efficiency *in vivo* can be low. Functionalization adds significant complexity and cost.	(Chen *et al*., 2024; Fatima *et al*., 2024)

Continued

Table 8.4. Continued

Aspect	Strategy/technology	Mechanism of action	Key benefits	Critical challenges & limitations	Citations
	Controlled release systems: engineered nanocarriers with specific release triggers.	Regulates the release rate of the encapsulated compound, maintaining therapeutic concentrations over a longer period.	Extends duration of action, reduces dosing frequency, and improves patient compliance.	Achieving precise, predictable, and consistent release kinetics *in vivo* is difficult.	(Barik *et al.*, 2022; Chen *et al.*, 2024)
Complementary enhancement strategies	Biofortification (foliar/ soil nutrients, genetic modification).	Increases the intrinsic concentration of nutraceuticals in the plant source.	Addresses the problem at the source, creating inherently more potent raw materials.	Does not directly address post-harvest bioavailability challenges. Consumer acceptance of GM biofortified plants is low.	(Martinez-Alvarado *et al.*, 2022)
	Co-administration with bioenhancers (e.g. piperine, quercetin).	Inhibits metabolic enzymes (e.g. cytochrome P450) or efflux pumps (e.g. P-glycoprotein) to reduce pre-systemic metabolism.	A simple, often natural, strategy to boost the bioavailability of a co-consumed nutraceutical.	Risk of unintended food–drug interactions if the consumer is on medication.	(Implied by text)
Overarching hurdles & future directions	Regulatory frameworks & standardization:	Establishing safety and quality controls for novel nutraceutical formulations, especially nano-enabled and GM-derived.	Essential for consumer safety and market trust.	Regulations for nano-nutraceuticals are underdeveloped and lag behind the technology. Standardizing complex plant extracts is inherently difficult.	(Singh *et al.*, 2024)
	Predictive modeling & integration:	Using bioinformatics to integrate omics data and predict *in vivo* bioavailability and efficacy.	Could accelerate rational design of next-generation nutraceuticals and reduce reliance on costly trial and error.	The biological system is highly complex; creating accurate predictive models is a monumental challenge.	(Dai *et al.*, 2022)

Nanoencapsulation addresses the volatility, instability, and aqueous insolubility of plant volatile organic compounds, enhancing their bioavailability and biological activity (Mun and Townley, 2021). By overcoming these physical limitations, nanoencapsulation promotes the controlled release and cellular absorption of volatile compounds, expanding their applications in medical, food, and agricultural fields (Mun and Townley, 2021). Despite the promising potential of nanoencapsulation, challenges remain, such as understanding the pharmacokinetic processes and interactions between nanoencapsulated phytochemicals and the host (Zheng *et al.*, 2022). The stability and safety of nanoformulations, particularly concerning potential toxicity with long-term use, require further investigation (Chen *et al.*, 2024). While nanoencapsulation offers significant advantages in enhancing the absorption and efficacy of medicinal plants, it is essential to consider the potential risks and limitations associated with this technology. The stability, safety, and long-term effects of nanoformulations need thorough investigation to ensure their safe and effective application in herbal medicine. Additionally, the scalability of production and understanding of working principles remain areas for further exploration. Despite these challenges, the integration of nanotechnology with traditional herbal medicine holds substantial promise for advancing therapeutic strategies and improving patient outcomes (Table 8.4 and Fig. 8.2).

8.5 Ginsenoside-enriched Ginseng for Cognitive Health

Ginsenoside-enriched ginseng has been extensively studied for its potential benefits for cognitive health, particularly in medicinal plants. Ginsenosides, the active compounds in ginseng, have shown promise in improving cognitive function through various mechanisms. These include modulating signaling pathways related to oxidative stress, inflammation, and neurogenesis, which are crucial in cognitive impairment and neurodegenerative diseases. Despite the promising preclinical and clinical findings, the evidence remains inconclusive, necessitating further research to establish the efficacy and mechanisms of ginsenosides in cognitive health. Ginsenosides have been shown to

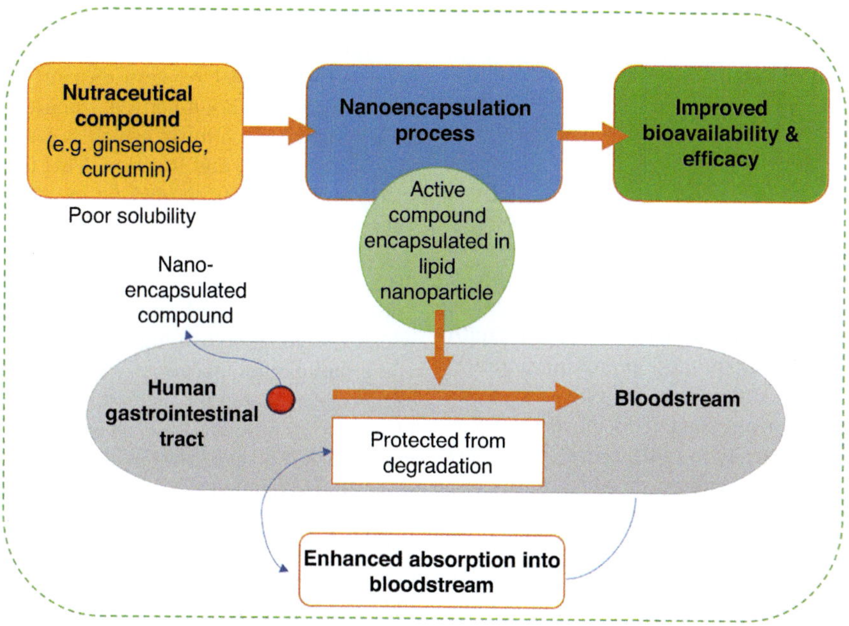

Fig. 8.2. Integrated development pipeline for plant-based nutraceuticals. (Author's own figure.)

improve cognitive function in both clinical and preclinical studies. They are believed to enhance memory and cognitive performance in healthy individuals and those with cognitive impairments such as mild cognitive impairment (MCI) and early Alzheimer's disease (Feng *et al.*, 2022; Lee *et al.*, 2024). Clinical studies have shown that ginseng can improve cognitive function in healthy individuals and those with MCI or early Alzheimer's disease dementia. However, the results are inconsistent, and more standardized trials are needed (Geng *et al.*, 2010; Zeng *et al.*, 2024). Preclinical studies have demonstrated that ginsenoside Rg1 significantly improves cognitive impairments in Alzheimer's disease models, suggesting its potential as a therapeutic agent (Liang *et al.*, 2021).

Ginseng has been shown to promote neuronal plasticity and improve short-term memory, indicating its potential benefits in aging-related cognitive decline (Lelong *et al.*, 2024). The cognitive benefits of ginsenosides are linked to their ability to modulate signaling pathways such as PI3K/Akt, CREB/BDNF, and NF-κB/NLRP3 inflammasome pathways. These pathways are involved in reducing oxidative stress, inflammation, and apoptosis, while promoting synaptic plasticity and neurogenesis (Jakaria *et al.*, 2018; Feng *et al.*, 2022). The use of smart farming techniques has been shown to enhance the bioactive compound content of ginseng, particularly ginsenosides Rb2 and Rd, which have been linked to anti-inflammatory effects and cognitive improvement (Huang *et al.*, 2023).

Ginsenosides compounds like Rg3, Rh1, Rh2, Rb1, Rd, Rg2, and Rb3 have been identified as particularly effective in enhancing cognitive function. Compound K, a metabolite of ginsenosides, also shows a promising pharmacokinetic profile and potential for neuroprotection (Smith *et al.*, 2014). Despite these promising findings, results from studies on ginseng's cognitive benefits are inconsistent. This inconsistency is attributed to variations in study design, ginseng dosage, and the specific ginsenosides used (Geng *et al.*, 2010; Smith *et al.*, 2014). There is a need for standardized clinical trials to address these inconsistencies. Properly conducted trials using standardized tests in healthy individuals and those with cognitive impairments are essential to validate the cognitive-enhancing effects of ginsenosides (Smith *et al.*, 2014). While ginseng is generally well tolerated, the long-term effects and potential side effects of ginsenoside supplementation need further investigation to ensure safety (Table 8.5; Geng *et al.*, 2010).

8.5.1 Omics-guided optimization of ginsenosides to target neurodegenerative diseases

The optimization of ginsenosides, active compounds derived from ginseng, through omics-guided approaches holds significant promise for targeting neurodegenerative diseases (NDDs). These compounds have been shown to exert neuroprotective effects through various molecular mechanisms, making them potential therapeutic agents for conditions such as Alzheimer's disease (AD) and Parkinson's disease (PD). The integration of omics technologies can enhance the understanding of these mechanisms and optimize the use of ginsenosides in treating NDDs. This section will explore the pathways influenced by ginsenosides, their specific roles in neuroprotection, and the potential of omics-guided strategies in optimizing their therapeutic effects.

Ginsenosides target the Akt (protein kinase B: PKB)/CREB (cAMP response element-binding protein)/BDNF (brain-derived neurotrophic factor) signaling pathway, which is crucial for neuronal survival and function. This pathway is implicated in neuroprotection by promoting neuronal growth and survival, thus offering a therapeutic target for NDDs (Zarneshan *et al.*, 2022; Ding *et al.*, 2023). Ginsenosides exhibit anti-inflammatory and antioxidant properties, which are vital in mitigating the oxidative stress and inflammation associated with NDDs. These effects are mediated through pathways such as PI3K/Akt, NF-κB, and Nrf2, which are involved in cellular stress responses (Lu *et al.*, 2022; Zhou *et al.*, 2024; Jiang *et al.*, 2025). The upregulation of BDNF and activation of the BDNF/TrkB pathway by ginsenosides is particularly significant in PD, as it supports neuronal growth and differentiation, countering the neurodegenerative processes (Ding *et al.*, 2023).

Table 8.5. Ginsenoside-enriched ginseng for cognitive health and NDDs. Author's own table.

Aspect	Key findings & mechanisms	Promising evidence	Critical Challenges & Limitations	Citations
Clinical efficacy & consistency	Shows potential to improve cognitive function in healthy adults, MCI, and early AD.	Some clinical trials report improvements in memory and cognitive performance.	Inconsistent results across studies due to variations in dosage, ginseng species, ginsenoside profiles, and study design. Lack of large-scale, definitive trials.	(Geng *et al*., 2010; Smith *et al*., 2014; Feng *et al*., 2022; Lee *et al*., 2024)
Neuroprotective mechanisms	Multi-target actions: modulates PI3K/Akt, CREB/BDNF, Nrf2, and NF-κB pathways. Reduces oxidative stress, inflammation, and apoptosis while promoting synaptic plasticity and neurogenesis.	Preclinical models (e.g. AD, PD) show clear mechanistic benefits, such as reduced amyloid-beta and improved neuronal survival.	The complexity of mechanisms makes it difficult to attribute benefits to a single pathway. Translation from animal models to human disease is uncertain.	(Jakaria *et al*., 2018; Feng *et al*., 2022; Ding *et al*., 2023; Jiang *et al*., 2025)
Key bioactive ginsenosides	Rg1, Rb1: neuroprotection via BDNF/TrkB and PI3K/Akt. Rd: anti-apoptotic and anti-inflammatory. Rg3, Ro: target amyloid-beta and tau pathology. Compound K: bioactive metabolite with a favorable pharmacokinetic profile.	Specific ginsenosides are linked to distinct pathways, allowing for targeted therapeutic development.	Bioavailability varies greatly among different ginsenosides. Many parent compounds are metabolized, and the active forms in the brain are not fully established.	(Smith *et al*., 2014; Li *et al*., 2022; Ding *et al*., 2023; Liu *et al*., 2023; Jiang *et al*., 2025)
Omics-guided optimization	Network pharmacology & bioinformatics: identifies multi-target interactions and predicts efficacy. Molecular docking: simulates ginsenoside-target binding.	Enables a systems biology approach, moving beyond single-target drugs to multi-target strategies suited for complex NDDs. Helps optimize ginseng cultivation for desired ginsenoside profiles.	Predictive models require experimental validation. Omics data is complex and correlative; establishing causation remains a challenge.	(Liu *et al*., 2023; Huang *et al*., 2023; He *et al*., 2024; Jiang *et al*., 2025)

Continued

Table 8.5. Continued

Aspect	Key findings & mechanisms	Promising evidence	Critical Challenges & Limitations	Citations
Production & standardization	Smart farming: can enhance the content of specific ginsenosides (e.g. Rb2, Rd).	Improves consistency and potency of the raw material, addressing one source of variability in clinical studies.	Lack of standardized extracts with consistent ginsenoside ratios makes it difficult to compare study results and ensure product quality.	(Huang *et al*., 2023; Smith *et al*., 2014)
Safety & long-term use	Generally well-tolerated in the short term.	A favorable safety profile makes it an attractive candidate for long-term preventive use.	Long-term safety data is limited. Potential for side effects or interactions with medications needs more comprehensive study.	(Geng *et al*., 2010)
Overarching critical gaps	The multi-target nature is a strength for complex diseases like AD and PD.	Offers a holistic therapeutic strategy unlike many failed single-target pharmaceutical drugs.	The "shotgun" approach makes it difficult to pinpoint the essential mechanisms and standardize dosing. Personalized response variability is not well understood.	(Smith *et al*., 2014; He *et al*., 2024)

Ginsenoside Rg1 and Rb1 have been identified as promising agents for PD and AD due to their ability to modulate key signaling pathways, including BDNF/TrkB and PI3K/Akt, thereby exerting neuroprotective effects (Jiang *et al.*, 2025). Ginsenoside Rd, known for its role in ameliorating ischaemic stroke and nerve injury, acts through mechanisms involving apoptosis, inflammation, and oxidative stress, making it a candidate for broader applications in NDDs (Li *et al.*, 2022). Ginsenoside Rg3 and Ro compounds have shown potential in reducing amyloid-beta deposition and tau hyperphosphorylation in AD, highlighting their role in addressing key pathological features of the disease (Liu *et al.*, 2023).

The network pharmacology approach allows for the identification of potential targets and pathways modulated by ginsenosides, facilitating the optimization of their therapeutic effects. By integrating data from various omics technologies, researchers can better understand the complex interactions and mechanisms involved in NDDs (He *et al.*, 2024; Jiang *et al.*, 2025). Bioinformatics and molecular docking tools are used to predict the interactions between ginsenosides and their targets, providing insights into their efficacy and potential side effects. This information is crucial for developing more effective ginsenoside-based therapies (Liu *et al.*, 2023). While ginsenosides show great promise in targeting NDDs, challenges remain in translating these findings into clinical applications. The complexity of NDDs and the variability in individual responses necessitate a personalized approach to treatment. Omics-guided strategies offer a pathway to overcome these challenges by providing a comprehensive understanding of the molecular underpinnings of ginsenosides' effects. However, further research is needed to fully realize their potential and to address any safety and efficacy concerns with regard to long-term use (Table 8.5 and Fig. 8.3).

8.6 Regulatory Frameworks for Nutraceutical Safety and Efficacy

The regulatory frameworks for nutraceutical safety and efficacy, particularly in medicinal plants, are complex and vary significantly across different regions. These frameworks are essential to ensure consumer safety, product quality, and efficacy, given the increasing global demand for plant-based medicinal products. The regulatory landscape is shaped by national and regional policies, with efforts toward harmonization to address the challenges posed by diverse definitions and standards. This section explores the current state of these frameworks, highlighting key aspects and challenges.

Different countries have adopted varied regulatory approaches for nutraceuticals and medicinal plants. For instance, the USA operates under the Dietary Supplement Health and Education Act (DSHEA), which allows for a relatively open market, whereas Canada requires premarket review under the Natural Health

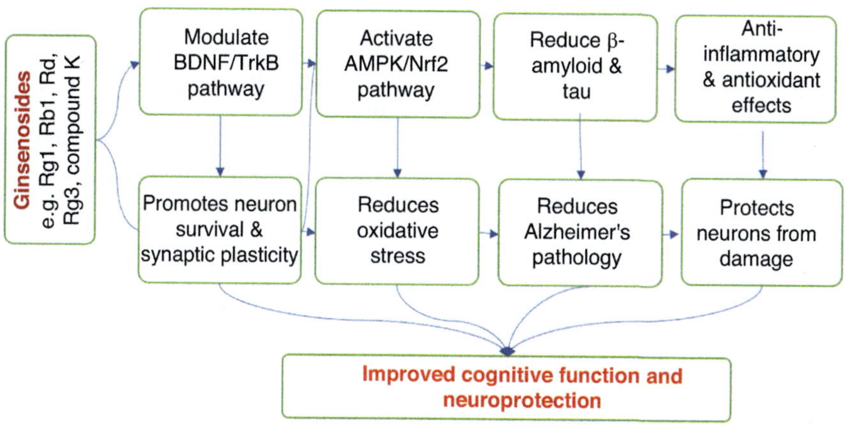

Fig. 8.3. Integrated development pipeline for plant-based nutraceuticals. (Author's own figure.)

Product Regulations (Charrette *et al.*, 2021). The European Union (EU) has established a harmonized framework through the Committee on Herbal Medicinal Products (HMPC), which develops Community Monographs to ensure safety and efficacy standards (Knöss, 2015). A significant challenge in the global regulatory landscape is the lack of consensus on definitions and standards for nutraceuticals. This lack of harmonization complicates international trade and consumer protection (Thakur *et al.*, 2024). The World Health Organization (WHO) plays a crucial role in promoting international standards, but achieving global harmonization remains a complex task (Picking, 2024).

In Africa, the regulatory framework for medicinal plants is still developing. The continent's rich biodiversity offers significant potential for the medicinal plant industry, but stringent regulations are needed to ensure product standardization and safety (Bahorun *et al.*, 2019). Asian countries like India and China have more established systems for regulating traditional herbal medicines, but challenges remain in ensuring consistent quality and safety across products (Dubale *et al.*, 2025). In North America, regulatory frameworks in the USA and Canada reflect a balance between consumer access and safety. However, the rise of illicit products, such as cannabidiol, is prompting regulatory changes (Charrette *et al.*, 2021).

The quality and safety of nutraceuticals is a primary concern. Many products lack sufficient data to guarantee their safety and efficacy, and adverse effects, although less frequent than with synthetic drugs, do occur (Fokunang *et al.*, 2020). The need for standardized quality control measures is critical to addressing these issues. The scientific evaluation of herbal medicines is essential for establishing safety and efficacy. The HMPC in the EU serves as a model for how scientific studies can be integrated with traditional knowledge to create robust regulatory standards (Knöss, 2015). There is a pressing need for more research on the safety and efficacy of medicinal plants. This includes developing new technologies for phytochemical delivery and improving the standardization of extracts (Bahorun *et al.*, 2019). Strengthening regulatory frameworks through better training, policy development, and collaboration between traditional healers and scientific communities is crucial for improving the regulation of traditional herbal medicines (Dubale *et al.*, 2025). While the regulatory frameworks for nutraceuticals and medicinal plants are evolving, significant challenges remain in achieving global harmonization and ensuring product safety and efficacy. The diverse approaches across regions reflect different priorities and levels of development, but there is a shared need for improved research, standardization, and international cooperation to protect consumers and support industry growth (Table 8.6).

8.6.1 Compliance with FDA/EFSA guidelines for health claims and quality control

Compliance with FDA/EFSA guidelines for health claims and quality control of medicinal plants is a multifaceted process that involves ensuring the safety, efficacy, and quality of herbal products. This process is crucial for integrating herbal medicines into mainstream healthcare systems and for gaining consumer trust. The guidelines set by regulatory bodies like the FDA and EFSA are designed to ensure that herbal products meet specific standards from cultivation to consumption. These standards include the authentication of plant materials, standardization of production processes, and rigorous quality control measures. The following sections detail the key aspects of compliance with these guidelines.

Standardization is essential for ensuring the consistency of herbal medicinal products (HMPs). It involves developing a body of information and controls that guarantee the constancy of composition and efficacy of HMPs (Wijayabandara, 2019). Quality control measures include the authentication of plant materials, detection of contaminants, and adherence to good manufacturing practices (GMP; Wang *et al.*, 2023). Analytical methods such as chromatography, spectroscopy, and DNA-based technologies are employed to ensure the identity, purity, and composition of herbal products (Efferth and Greten, 2012; Edo *et al.*, 2024).

WHO guidelines provide a comprehensive framework for ensuring the safety, efficacy, and

Table 8.6. Regulatory frameworks for nutraceutical safety and efficacy. Author's own table.

Aspect	Region/guideline	Key requirements & approaches	Critical challenges & limitations	Citations
Regional regulatory models	USA (FDA/DSHEA): regulates products as "dietary supplements."	Premarket approval is not required for safety and efficacy. Manufacturers are responsible for ensuring safety and substantiating claims; FDA acts post-market.	Reactive, not proactive. Creates a market where products can be sold with limited premarket scrutiny, placing the burden of proof on the FDA after consumer harm occurs.	(Charrette *et al.*, 2021)
	EU (EFSA/HMPC): has a harmonized framework for "herbal medicinal products."	Requires premarket authorization based on safety, efficacy, and quality. Community Monographs standardize approved uses and information.	High barrier to entry; process can be lengthy and expensive, potentially limiting product diversity and access to traditional remedies.	(Knöss, 2015)
	Canada (NHPD): regulates under "Natural Health Product Regulations."	Premarket review and licensing are required. Products must have a product license before sale, demonstrating safety, efficacy, and quality.	Strikes a balance but still faces challenges with a vast number of applications, leading to potential backlogs.	(Charrette *et al.*, 2021)
Quality control & standardization	Good practices (GACP, GMP): adherence from cultivation to manufacturing.	Ensures consistency, minimizes contamination, and guarantees the identity and purity of raw plant materials and finished products.	Inherent variability of plant materials makes batch-to-batch consistency difficult and expensive to achieve.	(Rawat and Tewari, 2015; Wang *et al.*, 2023)
	Advanced analytical methods: chromatography, spectroscopy, DNA barcoding.	Used for authentication, detecting adulterants, and standardizing the composition of complex herbal mixtures.	Techniques are sophisticated and costly, creating a high barrier for small producers. Standardizing multi-component extracts for "active" constituents is scientifically complex.	(Efferth and Greten, 2012; Wang *et al.*, 2023; Edo *et al.*, 2024)

Continued

Table 8.6. Continued

Aspect	Region/guideline	Key requirements & approaches	Critical challenges & limitations	Citations
Substantiating health claims	EFSA & FDA health claim approval	Requires robust scientific evidence, typically from human clinical trials, to substantiate any structure/function or disease-risk reduction claims.	The high cost of clinical trials for non-patentable natural products offers little incentive for manufacturers. Leads to vague, non-meaningful claims.	(Implied by text (FDA, EFSA guidelines))
Global harmonization & developing nations	WHO: provides international guidelines for herbal medicines.	Aims to promote global standardization and help countries, especially developing ones, establish their own regulatory frameworks.	Lack of global consensus on definitions (nutraceutical vs drug) and standards hinders international trade and consumer protection.	(Jain *et al.*, 2024; Picking, 2024; Thakur *et al.*, 2024)
	Africa & many Asian nations: frameworks are often underdeveloped.	Rich biodiversity offers great potential, but regulatory systems are fragmented or lacking.	Lack of stringent regulations leads to issues with product quality, safety, and adulteration. Struggles with capacity, enforcement, and integrating traditional knowledge.	(Bahorun *et al.*, 2019; Dubale *et al.*, 2025)
Overarching critical gaps	The "natural equals safe" fallacy	Regulatory frameworks must combat consumer misconception that herbal products are inherently safe due to their natural origin.	This perception leads to underreporting of adverse effects and potential risks of herb–drug interactions, which are not always adequately studied.	(Fokunang *et al.*, 2020)
	Integration of traditional knowledge	Modern regulations require scientific validation, which can be at odds with centuries of traditional use.	Bridging the gap between traditional evidence and modern evidence-based requirements is a significant challenge without dismissing valuable historical data.	(Knöss, 2015; Mukherjee, 2019; Wang *et al.*, 2023)

quality of herbal products. These guidelines help manufacturers minimize risks associated with contamination and variability in potency (Jain *et al.*, 2024). Good agricultural and collection practices (GACP) are recommended to ensure the quality of raw plant materials. These practices help in maintaining the consistency and reliability of herbal medicines (Rawat and Tewari, 2015). Regulatory harmonization is crucial for the global promotion of standardized herbal medicines. This involves aligning national regulations with international standards to facilitate the free movement of safe and effective herbal products (Mukherjee, 2019).

The complexity of herbal products, which are mixtures of various plant constituents, poses challenges in ensuring consistent quality and efficacy. This necessitates the use of advanced analytical techniques and rigorous quality control measures (Wang *et al.*, 2023). Innovations in phytotherapy research, such as the use of molecular biology and systems biology, are enhancing the quality control processes for herbal medicines (Efferth and Greten, 2012). The integration of traditional knowledge with modern scientific approaches is vital for achieving optimal quality control outcomes and fostering consumer trust (Wang *et al.*, 2023).

While compliance with FDA/EFSA guidelines is crucial for ensuring the safety and efficacy of herbal products, it is important to recognize the challenges posed by the inherent complexity of these products. The variability in plant materials and the presence of multiple active constituents can complicate the standardization and quality control processes. Additionally, the perception that herbal products are inherently safe can lead to underestimation of potential risks. Therefore, continuous efforts in research, innovation, and regulatory harmonization are necessary to address these challenges and ensure the responsible use of herbal medicines (Table 8.6).

8.7 Emerging Consumer Trends in Plant-based Functional Foods

The emergence of plant-based functional foods, particularly those derived from medicinal plants, is a significant trend in the food industry, driven by increasing consumer demand for health-enhancing products. These foods are valued not only for their nutritional content but also for their potential therapeutic benefits, which include managing chronic diseases and improving overall health. This trend is supported by a growing body of research and innovation in the field, highlighting the potential of medicinal plants to serve as functional food ingredients.

Medicinal plants have been traditionally used for their neuro-therapeutic properties, offering benefits such as antidepressant and anti-anxiety effects. The field of nutritional neuroscience is expanding, with evidence supporting the use of plant-based functional foods in managing neurological disorders (Ahmed, 2020). Functional foods incorporating medicinal plants are being developed to address metabolic disorders like diabetes. These foods can help manage blood glucose levels and improve insulin function, offering a natural alternative to conventional diabetes treatments (Devaraj and Mahalingam, 2021; Ansari *et al.*, 2023). The Lamiaceae family of plants, rich in bioactive compounds, is particularly noted for its potential in preventing chronic diseases. These plants are used in functional foods targeting metabolic health, cognitive function, and other health areas (Carović-Stanko *et al.*, 2016; El Boukhari and Fatimi, 2024).

The global market for functional foods is expanding rapidly, with projections reaching US$275.77 billion by 2025. This growth is fueled by consumer interest in health and wellness, particularly in the wake of the COVID-19 pandemic, which has increased demand for immune-boosting products (Chua and Nicholas, 2023). Companies are investing in the development of novel functional food products that meet consumer demands for health benefits. This includes enhancing the nutritional value of foods and developing new formulations that incorporate bioactive compounds from medicinal plants (Chala and Emire, 2022).

While many plant-based functional foods are claimed to offer health benefits, there is a need for more rigorous scientific validation to substantiate these claims. This includes understanding the bioactive components and their mechanisms of action (Chala and Emire, 2022). The incorporation of medicinal plants into functional foods requires advanced processing methods to retain the efficacy of bioactive compounds. This includes addressing challenges related to bioavailability and potential interactions between different ingredients (Chua and Nicholas, 2023). While the trend toward plant-based functional foods is promising, it is important to consider the challenges associated with their development and validation. The need for scientific evidence to support health claims and the complexities of processing medicinal plants into functional foods are significant hurdles. However, with continued research and innovation, these challenges can be addressed, paving the way for the broader adoption of plant-based functional foods in the market (Table 8.7).

8.7.1 Demand for adaptogen-infused products, personalized nutrition, and clean-label trends

The demand for adaptogen-infused products, personalized nutrition, and clean-label trends is growing, particularly in medicinal plants. Adaptogens, which are plant-derived substances, are known for their ability to enhance the body's resistance to stress and improve overall health. This trend is driven by consumer interest in natural health products and the therapeutic benefits of medicinal plants. The integration of adaptogens into functional foods and nutraceuticals aligns with the clean-label movement, which emphasizes transparency and natural ingredients. This demand is further fueled by the potential of personalized nutrition to cater to individual health needs, leveraging the diverse bioactive compounds found in medicinal plants.

Adaptogens like *Rhodiola rosea*, *Schisandra chinensis*, and *Eleutherococcus senticosus* have been extensively studied for their stress-protective and geroprotective properties (Sharma *et al.*, 2024). These plants are used to enhance physical endurance, mental alertness, and overall resilience against stressors (Agarwal, 2023). Adaptogen-infused products, such as gummies, offer convenience and higher bioavailability, making them popular among consumers seeking natural health solutions (Agarwal, 2023). Personalized nutrition involves tailoring dietary recommendations to individual health profiles, which can be enhanced by the use of adaptogens (Singh *et al.*, 2024). Medicinal plants provide a rich source of bioactive compounds, such as polyphenols and flavonoids, which can be used to address specific health concerns (Singh *et al.*, 2024). The integration of adaptogens into personalized nutrition plans can help manage stress-related disorders and improve quality of life (Ray *et al.*, 2021).

The clean-label trend emphasizes the use of natural, minimally processed ingredients, which aligns with the use of adaptogens in food products (Krylov *et al.*, 2024). Consumers are increasingly seeking products that are free from artificial additives and preservatives, making adaptogen-infused products appealing (Agarwal, 2023). The natural origin and low toxicity of adaptogens make them suitable for long-term use, further supporting their inclusion in clean-label products (Krylov *et al.*, 2024). Medicinal plants offer a sustainable source of nutraceuticals, providing health benefits beyond basic nutrition (Singh *et al.*, 2024). The use of adaptogens in nutraceuticals is supported by their ability to modulate immune function, neurotransmitters, and signaling pathways, promoting homeostasis (Agarwal, 2023). The therapeutic potential of medicinal plants in disease prevention and management is well documented, making them valuable in the development of functional foods (Ahmed, 2020).

While the demand for adaptogen-infused products and personalized nutrition is on the rise, it is important to consider the challenges associated with standardization and quality control of medicinal plant-based products. Ensuring the efficacy and safety of these products requires rigorous scientific validation and adherence to regulatory frameworks. Additionally, while adaptogens offer numerous health benefits, they may not be suitable for everyone, and potential interactions with other medications should be considered. As the field of nutraceutical research continues to evolve, addressing these challenges will be crucial to fully harnessing the potential of medicinal plants in promoting health and well-being (Table 8.7).

Table 8.7. Emerging consumer trends in plant-based functional foods. Author's own table.

Trend	Core concept & driver	Key market applications & examples	Critical challenges & considerations	Citations
Adaptogen-infused products	Concept: use of plants (e.g. Rhodiola, ashwagandha) that non-specifically enhance the body's resistance to physical, chemical, and biological stressors. Driver: rising consumer stress and demand for natural mental health and performance aids.	Stress-relief gummies, functional beverages (teas, coffees), and snack bars marketed for resilience, focus, and energy. Products like Rhodiola for endurance and Schisandra for mental alertness.	Lack of standardization: potency and phytochemical profiles can vary dramatically between sources. Over-hyped claims: marketing often outpaces robust clinical evidence for specific product formulations. Potential for misuse: not suitable for all health conditions; risk of herb–drug interactions.	(Agarwal, 2023; Krylov *et al.*, 2024; Sharma *et al.*, 2024)
Personalized nutrition	Concept: tailoring food and supplement recommendations to an individual's genetics, metabolism, microbiome, and lifestyle. Driver: advances in omics technologies (genomics, metabolomics) and consumer desire for targeted health solutions.	Use of bioactive compounds from medicinal plants (polyphenols, adaptogens) in formulations designed for specific genetic markers or health goals (e.g. stress management, metabolic support).	High cost & accessibility: omics testing and personalized products often come at a premium, creating equity issues. Data privacy: concerns over the use of genetic and health data. Scientific complexity: translating complex genetic data into simple, effective dietary advice is still in its infancy.	(Ray *et al.*, 2021; Singh *et al.*, 2024)
Clean-label trends	Concept: consumer preference for products with simple, recognizable, natural ingredients and minimal processing. Driver: distrust of artificial additives, preservatives, and complex chemical names; desire for "wholesome" foods.	Adaptogen-infused products that highlight "natural," "organic," or "whole-plant" ingredients. Elimination of artificial sweeteners, colors, and preservatives from functional food formulations.	Tension with technology: advanced processing (e.g. for bioavailability) or nanoencapsulation may be needed for efficacy but conflicts with "minimally processed" ideals. "Natural" fallacy: the assumption that "clean-label" equals safe, which may not be true for all botanicals (e.g. dose-dependent toxicity). Vague definition: "Clean-label" is a marketing term, not a regulated standard, leading to inconsistent interpretation.	(Agarwal, 2023; Krylov *et al.*, 2024)

Continued

Table 8.7. Continued

Trend	Core concept & driver	Key market applications & examples	Critical challenges & considerations	Citations
Broader functional food development	Concept: incorporating medicinal plants (e.g. from Lamiaceae family) into everyday foods to provide targeted health benefits for neurological, metabolic, and chronic diseases. Driver: growing scientific evidence and consumer proactivity in managing health through diet.	Functional foods for cognitive support (e.g. anti-anxiety), blood glucose management in diabetes, and general metabolic health.	Scientific validation gap: many health claims are based on traditional use or preliminary studies; lack of large-scale human trials for final food products. Bioavailability hurdle: ensuring active compounds survive processing and are absorbed effectively in the body. Regulatory hurdles: navigating health claim approvals (EFSA/FDA) for complex plant mixtures in food matrices.	(Carović-Stanko *et al.*, 2016; Ahmed, 2020; Chala and Emire, 2022; Ansari *et al.*, 2023)
Market growth & economic drivers	Concept: rapid expansion of the global functional food market, accelerated by post-COVID-19 health consciousness. Driver: corporate investment and consumer willingness to pay a premium for health-enhancing foods.	Projected market value of $275 billion+. Proliferation of novel product launches in categories like fortified beverages, snacks, and dairy alternatives.	Market saturation & “greenwashing”: risk of products with negligible functional benefits capitalizing on the trend. Cost vs benefit: the high price of validated functional foods may not be justified by their actual health impact, leading to consumer skepticism.	(Chala and Emire, 2022; Chua and Nicholas, 2023)

8.8 Challenges and Future Directions

The development of functional foods and nutraceuticals from medicinal plants faces multifaceted challenges spanning scientific, regulatory, and commercial domains. Technically, poor bioavailability of bioactive compounds (e.g. ginsenosides, polyphenols) remains a critical barrier, as their hydrophilic nature and molecular complexity limit absorption and therapeutic efficacy. Metabolic engineering efforts, while promising, contend with the intricate interplay of biosynthetic pathways; unintended consequences of pathway manipulation—such as altered plant growth or accumulation of toxic intermediates—demand rigorous genomic and metabolomic oversight. Scaling advanced delivery systems (e.g. nanoencapsulation) for industrial production also poses economic hurdles, particularly for nanoparticle-based formulations requiring sterile conditions and specialized equipment. Regulatory fragmentation further complicates progress: divergent FDA (USA), EFSA (EU), and global standards for health claims, safety testing, and GMO classifications create compliance bottlenecks, stifling innovation and international market access. This is exacerbated by inconsistent definitions of "nutraceuticals" across jurisdictions and sparse long-term safety data for novel formulations like lipid nanocarriers. Consumer acceptance challenges persist, with skepticism toward genetically modified crops conflicting with clean-label demands, while limited clinical validation for neuroprotective effects (e.g. ginseng's impact on AD) undermines market confidence. Additionally, supply chain vulnerabilities—including seasonal variability in medicinal plant yields and contamination risks—threaten ingredient standardization.

Future advancements hinge on integrated interdisciplinary strategies. Precision biotechnology, leveraging CRISPR-Cas9 coupled with multi-omics (genomics, metabolomics), could optimize phytochemical production without compromising plant viability, while AI-driven predictive modeling may unravel compound interactions and bioavailability barriers. Regulatory harmonization, guided by WHO frameworks and blockchain-enabled traceability, must streamline global standards for safety and labeling. Economically viable scaling could be achieved through "smart farming" (e.g. AI-monitored cultivation of ginseng to boost ginsenoside content) and circular-economy approaches, such as valorizing extraction waste into biodegradable packaging. To bridge scientific validation gaps, robust clinical trials focusing on personalized nutrition—using omics to tailor adaptogen blends (e.g. *Rhodiola* for stress resilience) to individual biomarkers—are essential. Concurrently, consumer education initiatives demystifying biofortification and metabolic engineering could foster broader acceptance, positioning plant-based nutraceuticals as credible solutions for global health challenges (Table 8.8).

8.9 Conclusion

Functional foods and nutraceuticals derived from medicinal plants represent a transformative frontier in bridging nutritional deficits and combating chronic diseases. Through targeted biofortification, staple crops like iron-enhanced beans and zinc-rich rice offer scalable solutions to micronutrient deficiencies, particularly in resource-limited regions. Metabolic engineering, empowered by CRISPR and omics technologies, has unlocked unprecedented potential to amplify health-boosting phytochemicals—from cancer-preventive lignans in flaxseed to neuroprotective ginsenosides in *Panax ginseng*. The latter exemplifies this synergy: omics-guided optimization reveals its efficacy in modulating BDNF/TrkB and AMPK/Nrf2 pathways, positioning it as a promising adjuvant for neurodegenerative conditions like AD and PD. Concurrently, nanoencapsulation and lipid-based delivery systems address historical bioavailability constraints, ensuring optimal absorption of hydrophobic compounds such as carotenoids and curcuminoids. Yet, the sector's success hinges on navigating persistent challenges. Regulatory coherence across FDA, EFSA, and emerging economies remains critical to validating health claims and ensuring product safety amidst rising demand for adaptogen-infused clean-label products. While consumer trends toward personalized nutrition

Table 8.8. Challenges and future directions in medicinal plant-based nutraceuticals. Author's own table.

Domain	Critical challenges	Proposed future directions & integrated strategies	Citations
Scientific & technical	• Poor bioavailability: hydrophilic nature and complex structure of bioactive compounds (e.g. ginsenosides, polyphenols) limit absorption and efficacy.	Advanced delivery systems: scale-up of nanoencapsulation and lipid-based carriers. Integration with omics to design targeted delivery for specific compounds.	(Chen *et al*., 2024; Singab, 2024)
	• Metabolic pathway complexity: unintended consequences of engineering (e.g. growth inhibition, toxic intermediates) due to interconnected biosynthetic networks.	Precision biotechnology: combine CRISPR-Cas9 with multi-omics (genomics, metabolomics) for fine-tuning pathways without affecting core plant functions. Use AI for predictive modeling of metabolic fluxes.	(Zhu *et al*., 2013; Sharma *et al*., 2019a; Arya *et al*., 2022)
	• Scalability of production: high cost and technical difficulty in industrially scaling advanced systems (e.g. nanoparticle synthesis, plant cell culture).	"Smart farming" & circular economy: use AI and Internet of Things (IoT) to optimize cultivation for enhanced phytochemical content (e.g. ginsenosides). Valorize extraction waste into bio-materials to improve economics.	(Nielsen *et al*., 2019; Huang *et al*., 2023)
Regulatory & safety	• Regulatory fragmentation: divergent standards for health claims, safety, and GMOs between FDA, EFSA, and other global bodies creates compliance bottlenecks.	International harmonization: advocate for globally aligned standards based on WHO frameworks. Implement blockchain for transparent, verifiable supply chain traceability.	(Knöss, 2015; Mukherjee, 2019; Charrette *et al*., 2021; Thakur *et al*., 2024)
	• Insufficient long-term data: lack of long-term safety and toxicology data for novel formulations (e.g. nanocarriers, engineered plants).	Proactive safety science: develop new regulatory science frameworks for complex products. Require post-market surveillance studies as a condition of approval.	(Fokunang *et al*., 2020; Chen *et al*., 2024)

Commercial & consumer	• Consumer acceptance & trust: skepticism toward GMOs conflicts with "clean-label" demand. Lack of robust clinical trials undermines confidence in health claims.	Transparent education & personalized nutrition: demystify technologies like biofortification through clear communication. Use omics to create personalized adaptogen blends (e.g. for stress), providing tangible, tailored benefits that build trust.	(Geng *et al*., 2010; Smith *et al*., 2014; Agarwal, 2023; Singh *et al*., 2024)
	• Supply chain vulnerabilities: seasonal variability in plant yields and contamination risks threaten ingredient standardization and quality.	Supply chain digitization & standardization: use blockchain and IoT sensors for real-time quality monitoring from farm to product. Enforce strict GACP.	(Rawat and Tewari, 2015; Wang *et al*., 2023)
Validation & evidence	• Translational gap: promising preclinical data (e.g. ginseng in Alzheimer's models) fails to consistently translate into definitive human clinical benefits.	Robust clinical trial design: conduct large-scale, long-term human studies with standardized, well-characterized extracts. Focus on personalized nutrition cohorts to identify responsive sub-populations.	(Smith *et al*., 2014; Feng *et al*., 2022; Lee *et al*., 2024)

underscore market viability, they also necessitate rigorous clinical substantiation—especially for cognitive health assertions. Moving forward, interdisciplinary collaboration among geneticists, nutritionists, and policy makers will be paramount. Prioritizing sustainable scaling through climate-resilient agriculture, ethical biotechnology deployment, and AI-driven supply chain optimization can democratize access to these innovations. Ultimately, medicinal plants, underpinned by cutting-edge science and equitable frameworks, offer not merely dietary supplements but foundational tools for redefining global public health paradigms.

Acknowledgments

The authors express their gratitude for the financial support received through the Distinguished High-Level Talents Research Grant from the Guizhou Science and Technology Corporation Platform Talents Fund (Grant nos.: [2017]5733-001 and CK-1130-002), high-level specialists project fund of Zunyi Medical University (1092/2000006)F-1092, the National Natural Science Foundation of China (82373981), and the support provided by Zunyi Medical University, China. Special appreciation is extended to all laboratory colleagues and research staff members for their valuable insights, constructive guidance, and assistance throughout this study.

References

Adeyanju, A.A., Uduehi, F.E. and Iranloye, Y.M. (2024) Enhancing the absorption of iron and zinc in plant-based foods with citric acid and ascorbic acid: A review. In: *2024 International Conference on Science, Engineering and Business for Driving Sustainable Development Goals (SEB4SDG)*, Institute of Electrical and Electronics Engineers, New York, pp. 1–6 (Vol. SEB4SDG). DOI: 10.1109/SEB4SDG60871.2024.10630323.

Agarwal, A. (2023) Adaptogenic plants within food supplement. *Annals of Innovation in Medicine* 1(1). DOI: 10.59652/aim.v1i1.29.

Ahmed, Md. (2020) Medicinal plant-based functional foods for the management of neurological health. DOI: 10.20944/preprints202006.0311.v1.

Ananga, A., Georgiev, V. and Tsolova, V. (2013) Manipulation and engineering of metabolic and biosynthetic pathway of plant polyphenols. *Current Pharmaceutical Design* 19(34), 6186–6206. DOI: 10.2174/1381612811319340011.

Ansari, P., Samia, J.F., Khan, J.T., Rafi, M.R., Rahman, M.S. *et al.* (2023) Protective effects of medicinal plant-based foods against diabetes: A review on pharmacology, phytochemistry, and molecular mechanisms. *Nutrients* 15(14), 3266. DOI: 10.3390/nu15143266.

Arya, A., Gautam, S., Goel, S., Grewal, S. and Bhattacharyya, M. (2022) Metabolic engineering for high-value bioactive compounds from medicinal plants. In: Swamy, M.K. and Kumar, A. (eds) *Phytochemical Genomics*. Springer Nature, Singapore, pp. 521–544.

Bahorun, T., Aruoma, O.I. and Neergheen-Bhujun, V.S. (2019) Phytomedicines, nutraceuticals, and functional foods regulatory framework. In: Bagchi, D. (ed.) *Nutraceutical and Functional Food Regulations in the United States and around the World*. Academic Press, Cambridge, Massachusetts, pp. 509–521.

Barik, R., Sugunan, S. and Pal, A. (2022) Nanotechnology-based drug delivery of phytotherapeutics: Herbal drugs and nanotechnology. In: Verma, S. and Kumar Verma, S. (eds) *Advancements in Controlled Drug Delivery Systems*. IGI Global Scientific Publishing, Hershey, Pennsylvaniapp, pp. 73–96.

Behera, S.R., Pandey, R., Dhathri, M., Negi, P., Jakhwal, R. *et al.* (2024) Agronomic biofortification of vegetable crops – A systematic review. *Journal of Advances in Biology & Biotechnology* 27(9), 942–956. DOI: 10.9734/jabb/2024/v27i91366.

Bhavsar, S. (2018) Application of nanotechnology for phyto constituents: Review. *Archives of Nanomedicine* 1(1). DOI: 10.32474/ANOAJ.2018.01.000103.

Białowąs, W., Blicharska, E. and Drabik, K. (2024) Biofortification of plant- and animal-based foods in limiting the problem of microelement deficiencies - A narrative review. *Nutrients* 16(10), 1481. DOI: 10.3390/nu16101481.

Carović-Stanko, K., Petek, M., Grdiša, M., Pintar, J., Bedeković, D. *et al.* (2016) Medicinal plants of the family lamiaceae as functional foods - A review. *Czech Journal of Food Sciences* 34(5), 377–390. DOI: 10.17221/504/2015-CJFS.

Chala, D.B. and Emire, S.A. (2022) Development of plant based functional food products: A review. The annals of the University Dunarea de jos of Galati. *Fascicle VI-Food Technology* (1), 169–183. DOI: 10.35219/foodtechnology.2022.1.13.

Charrette, A., Guthrie, N. and Hilmas, C.J. (2021) The regulatory framework for nutraceuticals: North America. In: Gupta, R.C., Lall, R. and Srivastava, A. (eds) *Nutraceuticals*. Academic Press, Cambridge, Massachusetts, pp. 1215–1239.

Chen, Y., Tang, Y., Li, Y., Rui, Y. and Zhang, P. (2024) Enhancing the efficacy of active pharmaceutical ingredients in medicinal plants through nanoformulations: A promising field. *Nanomaterials* 14(19), 1598. DOI: 10.3390/nano14191598.

Chowdhury, S., Nath, D., Das, S.R.C., Kar, K., Chakraborty, P. *et al.* (2024) Nanotechnology based herbal drug delivery system: Current insights and future prospects. *Current Nanomedicine* 15. DOI: 10.2174/0124681873349580241113081309.

Cho, Y.Y., Park, J.H., Lee, J.H. and Chung, S. (2024) Ginsenosides decrease β-amyloid production via potentiating capacitative calcium entry. *Biomolecules & Therapeutics* 32(3), 301–308. DOI: 10.4062/biomolther.2023.172.

Chua, H.P. and Nicholas, D. (2023) Processing of herbal-based natural products and functional foods: A review. *Sains Malaysiana* 52(9), 2587–2598. DOI: 10.17576/jsm-2023-5209-09.

Dai, R., Liu, M., Xiang, X., Li, Y., Xi, Z. *et al.* (2022) OMICS applications for medicinal plants in gastrointestinal cancers: Current advancements and future perspectives. *Frontiers in Pharmacology* 13, 842203. DOI: 10.3389/fphar.2022.842203.

Devaraj, A. and Mahalingam, G. (2021) Bioactive molecules from medicinal plants as functional foods (biscuits) for the benefit of human health as antidiabetic potential. In: Sharma, K., Mishra, K., Senapati, K.K. and Dancui, C. (eds) *Bioactive Compounds in Nutraceutical and Functional Food for Good Human Health*. IntechOpen, London, pp. 1–15.

Dhaliwal, S.S., Sharma, V., Shukla, A.K., Kaur, J., Verma, V. *et al.* (2021) Enrichment of zinc and iron micronutrients in lentil (*Lens culinaris* Medik.) through biofortification. *Molecules (Basel, Switzerland)* 26(24), 7671. DOI: 10.3390/molecules26247671.

Ding, Y., Botchway, B.O.A., Zhang, Y. and Liu, X. (2023) Ginsenosides can target brain-derived neurotrophic factor to improve parkinson's disease. *Food & Function* 14(12), 5537–5550. DOI: 10.1039/D2FO03484K.

Dubale, S., Usure, R.E., Mekasha, Y.T., Hasen, G., Hafiz, F. *et al.* (2025) Traditional herbal medicine legislative and regulatory framework: A cross-sectional quantitative study and archival review perspectives. *Frontiers in Pharmacology* 16, 1475297. DOI: 10.3389/fphar.2025.1475297.

Edo, G.I., Obasohan, P., Makia, R.S., Abiola, O.T., Umelo, E.C. *et al.* (2024) The use of quality control parameters in the evaluation of herbal drugs. A review. *Discover Medicine* 1(1), 168. DOI: 10.1007/s44337-024-00177-6.

Efferth, T. and Greten, H.J. (2012) Quality control for medicinal plants. *Medicinal & Aromatic Plants* 01, 07. DOI: 10.4172/2167-0412.1000e131.

El Boukhari, R. and Fatimi, A. (2024) Advancing functional food innovation: A patent landscape analysis of Lamiaceae bioactives through cooperative patent classification systems. In: *International Electronic Conference on Nutrients (IECN 2024)*, p. 1. DOI: 10.3390/blsf2024038001.

Erangika, L.H.H. and Kuruppu, A.I. (2024) Revolutionizing herbal extracts by nanotechnology approaches for enhanced delivery and efficacy. *Vidyodaya Journal of Science* 27(2). DOI: 10.31357/vjs.v27i02.7841.

Fatima, M., Saleem, M.Z., Zafar, S., Askri, S.M.H., Ayub, M.A. *et al.* (2022) Improving the quality of medicinal and aromatic plants through metabolic engineering. In: Aftab, T. and Hakeem, K.R. (eds) *Metabolic Engineering in Plants*. Springer Nature, Singapore, pp. 321–339.

Fatima, N., Parveen, M., Mahmood, I. and Rizvi, R. (2024) Nanotechnology-based drug delivery system for herbal medicines. In: Swamy, M.K. and Akhtar, M.S. (eds) *Metabolites of Medicinal Plants: Insightful Approaches*. Bentham Science Publishers, Singapore, pp. 232–243.

Feng, H., Xue, M., Deng, H., Cheng, S., Hu, Y. *et al.* (2022) Ginsenoside and its therapeutic potential for cognitive impairment. *Biomolecules* 12(9), 1310. DOI: 10.3390/biom12091310.

Fokunang, E.T., Fonmboh, D.J., Mballa, R.N., Nyuyki, A.B., Fokunang, L.B. *et al.* (2020) Pharmacovigilance of natural herbal medicines research for efficacy, safety and quality assurance of phytomedicine products. *Journal of Complementary and Alternative Medical Research* 12(1), 21–37. DOI: 10.9734/jocamr/2020/v12i130198.

Geng, J., Dong, J., Ni, H., Lee, M.S., Wu, T. *et al.* (2010) Ginseng for cognition. *The Cochrane Database of Systematic Reviews* 12, CD007769. DOI: 10.1002/14651858.CD007769.pub2.

Halder, M. and Roy, S. (2023) Medicinal plants: Biodiversity, biotechnology and conservation. In: Jha, S. and Halder, M. (eds) *Current Status of Metabolic Engineering of Medicinal Plants for Production of Plant-Derived Secondary Metabolites*. Springer Nature, Singapore, pp. 819–869.

He, S., Shi, J., Chai, H., Ma, L., Pei, H. *et al.* (2024) Mechanisms with network pharmacology approach of ginsenosides in Alzheimer's disease. *Heliyon* 10(5), e26642. DOI: 10.1016/j.heliyon.2024.e26642.

Huang, T., Lee, S., Lee, T., Yun, S., Kim, Y. *et al.* (2023) Smart farming enhances bioactive compounds content of Panax ginseng on moderating scopolamine-induced memory deficits and neuroinflammation 12(3), 640. DOI: 10.3390/plants12030640.

Jabeen, N. (2023) The role of biofortification in enhancing plant growth, development, yield, and quality. In: Hasanuzzaman, M., Tahir, M.S., Tanveer, M. and Shah, A.N. (eds) *Mineral Biofortification in Crop Plants for Ensuring Food Security*. Springer Nature, Singapore, pp. 37–94.

Jain, S.D., Shrivastava, S.K., Agrawal, A. and Gupta, A.K. (2024) WHO guidelines for quality control of herbal medicines: From cultivation to consumption. *International Journal of Pharmaceutical Chemistry and Analysis* 11(3), 212–225. DOI: 10.18231/j.ijpca.2024.031.

Jakaria, M., Haque, M.E., Kim, J., Cho, D.-Y., Kim, I.-S. *et al.* (2018) Active ginseng components in cognitive impairment: Therapeutic potential and prospects for delivery and clinical study. *Oncotarget* 9(71), 33601–33620. DOI: 10.18632/oncotarget.26035.

Jiang, M., Chi, J., Qiao, Y., Wang, J., Zhang, Z. *et al.* (2025) Ginsenosides Rg1, Rb1 and rare ginsenosides: Promising candidate agents for Parkinson's disease and Alzheimer's disease and network pharmacology analysis. *Pharmacological Research* 212, 107578. DOI: 10.1016/j.phrs.2025.107578.

Kayser, O. (2010) Metabolic engineering strategies for the optimization of medicinal and aromatic plants: Expectations and realities. *Acta Horticulturae* 860(860), 199–204. DOI: 10.17660/ActaHortic.2010.860.29.

Knöss, W. (2015) Harmonization of regulatory requirements in europe to ensure quality, safety and efficacy of herbal medicinal products. In: Mukherjee, P.K. (ed.) *Evidence-Based Validation of Herbal Medicine*. Elsevier, Amsterdam, pp. 201–216.

Krylov, I.A., Kubasova, E., Kubassov, R., Korelskaya, G.V., Korobitsyn, A.P. *et al.* (2024) Perspectives of medical use of herbal origin adaptogens. *Medical & Pharmaceutical Journal Pulse* 26(12), 53–59. DOI: 10.26787/nydha-2686-6838-2024-26-12-53-59.

Kumar, A., Mosa, K.A., Ji, L., Kage, U., Dhokane, D. *et al.* (2018) Metabolomics-assisted biotechnological interventions for developing plant-based functional foods and nutraceuticals. *Critical Reviews in Food Science and Nutrition* 58(11), 1791–1807. DOI: 10.1080/10408398.2017.1285752.

Kumar, B., Misra, A. and Srivastava, S. (2022) Bioactive phenolic compounds from indian medicinal plants for pharmaceutical and medical aspects. In: Badria, F.A. (ed.) *Phenolic Compounds-Chemistry, Synthesis, Diversity, Non-Conventional Industrial, Pharmaceutical and Therapeutic Applications*. IntechOpen, London, pp. 1–15.

Kumari, M., Checker, V.G., Kathpalia, R., Srivastava, V., Singh, I.K. *et al.* (2024) Metabolic engineering for enhanced terpenoid production: Leveraging new horizons with an old technique. *Plant Physiology and Biochemistry* 210, 108511. DOI: 10.1016/j.plaphy.2024.108511.

Kurdekar, A.K., Banu, A.M., Das, H., Rani, K.S., Rana, S. *et al.* (2023) A promising approach for addressing micronutrient deficiencies and enhancing nutritional quality in food crops through biofortification. *International Journal of Plant & Soil Science* 35(18), 2127–2134. DOI: 10.9734/ijpss/2023/v35i183503.

Lee, R., Kim, J.-H., Kim, W.-W., Hwang, S.-H., Choi, S.-H. *et al.* (2024) Emerging evidence that ginseng components improve cognition in subjective memory impairment, mild cognitive impairment, and early Alzheimer's disease dementia. *Journal of Ginseng Research* 48(3), 245–252. DOI: 10.1016/j.jgr.2024.02.002.

Lelong, C., Mandoux, A., Houyoux, N., Defrere, S. and Mariage, P.-A. (2024) Red Panax ginseng root promotes neuronal plasticity *in vitro* and improves cognitive function in aged animals. *Functional Foods in Health and Disease* 14(8), 574–588. DOI: 10.31989/ffhd.v14i8.1388.

Li, J., Huang, Q., Yao, Y., Ji, P., Mingyao, E. *et al.* (2022) Biotransformation, pharmacokinetics, and pharmacological activities of ginsenoside Rd against multiple diseases. *Frontiers in Pharmacology* 13, 909363. DOI: 10.3389/fphar.2022.909363.

Liang, H.-Y., Zhang, P.-P., Zhang, X.-L., Zheng, Y.-Y., Huang, Y.-R. *et al.* (2021) Preclinical systematic review of ginsenoside Rg1 for cognitive impairment in Alzheimer's disease. *Aging* 13(5), 7549–7569. DOI: 10.18632/aging.202619.

Liu, J., Yu, W., Ma, C., Li, T., Liang, Y. *et al.* (2023) Network pharmacology and mechanism studies of the protective effect of ginseng against Alzheimer's disease based on aβ pathogenesis. *Planta Medica* 89(10), 990–1000. DOI: 10.1055/a-2014-6061.

Lu, J., Wang, X., Wu, A., Cao, Y., Dai, X. *et al.* (2022) Ginsenosides in central nervous system diseases: Pharmacological actions, mechanisms, and therapeutics. *Phytotherapy Research* 36(4), 1523–1544. DOI: 10.1002/ptr.7395.

Manzoor, Z., Khan, A.A., Khurshid, M. and Hassan, Z. (2024) Molecular mechanistic approaches for iron and zinc biofortification in field plants. In: Azhar, M.T., Ahmad, M.Q., Rana, I.A. and Atif, R.M. (eds) *Biofortification of Grain and Vegetable Crops*. Academic Press, Cambridge, Massachusetts, pp. 307–327.

Martinez-Alvarado, L., Acosta-Estrada, B.A. and Marilena, A. (2022) Strategies for crop nutraceutical content enhancement, A review. *JSFA Reports* 2(4), 146–154. DOI: 10.1002/jsf2.38.

Misra, B. and Sharma, V.P. (2024) Functional foods and nutraceuticals: Biochemical and safety insights. *Advances in Clinical Toxicology* 9(4). DOI: 10.23880/act-16000322.

Mogali, S., Jaggal, L. and Sridevi, O. (2020) Biofortification of linseed as functional food for profitability and nutritional security. *Journal of Oilseeds Research* 37(Special issue). DOI: 10.56739/jor.v37iSpecialissue.139510.

Mukherjee, P.K. (2019) Regulatory harmonization and good quality practices for the development of herbal medicine. In: Mukherjee, P.K. (ed.) *Quality Control and Evaluation of Herbal Drugs*. Elsevier, Amsterdam, pp. 723–739.

Mun, H. and Townley, H.E. (2021) Nanoencapsulation of plant volatile organic compounds to improve their biological activities. *Planta Medica* 87(3), 236–251. DOI: 10.1055/a-1289-4505.

Naik, B., Kumar, V., Rizwanuddin, S., Mishra, S., Kumar, V. *et al.* (2024) Biofortification as a solution for addressing nutrient deficiencies and malnutrition. *Heliyon* 10(9), e30595. DOI: 10.1016/j.heliyon.2024.e30595.

Nazma, S., Hemalatha, M. and Sudha, T. (2023) Biofortification of iron and zinc in field crops through plant microbe interaction: A review. *Agricultural Reviews* 46(3), 393–400. DOI: 10.18805/ag.R-2606.

Nielsen, E., Temporiti, M.E.E. and Cella, R. (2019) Improvement of phytochemical production by plant cells and organ culture and by genetic engineering. *Plant Cell Reports* 38(10), 1199–1215. DOI: 10.1007/s00299-019-02415-z.

Picking, D. (2024) The global regulatory framework for medicinal plants. In: McCreath, S.B. and Clement, Y.N. (eds) *Pharmacognosy*. Academic Press, Cambridge, Massachusetts, pp. 769–782.

Ratheesh, G., Xiao, Y., Ezhilarasu, H., Sadiq, A., Devassy, G. *et al.* (2018) Nanodrug delivery system using medicinal plants. In: Sharma, C.P. (ed.) *Drug Delivery Nanosystems for Biomedical Applications*. Elsevier, Amsterdam, pp. 357–375.

Raut, D.A., Afrayeem, S., Singh, V., Jejal, A.D., Sachan, P. *et al.* (2024) Enhancing nutrition, crop resilience, and food security through biofortification. *International Journal of Environment and Climate Change* 14(2), 241–253. DOI: 10.9734/ijecc/2024/v14i23942.

Rawat, A.K.S. and Tewari, S.K. (2015) Quality assurance of medicinal and aromatic plants - good agricultural and collection practices. In: *Medicinal and Aromatic Plants of the World: Scientific, Production, Commercial and Utilization Aspects*. Springer, Dordrecht, Netherlands, pp. 273–303.

Ray, A., Gulati, K., Rehman, S., Rai, N. and Anand, R. (2021) Role of nutraceuticals as adaptogens. In: Gupta, R.C., Lall, R. and Srivastava, A. (eds) *Nutraceuticals*. Academic Press, Cambridge, Massachusetts, pp. 229–244.

Saad, B., Ghareeb, B. and Kmail, A. (2021) Metabolic and epigenetics action mechanisms of antiobesity medicinal plants and phytochemicals. *Evidence-Based Complementary and Alternative Medicine* 9995903. DOI: 10.1155/2021/9995903.

Sahu, J., Sen, P., Choudhury, M.D., Dehury, B., Barooah, M. *et al*. (2014) Rediscovering medicinal plants' potential with OMICS: Microsatellite survey in expressed sequence tags of eleven traditional plants with potent antidiabetic properties. *OMICS* 18(5), 298–309. DOI: 10.1089/omi.2013.0147.

Saini, P.K., Bhati, J., Kumar, K., Goyal, G., Tripathi, S.K. *et al*. (2022) Biofortification of iron, zinc, and iodine in crops using transgenic techniques. *OPSearch* 1(2), 48–61. DOI: 10.58811/opsearch.v1i2.13.

Sakure, A.A. (2022) Enhancement of nutritional, pharmaceutical and industrial value of crops through genetic modification with carotenoid pathway genes. In: Kumar, N. (ed.) *Biotechnology and Crop Improvement*. CRC Press, Boca Raton, Florida, pp. 63–77.

Samtiya, M., Aluko, R.E., Puniya, A.K. and Dhewa, T. (2021) Enhancing micronutrients bioavailability through fermentation of plant-based foods: A concise review. *Fermentation* 7(2), 63. DOI: 10.3390/fermentation7020063.

Sethi, A., Bhandawat, A. and Pati, P.K. (2022) Engineering medicinal plant-derived CYPs: A promising strategy for production of high-valued secondary metabolites. *Planta* 256(6), 119. DOI: 10.1007/s00425-022-04024-9.

Sharma, A. and Verma, R.K. (2019) Biofortification: A promising approach toward eradication of hidden hunger. In: Singh, D.P., Gupta, V.K. and Prabha, R. (eds) *Microbial Interventions in Agriculture and Environment*. Springer, Singapore, pp. 313–327.

Sharma, D., Dhama, K. and Sharma, D. (2024) Adaptogens and functional food and their impact on health. In: *Futuristic Trends in Agriculture Engineering & Food Sciences*, Vol. 3. Iterative International Publisher, Chikmagalur, Karnataka, India, pp. 110–122.

Sharma, M., Koul, A., Sharma, D., Kaul, S., Swamy, M.K. *et al*. (2019a) Metabolic engineering strategies for enhancing the production of bio-active compounds from medicinal plants. In: Akhtar, M.S., Swamy, M.K. and Sinniah, U.R. (eds) *Natural Bio-Active Compounds*. Springer Singapore, pp. 287–316.

Sharma, R., Kumar, S., Kumar, V. and Thakur, A. (2019b) Comprehensive review on nutraceutical significance of phytochemicals as functional food ingredients for human health management. *Journal of Pharmacognosy and Phytochemistry* 8(5), 385–395. DOI: 10.22271/phyto.2019.v8.i5h.9589.

She, L., Sun, J., Xiong, L., Li, A., Li, L. *et al*. (2024) Ginsenoside RK1 improves cognitive impairments and pathological changes in Alzheimer's disease via stimulation of the AMPK/NRF2 signaling pathway. *Phytomedicine* 122, 155168. DOI: 10.1016/j.phymed.2023.155168.

Shi, Z., Chen, H., Zhou, X., Yang, W. and Lin, Y. (2022) Pharmacological effects of natural medicine ginsenosides against Alzheimer's disease. *Frontiers in Pharmacology* 13, 952332. DOI: 10.3389/fphar.2022.952332.

Singab, A.N. (2024) Unveiling revealing nature's bounty: A comprehensive exploration of bioavailability in natural products. *Bioequivalence & Bioavailability International Journal* 8(1), 1–9. DOI: 10.23880/beba-16000226.

Singh, M., Chauhan, P.K. and Kumar, P. (2017) Bio-fortification of underutilized himalayan pear and plum — a next hotspot for nutritional strategies of the "neglected" micronutrients. *International Journal of Trend in Scientific Research and Development* 2(1), 762–769. DOI: 10.31142/ijtsrd7050.

Singh, V., Akansha, A., Islam, Z. and Shaida, B. (2024) Medicinal plants: Sustainable scope to nutraceuticals. In: Thakur, M. (ed.) *Sustainable Food Systems*. Cham, Switzerland, pp, SFS, pp. 205–236.

Smith, I., Williamson, E.M., Putnam, S., Farrimond, J. and Whalley, B.J. (2014) Effects and mechanisms of ginseng and ginsenosides on cognition. *Nutrition Reviews* 72(5), 319–333. DOI: 10.1111/nure.12099.

Thakur, S., Gupta, M.M. and Sharma, D. (2024) Nutraceuticals regulation: An overview of the regulatory frameworks in USA, EU, and Japan. In: Keservani, R.K., Kesharwani, R.K., Emerald, M. and Sharma, A.K. (eds) *Nutraceutical Fruits and Foods for Neurodegenerative Disorders*. Academic Press, Cambridge, Massachusetts, pp. 421–440.

Wang, H., Chen, Y., Wang, L., Liu, Q., Yang, S. *et al*. (2023) Advancing herbal medicine: Enhancing product quality and safety through robust quality control practices. *Frontiers in Pharmacology* 14, 1265178. DOI: 10.3389/fphar.2023.1265178.

Wijayabandara, J. (2019) Standardisation and quality control of herbal medicines. In: Ghosh, D. and Mukherjee, P.K. (eds) *Natural Medicines*. CRC Press, Boca Raton, Florida, pp. 13–24.

Zahra, W., Birla, H., Singh, S.S., Rathore, A.S., Dilnashin, H. *et al*. (2022) Medicinal plants and natural compounds as anti parkinsonian agents. In: Qidwai, T. (ed.) *Neurodegenerative Diseases: Multifactorial Degenerative Processes, Biomarkers and Therapeutic Approaches*. Bentham Science Publishers, Singapore, pp. 199–217.

Zarneshan, S.N., Fakhri, S. and Khan, H. (2022) Targeting akt/CREB/BDNF signaling pathway by ginsenosides in neurodegenerative diseases: A mechanistic approach. *Pharmacological Research* 177, 106099. DOI: 10.1016/j.phrs.2022.106099.

Zeng, M., Zhang, K., Yang, J., Zhang, Y., You, P. *et al.* (2024) Effects of ginseng on cognitive function: A systematic review and meta-analysis. *Phytotherapy Research* 38(12), 6023–6034. DOI: 10.1002/ptr.8359.

Zhang, W., Zeng, Y., Jiao, M., Ye, C., Li, Y. *et al.* (2023) Integration of high-throughput omics technologies in medicinal plant research: The new era of natural drug discovery. *Frontiers in Plant Science* 14, 1073848. DOI: 10.3389/fpls.2023.1073848.

Zhang, Y.Y., Stockmann, R., Ng, K. and Ajlouni, S. (2021) Opportunities for plant-derived enhancers for iron, zinc, and calcium bioavailability: A review. *Comprehensive Reviews in Food Science and Food Safety* 20(1), 652–685. DOI: 10.1111/1541-4337.12669.

Zheng, T., Yin, Z. and Huang, Q. (2022) Assessment of digestion, absorption, and metabolism of nanoencapsulated phytochemicals using *in vitro* and *in vivo* models: A perspective paper. *Journal of Agricultural and Food Chemistry* 70(15), 4548–4555. DOI: 10.1021/acs.jafc.1c07919.

Zhou, L., Tan, F., Zhang, X., Li, Y. and Yin, W. (2024) Neuroprotection and mechanisms of ginsenosides in nervous system diseases: Progress and perspectives. *IUBMB Life* 76(11), 862–882. DOI: 10.1002/iub.2862.

Zhu, C., Sanahuja, G., Yuan, D., Farré, G., Arjó, G. *et al.* (2013) Biofortification of plants with altered antioxidant content and composition: Genetic engineering strategies. *Plant Biotechnology Journal* 11(2), 129–141. DOI: 10.1111/j.1467-7652.2012.00740.x.

9 Omics-guided Conservation and Ethical Bioprospecting of Medicinal Plants

Surendra Sarsaiya, Archana Jain and Qihai Gong*
Key Laboratory of Basic Pharmacology and Joint International Research Laboratory of Ethnomedicine of Ministry of Education, Zunyi Medical University, Zunyi, Guizhou, China

9.1 Introduction

Medicinal plants have been integral to human health and traditional healing systems for millennia, serving as the foundation for numerous modern pharmaceuticals. However, the escalating demand for plant-derived bioactive compounds, coupled with habitat destruction and unsustainable harvesting practices, has placed immense pressure on these valuable biological resources. The conservation of medicinal plant species is no longer merely a botanical concern but a pressing global priority that intersects biodiversity preservation, public health, and socio-economic equity. In response to these challenges, the integration of advanced scientific methodologies—particularly omics technologies—has emerged as a transformative approach to both understanding and safeguarding medicinal plant diversity (Sharma and Sharon, 2025).

The term "omics" encompasses a suite of high-throughput technologies—including genomics, transcriptomics, proteomics, and metabolomics—that enable comprehensive molecular profiling of biological systems. These tools have revolutionized medicinal plant research by providing unprecedented insights into the genetic blueprint, regulatory mechanisms, and metabolic output of species of therapeutic interest. For instance, genomics facilitates the identification of genes involved in the biosynthesis of valuable secondary metabolites, while transcriptomics reveals how these genes are expressed under varying environmental conditions (Zhang *et al.*, 2023). Proteomics and metabolomics further elucidate the functional proteins and chemical compounds that define a plant's medicinal value (Yang *et al.*, 2023). Together, these technologies form a multi-omics framework that supports the discovery of novel bioactive compounds and informs strategies for sustainable use.

The application of omics technologies has significantly accelerated the discovery of novel bioactive compounds from medicinal plants. Genomics provides a foundational understanding of the genetic makeup, enabling researchers to decode entire genomes and identify gene clusters responsible for the synthesis of complex molecules such as alkaloids, terpenoids, and phenolics. Comparative genomics further

*Corresponding author: gqh@zmu.edu.cn

DOI: 10.1079/9781836994206.0009

reveals evolutionary relationships and genetic diversity among species, which is crucial for selecting candidates with high therapeutic potential (Jabeen *et al.*, 2024). Advances in sequencing technologies have made it feasible to generate high-quality genomes for non-model medicinal plants, opening new avenues for gene editing and metabolic engineering (Hussain *et al.*, 2023).

Transcriptomics complements genomics by capturing dynamic gene expression patterns in response to developmental cues, environmental stresses, or elicitor treatments. This approach helps pinpoint key regulatory genes and biosynthetic pathways activated during the production of target compounds (Yang *et al.*, 2023). Proteomics, on the other hand, delves into the functional entities—proteins and enzymes—that catalyze biosynthetic reactions. Through proteomic profiling, researchers can identify post-translational modifications and protein–protein interactions that modulate metabolic flux (Barbosa *et al.*, 2025). Metabolomics serves as the phenotypic endpoint of omics investigations, offering a direct readout of the metabolic composition of plant tissues. By integrating metabolomic data with other omics layers, scientists can reconstruct entire metabolic networks and identify previously unknown bioactive molecules (Guo *et al.*, 2024).

The true power of omics lies in the integration of multidimensional datasets to construct holistic models of plant physiology and biochemistry. Multi-omics integration enables the correlation of genetic variants with transcript abundance, protein expression, and metabolite accumulation, thereby uncovering the molecular mechanisms underpinning bioactive compound synthesis (Yang *et al.*, 2023). For example, combined genomics and metabolomics have been used to link gene families with the production of specific anti-cancer compounds in medicinal species (Das *et al.*, 2025). Such integrative approaches are increasingly supported by bioinformatics pipelines and artificial intelligence (AI) tools, which facilitate data mining, pattern recognition, and predictive modeling (Murmu *et al.*, 2024).

Despite these advances, the field faces significant challenges in data management and interpretation. The volume and complexity of omics data require robust computational infrastructure and sophisticated algorithms to extract biologically meaningful insights (Gupta *et al.*, 2024). Moreover, the inherent variability in plant growth conditions, genetic backgrounds, and metabolic plasticity necessitates careful experimental design and validation. Overcoming these hurdles will be essential for translating omics data into actionable conservation strategies and drug discovery outcomes.

As omics technologies enhance our capacity to identify and exploit medicinal plants, the ethical dimensions of bioprospecting become increasingly salient. Bioprospecting—the systematic search for commercially valuable biochemical and genetic resources—must be conducted in a manner that respects the rights and knowledge of Indigenous and local communities. Traditional ecological knowledge (TEK) has long guided the use of medicinal plants, and its integration with omics approaches can accelerate the discovery of natural products while ensuring cultural relevance and community benefit (Chele *et al.*, 2025). However, the historical exploitation of such knowledge, often termed biopiracy, underscores the need for ethical frameworks that prioritize informed consent, fair compensation, and legal recognition of intellectual property rights (Katu, 2025).

International agreements, such as the Nagoya Protocol on Access and Benefit-Sharing (ABS), provide a legal foundation for equitable partnerships between bioprospectors and source countries. Compliance with such frameworks ensures that benefits—whether monetary or non-monetary—are shared fairly with communities that have conserved and managed medicinal plant diversity for generations (Hao *et al.*, 2025a). Case studies from India and Kenya illustrate how collaborative, community-led bioprospecting initiatives can simultaneously advance conservation goals and improve local livelihoods (Katu, 2025).

The conservation of medicinal plants relies on a combination of *in situ* and *ex situ* strategies. *In situ* conservation involves protecting species in their natural habitats through the establishment of protected areas, conservation corridors, and sustainable harvesting protocols (Sharma and Sharon, 2025). *Ex situ* methods, such as seed banks, botanical gardens, and tissue culture repositories, serve as insurance against extinction and provide material for research and restoration (Chen *et al.*, 2016). Omics data can inform both approaches by identifying genetically distinct populations, guiding seed collection efforts, and optimizing cryopreservation protocols.

Biotechnological tools further enhance conservation outcomes. Micropropagation, synthetic seed technology, and cell suspension cultures enable the mass production of high-value medicinal plants without depleting wild populations (Gusain *et al.*, 2021). Genetic engineering, particularly CRISPR-Cas9-based (clustered regularly interspaced short palindromic repeats-associated protein 9) genome editing, offers the potential to enhance the production of bioactive compounds in cultivated plants, reducing pressure on wild stocks (Kielich *et al.*, 2024). When combined with omics-driven insights, these technologies form a powerful toolkit for sustainable utilization and long-term conservation.

The future of omics-guided conservation and ethical bioprospecting is poised at the intersection of technology, policy, and collaboration. Emerging trends include the adoption of pharmacophylomics—a field that integrates phylogenetics with metabolomics to predict chemodiversity across plant lineages—and the expanded use of AI and machine learning (ML) for predictive ecology and conservation prioritization (Hao *et al.*, 2025a). Synthetic biology will likely play an increasingly prominent role in engineering plant systems for enhanced compound production, while blockchain and other digital tools may improve transparency and traceability in bioprospecting agreements (Murmu *et al.*, 2024).

However, technical advancements must be matched by ethical vigilance and inclusive governance. The engagement of scientists, policymakers, Indigenous communities, and industry stakeholders is critical to developing frameworks that are both scientifically sound and socially just (Rose *et al.*, 2012). As the demand for medicinal plants continues to grow, the integration of omics technologies with ethical bioprospecting and conservation practices offers a promising pathway to reconcile human health needs with the preservation of plant biodiversity (Table 9.1 and Fig. 9.1).

9.2 Omics in Prioritizing Endangered Medicinal Plants: Informing IUCN Assessments with Genomic and Metabolomic Data

The escalating global demand for plant-based therapeutics, coupled with habitat loss and climate change, has pushed numerous medicinal plant species to the brink of extinction. In this critical scenario, the integration of omics technologies—genomics, transcriptomics, proteomics, and metabolomics—provides a powerful, data-driven framework for identifying and prioritizing endangered medicinal plants for conservation. These approaches move beyond traditional morphological assessments to offer a deep, molecular-level understanding of genetic diversity, metabolic potential, and adaptive capacity, which is essential for formulating effective and targeted conservation strategies (Rohini, 2020).

Genomic tools form the cornerstone of this effort. Techniques such as DNA barcoding and molecular fingerprinting are crucial for the authentic identification of medicinal species, helping to prevent adulteration in herbal trade and ensuring that conservation efforts are focused on the correct genetic entities (Rohini, 2020). Furthermore, the assessment of genetic diversity and population structure through molecular markers allows conservationists to gauge the genetic health of endangered populations. This information is vital for understanding species' evolutionary trajectories and their ability to adapt to environmental changes, thereby directly informing priority-setting in conservation programs (Rohini, 2020).

The application of multi-omics approaches significantly deepens this prioritization process. By integrating transcriptomics, proteomics, and metabolomics, researchers can gain a comprehensive understanding of the biosynthetic pathways responsible for producing valuable bioactive compounds (Yang *et al.*, 2023; Zhang *et al.*, 2023). This integrated perspective reveals the molecular intricacies and dynamic metabolic changes that occur in response to environmental stimuli. A species that demonstrates a high abundance and diversity of key therapeutic metabolites, or possesses unique biosynthetic pathways, may be assigned

Table 9.1. Omics technologies in medicinal plant conservation and ethical bioprospecting. Author's own table.

Omics technology	Core application in medicinal plants	Key advantages	Specific challenges & ethical considerations	Citations
Genomics	• Decoding entire genomes to identify biosynthetic gene clusters. • Assessing population genetics and genetic diversity for conservation prioritization.	• Provides a foundational blueprint for understanding metabolic potential. • Enables marker-assisted breeding and traceability.	• High cost and computational demands for non-model species. • Risk of "digital biopiracy" if genetic sequences are commercialized without benefit-sharing.	(Hussain *et al.*, 2023; Jabeen *et al.*, 2024; Sharma and Sharon, 2025)
Transcriptomics	• Analyzing gene expression patterns in response to environment/stress. • Identifying key regulatory genes in secondary metabolite pathways.	• Reveals dynamic molecular responses, guiding cultivation and harvest timing. • Pinpoints rate-limiting steps in valuable compound synthesis.	• RNA instability requires rapid, complex sample preservation. • Expression data is context-specific, limiting generalizability.	(Yang *et al.*, 2023; Zhang *et al.*, 2023)
Proteomics	• Profiling functional enzymes and proteins involved in biosynthesis. • Identifying post-translational modifications that regulate metabolic flux.	• Bridges the gap between genetic instruction and metabolic activity. • Directly targets the functional molecules catalyzing reactions.	• Technically challenging due to protein complexity and dynamic range. • Requires well-annotated genomes for accurate protein identification.	(Yang *et al.*, 2023; Barbosa *et al.*, 2025)
Metabolomics	• Comprehensive profiling of small molecules (alkaloids, terpenoids, etc.). • Non-destructive monitoring for sustainable wild harvesting.	• Directly characterizes medicinal value and chemo-diversity. • Informs optimal harvest times and authenticates raw materials.	• Metabolite levels are highly variable with environment and development. • Compound identification remains a major bottleneck in untargeted studies.	(Yang *et al.*, 2023; Guo *et al.*, 2024)

Continued

Table 9.1. Continued

Omics technology	Core application in medicinal plants	Key advantages	Specific challenges & ethical considerations	Citations
Multi-omics integration	• Constructing holistic models by correlating genotype to phenotype. • Identifying candidate genes for metabolic engineering (e.g. CRISPR).	• Provides a systems-level understanding of plant biology and resilience. • Powerful for discovering novel bioactive compounds and their pathways.	• Requires immense computational resources and sophisticated bioinformatics. • Lack of standardized protocols for data integration across studies and platforms.	(Yang *et al*., 2023; Murmu *et al*., 2024; Das *et al*., 2025)
Cross-cutting ethical & policy frameworks	• Ensuring equitable benefit-sharing with Indigenous and local communities. • Complying with international agreements like the Nagoya Protocol.	• Promotes sustainable and just partnerships in bioprospecting. • Helps prevent biopiracy and recognizes TEK.	• Power imbalances and lack of legal capacity in source communities/countries. • Ambiguity in regulating digital sequence information under current ABS frameworks.	(Chele *et al*., 2025; Hao *et al*., 2025a; Katu, 2025)

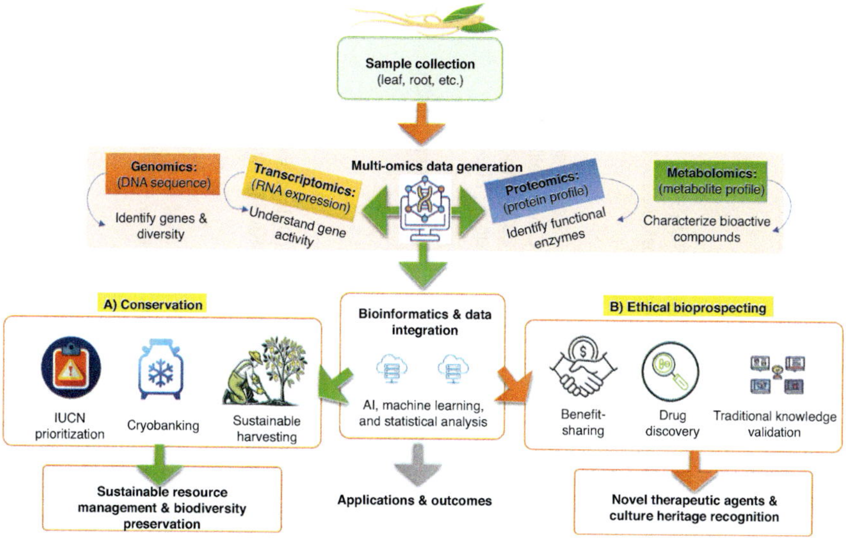

Fig. 9.1. Multi-omics integration pipeline for medicinal plant research. (Author's own figure.)

a higher conservation priority due to its irreplaceable pharmaceutical value (Zhang *et al.*, 2023). For instance, transcriptomic studies can identify highly expressed genes in critical pathways, while metabolomics provides a direct chemical fingerprint of the plant's medicinal potential, creating a powerful rationale for its conservation.

Bioinformatics serves as the essential engine that powers this omics-driven prioritization. Advanced computational tools are required to manage, analyze, and interpret the vast and complex datasets generated by high-throughput technologies (Gupta *et al.*, 2024). The development of specialized genomic and transcriptomic repositories offers invaluable insights into the secondary metabolism of endangered species, which is central to their therapeutic application (Kapoor *et al.*, 2021). By systematically analyzing this data, researchers can characterize pharmacological properties, predict chemodiversity, and model population genetics, all of which contribute to a more nuanced and scientifically robust framework for ranking species based on their risk of extinction and their potential value to human health.

Despite these significant advancements, challenges remain in fully leveraging omics for conservation prioritization. Issues of data accessibility, the high cost of technologies, and the need for interdisciplinary collaboration between bioinformaticians, conservation biologists, and ethnobotanists can hinder practical application, especially in resource-limited regions. Nonetheless, the synergy between omics technologies and TEK holds immense promise. By providing an unprecedented molecular understanding of genetic uniqueness and biochemical value, omics approaches are transforming how we identify which endangered medicinal plants to save first, ensuring that conservation resources are allocated to protect the most vulnerable and the most valuable species for future generations (Table 9.2 and Fig. 9.2).

9.2.1 Informing IUCN assessments with genomic diversity and metabolomic profiling

The integration of genomic diversity and metabolomic profiling into International Union for Conservation of Nature (IUCN) Red List assessments offers a transformative, multidimensional approach to evaluating the conservation status of endangered medicinal plants. Traditionally, assessments have relied heavily on population demographics and threat analysis, but incorporating these advanced molecular techniques

Table 9.2. Omics applications in prioritizing endangered medicinal plants for IUCN assessments. Author's own table.

Application area	Omics approach	Specific role in IUCN assessment & conservation prioritization	Key insights provided	Citations
Species authentication & population delineation	Genomics (DNA barcoding, ISSR, SSR, AFLP)	Correctly identifies species and delineates population structure, ensuring conservation targets are accurate. Prevents effort wastage on misidentified or hybrid taxa.	Reveals evolutionary significant units (ESUs) and management units (MUs) for targeted action.	(Rohini, 2020; Rai *et al.*, 2023)
Assessment of adaptive potential & resilience	Genomics (neutral markers)	Quantifies genetic diversity and inbreeding levels, directly informing Criterion C (small population size and decline) of IUCN assessments.	Populations with low heterozygosity and high inbreeding are less resilient to environmental change, indicating higher extinction risk.	(Ma *et al.*, 2023; Rai *et al.*, 2023)
Valuation of chemical utility & bioprospecting potential	Metabolomics (LC-MS, HPLC)	Identifies and quantifies key therapeutic compounds (e.g. paclitaxel in *Taxus*, berberine in *Coptis*), adding a "chemical value" dimension to prioritization.	Flags populations with high or unique concentrations of valuable metabolites as conservation priorities to prevent loss of pharmaceutical potential.	(Zhou *et al.*, 2019; Ćavar Zeljković *et al.*, 2024; Mohammed *et al.*, 2024)
Monitoring population health & stress response	Metabolomics (stress metabolome profiling)	Assesses physiological resilience by profiling stress-induced metabolites (e.g. antioxidants, phenolics).	Populations with a robust stress metabolome may be more suitable for *in situ* conservation, while those with a weak response may need urgent *ex situ* safeguarding.	(Salam *et al.*, 2023)
Understanding chemodiversity & ecotypic variation	Metabolomics	Documents intra-species chemical variation across different environments and genotypes.	Highlights chemically distinct populations that may represent unique adaptations or chemotypes, warranting separate conservation strategies.	(Dalmaris *et al.*, 2020; Karalija *et al.*, 2025)
Linking genotype to chemotype for predictive conservation	Integrative omics (GWAS, eQTL (expression quantitative trait locus) mapping)	Identifies genetic markers linked to the production of high-value metabolites through genome-wide association studies.	Enables non-destructive genetic screening for high-yield individuals and informs the strategic collection of germplasm for seed banks.	(Wei *et al.*, 2021; Wang *et al.*, 2024)
Holistic threat and value assessment	Integrated genomics & metabolomics	Provides a multidimensional view of extinction risk by combining data on genetic erosion and chemical erosion.	A population can be prioritized if it is both genetically unique *and* chemically valuable, or if it is genetically depleted but is the last source of a critical medicine.	(Alami *et al.*, 2022; Jeon *et al.*, 2024)

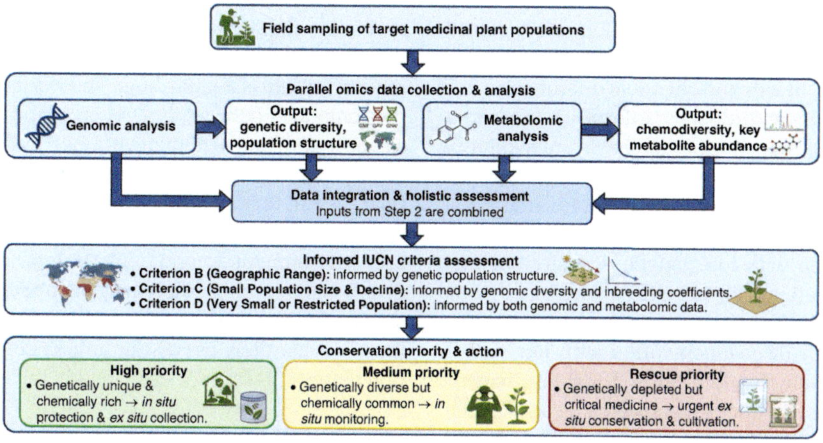

Fig. 9.2. Omics-guided workflow for IUCN Red List assessment and conservation prioritization. (Author's own figure.)

provides a deeper understanding of a species' evolutionary potential and intrinsic medicinal value, leading to more nuanced and effective conservation strategies.

Genomic diversity analysis is critical for predicting the long-term adaptability and resilience of endangered medicinal plants. By employing molecular markers such as SSR (simple sequence repeats), ISSR (inter-simple sequence repeats), and AFLP (amplified fragment length polymorphism), conservationists can quantify the genetic variation within and among populations. High genetic diversity is often correlated with a greater capacity to withstand environmental stressors, diseases, and climatic changes, which is essential information for a robust IUCN assessment (Rai *et al.*, 2023). Furthermore, genomic tools like DNA barcoding are vital for authentic species identification, preventing the conservation of misidentified or adulterated genetic lines and ensuring that conservation efforts are focused on the correct taxa (Rohini, 2020). This genetic insight allows for the formulation of targeted strategies, such as identifying priority populations for *in situ* protection or optimizing sampling for *ex situ* germplasm banks to capture the fullest possible genetic breadth, thereby directly informing the planning and action aspects of the IUCN framework (Ma *et al.*, 2023; Rai *et al.*, 2023).

Concurrently, metabolomic profiling addresses a unique and often overlooked aspect of conservation for medicinal species: the preservation of their chemical utility. The therapeutic value of a plant is derived from its specific profile of bioactive compounds, such as alkaloids and essential oils, which can exhibit significant intra-species variation (Karalija *et al.*, 2025). Metabolomic studies, using techniques like HPLC (high-performance liquid chromatography) and LC-MS (liquid chromatography-mass spectrometry), can map this chemical diversity, identifying populations that are not only genetically distinct but also rich in key medicinal metabolites (Mohammed *et al.*, 2024). For example, research on *Sinocalycanthus chinensis* has demonstrated how metabolomic data can reveal the plant's therapeutic potential and even guide sustainable harvesting practices by identifying alternative plant parts or optimal collection times (Tong *et al.*, 2025). Integrating this data into IUCN assessments helps prioritize conservation for populations that are chemically unique or possess high concentrations of valuable compounds, ensuring that the very reason for their medicinal use is not lost.

The synergistic combination of genomic and metabolomic data provides a holistic view of plant diversity that significantly strengthens the IUCN assessment process. This integrated approach allows for a more comprehensive evaluation of a species' extinction risk by considering not just its population size, but also its evolutionary health and chemical uniqueness (Jeon *et al.*, 2024). It directly informs conservation actions to mitigate threats like overexploitation and habitat destruction by providing a scientific basis for managing genetic resources and chemical traits (Cavers *et al.*, 2004). For instance, a population with low genetic diversity but high concentrations of a critical medicine would require a different conservation strategy than a genetically diverse population with low medicinal potency. By informing IUCN assessments with this dual-lens perspective, conservation planners can develop more precise and effective policies, ensuring that both the survival of the species and the preservation of its invaluable medicinal properties are secured for the future (Table 9.2 and Fig. 9.2).

9.2.2 Key metabolomic markers for conservation prioritization

The use of metabolomic markers moves beyond single-compound analysis to a systems-level understanding of a plant's biochemical value and stress status. Markers of economic and medicinal value are the direct targets of bioprospecting and harvesting. For *Taxus* species, the key markers are taxanes, most notably paclitaxel, and its biosynthetic precursors (Zhou *et al.*, 2019). LC-MS-based metabolomics can map the full spectrum and concentration of these compounds, identifying "high-yield" individuals or populations that are of utmost conservation priority. Similarly, for *Coptis*, the protoberberine alkaloids (berberine, palmatine, coptisine) serve as the primary metabolomic markers of value (Ćavar Zeljković *et al.*, 2024).

Metabolomes are dynamic and respond to environmental pressures. The variation in stress-induced metabolites, such as antioxidants, phenolic compounds, and pathogenesis-related proteins, can indicate population-level resilience (Salam *et al.*, 2023). A population of an endangered species showing a robust, diverse stress metabolome may be more resilient to climate change or pathogen outbreaks than a population with a blunted response, informing decisions about which populations are best suited for *in situ* conservation and which may require *ex situ* safeguarding. The variation of these markers across species is profound. While *Taxus wallichiana* (Himalayan yew) and *Taxus baccata* (European yew) both produce paclitaxel, their overall taxane profiles and concentrations vary significantly due to genetic differences and environmental conditions (Dalmaris *et al.*, 2020). This interspecific variation underscores the need for species-specific metabolomic baselines to accurately assess the unique conservation value and status of each taxon (Table 9.2).

9.2.3 Power of integrative omics for a comprehensive conservation strategy

While powerful individually, the true potential of omics is realized through the integration of genomics and metabolomics. This integrative approach bridges the gap between a plant's genetic blueprint (genotype) and its observable biochemical characteristics (chemo-phenotype), providing a mechanistic understanding of the factors driving its conservation status. An integrated omics approach can, for example, identify the genetic loci responsible for high-value metabolite production. By conducting genome-wide association studies (GWAS) on a metabolomically characterized population of an endangered medicinal plant, researchers can pinpoint the specific gene variants associated with high paclitaxel or berberine yield (Wang *et al.*, 2024). This knowledge is transformative: it allows for the genetic screening of wild populations to identify and protect high-genetic-value individuals without destructive chemical testing. Furthermore, it enables the intelligent design of *ex situ* collections (seed banks, living collections) that are curated not just for genetic diversity, but explicitly for the conservation of adaptive and economically significant traits (Wei *et al.*, 2021).

This synergy provides a more comprehensive understanding by linking neutral genetic diversity (a measure of evolutionary history) with adaptive potential (as reflected in the metabolome). A population might show moderate genomic diversity but possess a unique and valuable chemotype, or conversely, it might be genetically diverse but lack the key therapeutic compounds, thereby altering its conservation priority. Therefore, integrative omics moves conservation beyond a one-dimensional focus on genetics or simple presence/absence data, toward a multidimensional strategy that safeguards the full spectrum of diversity—genetic, functional, and chemical—that defines an endangered medicinal plant and secures its value for future generations (Table 9.2 and Fig. 9.2; Alami *et al.*, 2022).

9.3 Sustainable Wild Harvesting: Metabolomics for Non-destructive Sampling

The sustainable management of wild-harvested medicinal plants presents a critical conservation challenge, as traditional harvesting methods often threaten the survival of source populations. Metabolomics, defined as the comprehensive study of small-molecule metabolites, has emerged as a transformative approach through its application in non-destructive sampling. This methodology focuses on using omics technologies to monitor metabolite levels without damaging wild populations, thereby aligning resource utilization with long-term conservation goals (Skubel *et al.*, 2018). By enabling detailed biochemical profiling from minimal or renewable tissues, metabolomics provides essential data for informed harvesting decisions while preserving plant integrity and ecosystem function (Table 9.3 and Fig. 9.3; Świątły-Błaszkiewicz *et al.*, 2025).

9.3.1 Analytical platforms for minimal impact sampling

Advanced analytical technologies form the foundation of non-destructive metabolomic approaches. Spectroscopic methods such as attenuated total reflectance-Fourier transform infrared (ATR-FTIR) and energy dispersive-X-ray fluorescence (ED-XRF) have demonstrated particular utility in monitoring metabolic changes in medicinal plants like *Scutellaria baicalensis* without significant tissue destruction (Świątły-Błaszkiewicz *et al.*, 2025). These techniques enable the detection of seasonal variations in bioactive compounds while maintaining plant viability. Complementary approaches including nuclear magnetic resonance (NMR) spectroscopy and mass spectrometry (MS) provide detailed metabolomic data from minimal sample masses, significantly reducing the ecological impact of phytochemical research (Razzaq *et al.*, 2019).

9.3.2 Field-based technologies and sampling methodologies

The development of field-deployable technologies represents a crucial advancement for practical implementation. The RApid Metabolome Extraction and Storage (RAMES) system enables on-site metabolite extraction and stabilization from tiny tissue samples, facilitating comprehensive metabolomic library development from remote ecosystems with minimal population impact (Skubel *et al.*, 2018). This approach is particularly valuable for protected areas where repeated sampling is necessary for monitoring. The entire workflow requires careful standardization, including controlled collection timing and sample pooling to account for diurnal and developmental metabolic variations, ensuring data reproducibility while maintaining conservation priorities (Karakas *et al.*, 2023).

9.3.3 Applications in sustainable harvest management

Non-destructive metabolomics enables evidence-based harvest management through temporal monitoring of bioactive compounds. Research on endangered species such as *S. chinensis* has revealed significant seasonal fluctuations in medicinal metabolites, allowing managers to identify optimal harvest periods that maximize yield while minimizing ecological impact (Tong *et al.*, 2025). This temporal profiling provides scientific basis for establishing sustainable yield limits that prevent population decline. Additionally, the integration of metabolomic data with morphological and ecological parameters enables the

Table 9.3. Applications and challenges of non-destructive metabolomics in sustainable wild harvesting. Author's own table.

Aspect	Technology/method	Key application	Benefits	Limitations/challenges	Citations
Field-based sampling	RAMES	Enables on-site metabolite stabilization from tiny tissue samples (e.g. single leaf punches) in remote areas.	Minimal impact on plant viability; allows for longitudinal monitoring of individual plants.	Requires immediate field processing; limited by sample size for deep metabolomic coverage.	(Skubel *et al.*, 2018)
Non-/minimally invasive analysis	ATR-FTIR, ED-XRF spectroscopy	Provides chemical profiles directly from intact plant surfaces or very small samples.	Truly non-destructive or minimally invasive; rapid and portable options available.	Lower sensitivity and metabolite coverage compared to MS; limited to abundant compounds.	(Świątły-Błaszkiewicz *et al.*, 2025)
High-resolution metabolite profiling	LC-MS, NMR (with minimal sampling)	Detailed characterization of complex metabolite mixtures from small, renewable tissue samples.	High sensitivity and ability to identify novel bioactive compounds.	Still requires destructive sampling, albeit minimal; expensive equipment and expertise needed.	(Razzaq *et al.*, 2019)
Temporal harvest optimization	Seasonal metabolite monitoring	Tracks changes in key bioactive compound levels (e.g. in *S. chinensis*) over time.	Identifies optimal harvest windows to maximize yield and minimize harvest frequency/impact.	Metabolite levels are highly dynamic and influenced by micro-environmental factors.	(Tong *et al.*, 2025)
Quality control & authentication	Metabolomics + DNA barcoding	Combines chemical and genetic fingerprints to authenticate species and detect adulterants in trade.	Protects both consumers and wild populations from misidentification and illegal trade.	Requires comprehensive reference databases for both metabolites and DNA sequences.	(Sánchez *et al.*, 2020)

Precision conservation	Integrated genomic-metabolomic analysis	Identifies populations with unique, high-value chemotypes (e.g. *G. uralensis*) for targeted protection.	Enables conservation of both genetic diversity and specific chemical traits.	Complex data integration; requires significant bioinformatics capacity.	(Zhong *et al.*, 2022)
Data standardization & analysis	ML, AI	Automates metabolite annotation and models metabolic responses to environmental stressors.	Unlocks patterns from complex datasets; predicts plant resilience and compound yield.	Lack of standardized protocols and shared libraries hinders reproducibility and comparison.	(Tsugawa *et al.*, 2021; Shen *et al.*, 2023)
Future field monitoring	Nano/micro-sensors, spatial metabolomics	Aims for real-time, *in planta* metabolite sensing and mapping of compound distribution in intact tissues.	Potential for continuous, truly non-destructive monitoring of plant chemical status.	Largely in developmental stages; challenges with sensor stability, specificity, and deployment.	(Ang and Lew, 2022; Ali *et al.*, 2025)
Accessibility & equity	Cost and technical expertise	Implementing advanced metabolomics in resource-limited regions where medicinal plants are often harvested.	Empowers local communities with data for sustainable management.	High cost of equipment and need for specialized training create significant barriers to global use.	(Karakas *et al.*, 2023; Hao *et al.*, 2025b)

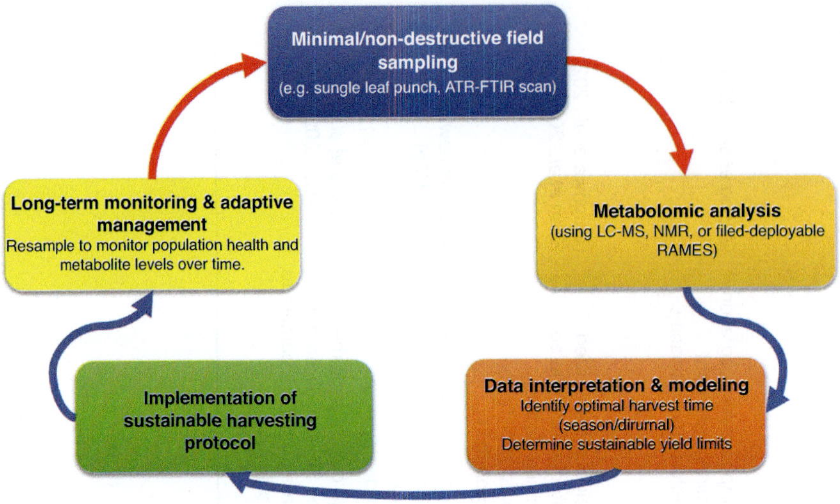

Fig. 9.3. Cycle of non-destructive metabolomics for sustainable wild harvesting. (Author's own figure.)

development of comprehensive conservation strategies that address both chemical resource preservation and species survival (Table 9.3).

9.3.4 Authentication and multi-omics integration

The combination of metabolomic profiling with DNA-based authentication creates a powerful tool for conservation quality control. This integrated approach ensures correct species identification in field collections and detects market adulterants, protecting both consumers and wild populations from misidentification consequences (Sánchez *et al.*, 2020). When expanded to multi-omics frameworks, metabolomics contributes to systems-level understanding of plant physiology. Studies on medicinal species like *Glycyrrhiza uralensis* demonstrate how integrated genomic-metabolomic analyses can elucidate biosynthetic pathways and environmental adaptation mechanisms, enabling identification of populations with unique chemotypes for targeted conservation (Zhong *et al.*, 2022).

Despite its potential, non-destructive metabolomics faces significant implementation barriers. The absence of standardized protocols for cross-species applications and technical challenges in metabolite identification from complex spectra limit comparative analyses and database development (Shen *et al.*, 2023). Furthermore, the high cost and technical requirements of field-deployable equipment present accessibility challenges for conservation programs in developing regions, where many medicinal species face the greatest harvest pressures (Hao *et al.*, 2025b). There remains a critical need for validation studies comparing non-destructive methods with conventional approaches across diverse plant families and growth forms.

Emerging technologies offer promising solutions to current limitations. Micro- and nanoengineered sensing devices are being developed for real-time, *in planta* metabolite monitoring, potentially revolutionizing field data collection (Ali *et al.*, 2025). Computational approaches using ML are increasingly employed for automated metabolite annotation and predictive modeling of metabolic responses to environmental factors (Tsugawa *et al.*, 2021). Spatial metabolomics technologies, which enable the mapping of compound distributions within intact tissues, provide new opportunities for understanding plant biochemical ecology without destructive sampling (Ang and Lew, 2022).

Non-destructive metabolomics represents a paradigm shift in sustainable wild plant management, offering sophisticated tools to balance conservation with resource use. As technological innovations address current challenges, these approaches will increasingly enable precise, adaptive management of medicinal plant populations. The continued development of accessible field technologies and standardized methodologies will be crucial for global implementation, ensuring that metabolomic insights can effectively contribute to preserving both biological and chemical diversity in threatened plant species worldwide (Table 9.3 and Fig. 9.3).

9.4 Bioprospecting Ethics: Omics in Benefit-sharing with Indigenous Communities

The rapid advancement of omics technologies—including genomics, metabolomics, and proteomics—has created new frontiers in bioprospecting while simultaneously raising critical ethical questions about benefit-sharing with Indigenous communities. These powerful analytical tools can validate traditional knowledge by identifying active compounds and genetic markers in biologically significant species, yet they also create potential pathways for the appropriation of Indigenous knowledge without adequate compensation or recognition. The case of the San people and *Hoodia gordonii* represents a pivotal moment in this ongoing dialogue, illustrating both the potential for equitable partnerships and the persistent challenges in achieving truly ethical bioprospecting frameworks (Wynberg, 2023). This case serves as a crucial reference point for understanding how omics research can either perpetuate colonial patterns of resource extraction or help establish new paradigms of respectful collaboration and fair benefit distribution (Table 9.4 and Fig. 9.4).

9.4.1 The *H. gordonii* case: Traditional knowledge and scientific validation

H. gordonii, a succulent plant native to the Kalahari Desert, has been used for generations by the San people as an appetite and thirst suppressant during long hunting expeditions. This traditional knowledge attracted scientific interest in the 1990s when the South African Council for Scientific and Industrial Research (CSIR) isolated the plant's active compound (P57) and patented it without initial consultation with or compensation for the San community (Amusan, 2016). The subsequent application of metabolomic technologies to validate and refine the understanding of *Hoodia*'s biochemical properties created a situation where advanced scientific methods were being used to study—and potentially commercialize—knowledge that had been developed and preserved by Indigenous peoples without their meaningful participation in the research process or equitable sharing of the potential benefits (Van Heerden, 2008; Amusan, 2016). This scenario exemplifies the ethical tensions that arise when modern scientific capabilities interface with traditional knowledge systems that have historically been excluded from formal intellectual property protections and scientific recognition.

9.4.2 From conflict to agreement: The San-CSIR benefit-sharing model

Following international pressure and advocacy, the CSIR and the San people reached a landmark benefit-sharing agreement in 2003, which acknowledged the San as holders of traditional knowledge related to *Hoodia* and entitled them to a percentage of royalties from commercial products derived from the plant (Amusan, 2016). This agreement was widely celebrated as a progressive model for recognizing Indigenous contributions to drug discovery and establishing mechanisms for financial compensation. However, critical analysis reveals significant limitations in both the process and outcomes of this arrangement. The San's perspectives on the agreement highlighted how definitions of fairness were largely shaped by more powerful commercial and scientific actors, failing to fully address historical power imbalances or ensure truly equitable participation in decision-making processes (Vermeylen, 2007). Furthermore,

Table 9.4. Ethical dimensions of omics in bioprospecting and benefit-sharing with Indigenous communities. Author's own table.

Ethical dimension	Key concepts & frameworks	Applications & case studies	Challenges & barriers	Future directions & solutions	Citations
Recognition & validation	Using omics to scientifically validate traditional knowledge (TK).	Metabolomics identified the active compound (P57) in *H. gordonii*, confirming San people's traditional use.	Risk of legitimizing knowledge for commercial use without recognizing its Indigenous origins, leading to biopiracy.	Use omics data to strengthen legal claims of prior art and support Indigenous intellectual property (IP) rights.	(Amusan, 2016; Wynberg, 2023)
Consent & governance	Free, prior, and informed consent (FPIC); community-led governance.	The San-*Hoodia* case initially proceeded without FPIC, leading to conflict and a later, albeit imperfect, agreement.	Power imbalances; FPIC is often treated as a one-time signature rather than an ongoing process.	Implement the CARE Principles; establish community review boards for project oversight.	(Vermeylen, 2007; Carroll *et al*., 2022; Astolfi *et al*., 2025)
Benefit-sharing models	Monetary and non-monetary benefit-sharing as per the Nagoya Protocol.	The 2003 San-CSIR agreement promised royalties, but commercial failures limited financial benefits.	Over-reliance on monetary benefits from market success; lack of immediate, guaranteed non-monetary benefits.	Develop agreements with upfront benefits (e.g. capacity building, infrastructure) regardless of commercial outcome.	(Amusan, 2016; Thaldar and Shozi, 2023)
Digital biopiracy	DSI; data sovereignty.	Omics data from genetic resources can be shared globally in databases, circumventing ABS agreements on physical material.	The Nagoya Protocol does not clearly regulate DSI, creating a major legal loophole for the avoidance of benefit-sharing.	Develop multilateral systems for DSI benefit-sharing; implement digital tracking and metadata tagging for provenance.	(Wynberg, 2023; Astolfi *et al*., 2025)
Legal & structural barriers	IP laws favoring individual/ institutional ownership.	Western IP systems struggle to protect communally held, intergenerational TK, as seen in the *Hoodia* patent dispute.	Legal systems are misaligned with Indigenous worldviews, creating inherent inequity in negotiations.	Promote *sui generis* IP systems, geographical indications, and contractual agreements that recognize collective rights.	(Foster, 2016; Thaldar and Shozi, 2023)

Continued

Table 9.4. Continued

Ethical dimension	Key concepts & frameworks	Applications & case studies	Challenges & barriers	Future directions & solutions	Citations
Equitable partnerships	Co-design, co-ownership, and capacity building.	Moving from a model where communities are "subjects" to one where they are "partners" in omics research.	Lack of trust from historical exploitation; disparities in technical and legal capacity.	Fund and support community-led research priorities; invest in local omics infrastructure and training.	(Claw *et al*., 2018; Soares *et al*., 2023)
Data sovereignty & ethics	Indigenous data sovereignty; FAIR and CARE principles.	Ensuring Indigenous communities control how their associated genetic and TK data is stored, used, and shared.	Open-access data mandates (FAIR) can conflict with Indigenous rights to control and restrict data (CARE).	Integrate CARE Principles with FAIR; develop data management plans that respect community-defined restrictions.	(Claw *et al*., 2018; Carroll *et al*., 2022)

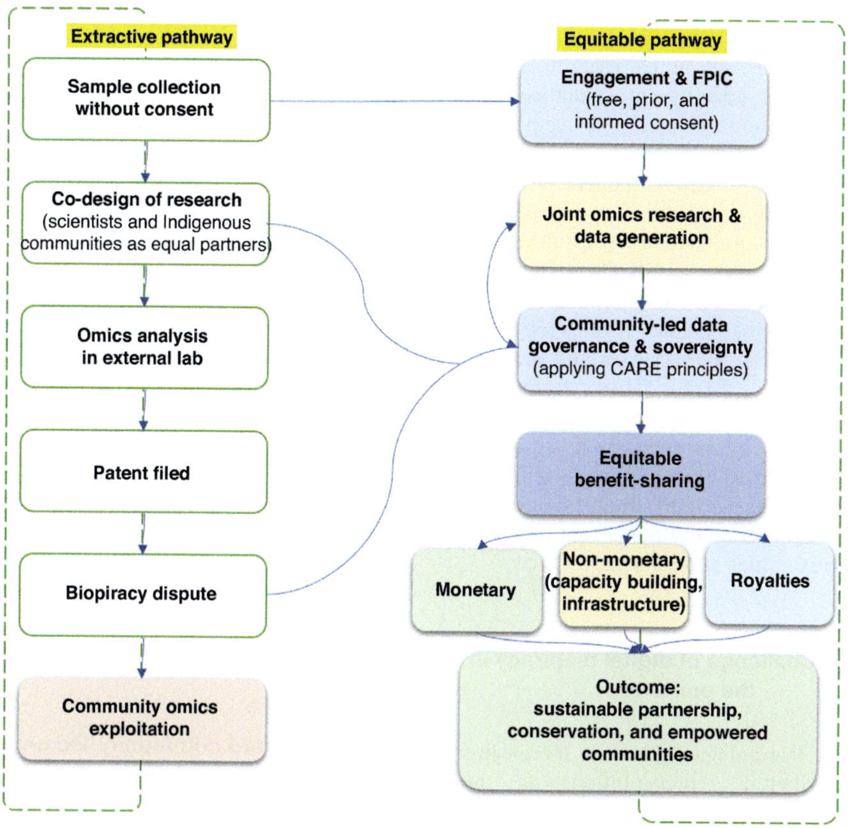

Fig. 9.4. Pathway from extractive to equitable omics bioprospecting. (Author's own figure.)

the agreement faced practical challenges as *Hoodia* products encountered regulatory hurdles and market limitations, ultimately generating fewer financial benefits than initially anticipated, thereby demonstrating the vulnerabilities of benefit-sharing models that depend heavily on commercial success.

9.4.3 Ethical frameworks for omics research in Indigenous contexts

The *Hoodia* case and similar bioprospecting initiatives have stimulated the development of more robust ethical frameworks for omics research involving Indigenous knowledge and biological resources. Central to these frameworks is the principle of free, prior, and informed consent (FPIC), which ensures that Indigenous communities maintain the right to approve or decline research and commercialization projects affecting their resources and knowledge systems (Astolfi *et al.*, 2025). Beyond FPIC, ethical omics research requires community-led governance structures that position Indigenous peoples as active partners rather than merely research subjects. The CARE Principles (collective benefit, authority to control, responsibility, and ethics) have emerged as a complementary framework to the FAIR principles, specifically designed to ensure that genomic and related research delivers collective benefits while recognizing Indigenous authority over data and resources (Carroll *et al.*, 2022). These frameworks emphasize that benefit-sharing must extend beyond monetary compensation to include capacity building, infrastructure development, knowledge exchange, and support for environmental conservation initiatives that align with community values and priorities.

9.4.4 Challenge of digital biopiracy in the omics era

As omics technologies generate increasingly detailed digital representations of biological resources, new ethical challenges have emerged around what has been termed “digital biopiracy” or “digital sequence information.” This concept refers to the extraction, digitization, and potential commercialization of genetic and biochemical data without adequate benefit-sharing arrangements with the source communities (Astolfi *et al.*, 2025). The very nature of omics data—which can be easily stored, shared, and analyzed digitally—creates separation between the physical biological material and the valuable information derived from it, potentially bypassing ABS regulations that typically govern physical biological resources. This technological evolution has raised urgent questions about whether existing legal frameworks, such as the Nagoya Protocol, adequately address the complexities of digital sequence information (DSI) and its implications for Indigenous rights and benefit-sharing (Wynberg, 2023).

9.4.5 Legal and structural barriers to equitable benefit-sharing

Despite international frameworks supporting benefit-sharing, significant legal and structural barriers often impede implementation. In the South African context, research has questioned whether benefit-sharing with research participants is even lawful under current legislation, creating regulatory uncertainty that can disadvantage Indigenous communities (Thaldar and Shozi, 2023). More broadly, intellectual property regimes typically prioritize individual or institutional ownership rather than collective rights, creating fundamental incompatibilities with Indigenous knowledge systems that are often held communally and transmitted intergenerationally (Foster, 2016). These structural inequities are compounded by power imbalances in negotiating processes, where Indigenous communities may lack the legal expertise, financial resources, and scientific capacity to engage as equal partners with well-resourced research institutions and corporations.

9.4.6 Toward community-led omics research and equitable partnerships

Moving beyond extractive research models requires a fundamental reorientation toward community-led omics research that centers Indigenous sovereignty and self-determination. This approach involves establishing Indigenous research governance structures, such as community review boards and research agreements that clearly define data ownership, access conditions, and benefit-sharing mechanisms before research commences (Claw *et al.*, 2018). Successful models emphasize long-term partnerships built on trust and mutual respect, where omics research addresses questions prioritized by Indigenous communities themselves and contributes to community-defined goals for health, economic development, and cultural preservation (Soares *et al.*, 2023). These partnerships recognize that Indigenous communities are not merely sources of data but essential collaborators with expertise in their own right, including deep ecological knowledge and understanding of traditional uses of biological resources.

9.4.7 Integrating ethics and omics in future bioprospecting

The integration of omics technologies into bioprospecting presents both ethical challenges and opportunities for establishing more equitable relationships between research institutions and Indigenous communities. The *H. gordonii* case illustrates that while benefit-sharing agreements represent important steps forward, they often fall short of addressing deeper structural inequities or ensuring meaningful community control over research processes and outcomes. Future ethical omics research must prioritize Indigenous data sovereignty, implement robust governance structures that recognize Indigenous authority, and ensure that benefits align with community-defined needs and values. As omics technologies continue to advance in sophistication and accessibility, maintaining focus on these ethical foundations will be essential for ensuring that scientific progress does not come at the cost of perpetuating historical injustices against Indigenous peoples. Instead, ethically conducted omics research has the potential to create new models of collaboration that respect Indigenous rights while contributing to scientific knowledge and therapeutic development.

9.5 Conservation Omics: Cryobanking and DNA Barcoding of Medicinal Germplasm

Medicinal plants represent an invaluable resource for global health care, with approximately 75% of the world's population relying on plant-based remedies and over 21,000 species documented for their therapeutic properties. However, this biological treasure faces unprecedented threats from overharvesting, habitat destruction, climate change, and unsustainable agricultural practices. The genetic diversity of rare medicinal species, such as *Nardostachys jatamansi*, is particularly vulnerable, with many taxa facing extinction before their full pharmacological potential can be understood or utilized. Conservation omics—integrating cryobanking and DNA barcoding technologies—has emerged as a critical approach for preserving these genetic resources against accelerating biodiversity loss (Zeng *et al.*, 2024). This integrated framework provides scientific tools not only for safeguarding genetic diversity but also for ensuring the authentic identification and sustainable utilization of medicinal species for future generations (Table 9.5 and Fig. 9.5).

9.5.1 Cryobanking: Principles and technical foundations

Cryobanking operates on the principle of storing plant germplasm—including seeds, embryos, shoot tips, or pollen—at ultra-low temperatures, typically in liquid nitrogen at −196°C. At these temperatures, all metabolic processes effectively cease, allowing for the long-term preservation of genetic material without significant deterioration or genetic drift. This approach is especially valuable for rare, endangered, or recalcitrant species that do not produce seeds suitable for conventional seed banking (Nagel *et al.*, 2024). The success of cryopreservation depends critically on species-specific protocols that account for variations in tissue type, moisture content, and sensitivity to freezing damage. Orthodox seeds, which tolerate dehydration, can often be preserved through direct freezing, while recalcitrant seeds and vegetative tissues typically require more

Table 9.5. Conservation omics: Integrating cryobanking and DNA barcoding for medicinal germplasm preservation. Author's own table.

Conservation component	Technology/method	Primary application & value	Key challenges	Future directions	Citations
Long-term germplasm preservation	Cryobanking (seeds, embryos, shoot tips at −196°C)	Provides an "insurance policy" against extinction for rare species (e.g. *N. jatamansi*); halts genetic drift.	Lack of optimized protocols for recalcitrant seeds and vegetatively propagated species; risk of somaclonal variation.	Develop universal vitrification protocols; explore cryobionics to understand molecular stress responses.	(El Merzougui *et al*., 2023; Nagel *et al*., 2024; Zeng *et al*., 2024)
Genetic fidelity assurance	Molecular markers (RAPD, ISSR, SSR, AFLP)	Post-thaw verification of genetic stability in regenerated plants (e.g. *Gentiana kurroo*, *Dioscorea deltoidea*).	Some protocols may induce genetic or epigenetic changes not detected by standard markers.	Implement epigenetic stability monitoring (DNA methylation) as a standard check.	(Mujib *et al*., 2022; Sharma *et al*., 2022; Aryakia, 2025)
Species authentication & quality control	DNA barcoding (ITS, *matK*, *rbcL*)	Authenticates germplasm before cryopreservation and detects adulteration in herbal trade; crucial for regulatory compliance.	Incomplete reference libraries for many medicinal species, especially in biodiverse regions.	Expand barcode libraries using high-throughput sequencing; promote data sharing via global networks.	(Cahyaningsih *et al*., 2022; Zhu *et al*., 2022; Chen *et al*., 2023)
Conservation prioritization & planning	DNA barcoding for phylogenetics	Identifies evolutionarily significant units (ESUs) and cryptic diversity to guide targeted *in situ* and *ex situ* efforts.	Difficulty in translating genetic data into immediate, on-the-ground conservation action.	Integrate barcoding data with GIS (geographic information system) and ecological modeling for spatial conservation planning.	(Hernández-Amasifuen *et al*., 2025; Jiamahate *et al*., 2025)

Integrated workflow for holistic conservation	Cryobanking + DNA barcoding	Creates a verified, traceable pipeline from field collection to long-term storage and regeneration.	High cost and expertise required for integrated workflow; logistical hurdles in remote areas.	Develop portable, low-cost barcoding tools for field use; establish international cryobank networks for resource sharing.	(Kaviani and Kulus, 2022; Parasher *et al.*, 2024; Aryakia, 2025)
Biochemical fidelity check	Metabolomic profiling of cryo-derived plants	Confirms post-thaw plants maintain production of key medicinal compounds (e.g. vinca alkaloids in *Catharanthus roseus*).	Metabolite levels can be influenced by *in vitro* conditions and not just genetic stability.	Standardize metabolomic screening as part of the regeneration and quality control process.	(Mujib *et al.*, 2022)
Protocol development & optimization	Species-specific cryopreservation	Tailors techniques (e.g. vitrification, encapsulation) to the specific biology of each medicinal plant.	Process is time-consuming, expensive, and not scalable for thousands of endangered species.	Utilize AI and ML to predict optimal cryopreservation parameters from plant traits.	(Roque-Borda *et al.*, 2021; Nits *et al.*, 2024)

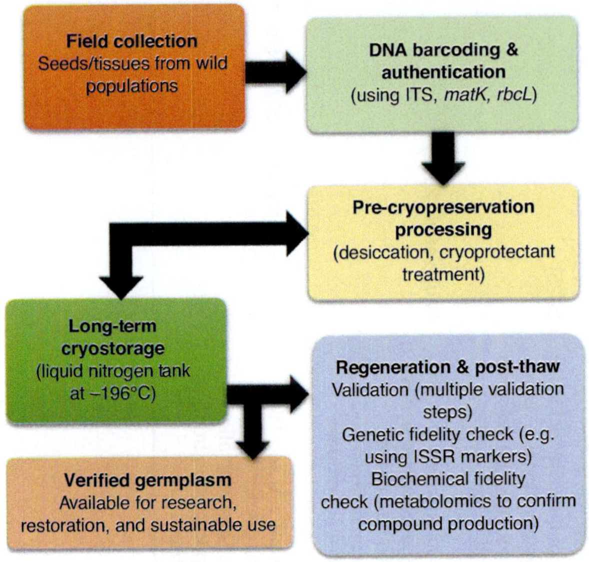

Fig. 9.5. Integrated pipeline for cryobanking and DNA barcoding of medicinal germplasm. (Author's own figure.)

complex approaches such as vitrification or encapsulation-dehydration to prevent ice crystal formation that would otherwise damage cellular structures (Fig. 9.5; Roque-Borda *et al.*, 2021).

9.5.2 Protocol optimization and genetic stability assessment

The development of effective cryopreservation protocols requires meticulous optimization of multiple variables, including preculture conditions, cryoprotectant composition, cooling rates, and thawing procedures. Recent research has demonstrated that under optimized conditions, cryopreserved germplasm can retain high viability and genetic fidelity, as verified through molecular markers including RAPD (random amplified polymorphic DNA), ISSR, SSR, and AFLP (Aryakia, 2025). Studies on critically endangered medicinal plants such as *G. kurroo* and *D. deltoidea* have achieved regrowth rates exceeding 70% post-cryopreservation, with genetic stability assessments confirming the maintenance of original genetic profiles across multiple regeneration cycles (Sharma *et al.*, 2021, 2022). However, some species exhibit reduced germination rates or somaclonal variation after extended storage, highlighting the need for continued protocol refinement and the development of species-specific recovery strategies (Nits *et al.*, 2024).

9.5.3 DNA barcoding: Technology and authentication applications

DNA barcoding utilizes short, standardized genetic markers from specific genome regions to enable accurate species identification and authentication. For plants, the most commonly used barcode regions include internal transcribed spacer (ITS), maturase K (matK), ribulose-bisphosphate carboxylase (rbcL), and tRNA-Leu (trnL) (Chen *et al.*, 2023). This technology has revolutionized the authentication of medicinal materials by providing a reliable method to detect adulteration and substitution in herbal products, which represents a significant challenge in quality control and safety assurance. The construction of comprehensive DNA barcode reference libraries has been particularly impactful, with major pharmacopeias now incorporating barcoding standards that cover over 95% of commonly traded medicinal materials (Zhu *et al.*, 2022). These reference databases enable regulatory agencies, researchers, and industry stakeholders to verify the taxonomic identity of medicinal plants throughout the supply chain, from wild collection to finished products.

9.5.4 Conservation applications of DNA barcoding

Beyond authentication, DNA barcoding serves crucial functions in conservation planning and prioritization. By revealing cryptic diversity and population structure, barcoding data can identify evolutionarily significant units and genetically distinct populations that warrant targeted conservation efforts (Jiamahate *et al.*, 2025). This application is particularly valuable for rare medicinal species with limited distribution or those facing high exploitation pressure. For example, barcoding studies of endangered *Solanum corymbosum* in Peru have not only confirmed taxonomic identity but also supported subsequent *in vitro* propagation and reintroduction programs (Hernández-Amasifuen *et al.*, 2025). Large-scale barcoding initiatives across biodiversity hotspots have enabled the systematic assessment of conservation status for numerous medicinal plants, facilitating data-driven decision making for resource allocation and protection measures (Cahyaningsih *et al.*, 2022).

9.5.5 Integration of cryobanking and DNA barcoding in conservation programs

The synergy between cryobanking and DNA barcoding creates a powerful framework for holistic conservation of medicinal germplasm. DNA barcoding ensures the correct taxonomic identification and genetic characterization of accessions before they enter cryostorage, preventing the misidentification that could compromise the value of cryobanked collections (Aryakia, 2025). Post-thaw molecular analysis further verifies the genetic integrity of regenerated plants, confirming that cryopreservation has not induced significant genetic alterations (Kaviani and Kulus, 2022). This integrated approach is exemplified by successful conservation programs for endangered medicinal species such as *Acorus calamus*, where barcoding authenticated source material before cryopreservation and molecular markers confirmed genetic stability after regeneration (Parasher *et al.*, 2024). Similarly, studies on *C. roseus* demonstrated that cryo-derived plants maintained normal genome size and production of valuable vinca alkaloids, confirming both genetic and biochemical fidelity after preservation (Mujib *et al.*, 2022).

Despite significant advances, the implementation of conservation omics faces several substantial challenges. Many rare medicinal species lack optimized cryopreservation protocols, particularly those with recalcitrant seeds or complex tissue structures that are difficult to preserve (El Merzougui *et al.*, 2023). Technical barriers remain especially pronounced for geophytes and vegetatively propagated species that cannot be preserved through conventional seed banking approaches. Similarly, DNA barcoding efforts are hampered by incomplete reference libraries, with many medicinal species from biodiverse regions remaining poorly represented in public databases (Zeng *et al.*, 2024). Practical constraints include the high costs of establishing and maintaining cryobank infrastructure, the technical expertise required for protocol development, and the logistical challenges of collecting germplasm from remote or politically unstable regions where many endangered medicinal species are found.

The future of conservation omics lies in the integration of emerging technologies that enhance both preservation and analysis capabilities. High-throughput sequencing approaches are enabling more comprehensive genetic characterization of cryobanked collections, moving beyond traditional barcoding regions to encompass whole plastid genomes and nuclear markers that provide richer phylogenetic resolution (Sarwat and Yamdagni, 2016). The developing field of cryobionics explores the molecular responses to cryopreservation stress, seeking to identify biomarkers that predict preservation success and guide protocol optimization (Kaviani and Kulus, 2022). Epigenetic stability monitoring represents another frontier, assessing whether cryopreservation induces heritable changes in DNA methylation patterns that might affect gene expression in regenerated plants (El Merzougui *et al.*, 2023).

International collaboration through networks such as the International Society for Biological and Environmental Repositories and the Millennium Seed Bank Partnership is crucial for standardizing methods, sharing resources, and building capacity in biodiversity-rich but resource-limited regions.

Conservation omics, through the strategic integration of cryobanking and DNA barcoding, provides robust scientific foundations for preserving the genetic diversity of rare medicinal species. These complementary approaches enable not only the long-term safeguarding of genetic resources but also their accurate identification, quality control, and sustainable utilization. As threats to medicinal plant diversity intensify through habitat loss, climate change, and overexploitation, the implementation of comprehensive conservation omics programs becomes increasingly urgent. Through continued technological innovation, international collaboration, and capacity building, these approaches offer hope for maintaining the rich tapestry of medicinal plant diversity that sustains both global healthcare systems and ecosystem functioning. The successful conservation of species such as *N. jatamansi* and other threatened medicinal plants will depend on our ability to leverage these powerful omics tools within holistic conservation strategies that integrate both *ex situ* and *in situ* approaches (Table 9.5 and Fig. 9.5).

9.6 Case Study: Neem (*Azadirachta indica*)—Omics-driven Global Access and Equity

The neem tree (*Azadirachta indica*) stands as a powerful symbol in the global discourse on biopiracy, traditional knowledge protection, and equitable benefit-sharing. For centuries, this plant has been integral to agricultural, medicinal, and cultural practices across South Asia and Africa, with its uses documented in ancient Sanskrit texts. However, the late 20th century witnessed a surge in patent claims by multinational corporations on neem-based products, triggering high-profile legal battles that exposed fundamental tensions between Western intellectual property regimes and Indigenous knowledge systems (Gupta *et al.*, 2017). The emergence of omics technologies—genomics and metabolomics—has created new opportunities to address these historical inequities by providing scientific frameworks that can authenticate traditional knowledge, ensure product traceability, and support fair commercialization models that recognize the contributions of source communities (Table 9.6).

9.6.1 Historical context: The Neem biopiracy controversy

The neem biopiracy controversy reached its zenith in the 1990s when over 130 patents were filed on neem-derived products and processes, primarily by European and American entities. The most notable case involved a European patent granted for a method of controlling fungi on plants using neem oil, which was successfully challenged by Indian activists and organizations who demonstrated that this application represented prior art deeply embedded in traditional practices (Gupta *et al.*, 2017). This case highlighted the systemic inadequacies of patent systems in recognizing traditional knowledge and sparked global debates that ultimately influenced the development of international agreements like the Convention on Biological Diversity and the Nagoya Protocol, which established principles for ABS. The neem controversy thus became a landmark in the movement to protect Indigenous knowledge from misappropriation while simultaneously raising complex questions about how to fairly commercialize traditionally derived innovations (Table 9.6).

9.6.2 Genomic characterization of neem diversity and origins

Genomic approaches have provided critical insights into neem's extensive genetic diversity, which forms the foundation for both conservation strategies and equitable commercialization. Recent research characterizing 170 neem germplasm accessions has revealed significant variability in key traits, including oil content, seed weight, azadirachtin concentration, and kernel yield, with oil content ranging from 31.2% to 45.6% across different accessions (Rajarajan

Table 9.6. Omics-driven resolution of the neem biopiracy case and path to equity. Author's own table.

Aspect	Omics contribution	Impact on bioprospecting & equity	Persisting challenges	Future directions	Citations
Historical biopiracy context	N/A (Pre-omics baseline)	Over 130 patents filed on neem-based products without recognition of prior traditional knowledge (TK).	Western IP systems systematically excluded communal, intergenerational TK from recognition.	Use omics data to retroactively challenge unjust patents and inform new ethical frameworks.	(Gupta *et al.*, 2017)
Genetic resource traceability & value assessment	Genomics (characterization of 170 accessions)	Revealed vast diversity in traits (oil content: 31.2–45.6%; azadirachtin levels), enabling traceability to geographic origin.	Prevents "biopiracy by anonymity"; provides scientific basis for claiming uniqueness of regional germplasm.	Create certified supply chains based on genetic identity; use data for geographical indications (GIs).	(Rajarajan *et al.*, 2025)
Traditional knowledge (TK) validation	Metabolomics (LC-MS, NMR)	Identified & quantified 300 + bioactive compounds (e.g. azadirachtin, nimbin), scientifically validating specific traditional uses (e.g. contraceptive).	Transforms TK from "anecdotal" to "evidence-based," strengthening legal standing in IP and ABS disputes.	Build comprehensive "chemotype maps" linked to traditional uses to document prior art.	(Islas *et al.*, 2020; Patil *et al.*, 2021; Wylie and Merrell, 2022)
Legal evidence for patent challenges	Integrated genomics & metabolomics	Provides objective data to prove "prior art" by linking unique chemoprofiles/ genotypes to specific traditional preparations and regions.	Was successfully used to revoke the European neem oil patent. Legal systems are still adapting to this type of evidence.	Standardize the use of omics data as admissible evidence in international patent law and ABS disputes.	(Gupta *et al.*, 2017; Rajarajan *et al.*, 2025)
Equitable commercialization models	Omics for authentication & certification	Enables creation of transparency schemes and certification labels (e.g. "Equitably Sourced Neem") based on verifiable genetic and chemical data.	Power imbalances persist; large corporations can still dominate markets despite certification.	Support community-based enterprises that use omics for quality control and brand storytelling.	(Rajarajan *et al.*, 2025)

Continued

Table 9.6. Continued

Aspect	Omics contribution	Impact on bioprospecting & equity	Persisting challenges	Future directions	Citations
DSI challenge	Genomic & metabolomic data as DSI	The very omics data used for protection can be copied and used commercially without benefit-sharing, creating a new loophole ("digital biopiracy").	The Nagoya Protocol does not clearly regulate non-physical DSI, undermining ABS frameworks.	Develop new multilateral systems for DSI benefit-sharing that are integrated with omics databases.	(Rajarajan *et al*., 2025)
Capacity & sovereignty barriers	Need for local omics capacity	Source countries often lack infrastructure to generate and interpret their own omics data, creating dependency on foreign labs.	Perpetuates inequity; local communities cannot leverage the technology for their own benefit.	Invest in building omics capacity and bioinformatics expertise in source countries and communities.	(Gupta *et al*., 2017)
Paradigm shift	Omics as a bridge	Shifts the role of science from a tool of extraction to a tool for validation, transparency, and justice.	Requires a fundamental rethinking of research partnerships to be co-designed and community-led.	Apply the "neem model" to other high-value medicinal species with histories of appropriation.	(Gupta *et al*., 2017; Rajarajan *et al*., 2025)

et al., 2025). This genetic mapping enables the identification of superior genotypes for breeding programs while simultaneously creating a reference database that can trace commercial neem products back to their geographic origins. Such traceability is essential for implementing benefit-sharing mechanisms that ensure communities receive appropriate recognition and compensation for their stewardship of specific genetic resources. Genomic data also helps document the historical pathways of neem dissemination, providing scientific evidence of traditional knowledge networks that have maintained and distributed this species across generations.

9.6.3 Metabolomic profiling of bioactive compounds and quality control

Metabolomic studies have comprehensively characterized neem's complex phytochemical profile, identifying over 300 bioactive compounds, including limonoids, flavonoids, and triterpenoids with diverse pharmacological activities. Key compounds such as azadirachtin, nimbin, nimbidin, and nimbolide have been documented for their antimicrobial, anti-inflammatory, antimalarial, and contraceptive properties (Islas *et al.*, 2020; Wylie and Merrell, 2022). Metabolomic fingerprinting enables quality control and standardization of neem products, addressing significant issues of adulteration and variable potency that have historically undermined both efficacy and consumer trust. Furthermore, metabolomic analyses have validated specific traditional uses, such as neem's contraceptive applications, by identifying the compounds responsible for these effects and elucidating their mechanisms of action (Patil *et al.*, 2021). This scientific validation strengthens the case for recognizing traditional knowledge within regulatory and intellectual property frameworks.

9.6.4 Omics as tools for resolving biopiracy claims

The integration of genomic and metabolomic data provides powerful evidence for challenging questionable patent claims and supporting legitimate benefit-sharing arrangements. Omics technologies can establish clear links between commercial products and specific traditional knowledge systems by demonstrating chemical profiles unique to traditionally used preparations or genetic markers specific to regions where particular uses originated (Gupta *et al.*, 2017; Rajarajan *et al.*, 2025). In legal contexts, this data can demonstrate prior art by showing that compounds or applications claimed in patents were previously documented in traditional practices, as was successfully argued in the revocation of the European neem oil patent. Additionally, omics-based authentication creates supply chain transparency, allowing regulators and consumers to verify product origins and ensuring that benefit-sharing commitments are honored throughout commercial networks. This evidentiary capacity positions omics as crucial tools for implementing the Nagoya Protocol's provisions regarding access to genetic resources and the fair and equitable sharing of benefits arising from their utilization.

9.6.5 Toward equitable commercialization frameworks

The application of omics data to neem commercialization has facilitated more equitable business models that acknowledge traditional knowledge contributions. By providing scientific documentation of neem's biochemical complexity and genetic diversity, omics research supports the development of geographical indications and other forms of intellectual property that recognize collective rather than individual innovation (Rajarajan *et al.*, 2025). Several initiatives in India have leveraged omics data to establish community-based enterprises that maintain control over neem genetic resources while partnering with ethical commercial entities. These models often include benefit-sharing components such as royalty payments, technology transfer, capacity building, and support for conservation activities that preserve neem biodiversity. The transparency afforded by omics technologies also enables certification schemes that allow consumers to identify products developed through equitable partnerships, creating market incentives for ethical commercialization practices.

9.6.6 Persisting challenges and implementation barriers

Despite these advances, significant challenges remain in translating omics insights into consistently equitable outcomes. Power imbalances between multinational corporations and local communities persist, often manifested in unequal negotiating capacity and legal expertise (Gupta *et al.*, 2017). Many source countries lack the technical infrastructure and scientific capacity to independently conduct omics research, creating dependency on external institutions that may not prioritize local interests. Legal frameworks for protecting traditional knowledge remain fragmented internationally, with uncertain enforcement mechanisms for benefit-sharing agreements. Additionally, the DSI generated through omics research presents new regulatory challenges, as current ABS regimes primarily focus on physical genetic resources rather than the data derived from them. These limitations highlight the need for complementary policy interventions alongside technological solutions.

The neem case study offers important lessons for applying omics technologies to other traditionally used species facing similar biopiracy and equity challenges. Future efforts should focus on building omics capacity in source countries, developing standardized protocols for documenting and protecting traditional knowledge, and creating international databases that respect Indigenous data sovereignty principles (Rajarajan *et al.*, 2025). The integration of omics with other data sources, including ethnobotanical records and historical texts, can create comprehensive digital repositories that preserve traditional knowledge while making it accessible for legitimate research and innovation. For neem specifically, ongoing omics research continues to reveal new bioactive compounds and applications, underscoring the importance of maintaining genetic diversity and supporting the communities that have conserved this species through generations.

The application of omics technologies to neem represents a paradigm shift in addressing historical biopiracy concerns while enabling scientifically grounded, equitable commercialization. By providing transparent, verifiable data on genetic diversity and biochemical composition, omics approaches have helped rebalance power dynamics in the commercialization of traditionally derived products. The neem case demonstrates that modern science need not operate in opposition to traditional knowledge systems but can instead serve as a bridge that validates and protects Indigenous innovations. As omics technologies continue to advance, their thoughtful application to other species with similar histories of appropriation offers promise for creating more just and sustainable models of biodiscovery—models that recognize the intellectual contributions of traditional knowledge holders while ensuring that benefits are shared fairly with source communities (Table 9.6).

9.7 Challenges and Future Perspectives

9.7.1 Omics in prioritizing endangered medicinal plants

The integration of genomics and metabolomics into IUCN Red List assessments represents a transformative approach for evaluating the conservation status of endangered medicinal plants like *Taxus* (yew) and *Coptis* (goldthread). However, significant challenges impede implementation. Data integration remains problematic due to inconsistent formats and lack of standardized protocols across studies, complicating comparative analyses of genetic diversity and specialized metabolite profiles (Wang *et al.*, 2024). The dynamic nature of plant metabolism further complicates assessments, as metabolite concentrations fluctuate with developmental stage, season, and environmental conditions, creating difficulties in establishing stable baseline profiles for threat evaluation (Zhang *et al.*, 2023). Resource limitations present additional barriers, as high-throughput omics requires substantial computational infrastructure and technical expertise that are often unavailable in biodiversity-rich regions where many endangered medicinal species are found (Yang *et al.*, 2023).

Future progress depends on developing robust bioinformatics platforms that standardize data collection and analysis, enabling reliable cross-species comparisons. Advanced modeling approaches incorporating ML and AI show particular promise for predicting metabolic pathways and identifying key regulatory nodes that could serve as conservation priorities (Cembrowska-Lech *et al.*, 2023). Establishing international collaborative networks will be essential for building comprehensive omics datasets that reflect global medicinal plant diversity, while capacity-building initiatives in source countries can address existing technical and resource disparities (Table 9.7; Hao *et al.*, 2025b).

9.7.2 DSI and medicinal plant sovereignty

The emergence of DSI from omics research has created fundamental challenges for existing ABS frameworks. Legal ambiguity represents the primary obstacle, as current international agreements like the Nagoya Protocol lack clear definitions and governance mechanisms for DSI, creating uncertainty about access, ownership, and benefit-sharing obligations (Maia and Bourgeois-Gironde, 2025). This regulatory gap creates tension between the scientific community's need for open data access and provider countries' rights to maintain sovereignty over their genetic resources, potentially enabling new forms of "digital biopiracy" where sequence data is commercialized without appropriate benefit-sharing (Bond and Scott, 2020). Enforcement presents additional complications, as the non-physical nature of DSI allows for global replication and sharing without the physical transfer of materials that traditional ABS frameworks were designed to regulate (Akpoviri *et al.*, 2023).

Future solutions will require evolving international agreements to explicitly include DSI within their scope, with multilateral benefit-sharing mechanisms that recognize the global nature of genetic sequence data (Scholz *et al.*, 2023). Inclusive governance models that engage provider countries and Indigenous communities in DSI policy development are essential for establishing equitable frameworks (Cowell *et al.*, 2022). Technological innovations such as blockchain and digital tracking tools offer potential for monitoring DSI use and ensuring compliance with benefit-sharing agreements, though these approaches require further development and standardization (Table 9.7; Aubry, 2019).

9.7.3 Sustainable wild harvesting: Metabolomics for non-destructive sampling

Non-destructive metabolomic approaches face several implementation challenges that limit their current effectiveness for sustainable wild harvesting. Sampling complexity represents a significant barrier, as these methods require sensitive analytical techniques and careful experimental design to obtain meaningful data from minimal tissue samples without harming plant populations (Sharma and Yadav, 2022). Environmental variability further complicates monitoring efforts, as metabolite levels fluctuate significantly with microhabitat conditions, seasonal changes, and plant developmental stages, making it difficult to establish reliable harvesting guidelines (Wang *et al.*, 2024). Technical barriers are equally problematic, as many regions with high medicinal plant diversity lack access to the advanced metabolomics infrastructure and expertise required for implementation (Zhang *et al.*, 2023).

Future advancements will depend on developing portable technologies such as handheld MS devices that enable real-time, field-based metabolomic monitoring (Sharma and Yadav, 2022). Establishing standardized protocols for sampling and analysis will improve data comparability and reliability across different populations and harvesting regimes (Wang *et al.*, 2024). Most importantly, metabolomic data must be effectively integrated into conservation policy through adaptive management

Table 9.7. Synthesis of challenges and future perspectives in omics-guided conservation and bioprospecting. Author's own table.

Domain	Key challenges	Specific barriers	Future perspectives & solutions	Critical citations
Omics for IUCN prioritization	• Data integration & standardization. • Dynamic metabolite profiles. • Resource limitations in biodiverse regions.	Inconsistent data formats; seasonal metabolic variation; lack of computational infrastructure & expertise.	• Develop standardized bioinformatics platforms. • Apply AI/ML for predictive modeling. • Establish international collaborative networks & capacity building.	(Cembrowska-Lech *et al.*, 2023; Zhang *et al.*, 2023; Wang *et al.*, 2024)
DSI	• Legal ambiguity in ABS frameworks. • Digital biopiracy risks. • Enforcement difficulties.	Nagoya Protocol doesn't clearly regulate DSI; non-physical data easily replicated & shared globally.	• Evolve international agreements for multilateral benefit-sharing. • Implement inclusive DSI governance. • Develop blockchain/DLT (distributed ledger technology) for tracking & transparency.	(Bond and Scott, 2020; Scholz *et al.*, 2023; Maia and Bourgeois-Gironde, 2025)
Non-destructive metabolomics	• Sampling complexity. • Environmental variability. • Technical & infrastructure barriers.	Sensitive techniques required; metabolite fluctuation with microhabitat; lack of field-deployable equipment.	• Develop portable technologies (e.g. handheld MS). • Establish standardized protocols. • Integrate data into adaptive management & certification schemes.	(Sharma and Yadav, 2022; Wang *et al.*, 2024; Hao *et al.*, 2025b)
Bioprospecting ethics	• Traditional knowledge recognition. • Benefit-sharing complexity. • Historical mistrust.	Use of omics without compensation; legal/ bureaucratic hurdles; extractive research legacy.	• Shift to co-design & co-ownership models. • Invest in capacity building. • Develop stronger ethical-legal frameworks for non-monetary benefits.	(Bond and Scott, 2020; Cowell *et al.*, 2022; Maia and Bourgeois-Gironde, 2025)

Conservation omics	• Technical protocol limitations. • Incomplete reference libraries. • Genetic erosion & resource constraints.	Lack of cryopreservation methods for recalcitrant species; incomplete barcode databases; funding limitations.	• Prioritize universal protocol development. • Adopt integrated *in/ex situ* approaches. • Establish global cryobank & barcode networks.	(Yang *et al.*, 2023; Zhang *et al.*, 2023; Wang *et al.*, 2024)
Equity in commercialization (neem case)	• Persistent biopiracy risks. • Regulatory gaps for digital biopiracy. • Capacity disparities.	Omics data used for patents without benefit-sharing; ABS frameworks inadequate; resource imbalances.	• Create transparent patent systems using omics as prior art. • Develop equitable commercialization policies. • Build local R&D capacity in source countries.	(Cowell *et al.*, 2022; Zhang *et al.*, 2023; Wang *et al.*, 2024)
Cross-cutting themes	• Technical: data standardization, protocol development. • Legal: DSI governance, ABS implementation. • Ethical: equity, consent, benefit-sharing. • Resource: infrastructure, capacity, funding.	Interconnected barriers requiring multidisciplinary solutions.	• Holistic approach: integrate technical solutions with ethical frameworks and policy development. • Global collaboration: international networks for data sharing, capacity building, and resource allocation • Community-centered models: prioritize Indigenous sovereignty and local engagement in all initiatives.	(All synthesized from above)

frameworks and certification schemes that directly link biochemical monitoring to sustainable harvesting practices (Table 9.7; Hao *et al.*, 2025b).

9.7.4 Bioprospecting ethics: Omics in benefit-sharing with Indigenous communities

The application of omics technologies in bioprospecting raises critical ethical challenges regarding the recognition and compensation of traditional knowledge. A fundamental issue involves the continued use of Indigenous knowledge without proper acknowledgment or benefit-sharing, as omics data may validate traditional uses while failing to ensure that communities receive appropriate recognition or compensation (Cowell *et al.*, 2022). The complexity of benefit-sharing mechanisms presents additional obstacles, as translating omics discoveries into tangible community benefits is often hampered by legal and bureaucratic hurdles that delay or prevent implementation (Maia and Bourgeois-Gironde, 2025). Historical exploitation has created deep-seated mistrust that limits community engagement in omics research, particularly when communities perceive such research as extractive rather than collaborative (Akpoviri *et al.*, 2023).

Future ethical bioprospecting requires a fundamental shift toward co-design and co-ownership models where Indigenous communities participate as equal partners in research planning, implementation, and data governance (Bond and Scott, 2020). Capacity-building initiatives that invest in local training and infrastructure are essential for empowering communities to engage meaningfully with omics research and benefit from its applications (Wang *et al.*, 2024). Stronger ethical guidelines and legal frameworks must be developed to ensure culturally appropriate benefit-sharing that recognizes both monetary and non-monetary benefits, including knowledge exchange, infrastructure development, and environmental conservation (Table 9.7; Cowell *et al.*, 2022).

9.7.5 Conservation omics: Cryobanking and DNA barcoding of medicinal germplasm

Technical limitations present significant challenges for conservation omics approaches. Many rare medicinal species lack optimized cryopreservation protocols, particularly those with recalcitrant seeds or complex tissue structures that are difficult to preserve (Yang *et al.*, 2023). Simultaneously, DNA barcoding reference libraries remain incomplete for many medicinal plants, limiting the effectiveness of authentication and monitoring efforts (Zhang *et al.*, 2023). Genetic erosion compounds these technical challenges, as overharvesting and habitat loss continue to diminish wild populations faster than conservation efforts can preserve their diversity (Chen *et al.*, 2016). Resource constraints further limit implementation, as cryobanking and barcoding require sustained funding, technical expertise, and infrastructure that are often unavailable in regions with high medicinal plant diversity (Hao *et al.*, 2025b).

Future progress depends on prioritizing protocol development to establish universal cryopreservation and barcoding methods applicable across diverse medicinal species (Wang *et al.*, 2024). Integrated conservation approaches that combine *in situ* protection with *ex situ* preservation and omics-based monitoring will maximize genetic resource preservation (Zhang *et al.*, 2023). Establishing global networks for cryobanking and barcode data sharing will facilitate coordinated conservation efforts and prevent duplication of limited resources (Table 9.7; Yang *et al.*, 2023).

9.7.6 Case study: Neem (*A. indica*)—Omics-driven global access and equity

The neem case illustrates persistent challenges in achieving equitable benefit-sharing from omics research. Biopiracy risks remain substantial, as genomic and metabolomic data can be used to support patent claims without ensuring fair recognition or benefit-sharing with source countries and traditional knowledge holders (Zhang *et al.*,

2023). Regulatory gaps in existing ABS frameworks struggle to address "digital biopiracy" and ensure equitable commercialization of omics-derived products (Cowell *et al.*, 2022). Capacity disparities create additional inequities, as source countries often lack the resources to leverage omics data for their own benefit, perpetuating historical power imbalances in bioprospecting relationships (Akpoviri *et al.*, 2023).

Future solutions must include transparent patent systems that utilize omics data to support prior art claims and prevent unjust patents based on traditional knowledge (Bond and Scott, 2020). Policies must be developed to ensure that omics-enabled products deliver tangible benefits to source countries and communities through equitable commercialization models (Maia and Bourgeois-Gironde, 2025). Most importantly, supporting local research and development capacity will help source countries move up the value chain and capture greater benefits from their genetic resources and associated traditional knowledge (Wang *et al.*, 2024).

The integration of omics technologies into medicinal plant conservation and bioprospecting faces multifaceted challenges spanning technical, legal, and ethical domains. Addressing these challenges requires coordinated efforts to develop standardized protocols, establish robust legal frameworks, implement inclusive benefit-sharing mechanisms, and build global collaborative networks. By addressing these issues systematically, the scientific community can ensure that omics advances translate into sustainable and equitable outcomes that benefit both biodiversity conservation and human well-being, while respecting the rights and contributions of Indigenous peoples and local communities who have stewarded these resources for generations (Table 9.7).

9.8 Conclusion

The journey through omics-guided conservation and ethical bioprospecting reveals a field at a critical inflection point. The powerful suite of omics technologies provides an unparalleled capacity to understand, value, and preserve medicinal plants, moving conservation science beyond mere population counts to a sophisticated appreciation of genetic potential and biochemical wealth. From prioritizing endangered species for IUCN assessments to enabling sustainable harvests and safeguarding germplasm in cryobanks, omics offers tangible solutions to pressing conservation challenges. However, this technological prowess must be guided by a steadfast ethical compass. The cases of *Hoodia* and neem serve as potent reminders that scientific advancement without equity risks perpetuating historical injustices and biopiracy. The path forward, therefore, demands an integrated approach where technical innovation walks hand in hand with ethical commitment. This entails developing standardized omics protocols and robust bioinformatics, while also resolving the legal ambiguities surrounding DSI and implementing inclusive, community-led governance models. Building capacity in source countries, fostering global collaborations, and ensuring that benefits—both monetary and non-monetary—flow fairly to Indigenous and local communities are not merely ethical imperatives but prerequisites for long-term success. By harmonizing cutting-edge science with deep respect for traditional knowledge and rights, we can forge a future where the immense value of medicinal plants continues to enrich human health, while the biological and cultural systems that sustain them are preserved for generations to come.

Acknowledgments

The authors express their gratitude for the financial support received through the Distinguished High-Level Talents Research Grant from the Guizhou Science and Technology Corporation Platform Talents Fund (Grant nos.: [2017]5733-001 and CK-1130-002), high-level specialists project fund of Zunyi Medical University (1092/2000006)F-1092, the National Natural Science Foundation of China (82373981), and the support provided by Zunyi Medical University, China. Special appreciation is extended to all laboratory colleagues and research staff members for their valuable insights, constructive guidance, and assistance throughout this study.

References

Akpoviri, F.I., Baharum, S.N. and Zainol, Z.A. (2023) Digital sequence information and the access and benefit-sharing obligation of the convention on biological diversity. *Nanoethics* 17(1), 1. DOI: 10.1007/s11569-023-00436-3.

Alami, M.M., Ouyang, Z., Zhang, Y., Shu, S., Yang, G. *et al.* (2022) The current developments in medicinal plant genomics enabled the diversification of secondary metabolites' biosynthesis. *International Journal of Molecular Sciences* 23(24), 15932. DOI: 10.3390/ijms232415932.

Ali, S., Amin, A., Akhtar, M.S. and Zaman, W. (2025) Micro- and nanoengineered devices for rapid chemotaxonomic profiling of medicinal plants. *Nanomaterials (Basel, Switzerland)* 15(12), 899. DOI: 10.3390/nano15120899.

Amusan, L. (2016) Politics of biopiracy: An adventure into Hoodia/Xhoba patenting in Southern Africa. *African Journal of Traditional, Complementary, and Alternative Medicines* 14(1), 103–109. DOI: 10.21010/ajtcam.v14i1.11.

Ang, M.C.-Y. and Lew, T.T.S. (2022) Non-destructive technologies for plant health diagnosis. *Frontiers in Plant Science* 13, 884454. DOI: 10.3389/fpls.2022.884454.

Aryakia, E. (2025) Cryobanking of rare and endangered plant species: Key propagules and genetic stability assessments. *Journal for Nature Conservation* 84(1), 126844. DOI: 10.1016/j.jnc.2025.126844.

Astolfi, M.C.T., Flores, W., Perez, R., Espinoza, U.J., Zimring, T.B. *et al.* (2025) Partnerships with indigenous peoples for an ethical bioeconomy. *Nature Communications* 16(1), 3010. DOI: 10.1038/s41467-025-57935-y.

Aubry, S. (2019) The future of digital sequence information for plant genetic resources for food and agriculture. *Frontiers in Plant Science* 10, 1046. DOI: 10.3389/fpls.2019.01046.

Barbosa, J.C.J., Barbieri, H.B., Fill, T.P. and Crnkovic, C.M. (2025) Avanços em técnicas ômicas na descoberta e no estudo de produtos naturais. *Química Nova*. DOI: 10.21577/0100-4042.20250190.

Bond, M.R. and Scott, D. (2020) Digital biopiracy and the (dis)assembling of the Nagoya Protocol. *Geoforum; Journal of Physical, Human, and Regional Geosciences* 117, 24–32. DOI: 10.1016/j.geoforum.2020.09.001.

Cahyaningsih, R., Compton, L.J., Rahayu, S., Magos Brehm, J. and Maxted, N. (2022) DNA barcoding medicinal plant species from Indonesia. *Plants (Basel, Switzerland)* 11(10), 1375. DOI: 10.3390/plants11101375.

Carroll, S.R., Plevel, R., Jennings, L.L., Garba, I., Sterling, R. *et al.* (2022) Extending the CARE principles from tribal research policies to benefit sharing in genomic research. *Frontiers in Genetics* 13, 1052620. DOI: 10.3389/fgene.2022.1052620.

Ćavar Zeljković, S., De Diego, N., Drašar, L., Nisler, J., Havlíček, L. *et al.* (2024) Comprehensive LC-MS/MS analysis of nitrogen-related plant metabolites. *Journal of Experimental Botany* 75(17), 5390–5411. DOI: 10.1093/jxb/erae129.

Cavers, S., Navarro, C. and Lowe, A.J. (2004) Targeting genetic resource conservation in widespread species: A case study of *Cedrela odorata* L. *Forest Ecology and Management* 197(1–3), 285–294. DOI: 10.1016/j.foreco.2004.05.019.

Cembrowska-Lech, D., Krzemińska, A., Miller, T., Nowakowska, A., Adamski, C. *et al.* (2023) An integrated multi-omics and artificial intelligence framework for advance plant phenotyping in horticulture. *Biology* 12(10), 1298. DOI: 10.3390/biology12101298.

Chele, K.H., Piater, L.A., Van der Hooft, J.J.J. and Tugizimana, F. (2025) Bridging ethnobotanical knowledge and multi-omics approaches for plant-derived natural product discovery. *Metabolites* 15(6), 362. DOI: 10.3390/metabo15060362.

Chen, S., Yin, X., Han, J., Sun, W., Yao, H. *et al.* (2023) DNA barcoding in herbal medicine: Retrospective and prospective. *Journal of Pharmaceutical Analysis* 13(5), 431–441. DOI: 10.1016/j.jpha.2023.03.008.

Chen, S.-L., Yu, H., Luo, H.-M., Wu, Q., Li, C.-F. *et al.* (2016) Conservation and sustainable use of medicinal plants: Problems, progress, and prospects. *Chinese Medicine* 11(1), 37. DOI: 10.1186/s13020-016-0108-7.

Claw, K.G., Anderson, M.Z., Begay, R.L., Tsosie, K.S., Fox, K. *et al.* (2018) A framework for enhancing ethical genomic research with indigenous communities. *Nature Communications* 9(1), 2957. DOI: 10.1038/s41467-018-05188-3.

Cowell, C., Paton, A., Borrell, J.S., Williams, C., Wilkin, P. *et al.* (2022) Uses and benefits of digital sequence information from plant genetic resources: Lessons learnt from botanical collections. *Plants, People, Planet* 4(1), 33–43. DOI: 10.1002/ppp3.10216.

Dalmaris, E., Avramidou, E.V., Xanthopoulou, A. and Aravanopoulos, F.A. (2020) Dataset of targeted metabolite analysis for five taxanes of Hellenic *Taxus baccata* L. populations. *Data* 5(1), 22. DOI: 10.3390/data5010022.

Das, A., Pramanik, R., Chongder, D.B., Bania, P. and Debnath, S. (2025) Integrated molecular strategies for enhanced production of anti-cancer compounds from medicinal plants. In: Chen, J.-T. (ed.) *Biotechnology, Multiple Omics, and Precision Breeding in Medicinal Plants*. CRC Press, Boca Raton, Florida, pp. 270–286.

El Merzougui, S., Benelli, C., El Boullani, R. and Serghini, M.A. (2023) The cryopreservation of medicinal and ornamental geophytes: Application and challenges. *Plants (Basel, Switzerland)* 12(11), 2143. DOI: 10.3390/plants12112143.

Foster, L.A. (2016) Decolonizing patent law: Postcolonial technoscience and indigenous knowledge in South Africa. *Feminist Formations* 28(3), 148–173. DOI: 10.1353/ff.2016.0047.

Guo, M., Lv, H., Chen, H., Dong, S., Zhang, J. *et al.* (2024) Strategies on biosynthesis and production of bioactive compounds in medicinal plants. *Chinese Herbal Medicines* 16(1), 13–26. DOI: 10.1016/j.chmed.2023.01.007.

Gupta, G., Khatoon, S., Singh, R., Gupta, S. and Rai, D. (2024) Exploring bioinformatics opportunities for identification and study of medicinal plants: A comprehensive review. *Biology and Life Sciences*. DOI: 10.20944/preprints202405.0752.v1.

Gupta, S.C., Prasad, S., Tyagi, A.K., Kunnumakkara, A.B. and Aggarwal, B.B. (2017) Neem (*Azadirachta indica*): An Indian traditional panacea with modern molecular basis. *Phytomedicine* 34, 14–20. DOI: 10.1016/j.phymed.2017.07.001.

Gusain, P., Uniyal, D.P. and Joga, R. (2021) Conservation and sustainable use of medicinal plants. In: Egbuna, C., Mishra, A.P. and Goyalm, M.R. (eds) *Preparation of Phytopharmaceuticals for the Management of Disorders*. Academic Press, Cambridge, Massachusetts, pp. 409–427.

Hao, D.-C., Spjut, R.W., Deng, D., He, C.-N. and Yao, R. (2025a) Editorial: Plant metabolites in drug discovery: The prism perspective between plant phylogeny, chemical composition, and medicinal efficacy, volume IV. *Frontiers in Pharmacology* 16, 1666900. DOI: 10.3389/fphar.2025.1666900.

Hao, Y., Zhang, Z., Luo, E., Yang, J. and Wang, S. (2025b) Plant metabolomics: Applications and challenges in the era of multi-omics big data. *aBIOTECH* 6(1), 116–132. DOI: 10.1007/s42994-024-00194-0.

Hernández-Amasifuen, A.D., Pineda-Lázaro, A.J., Delgado-Mera, E., Vega-Herrera, S.S., Fernandes Pereira, C. *et al.* (2025) *In vitro* propagation and DNA barcode analysis of the threatened herb *Solanum corymbosum* from La Joya Desert, Peru. *International Journal of Plant Biology* 16(1), 1–15. DOI: 10.3390/ijpb16010015.

Hussain, S., Sagar, T., Kaur, S., Nipunta, Kapoor, N. *et al.* (2023) Omics approaches to study the biosynthesis of bioactive compounds in medicinal and aromatic plants. In: Kumar, N. and Singh, R.S. (eds) *Biosynthesis of Bioactive Compounds in Medicinal and Aromatic Plants: Manipulation by Conventional and Biotechnological Approaches*. Springer Nature, Cham, Switzerland, pp. 41–64.

Islas, J.F., Acosta, E., G-Buentello, Z., Delgado-Gallegos, J.L., Moreno-Treviño, M.G. *et al.* (2020) An overview of neem (*Azadirachta indica*) and its potential impact on health. *Journal of Functional Foods* 74, 104171. DOI: 10.1016/j.jff.2020.104171.

Jabeen, F., Khursheed, N., Ahmad, R., Rasool, S., Ali, A. *et al.* (2024) Crosstalk of multiomics approaches with medicinal plants of therapeutic importance. In: Ali, A., Rehman, M.U., Ahmad, S.B. and Arafah, A. (eds) *Biological Insights of Multi-Omics Technologies in Human Diseases*. Academic Press, Cambridge, Massachusetts, pp. 21–51.

Jeon, J.Y., Black, A.N., Heenkenda, E.J., Mularo, A.J., Lamka, G.F. *et al.* (2024) Genomic diversity as a key conservation criterion: Proof-of-concept from mammalian whole-genome resequencing data. *Evolutionary Applications* 17(9), e70000. DOI: 10.1111/eva.70000.

Jiamahate, A., Bozorov, T.A., Wang, J., Zhang, J., Zhang, H. *et al.* (2025) Insights from DNA barcodes-based phylogenetic analysis of medicinal plants and estimation of their conservation status: A case study in the Tianshan Wild Forest, China. *Plants (Basel, Switzerland)* 14(1), 99. DOI: 10.3390/plants14010099.

Kapoor, B., Kumar, A. and Kumar, P. (2021) Transcriptome repository of North-Western Himalayan endangered medicinal herbs: A paramount approach illuminating molecular perspective of phytoactive molecules and secondary metabolism. *Molecular Genetics and Genomics* 296(6), 1177–1202. DOI: 10.1007/s00438-021-01821-x.

Karakas, E., Ferrante, P., Schafleitner, R., Giuliano, G., Fernie, A.R. *et al.* (2023) Plant sample collection and shipment for multi-omic analyses and phytosanitary evaluation. *Current Protocols* 3(12), e952. DOI: 10.1002/cpz1.952.

Karalija, E., Macanović, A. and Ibragić, S. (2025) Revisiting traditional medicinal plants: Integrating multi-omics, *in vitro* culture, and elicitation to unlock bioactive potential. *Plants (Basel, Switzerland)* 14(13), 2029. DOI: 10.3390/plants14132029.

Katu, A.H. (2025) The future of medicinal plants: Innovations in research and community health strategies. *Research Output Journal of Public Health and Medicine* 5(2), 75–83. DOI: 10.59298/ROJPHM/2025/527583.

Kaviani, B. and Kulus, D. (2022) Cryopreservation of endangered ornamental plants and fruit crops from tropical and subtropical regions. *Biology* 11(6), 847. DOI: 10.3390/biology11060847.

Kielich, N., Mazur, O., Musidlak, O., Gracz-Bernaciak, J. and Nawrot, R. (2024) Herbgenomics meets papaveraceae: A promising-omics perspective on medicinal plant research. *Briefings in Functional Genomics* 23(5), 579–594. DOI: 10.1093/bfgp/elad050.

Maia, B.G. and Bourgeois-Gironde, S. (2025) Analysis of the Working Group's recommendations and COP-16 decision 16/2 on digital sequence information. *Global Policy* 16(4), 751–761. DOI: 10.1111/1758-5899.70013.

Ma, Q., Wu, G., Li, W., Yuzuak, S., Guan, F. *et al.* (2023) Research advances and perspectives of conservation genomics of endangered plants. DOI: 10.5772/intechopen.112281.

Mohammed, M.A., Elzefzafy, N., El-Khadragy, M.F., Alzahrani, A., Yehia, H.M. *et al.* (2024) Comprehensive tools of alkaloid/volatile compounds-metabolomics and DNA profiles: Bioassay-role-guided differentiation process of six *Annona* sp. grown in Egypt as anticancer therapy. *Pharmaceuticals (Basel, Switzerland)* (1), 103. DOI: 10.3390/ph17010103.

Mujib, A., Fatima, S. and Malik, M.Q. (2022) Cryo-derived plants through embryogenesis showed same levels of vinblastine and vincristine (anticancer) in *Catharanthus roseus* and had normal genome size. *Scientific Reports* 12(1), 16635. DOI: 10.1038/s41598-022-20993-z.

Murmu, S., Sinha, D., Chaurasia, H., Sharma, S., Das, R. *et al.* (2024) A review of artificial intelligence-assisted omics techniques in plant defense: Current trends and future directions. *Frontiers in Plant Science* 15, 1292054. DOI: 10.3389/fpls.2024.1292054.

Nagel, M., Pence, V., Ballesteros, D., Lambardi, M., Popova, E. *et al.* (2024) Plant cryopreservation: Principles, applications, and challenges of banking plant diversity at ultralow temperatures. *Annual Review of Plant Biology* 75(1), 797–824. DOI: 10.1146/annurev-arplant-070623-103551.

Nits, O.S., Sementsova, M.V., Osipova, E.S., Tereshonok, D.V. and Gladkov, E.A. (2024) IPPRAS cryobank for the conservation of orthodox seeds of rare, endangered, medicinal, and ornamental plant species-current research. *Plants (Basel, Switzerland)* 13(10), 1354. DOI: 10.3390/plants13101354.

Parasher, K., Sharma, C., Sharma, S., Shrishti, E.Y. and Mukherjee, P. (2024) Cryopreservation and assessment of genetic fidelity *Acorus calamus* Linn., an endangered medicinal plant. *Cryo Letters* 45(2), 122–133. DOI: 10.54680/fr24210110412.

Patil, S.M., Shirahatti, P.S., VB, C.K., Ramu, R., Prasad, N. *et al.* (2021) *Azadirachta indica* A. *Juss* (neem) as a contraceptive: An evidence-based review on its pharmacological efficiency. *Phytomedicine* 88, 153596. DOI: 10.1016/j.phymed.2021.153596.

Rai, S., Omar, A.F., Rehan, M., Al-Turki, A., Sagar, A. *et al.* (2023) Crop microbiome: Their role and advances in molecular and omic techniques for the sustenance of agriculture. *Planta* 257(2), 27. DOI: 10.1007/s00425-022-04052-5.

Rajarajan, K., Sahu, S., Gangwar, N., Arunachalam, A., Handa, A.K. *et al.* (2025) Phenotypic characterization of neem (*Azadirachta indica* A. *Juss*) germplasm accessions for high oil yield and associated traits. *Plant Genetic Resources* 23(4), 271–274. DOI: 10.1017/S147926212500005X.

Razzaq, A., Sadia, B., Raza, A., Khalid Hameed, M. and Saleem, F. (2019) Metabolomics: A way forward for crop improvement. *Metabolites* 9(12), 303. DOI: 10.3390/metabo9120303.

Rohini, M.R. (2020) Genomics and molecular characterization of threatened medicinal plants. In: Rajasekharan, P.E. and Wani, S.H. (eds) *Conservation and Utilization of Threatened Medicinal Plants*. Springer International Publishing, Cham, Switzerland, pp. 317–346.

Roque-Borda, C.A., Kulus, D., Vacaro de Souza, A., Kaviani, B. and Vicente, E.F. (2021) Cryopreservation of agronomic plant germplasm using vitrification-based methods: An overview of selected case studies. *International Journal of Molecular Sciences* 22(11), 6157. DOI: 10.3390/ijms22116157.

Rose, J., Quave, C.L. and Islam, G. (2012) The four-sided triangle of ethics in bioprospecting: Pharmaceutical business, international politics, socio-environmental responsibility and the importance of local stakeholders. *Ethnobiology and Conservation* 1(3). DOI: 10.15451/ec2012-8-1.3-1-25.

Salam, U., Ullah, S., Tang, Z.-H., Elateeq, A.A., Khan, Y. *et al.* (2023) Plant metabolomics: An overview of the role of primary and secondary metabolites against different environmental stress factors. *Life (Basel, Switzerland)* 13(3), 706. DOI: 10.3390/life13030706.

Sánchez, M., González-Burgos, E., Divakar, P.K. and Gómez-Serranillos, M.P. (2020) DNA-based authentication and metabolomics analysis of medicinal plants samples by DNA barcoding and ultra-high-performance liquid chromatography/triple quadrupole mass spectrometry (UHPLC-MS). *Plants (Basel, Switzerland)* 9(11), 1601. DOI: 10.3390/plants9111601.

Sarwat, M. and Yamdagni, M.M. (2016) DNA barcoding, microarrays and next generation sequencing: Recent tools for genetic diversity estimation and authentication of medicinal plants. *Critical Reviews in Biotechnology* 36(2), 191–203. DOI: 10.3109/07388551.2014.947563.

Scholz, A.H., Nunez-Vega, G., Weissgold, L. and Wussmann, K. (2023) The future of access and benefit-sharing: What next after the adoption of the global biodiversity framework and decision on digital sequence information? *Diversity* 16(1), 27. DOI: 10.3390/d16010027.

Sharma, B. and Yadav, D.K. (2022) Metabolomics and network pharmacology in the exploration of the multi-targeted therapeutic approach of traditional medicinal plants. *Plants (Basel, Switzerland)* 11(23), 3243. DOI: 10.3390/plants11233243.

Sharma, N., Gowthami, R., Devi, S.V., Malhotra, E.V., Pandey, R. *et al.* (2021) Cryopreservation of shoot tips of *Gentiana kurroo* Royle – A critically endangered medicinal plant of India. *Plant Cell, Tissue and Organ Culture (PCTOC)* 144(1), 67–72. DOI: 10.1007/s11240-020-01879-2.

Sharma, P. and Sharon, M. (2025) A review on medicinal plants the nature's pharmacy: Challenges and progress in conservation. *Asian Plant Research Journal* 13(5), 65–81. DOI: 10.9734/aprj/2025/v13i5331.

Sharma, N., Malhotra, E.V., Chandra, R., Gowthami, R., Sultan, S.M. *et al.* (2022) Cryopreservation and genetic stability assessment of regenerants of the critically endangered medicinal plant *Dioscorea deltoidea* Wall. ex Griseb. for cryobanking of germplasm. *In Vitro Cellular & Developmental Biology - Plant* 58(4), 521–529. DOI: 10.1007/s11627-022-10267-8.

Shen, S., Zhan, C., Yang, C., Fernie, A.R. and Luo, J. (2023) Metabolomics-centered mining of plant metabolic diversity and function: Past decade and future perspectives. *Molecular Plant* 16(1), 43–63. DOI: 10.1016/j.molp.2022.09.007.

Skubel, S.A., Dushenkov, V., Graf, B.L., Niu, Q., Poulev, A. *et al.* (2018) Rapid, field-deployable method for collecting and preserving plant metabolome for biochemical and functional characterization. *PLOS ONE* 13(9), e0203569. DOI: 10.1371/journal.pone.0203569.

Soares, G.H., Hedges, J., Sethi, S., Poirier, B. and Jamieson, L. (2023) From *biocolonialism* to emancipation: Considerations on ethical and culturally respectful omics research with Indigenous Australians. *Medicine, Health Care, and Philosophy* 26(3), 487–496. DOI: 10.1007/s11019-023-10151-1.

Świątły-Błaszkiewicz, A., Gębalski, J., Kaczmarek, T., Załuski, D. and Kupcewicz, B. (2025) Exploring ED-XRF and ATR-FTIR spectroscopic techniques in medicinal plant research: A pilot study with *Scutellaria baicalensis* roots. *Journal of Pharmaceutical and Biomedical Analysis* 265, 117008. DOI: 10.1016/j.jpba.2025.117008.

Thaldar, D. and Shozi, B. (2023) Is benefit sharing with research participants lawful in South Africa? An unexplored question in the governance of genomics research. *Journal of Law and the Biosciences* 10(1), lsad018. DOI: 10.1093/jlb/lsad018.

Tong, Y., Li, X., Wan, J., Zhou, Q., Jiang, C. *et al.* (2025) Integrating ultra-high-performance liquid chromatography and orbitrap high-resolution mass spectrometry, feature-based molecular networking, and network medicine to unlock harvesting strategies for endangered *Sinocalycanthus chinensis*. *Journal of Separation Science* 48(1), e70072. DOI: 10.1002/jssc.70072.

Tsugawa, H., Rai, A., Saito, K. and Nakabayashi, R. (2021) Metabolomics and complementary techniques to investigate the plant phytochemical cosmos. *Natural Product Reports* 38(10), 1729–1759. DOI: 10.1039/d1np00014d.

Van Heerden, F.R. (2008) *Hoodia gordonii*: A natural appetite suppressant. *Journal of Ethnopharmacology* 119(3), 434–437. DOI: 10.1016/j.jep.2008.08.023.

Vermeylen, S. (2007) Contextualizing 'fair' and 'equitable': The San's reflections on the Hoodia benefit-sharing agreement. *Local Environment* 12(4), 423–436. DOI: 10.1080/13549830701495252.

Wang, M., Zhang, S., Li, R. and Zhao, Q. (2024) Unraveling the specialized metabolic pathways in medicinal plant genomes: A review. *Frontiers in Plant Science* 15, 1459533. DOI: 10.3389/fpls.2024.1459533.

Wei, W., Li, S., Wang, Y., Wang, B., Fan, G. *et al.* (2021) Metabolome-based genome-wide association study provides genetic insights into the natural variation of foxtail millet. *Frontiers in Plant Science* 12, 665530. DOI: 10.3389/fpls.2021.665530.

Wylie, M.R. and Merrell, D.S. (2022) The antimicrobial potential of the neem tree *Azadirachta indica*. *Frontiers in Pharmacology* 13, 891535. DOI: 10.3389/fphar.2022.891535.

Wynberg, R. (2023) Biopiracy: Crying wolf or a lever for equity and conservation. *Research Policy* 52(2), 104674. DOI: 10.1016/j.respol.2022.104674.

Yang, L., Yang, Y., Huang, L., Cui, X. and Liu, Y. (2023) From single- to multi-omics: Future research trends in medicinal plants. *Briefings in Bioinformatics* 24(1), bbac485. DOI: 10.1093/bib/bbac485.

Zeng, L., Sun, Z., Fu, L., Gu, Y., Li, R. *et al.* (2024) Cryopreservation of medicinal plant seeds: Strategies for genetic diversity conservation and sustainability. *Plants* 13(18), 2577. DOI: 10.3390/plants13182577.

Zhang, W., Zeng, Y., Jiao, M., Ye, C., Li, Y. *et al.* (2023) Integration of high-throughput omics technologies in medicinal plant research: The new era of natural drug discovery. *Frontiers in Plant Science* 14, 1073848. DOI: 10.3389/fpls.2023.1073848.

Zhong, C., Chen, C., Gao, X., Tan, C., Bai, H. *et al.* (2022) Multi-omics profiling reveals comprehensive microbe-plant-metabolite regulation patterns for medicinal plant *Glycyrrhiza uralensis* Fisch. *Plant Biotechnology Journal* 20(10), 1874–1887. DOI: 10.1111/pbi.13868.

Zhou, T., Luo, X., Zhang, C., Xu, X., Yu, C. *et al.* (2019) Comparative metabolomic analysis reveals the variations in taxoids and flavonoids among three *Taxus* species. *BMC Plant Biology* 19(1), 529. DOI: 10.1186/s12870-019-2146-7.

Zhu, S., Liu, Q., Qiu, S., Dai, J. and Gao, X. (2022) DNA barcoding: An efficient technology to authenticate plant species of traditional Chinese medicine and recent advances. *Chinese Medicine* 17(1), 112. DOI: 10.1186/s13020-022-00655-y.

10 Frontier Technologies: AI, Synthetic Biology, and Omics Convergence

Surendra Sarsaiya, Archana Jain and Qihai Gong*

Key Laboratory of Basic Pharmacology and Joint International Research Laboratory of Ethnomedicine of Ministry of Education, Zunyi Medical University, Zunyi, Guizhou, China

10.1 Introduction

The dawn of the 21st century has witnessed a profound transformation in the way medicinal plant research is conceptualized and executed, driven by the convergence of artificial intelligence (AI), synthetic biology, and multi-omics technologies. This chapter explores the vanguard of these intersecting fields, charting a course beyond traditional predictive modeling and metabolic engineering to a future where next-generation AI, cell-free systems, and omics-guided synthetic biology collectively enable the rational, rapid, and scalable engineering of plant-derived therapeutics.

The integration of AI with multi-omics data has emerged as a cornerstone for deciphering the intricate genotype–environment–phenotype relationships that underlie plant metabolism and bioactive compound production. Recent advances in deep learning, particularly transformer-based neural networks and graph machine learning, have enabled the *de novo* prediction of biosynthetic pathways, enzyme functions, and protein–protein interactions with unprecedented accuracy and scope (Zheng *et al.*, 2022; Gainza *et al.*, 2025; Wu and Xie, 2025). These models transcend the limitations of pattern recognition, facilitating the design of novel metabolic routes and the discovery of previously uncharacterized enzymes, thus expanding the chemical space accessible for drug discovery and natural product biosynthesis (Yeh *et al.*, 2023). The application of large language models and geometric deep learning further augments the capacity to generate hypotheses, design experiments, and interpret complex multi-modal datasets, accelerating the pace of scientific discovery in plant biotechnology (Mathema *et al.*, 2023; Wang *et al.*, 2023).

Parallel to these computational advances, cell-free synthetic biology platforms have revolutionized the prototyping and optimization of biosynthetic pathways. By decoupling pathway assembly from the constraints of living cells, cell-free systems enable rapid, high-throughput testing of genetic circuits and metabolic modules, offering a flexible alternative to the cellular chassis systems (Laohakunakorn *et al.*, 2020; Bogart *et al.*, 2021). The integration of multi-omics feedback—encompassing genomics, transcriptomics, proteomics, and metabolomics—into these platforms allows for iterative refinement of pathway designs, overcoming challenges related to metabolic complexity, toxicity, and resource allocation that often

*Corresponding author: gqh@zmu.edu.cn

DOI: 10.1079/9781836994206.0010

hinder cellular metabolic engineering (Jamil *et al.*, 2020; Valous *et al.*, 2024). Recent developments in mechanism-based and data-driven modeling further enhance the predictive power and efficiency of cell-free systems, streamlining the design-build-test-learn cycle and facilitating the bottom-up construction of complex biomolecular systems (Yurchenko *et al.*, 2024).

The advent of CRISPR-based (clustered regularly interspaced short palindromic repeats) genome and epigenome editing, when coupled with multi-omics data, has ushered in a new era of precision metabolic engineering that far exceeds the scope of basic gene editing. Advanced CRISPRa/i systems, informed by transcriptomic and epigenomic landscapes, enable the targeted modulation of gene expression and regulatory networks, allowing for the fine-tuning of biosynthetic pathways in plants (Badia-i-Mompel *et al.*, 2023; Parthiban *et al.*, 2023; Barbadilla-Martínez *et al.*, 2025). This omics-guided approach provides a systems-level understanding of regulatory cross-talk and feedback mechanisms, thereby facilitating the rational engineering of complex metabolic traits and the development of crops with enhanced therapeutic potential (Alamillo *et al.*, 2023; Kim *et al.*, 2023a).

Spatial multi-omics technologies, such as multiplexed error-robust fluorescence *in situ* hybridization (MERFISH) and matrix-assisted laser desorption/ionization imaging mass spectrometry (MALDI-imaging MS), have further expanded the analytical toolkit available to plant scientists. These methods provide high-resolution, three-dimensional maps of metabolite distributions and gene expression patterns within plant tissues, revealing tissue- and cell-type-specific localization of biosynthetic pathways (Liu *et al.*, 2020; Baysoy *et al.*, 2023; Vandereyken *et al.*, 2023). The integration of spatial and single-cell omics data offers a holistic view of plant metabolic architecture, enabling the identification of key regulatory nodes, metabolic bottlenecks, and spatially resolved pathway activities that inform targeted engineering strategies (Rao *et al.*, 2021; Nobori, 2025).

Looking to the horizon, quantum computing represents a transformative leap in the computational analysis of omics data and the simulation of complex metabolic networks. While still in its infancy, quantum algorithms hold the potential to solve optimization problems and network simulations that are currently intractable with classical computing, thereby accelerating pathway design, protein structure prediction, and systems-level modeling in synthetic biology (Selvarajoo and Maurer-Stroh, 2024). The integration of quantum computing with AI and multi-omics analysis promises to unlock new frontiers in the rational design of plant-derived therapeutics and the exploration of biological complexity.

The synergy of these frontier technologies is exemplified in advanced case studies, such as the engineering of cannabinoid and progesterone biosynthetic pathways in yeast, where the combined application of AI-driven pathway prediction, CRISPR-omics regulatory control, and cell-free prototyping has enabled the tailor-made biosynthesis of complex natural products with desired pharmacological properties (Subramanian *et al.*, 2020; Li *et al.*, 2025). This integrated approach not only accelerates the discovery and production of high-value medicinal compounds but also sets a precedent for the engineering of other complex natural products, moving beyond the paradigms established for paclitaxel and essential oils. In summary, the convergence of next-generation AI, cell-free systems, omics-guided genome editing, spatial multi-omics, and quantum computing is redefining the landscape of medicinal plant biotechnology. These advances collectively enable the rational, rapid, and scalable engineering of plant-derived therapeutics, paving the way for a new era of precision natural product discovery and sustainable drug development. As these technologies continue to mature and integrate, they promise to transform the future of medicine, agriculture, and biotechnology, offering solutions to some of the most pressing challenges in health and sustainability (Table 10.1).

10.2 Deep Learning for *de novo* Biosynthetic Pathway Prediction

Natural products (NPs) derived from medicinal plants have been a cornerstone of drug discovery for centuries, yielding compounds with unparalleled structural diversity and potent biological activities. However, a significant

Table 10.1. Frontier technologies in medicinal plant biotechnology: Core concepts, applications. Author's own table.

Technology domain	Core concept	Key applications in medicinal plants	Citations
AI & deep learning	Uses neural networks (e.g. transformers, graph neural networks—GNNs) to learn from biological data and generate novel hypotheses and designs.	• *De novo* prediction of biosynthetic pathways. • Novel enzyme discovery and functional annotation • Generative design of proteins and small molecules.	(Zheng *et al.*, 2022; Yeh *et al.*, 2023; Gainza *et al.*, 2025; Wu and Xie, 2025)
Cell-free synthetic biology	Reconstitutes metabolic pathways *in vitro* using cellular extracts or purified enzymes, decoupling prototyping from living cells.	• Rapid, high-throughput pathway prototyping. • Engineering of toxic pathways. • Optimization of enzyme combinations and reaction conditions.	(Jamil *et al.*, 2020; Laohakunakorn *et al.*, 2020; Bogart *et al.*, 2021; Valous *et al.*, 2024)
CRISPR-omics	Integrates multi-omics data (epigenomics, transcriptomics) with precision CRISPR tools (CRISPRa/i) to fine-tune gene regulatory networks.	• Fine-tuning of biosynthetic pathway flux. • Targeted activation of silent gene clusters. • Rewiring of regulatory networks for enhanced production.	(Alamillo *et al.*, 2023; Badia-i-Mompel *et al.*, 2023; Parthiban *et al.*, 2023; Barbadilla-Martínez *et al.*, 2025)
Spatial multi-omics	Provides high-resolution maps of the spatial localization of molecules (transcripts, metabolites) within intact plant tissues.	• Mapping tissue- and cell-type-specific pathway activity. • Identifying metabolic bottlenecks *in situ*. • Understanding compartmentalization of biosynthesis.	(Liu *et al.*, 2020; Baysoy *et al.*, 2023; Vandereyken *et al.*, 2023; Nobori, 2025)
Quantum computing	Leverages quantum mechanical principles (superposition, entanglement) to solve complex optimization and simulation problems.	• Simulation of complex metabolic networks. • Optimization of pathway fluxes. • Accelerated protein-ligand docking and drug discovery.	(Selvarajoo and Maurer-Stroh, 2024)
Convergent workflow (case study)	Synergistic integration of the above technologies into a unified, iterative design-build-test-learn cycle.	• Tailor-made production of cannabinoids, progesterone, etc. • Accelerated discovery and engineering of high-value natural products.	(Subramanian *et al.*, 2020; Li *et al.*, 2025)

bottleneck in their sustainable production and therapeutic application lies in the fact that their biosynthetic pathways are often unknown or only partially characterized. Traditional methods for pathway elucidation, which are largely template-based and rely on known enzymatic reaction rules and homology, struggle to extrapolate to novel compounds or complex pathways, particularly within the intricate landscape of plant secondary metabolism. These approaches are inherently limited by the scope of existing biochemical knowledge, making them inadequate for the *de novo* design of pathways for newly discovered or engineered molecules. The advent of deep learning, with its profound capacity to model complex, nonlinear relationships from large-scale data, offers a transformative, template-free solution to this long-standing challenge. By learning the fundamentals of biochemistry directly from data, deep learning models are now enabling the *de novo* prediction of biosynthetic pathways, the discovery of novel enzymes, and the rational engineering of metabolic routes with unprecedented speed and accuracy, thereby accelerating the development of plant-derived therapeutics and expanding the frontiers of synthetic biology (Zheng *et al.*, 2022; Ryu *et al.*, 2023; Xie *et al.*, 2024).

The promise of these AI-driven approaches is to move beyond retro-biosynthesis—the reconstruction of known pathways—and into genuine *de novo* design, where plausible biosynthetic routes can be generated for any target molecule, even those not found in nature. This paradigm shift is powered by architectures like transformer networks and GNNs, which can navigate the vast combinatorial space of biochemical reactions. However, this endeavor is not without its challenges. The unique complexities of plant biosynthetic pathways, such as enzyme promiscuity, pathway branching, compartmentalization, and intricate regulatory feedback loops, present formidable hurdles for accurate computational modeling. Furthermore, the success of these models is critically dependent on the quality, diversity, and scale of the training data. Overcoming these limitations requires continuous innovation in model architectures, sophisticated data curation strategies, and the integration of multi-modal biological information, setting the stage for a new era of intelligent metabolic engineering (Table 10.2 and Fig. 10.1; Goshisht, 2024).

10.2.1 Data foundations: Databases and dataset curation for AI-driven pathway prediction

The performance and generalizability of any deep learning model are fundamentally constrained by the data on which it is trained. In the domain of biosynthetic pathway prediction, the construction of comprehensive, high-quality, and well-curated datasets is a prerequisite for success. The ecosystem of biochemical data is vast, encompassing several key repositories that serve as the foundational pillars for model development. General-purpose databases such as KEGG (Kyoto Encyclopedia of Genes and Genomes) and Reactome provide extensively annotated information on metabolic pathways, reactions, and enzymes, while UniProt offers a comprehensive resource for protein sequence and functional data. To expand the chemical space available for learning, large chemical databases like Reaxys, SciFinder, and PubChem provide rich structural and property information for millions of small molecules, enabling models to learn the relationship between molecular structure and potential reactivity (Zheng *et al.*, 2022; Xie *et al.*, 2024).

For applications specifically targeting plant NPs, specialized datasets are indispensable. Resources like the PMN offer plant-specific pathway annotations, and various curated collections of secondary metabolites provide the specialized data needed to train models capable of navigating the unique chemical landscapes of plant biochemistry. A significant and pervasive challenge in this field is the issue of unbalanced and multi-label datasets. Certain pathway classes or compound families are vastly over-represented in public databases, while others, often the most novel and interesting from a drug discovery perspective, are critically under-represented. This imbalance can lead to models that are biased toward predicting common pathways while failing to generalize to rare or novel ones. To mitigate this, careful dataset curation is essential, often involving techniques such as data augmentation, strategic

Table 10.2. Deep learning for *de novo* biosynthetic pathway prediction: Approaches, tools, and applications. Author's own table.

Key area	Core concept/methodology	Representative tools/models	Specific applications in plant natural products	Citations
Data foundations	Curation of comprehensive, high-quality biochemical datasets for model training. Utilizes general and specialized databases, with strategies to handle data imbalance.	KEGG, Reactome, UniProt, PubChem, Plant Metabolic Network (PMN).	• Training models on plant-specific pathway annotations. • Using transfer learning to generalize from general biochemistry to rare plant metabolites.	(Zheng *et al.*, 2022; Bao *et al.*, 2023; Xie *et al.*, 2024)
Deep learning architectures	Application of neural networks designed for sequential and graph-structured data to learn the "language" and "structure" of biochemistry.	Transformers: BioNavi-NP. GNNs: GTC model, MotifMol3D. Hybrid Models: GATs and CNNs.	• *De novo* multi-step pathway prediction for complex NPs. • Predicting pathways by capturing global relational context and local molecular motifs.	(Zheng *et al.*, 2022; Bao *et al.*, 2023; Xie *et al.*, 2024; Hu *et al.*, 2025; Wang *et al.*, 2025)
Pathway prediction tasks	Decomposing the challenge into single-step reaction prediction and multi-step pathway planning, often using retrosynthetic analysis.	Single-step: sequence-to-sequence models, graph-to-graph models. Multi-step: BioNavi-NP, READRetro.	• Validating individual enzymatic steps. • Designing complete biosynthetic routes from precursors to target NPs.	(Zheng *et al.*, 2022; Kim *et al.*, 2024)
Enzyme discovery & engineering	Using DL to predict enzyme function from sequence/structure and to guide protein engineering for novel catalytic activities.	DeepECtransformer, ALDELE (All-Purpose Deep Learning Toolkits for Predicting the Biocatalytic Activities of Enzymes), EnzRank.	• Functional annotation of "hypothetical proteins" in plant genomes. • *In silico* screening of enzyme variants for improved properties.	(Kim *et al.*, 2023b; Upadhyay *et al.*, 2023; Wang *et al.*, 2024)
Generative & language models	Moving beyond prediction to the generative design of novel biological parts, such as enzymes and small molecules.	ProGen, chemical language models (CLMs), deep reinforcement learning (DRL).	• Generating novel enzyme sequences with desired functions. • Designing new-to-nature reaction steps and NP analogs.	(Olivecrona *et al.*, 2017; Madani *et al.*, 2023; Moret *et al.*, 2023; Nana Teukam *et al.*, 2024)
Validation & interpretability	Techniques to explain model predictions and the critical step of experimental validation to close the design loop.	Integrated gradients, motif analysis, wet-lab synthesis & testing.	• Identifying key residues for enzyme function. • Confirming the catalytic activity of AI-designed enzymes (e.g. luciferases).	(Bao *et al.*, 2023; Yeh *et al.*, 2023; Hu *et al.*, 2025)

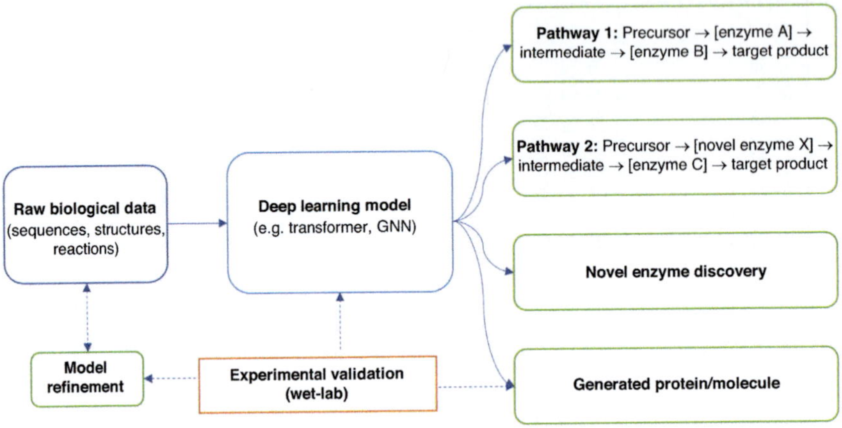

Fig. 10.1. Deep learning for *de novo* biosynthetic pathway prediction. (Author's own figure.)

sampling, and, most importantly, transfer learning. By pre-training a model on a large, general biochemical dataset and then fine-tuning it on a smaller, targeted plant-specific dataset, it is possible to imbue the model with broad biochemical knowledge while specializing its capabilities for the task at hand, thereby significantly improving generalization and predictive performance on under-represented classes (Bao *et al.*, 2023; Xie *et al.*, 2024).

10.2.2 Deep learning architectures: From sequence models to graph neural networks

The choice of neural network architecture is critical in determining how a model perceives and reasons about biochemical data. The field has largely converged on two powerful families of architectures: sequence-based models and graph-based models, with hybrid approaches combining their strengths now at the forefront of research. Following their revolutionary success in natural language processing (NLP), transformer neural networks have been powerfully adapted for biosynthetic pathway design. The underlying analogy is potent: just as a sentence is a sequence of words, a metabolic pathway can be viewed as a sequence of chemical transformations, and a molecular structure can be represented as a string (e.g. a SMILES (simplified molecular input line entry system) string). Transformer models, with their self-attention mechanisms, excel at learning the long-range dependencies and complex contextual relationships within these sequences. They have been successfully deployed for tasks ranging from single-step reaction prediction—identifying the most likely enzyme to catalyze a given biochemical transformation—to multi-step retrosynthetic analysis, which involves working backward from a target NP to plausible precursor molecules. For instance, the BioNavi-NP toolkit employs end-to-end transformer models to navigate the biosynthetic space of NPs, systematically predicting multi-step pathways with high accuracy and coverage, significantly outperforming traditional rule-based systems (Zheng *et al.*, 2022; Xie *et al.*, 2024; Wang *et al.*, 2025). These models effectively learn the "grammar" and "syntax" of biochemistry, enabling them to propose novel and biochemically plausible reaction steps that may not be present in existing rule sets.

While sequence models are powerful, the intrinsic structure of a molecule is inherently graph-like, with atoms as nodes and bonds as edges. Graph neural networks (GNNs) are specifically designed to operate on such data structures, allowing them to directly capture topological and relational information that can be lost in string-based representations. GNNs can learn to aggregate information from a node's local neighborhood, making them exceptionally well-suited for predicting molecular properties, simulating molecular interactions, and modeling entire metabolic networks. Hybrid models that combine the strengths of GNNs and other architectures are proving to be particularly effective. For example, models integrating graph attention networks (GATs) with convolutional neural networks (CNNs) can simultaneously process the structural graph of a molecule and other textual or spatial features. The GTC (graph transformer and CNN) model, for instance, leverages this hybrid approach to achieve high accuracy in predicting plant secondary metabolic pathways by capturing both the global relational context of the metabolic graph and local structural motifs (Yang *et al.*, 2022; Bao *et al.*, 2023). Another advanced model, MotifMol3D, further enhances this by integrating motif information and 3D molecular structure into a GNN framework, leading to improved interpretability and precision in pathway category prediction (Hu *et al.*, 2025). These hybrid architectures, when pre-trained on large, diverse biochemical datasets and fine-tuned on specific tasks, represent the state of the art in robust and generalizable pathway prediction.

10.2.3 Single-step and multi-step pathway prediction

The task of *de novo* pathway prediction can be conceptually divided into two levels of complexity: single-step reaction prediction and multi-step pathway design, each requiring distinct yet complementary modeling strategies. Single-step reaction prediction is a fundamental building block, focusing on identifying the specific enzyme or type of transformation that can convert a given substrate into a product. This task is well suited to both sequence-to-sequence models (treating substrate and product SMILES strings as input and output sequences) and graph-to-graph models, which directly map the molecular graph of the substrate to that of the product. Accurate single-step prediction is crucial for validating individual reaction steps within a larger pathway and for enzyme function annotation.

Multi-step pathway design is a more complex planning problem that involves stringing together a series of plausible single-step reactions to connect readily available building blocks to a target complex molecule. This often involves retrosynthetic analysis, working backward from the target. Models like BioNavi-NP combine deep learning with search algorithms, such as AND-OR tree-based planning, to efficiently explore the vast tree of possible synthetic routes. The AI model prunes this tree by scoring the plausibility of each reaction step, guiding the search toward the most feasible pathways. This approach has demonstrated remarkable success, with BioNavi-NP capable of identifying biosynthetic pathways for over 90% of test compounds and accurately recovering known biosynthetic building blocks with high fidelity (Zheng *et al.*, 2022). Further advancing this field, retrieval-augmented models like READRetro enhance pathway prediction for plant NPs by integrating ensemble deep learning architectures with retrievers that fetch biochemically relevant analogs from a database, thereby providing additional context and constraints to guide the pathway generation process (Kim *et al.*, 2024). These multi-step systems represent a critical step toward the fully automated design of synthetic pathways for novel therapeutic compounds.

10.2.4 Enzyme discovery and functional annotation

A major bottleneck in the practical implementation of computationally designed pathways is the identification and characterization of enzymes capable of catalyzing the proposed novel reactions. A significant portion of genes in any plant genome, especially those involved in secondary metabolism, are annotated as “hypothetical proteins” or have unknown functions. Deep learning is playing an increasingly vital role in bridging this annotation gap. Models are now capable of predicting enzyme function, typically represented by Enzyme Commission (EC) numbers, directly from amino acid sequence or protein structure data. Transformer-based architectures, such as DeepECtransformer, have shown high accuracy in this task by treating protein sequences as textual data and using self-attention to identify key functional motifs and domains critical for catalytic activity (Kim *et al.*, 2023b).

Beyond annotating existing enzymes, these models are instrumental in *discovering* new biocatalysts. Tools like ALDELE leverage hybrid models that integrate sequence and structural information to predict the catalytic activities of uncharacterized enzymes, effectively expanding the toolbox available for metabolic engineering (Wang *et al.*, 2024). Furthermore, deep learning guides enzyme engineering efforts. Models can predict the functional consequences of mutations, allowing for the *in silico* screening of vast virtual libraries of enzyme variants to identify those with improved catalytic activity, altered substrate specificity, or enhanced stability. Tools like EnzRank use CNNs to rank-order known enzymes based on their potential for successful re-engineering toward new substrate activities, providing a prioritized list of starting points for protein engineers (Upadhyay *et al.*, 2023). This synergy between AI-driven discovery and engineering is dramatically accelerating the process of matching novel enzymatic reactions to capable catalysts.

10.2.5 Generative and language models for pathway and enzyme design

The most frontier application of deep learning in this field moves beyond prediction and into generative design, where models create entirely novel biological components. Large language models (LLMs) and other generative deep learning architectures are at the heart of this revolution. Just as LLMs like GPT-4 can generate coherent and contextually relevant text, their biological counterparts are being trained to generate functional protein sequences and novel molecular structures.

Models like ProGen have demonstrated the ability to generate functional protein sequences across diverse families, conditioned on desired properties or functions specified by the user (Madani *et al.*, 2023). Similarly, hybrid CLMs can design novel small molecules with optimized bioactivity or propose new-to-nature reaction steps by learning from the combined language of chemistry and biology (Moret *et al.*, 2023). These generative approaches are being further enhanced by integration with other AI paradigms. For example, the combination of LLMs with genetic algorithms creates a powerful iterative cycle for enzyme design, where the LLM proposes novel sequences and the genetic algorithm optimizes them for stability and function (Nana Teukam *et al.*, 2024). DRL represents another powerful generative framework, where an agent learns to make decisions (e.g. which molecular building block to add next) by maximizing a reward signal (e.g. a desired molecular property). DRL has been successfully applied to *de novo* molecular design, enabling the exploration of vast chemical spaces to discover compounds with tailored pharmaceutical activities (Olivecrona *et al.*, 2017; Popova *et al.*, 2018). Together, these technologies are shifting the paradigm from discovering biology to designing it from first principles.

10.2.6 Interpretability, explainability, and model validation

As deep learning models for pathway prediction grow in complexity and are increasingly used to guide costly experimental efforts, the "black box" problem becomes a critical concern. It is no longer sufficient for a model to be accurate; it must also be interpretable, providing insights into *why* it made a particular prediction. This interpretability is essential for building trust among biologists and biochemists, for generating testable hypotheses, and for debugging and improving the models themselves. A suite of techniques is being deployed to open this black box. Methods like integrated gradients can identify which specific amino acids in a protein sequence or which atoms in a substrate molecule were most influential in the model's prediction of enzyme function or reaction outcome (Bao *et al.*, 2023). Motif analysis can reveal conserved structural or sequence patterns that the model has learned to associate with specific catalytic activities, providing a biologically meaningful explanation for its outputs (Hu *et al.*, 2025).

However, computational interpretability is only one part of the validation pipeline. The ultimate test of any AI-predicted pathway or enzyme is rigorous experimental validation. This is exemplified by high-profile successes in the field, such as the use of deep learning for the *de novo* design of luciferases, where computationally generated enzyme sequences were synthesized and experimentally confirmed to possess the predicted catalytic activity (Yeh *et al.*, 2023). Similarly, predictions from tools like DeepECtransformer require wet-lab validation to confirm the assigned EC numbers for previously uncharacterized genes. This continuous cycle of prediction, experimental testing, and model refinement based on feedback is the cornerstone of building robust, reliable, and ultimately transformative AI tools for synthetic biology.

In summary, deep learning has fundamentally transformed the field of biosynthetic pathway prediction, ushering in a new era of *de novo* metabolic design. Through the application of sophisticated architectures like transformers and GNNs, AI is no longer merely a tool for analyzing existing biological data but has become a creative partner in designing novel biochemical pathways and enzymes. From navigating the complex data landscape and predicting single enzymatic steps to generating entire pathways and designing functional proteins from scratch, these technologies are dramatically accelerating the discovery and sustainable production of high-value plant-derived therapeutics. While challenges in interpretability, biological complexity, and data scarcity persist, the trajectory of progress is clear. The continued convergence of deep learning with synthetic biology and multi-omics technologies promises to unlock unprecedented capabilities in rational metabolic engineering, paving the way for a future where the design of bespoke NPs for medicine and agriculture is both rapid and routine, thereby expanding the frontiers of what is biologically possible (Table 10.2).

10.3 Cell-free Biosynthesis and Omics-enabled Prototyping

The quest to harness the chemical diversity of medicinal plants for drug discovery and sustainable production is a central pursuit in modern biotechnology. However, traditional metabolic engineering in plants faces formidable challenges, including slow growth cycles, complex and often cryptic regulatory networks, recalcitrant genetics, and the sheer difficulty of transforming many medicinal species. These constraints render the classic design-build-test-learn (DBTL) cycle painfully slow, often stretching to months or years for a single iteration. Cell-free biosynthesis has emerged as a transformative paradigm that circumvents these limitations by decoupling pathway prototyping from the confines of the living cell. This approach utilizes crude cell extracts or purified enzyme systems to reconstitute metabolic pathways *in vitro*, offering an open and controllable environment for rapid experimentation. When this methodology is synergistically combined with multi-omics feedback—integrating genomics, transcriptomics, proteomics, and metabolomics—it creates a powerful, data-driven platform for the iterative optimization of biosynthetic pathways. This chapter critically examines the principles, applications, and future trajectory of cell-free biosynthesis and omics-enabled prototyping, with a specific focus on unlocking the biosynthetic potential of medicinal plants, thereby establishing a novel and accelerated workflow distinct from traditional cellular engineering (Table 10.3 and Fig. 10.2; Karim and Jewett, 2016; Jeung *et al.*, 2025).

10.3.1 Foundational principles of cell-free biosynthetic systems

At its core, a cell-free biosynthetic system is a biochemical platform that harnesses the catalytic machinery of the cell—including the transcriptional and translational apparatus, metabolic enzymes, and cofactors—without the barrier of a cell membrane. These systems are typically prepared as lysates from selected source organisms, such as *Escherichia coli*, yeast, or even plant tissues, which contain the necessary components for protein synthesis and metabolic activity. Alternatively, fully defined systems like the PURE (Protein synthesis Using Recombinant Elements) system offer a minimalist environment comprising purified components, granting maximum control and reproducibility at a higher cost (Sawasaki *et al.*, 2002; Liao *et al.*, 2025a).

The advantages of this *in vitro* approach are profound and directly address the key bottlenecks in plant pathway engineering. Speed and flexibility are paramount; DBTL cycles can be compressed from months to days, as there is no need for stable transformation, clonal selection,

Table 10.3. Cell-free biosynthesis and omics-enabled prototyping: Principles, platforms, and applications. Author's own table.

Key area	Core concept/methodology	Representative platforms/ systems	Specific applications & advantages	Citations
Foundational principles	Biochemical platform using cell lysates or purified systems to reconstitute pathways *in vitro*, decoupling from living cells.	Crude lysates (*E. coli*, yeast), PURE system	• Speed & flexibility: DBTL cycles in days. • Open system control: direct manipulation of reaction conditions. • Toxicity tolerance: study of pathways with lethal intermediates.	(Sawasaki *et al*., 2002; Karim and Jewett, 2018; Dudley *et al*., 2020; Liao *et al*., 2025a)
High-throughput platforms	Integrated systems for rapid assembly, testing, and analysis of pathway variants.	CFPS-ME, iPROBE, robotic automation, droplet microfluidics	• Single-pot testing of DNA-encoded pathways. • Ultra-high-throughput screening of enzyme libraries. • Nanoscale reactions for cost-effective prototyping.	(Karim and Jewett, 2016; Karim *et al*., 2020; Gan *et al*., 2022)
Multi-omics integration	Using systems-level data to diagnose bottlenecks and guide iterative optimization.	Proteomics, metabolomics, computational modeling	• Proteomics: verifies enzyme expression and stability. • Metabolomics: identifies kinetic bottlenecks and off-target products. • Data-driven design: informs rational re-engineering.	(Costello and Martin, 2018; Rasor *et al*., 2023; Yi *et al*., 2023; Poole *et al*., 2025)
Applications in medicinal plants	Prototyping complex plant-derived NP pathways.	Terpenoid, alkaloid, and glycosylated pathway reconstitution	• Limonene biosynthesis: screening synthase homologs. • P450-dependent pathways: co-expression with partners in supported membranes. • Non-model systems: use of *Streptomyces* lysates for plant-like metabolism.	(Kightlinger *et al*., 2019; Dudley *et al*., 2020; Ji *et al*., 2025; Rice *et al*., 2025)

Comparison with cellular engineering	Critical evaluation of cell-free vs cellular approaches for pathway development.	N/A	• Speed: days (cell-free systems) vs months (cellular). • Control & flexibility: high (cell-free systems) vs limited (cellular). • Predictive power: strong correlation for de-risking scale-up.	(Rasor *et al.*, 2021; Vögeli *et al.*, 2022)
Challenges & future directions	Identifying current limitations and key areas for innovation.	N/A	• Cost & scalability: for industrial production. • Reaction stability: extending operational lifetime. • Plant-specific systems: developing lysates from medicinal plants.	(Dudley *et al.*, 2015; Jeung *et al.*, 2025)

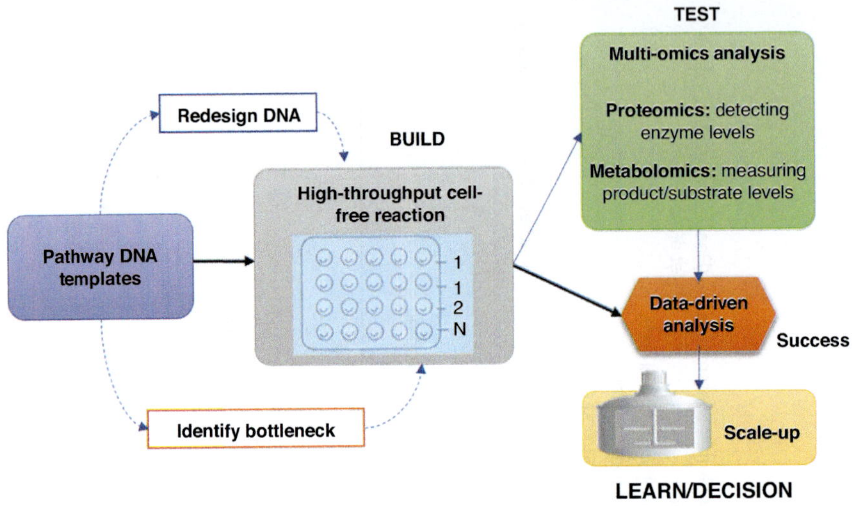

Fig. 10.2. Cell-free biosynthesis and omics-enabled prototyping. (Author's own figure.)

or lengthy cultivation periods. Pathway variants can be assembled and tested in a matter of hours. The Open System Control allows for precise manipulation of the reaction milieu—researchers can directly adjust substrate concentrations, replenish energy cofactors like ATP (adenosine triphosphate) and NADPH (nicotinamide adenine dinucleotide phosphate), and remove inhibitory by-products without concern for cellular homeostasis. This is particularly critical for pathways that produce toxic intermediates or final products that would be lethal to a living host, enabling the study and optimization of routes that are intractable *in vivo*. Furthermore, the inherent modularity of cell-free systems supports a "mix-and-match" approach, where enzymes produced separately via cell-free protein synthesis (CFPS) can be combinatorially assembled to rapidly test different pathway architectures and enzyme homologs (Karim and Jewett, 2018; Tinafar *et al.*, 2019; Dudley *et al.*, 2020).

10.3.2 Key platforms for high-throughput pathway prototyping

The practical application of cell-free biosynthesis has been greatly advanced by the development of specialized platforms designed for high-throughput experimentation. The CFPS-ME (cell-free protein synthesis–metabolic engineering) platform integrates protein synthesis and metabolic pathway operation in a single pot, allowing for the direct testing of DNA templates encoding pathway enzymes without the need for prior protein purification. This enables the rapid screening of enzyme libraries and regulatory parts. Building on this, the iPROBE (*in vitro* prototyping and rapid optimization of biosynthetic enzymes) platform provides a structured workflow for the high-throughput assembly and analysis of numerous pathway variants, facilitating the systematic exploration of design space (Karim and Jewett, 2016; Karim *et al.*, 2020).

These platforms are increasingly being coupled with automation technologies. Robotic liquid handlers and microfluidic devices enable the setup of thousands of parallel cell-free reactions with nanoliter volumes, ensuring high reproducibility while minimizing reagent costs. For example, droplet microfluidics has been used to encapsulate cell-free reactions for the ultra-high-throughput screening of enzymatic activities and pathway outputs (Gan *et al.*, 2022). This technological convergence transforms pathway prototyping into a rapid, scalable, and automated process, a stark contrast to the low-throughput, labor-intensive nature of traditional plant transformation and analysis.

10.3.3 Integrative power of multi-omics feedback

While cell-free systems simplify the biological context, the pathways within them remain complex networks. To fully understand and optimize these networks, a systems-level perspective is essential, and this is provided by multi-omics technologies. The integration of omics data creates a rich feedback loop that guides the DBTL cycle with unprecedented depth and precision. Proteomics is used to quantitatively monitor the expression, stability, and post-translational modification of every enzyme in the reconstituted pathway, confirming that each component is synthesized correctly and identifying potential issues with solubility or degradation that could limit flux (Rasor *et al.*, 2023).

Metabolomics provides a dynamic, comprehensive snapshot of the entire metabolic landscape of the reaction. It allows researchers to track the depletion of substrates, the accumulation of the desired final product, and, crucially, the formation of any off-target or inhibitory by-products that may be creating bottlenecks. This insight is invaluable for diagnosing pathway inefficiencies. For instance, the transient accumulation of a specific intermediate, revealed by time-course metabolomics, directly implicates the downstream enzyme as a kinetic bottleneck, prompting its re-engineering or replacement in the next design cycle (Costello and Martin, 2018; Poole *et al.*, 2025). This data-driven design moves engineering beyond guesswork, enabling rational optimization based on a holistic understanding of pathway dynamics. Computational tools then integrate these proteomic and metabolomic datasets to build predictive models that can forecast the outcome of further perturbations, accelerating the path to an optimally performing pathway (Yi *et al.*, 2023).

10.3.4 Applications in medicinal plant pathway engineering

The application of cell-free prototyping to medicinal plant pathways is a rapidly advancing frontier, demonstrating the practical utility of this approach for complex NP biosynthesis. Several case studies highlight its potential. The reconstitution of terpenoid pathways has been a major success. For example, the biosynthesis of limonene, a valuable monoterpene, was rapidly prototyped in a cell-free system, enabling the screening of multiple geranyl diphosphate synthase and limonene synthase homologs to identify the highest-yielding combination before any genetic modification of a host organism (Dudley *et al.*, 2020). Similarly, cell-free systems have been used to tackle the challenges of alkaloid and glycosylated NP biosynthesis. These pathways often involve membrane-associated cytochrome P450 enzymes and glycosyltransferases that are difficult to express functionally in heterologous hosts. Cell-free platforms allow for the co-expression of these enzymes with their necessary partners (e.g. cytochrome P450 reductases) and the addition of lipid membranes or other cofactors to support their activity, facilitating their characterization and optimization *in vitro* (Kightlinger *et al.*, 2019; Rice *et al.*, 2025). Furthermore, the development of cell-free systems from non-model organisms, such as *Streptomyces* or *Pseudomonas*, which possess metabolic features more akin to plants, provides a more native-like environment for prototyping plant-specific pathways, enabling the study of unique enzymes and regulatory elements relevant to plant secondary metabolism (Moore *et al.*, 2021; Ji *et al.*, 2025)

10.3.5 A critical comparison with cellular metabolic engineering

The cell-free approach offers distinct advantages that make it a superior choice for the initial prototyping phase compared to traditional cellular metabolic engineering. The most significant differentiator is speed. While a single DBTL cycle in a plant or microbial chassis can take weeks to months, the cell-free counterpart can be completed in days. This acceleration is due to the elimination of cell growth, transformation, and cloning steps. Control and flexibility are also greatly enhanced; the open system allows for direct intervention and tuning that is impossible within a living cell constrained by viability and homeostasis. Cell-free systems also exhibit a much higher tolerance for toxicity, allowing for the production and study of compounds that would kill or severely inhibit a cellular host. This is invaluable for engineering pathways that involve reactive intermediates. Finally, the predictive power of a well-optimized cell-free system is strong; insights gained *in vitro* on enzyme kinetics, cofactor balancing, and pathway bottlenecks have been shown to correlate well with subsequent *in vivo* performance, thereby de-risking the transition to cellular production hosts (Rasor *et al.*, 2021; Vögeli *et al.*, 2022). It is crucial to view these approaches as complementary: cell-free systems excel at rapid, high-throughput debugging and optimization of pathway designs, while cellular systems are ultimately required for scalable, sustainable production.

10.3.6 Current challenges and future research directions

Despite its considerable promise, the field of cell-free biosynthesis for medicinal plants is not without its challenges. Cost and scalability remain significant hurdles; the preparation of high-quality cell extracts and the reagents for large-scale reactions can be expensive, limiting its use to industrial-scale production rather than prototyping. The stability of cell-free reactions is also finite, typically lasting only a few hours to a day, which can be a constraint for very slow, multi-enzyme cascades. A major research gap is the underdevelopment of plant-specific cell-free systems. While microbial lysates are well established, robust CFPS systems derived from medicinal plants themselves are rare. Their development would be revolutionary, as they would inherently contain the specific cofactors, chaperones, and lipid environments necessary for the functional expression of complex plant biosynthetic enzymes. Future research must also focus on improved multi-omics integration, developing standardized computational pipelines to seamlessly translate omics data from cell-free reactions into actionable design rules. Furthermore, comparative cost–benefit analyses are needed to clearly delineate the economic trade-offs between cell-free prototyping and direct cellular engineering for different classes of plant-derived compounds (Dudley *et al.*, 2015).

Cell-free biosynthesis, powerfully augmented by multi-omics feedback, represents a paradigm shift in the engineering of medicinal plant metabolic pathways. By providing a rapid, flexible, and information-rich environment for prototyping, it effectively bypasses the major bottlenecks of slow growth and genetic complexity that have long hampered progress in plant biotechnology. This approach does not seek to replace cellular metabolic engineering but rather to create a more efficient and intelligent front end for it. The ability to rapidly test hundreds of pathway variants, diagnose failures with systems-level omics data, and rationally design optimized enzyme sets *in vitro* before committing to stable strain construction dramatically accelerates the entire biomanufacturing pipeline. As research addresses the current challenges of cost, scalability, and plant-specific system development, cell-free biosynthesis and omics-enabled prototyping are poised to become indispensable tools, unlocking the full therapeutic potential of medicinal plants and paving the way for a new generation of sustainably produced plant-derived NPs (Table 10.3).

10.4 CRISPR-omics: Editing Regulatory Networks via Multi-omics Feedback in Medicinal Plants

The discovery of the CRISPR-Cas system has revolutionized genetic engineering, offering an unprecedented ability to precisely alter DNA sequences. However, the initial focus on gene knockouts, while powerful, represents a relatively blunt instrument for the nuanced task of optimizing complex metabolic pathways in medicinal plants. These pathways are not merely strings of enzymes but are embedded within intricate regulatory networks governed by transcription factors, epigenetic modifications, and complex feedback loops. Merely deleting a gene can disrupt this delicate balance, leading to unforeseen consequences, metabolic bottlenecks, or even cell death. This limitation has given rise to a more sophisticated paradigm: CRISPR-omics. This approach integrates advanced CRISPR tools—particularly CRISPR activation and interference (CRISPRa/i)—with multi-omics data to move beyond simple gene editing toward the precise engineering of entire regulatory networks. By leveraging epigenomic and transcriptomic maps, CRISPR-omics allows for the targeted upregulation or downregulation of key regulatory nodes within a plant's metabolic circuitry. This enables the fine-tuning of biosynthetic pathways to enhance the production of valuable medicinal compounds, such as alkaloids, terpenoids, and phenylpropanoids, without the disruptive effects of gene knockouts. It represents a shift from a disruptive, bottom-up engineering strategy to a holistic, top-down recalibration of the plant's innate metabolic program, paving the way for a new era of precision metabolic engineering in medicinal plants (Table 10.4 and Fig. 10.3; Badia-i-Mompel *et al.*, 2023; Parthiban *et al.*, 2023; Barbadilla-Martínez *et al.*, 2025).

10.4.1 The CRISPRa/i toolkit: Precision dials for gene expression

At the heart of the CRISPR-omics approach are the CRISPRa and CRISPRi systems. Unlike the standard CRISPR-Cas9 system that creates double-strand breaks in DNA, CRISPRa/i uses a catalytically "dead" Cas9 (dCas9) that retains its ability to bind specific DNA sequences guided by a single-guide RNA (sgRNA) but does not cut the DNA. This dCas9 protein serves as a programmable platform that can be fused to various effector domains to modulate gene expression. CRISPR interference (CRISPRi) involves fusing dCas9 to a transcriptional repressor domain, such as the KRAB (Krüppel-associated box) domain. When this complex is targeted to a promoter region, it physically blocks the binding of RNA polymerase or recruits chromatin-remodeling complexes that promote a closed, transcriptionally silent chromatin state, effectively repressing gene expression. Conversely, CRISPR activation (CRISPRa) fuses dCas9 to transcriptional activator domains like VP64, p65, or EDLL. By targeting these complexes to promoter regions, they can recruit the cellular transcription machinery and open the chromatin structure, leading to a robust upregulation of the target gene.

The power of these systems lies in their programmability and reversibility. They act as "dials" for gene expression rather than "switches," allowing for graded control over metabolic flux. This is particularly advantageous in medicinal plants, where the constitutive overexpression of a pathway enzyme might lead to metabolic imbalance or the accumulation of toxic intermediates, while its complete knockout could halt the pathway entirely. CRISPRa/i enables the subtle tuning of expression levels to optimize flux through a desired pathway, making it an ideal tool for the rational engineering of complex metabolic traits (Xu *et al.*, 2020; Alamillo *et al.*, 2023; Kim *et al.*, 2023a).

10.4.2 Guiding CRISPRa/i with multi-omics maps

The strategic application of CRISPRa/i requires precise knowledge of which genes to target and when to target them. This is where multi-omics data becomes the essential guide. Relying on CRISPRa/i without the guidance of omics data is akin to navigating a complex circuit board without a schematic; one might get lucky, but systemic optimization is improbable. Epigenomics and transcriptomics provide this critical schematic of the plant's regulatory landscape. Transcriptomics, through techniques like RNA-Seq, provides a snapshot of the entire gene expression profile of a cell or tissue under specific conditions. By comparing transcriptomes of high- and low-producing plant varieties, or of plants subjected to different elicitors, researchers can identify key genes that are co-expressed with the biosynthetic pathways of interest. These genes may encode rate-limiting enzymes, positive transcription factors, or even negative regulators whose suppression could enhance flux. This gene list provides the initial set of candidates for CRISPRa/i targeting. Epigenomics adds a deeper layer of regulatory information.

Table 10.4. CRISPR-omics for engineering regulatory networks in medicinal plants. Author's own table.

Key area	Core concept/ methodology	Tools/techniques	Applications & strategic goals	Citations
CRISPRa/i toolkit	Uses catalytically inactive dCas9 fused to effector domains for precise gene regulation without DNA cleavage.	CRISPRi: dCas9-KRAB CRISPRa: dCas9-VP64/p65/EDLL	• Fine-tuning metabolic flux. • Graded control of gene expression as "dials" vs "switches." • Avoiding metabolic imbalance from constitutive overexpression.	(Xu *et al*., 2020; Alamillo *et al*., 2023; Kim *et al*., 2023a)
Multi-omics guidance	Leverages transcriptomic and epigenomic data to identify key regulatory nodes for targeted intervention.	RNA-Seq, ChIP-Seq (chromatin immunoprecipitation sequencing), ATAC-Seq (assay for transposase-accessible chromatin using sequencing)	• Identifying rate-limiting enzymes, master transcription factors, and negative regulators. • Targeting epigenetically silenced loci for activation. • Strategic network-level engineering.	(Liu *et al*., 2020; Badia-i-Mompel *et al*., 2023; Nobori, 2025)
Pathway engineering applications	Applying CRISPRa/i for targeted enhancement of medicinal compound biosynthesis.	Targeted sgRNAs for promoters/ enhancers	• Taxol (paclitaxel): upregulating bottleneck P450 enzymes. • Terpenoid indole alkaloids: activating master regulators (ORCA2/3—octadecanoid-responsive *Catharanthus* AP2-like 2 and 3). • Precursor channeling: repressing competing pathways.	(Parthiban *et al*., 2023; Yang *et al*., 2025)

Iterative DBTL cycle	A closed-loop, data-driven workflow for continuous optimization of metabolic networks.	Multi-omics phenotyping, computational modeling	• Design: target selection from omics maps. • Build: construct CRISPRa/i vectors. • Test: omics analysis of perturbations. • Learn: refine network model and targets.	(Wang *et al.*, 2023; Valous *et al.*, 2024)
Technical challenges	Key obstacles limiting the widespread application of CRISPR-omics.	N/A	• Delivery: transformation recalcitrance in medicinal plants. • Off-target effects: unintended gene modulation. • Phenotypic stability: context dependency and epigenetic drift. • Data integration: complexity of multi-omics analysis.	(Alamillo *et al.*, 2023; Goshisht, 2024)
Future directions	Emerging technologies to overcome current limitations and enhance capabilities.	Compact Cas systems, epigenome editors, spatial omics	• Improved delivery: Cas12/Cas13 systems. • Advanced editing: direct epigenetic reprogramming. • Specificity: tissue-specific optimization via spatial omics.	(Baysoy *et al.*, 2023; Vandereyken *et al.*, 2023)

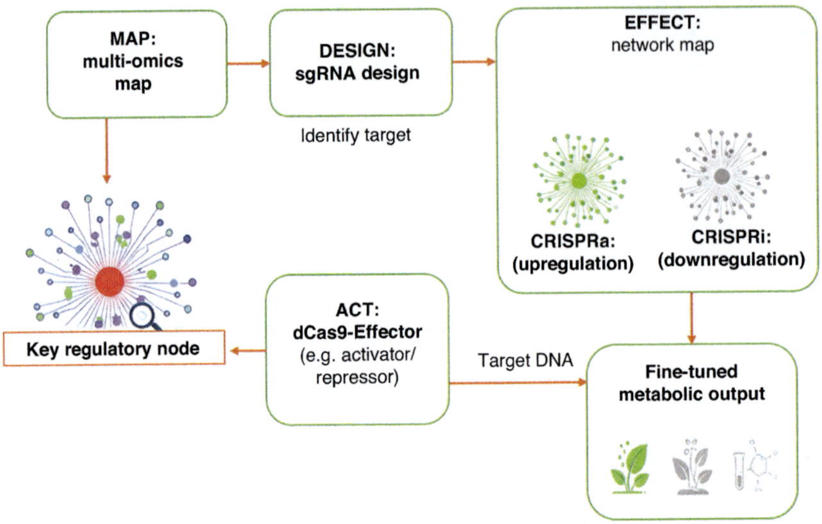

Fig. 10.3. CRISPR-omics: Editing regulatory networks via multi-omics feedback. (Author's own figure.)

Techniques such as ChIP-Seq (for histone modifications like H3K27ac, a mark of active enhancers) or ATAC-Seq (for mapping open chromatin regions) reveal the epigenetic "on" and "off" switches that control gene expression potential. For instance, a transcription factor gene that is crucial for a medicinal pathway might itself be under tight epigenetic control. Identifying such a locus—poised for activation but epigenetically silenced—provides a prime target for CRISPRa. By targeting the dCas9-activator to that specific regulatory region, it is possible to overcome the native epigenetic repression and unlock the expression of the master regulator, thereby activating the entire downstream pathway. This omics-guided approach ensures that CRISPRa/i interventions are not random but are strategically aimed at the most influential nodes within the regulatory network, maximizing the impact on the target metabolic pathway (Liu *et al.*, 2020; Badia-i-Mompel *et al.*, 2023; Nobori, 2025).

10.4.3 Applications in medicinal plant pathway engineering

The integration of CRISPR-omics holds immense potential for enhancing the production of a wide array of medicinal compounds. In many pathways, one or two enzymes act as key flux control points. Transcriptomic data can readily identify these bottlenecks. Instead of traditional overexpression, which can lead to protein aggregation or metabolic imbalance, CRISPRa can be used to subtly upregulate the gene encoding the rate-limiting enzyme, precisely adjusting its expression to a level that maximizes flux without causing cellular stress. For example, in the biosynthesis of the anti-cancer compound paclitaxel (Taxol), several cytochrome P450 enzymes are known bottlenecks. CRISPRa could be targeted to their promoters to enhance their expression based on transcriptomic data indicating their low native levels. Many plant secondary metabolic pathways are controlled by hierarchical transcription factors. In *Catharanthus roseus*, the AP2/ERF (APETALA2/ethylene-responsive factor) family transcription factors ORCA2 and ORCA3 are master regulators of terpenoid indole alkaloid biosynthesis. Epigenomic data might reveal that the loci of these regulators are in a repressed state in certain tissues or conditions. CRISPRa could be deployed to specifically activate these master regulators, thereby coordinately upregulating the entire suite of genes in the pathway, leading to a significant boost in the production of vinblastine and vincristine. Often, metabolic precursors are shunted into competing pathways. Transcriptomics can identify these competing routes. CRISPRi can be used to specifically repress key genes in these competing pathways, thereby channeling more carbon and energy toward the desired medicinal compound. Furthermore, some pathways are under the control of repressor proteins. Identifying these negative regulators via transcriptomic network analysis and then using CRISPRi to silence them can "release the brake" on biosynthesis, leading to enhanced compound accumulation (Parthiban *et al.*, 2023; Yang *et al.*, 2025).

10.4.4 Iterative DBTL cycle with multi-omics feedback

The true power of CRISPR-omics is realized when it is embedded within an iterative DBTL cycle, fueled by multi-omics feedback. This cycle begins with the *design* phase, where initial targets for CRISPRa/i are selected based on existing transcriptomic and epigenomic maps of the medicinal plant. The *build* phase involves constructing the CRISPRa/i vectors with sgRNAs targeted to the identified regulatory elements and transforming them into the plant host. The *test* phase is where the multi-omics feedback is critical. After applying the CRISPR-omics intervention, the plants are subjected to deep phenotyping, which includes not only measuring the final yield of the target medicinal compound but, more importantly, conducting a new round of transcriptomic and epigenomic analysis. This post-intervention omics profiling reveals the systemic consequences of the genetic perturbation. Did the targeted gene expression change as expected? What were the downstream effects on the rest of the pathway and the broader metabolic network? Were there any unexpected compensatory changes or activation of stress responses?

The *learn* phase involves computationally integrating this new omics data to understand the outcome. This feedback is used to refine the model of the regulatory network. Perhaps the initial intervention was insufficient, or maybe it activated a negative feedback loop. This new knowledge directly informs the next *design* phase, leading to a refined set of CRISPRa/i targets—for example, simultaneously activating a transcription factor and repressing a competing pathway enzyme. This closed-loop, data-driven cycle allows for the continuous and intelligent optimization of the plant's metabolic network, moving steadily toward a high-production phenotype (Wang *et al.*, 2023; Valous *et al.*, 2024).

10.4.5 Technical and biological challenges

Despite its transformative potential, the widespread implementation of CRISPR-omics in medicinal plants faces several significant challenges. Delivery and transformation remain a primary bottleneck. Many high-value medicinal plants are recalcitrant to genetic transformation, making the introduction of the relatively large dCas9-effector fusion constructs difficult. Developing efficient delivery methods, such as nanoparticle-based or viral vector systems, is a critical area of ongoing research. Off-target effects are a concern, as dCas9-sgRNA complexes could bind to and modulate genes with similar sequences, leading to unintended phenotypic consequences. While improved sgRNA design algorithms and high-fidelity Cas9 variants can mitigate this, rigorous off-target assessment using whole-genome sequencing and transcriptomics is essential.

Context dependency and phenotypic stability pose another challenge. The effect of a CRISPRa/i intervention can vary depending on the cell type, developmental stage, and environmental conditions. Furthermore, epigenetic states are dynamic and can change over time, potentially leading to the reversal of the engineered phenotype. Ensuring stable and consistent expression modulation across different generations is a non-trivial task. Finally, the complexity of data integration is a major hurdle. The sheer volume and complexity of multi-omics data require sophisticated bioinformatics expertise and computational resources for accurate interpretation and network modeling. Developing user-friendly software platforms that can translate omics data directly into actionable CRISPRa/i target recommendations will be crucial for democratizing this technology (Alamillo *et al.*, 2023; Goshisht, 2024).

The future of CRISPR-omics in medicinal plants is bright and will be shaped by several converging technological advances. The development of more compact and versatile CRISPR systems, such as those based on Cas12 or Cas13, could ease delivery constraints. The emergence of epigenome editors—fusions of dCas9 to enzymes that can directly write or erase DNA methylation or histone marks—offers an even more direct way to reprogram regulatory networks. Furthermore, the integration of spatial omics technologies will allow researchers to apply CRISPR-omics with tissue-specific resolution, optimizing metabolite production in the specific cells or glandular trichomes where they are naturally synthesized and stored (Baysoy *et al.*, 2023; Vandereyken *et al.*, 2023).

In conclusion, CRISPR-omics represents a paradigm shift from editing genes to engineering networks. By marrying the precision of CRISPRa/i with the insightful guidance of epigenomics and transcriptomics, this approach provides a powerful, rational framework for reprogramming medicinal plants into highly efficient biofactories. It moves metabolic engineering beyond disruptive gene knockouts and brute-force overexpression toward a more nuanced, systems-level control of metabolism. While challenges in delivery, stability, and data integration persist, the ongoing convergence of genomics, synthetic biology, and data science promises to overcome these barriers. As the technology matures, CRISPR-omics is poised to unlock the full metabolic potential of the plant kingdom, revolutionizing the sustainable production of next-generation plant-derived therapeutics and high-value NPs (Table 10.4).

10.5 Spatial Multi-omics: Mapping Metabolites in 3D Plant Architectures

Spatial multi-omics is fundamentally reshaping the landscape of plant biology by providing the tools to visualize and quantify the intricate spatial relationships between molecules within the complex three-dimensional (3D) architecture of plant tissues. This approach moves beyond homogenized tissue extracts, preserving the crucial context of where metabolites, transcripts, and proteins are localized, which is essential for understanding their functional roles (Zhang *et al.*, 2024a). At the forefront of this revolution are advanced techniques like MERFISH and MALDI-imaging MS. These technologies offer unprecedented spatial and molecular resolution, enabling researchers to deconstruct plant metabolic networks with a clarity previously unattainable. By mapping the precise locations of thousands of molecules simultaneously, spatial multi-omics allows for the correlation of metabolic activity with gene expression patterns directly *in situ*, opening new avenues for exploring tissue-specific pathway resolution, developmental programming, and adaptive responses to environmental stimuli (Table 10.5 and Fig. 10.4; Yu *et al.*, 2023).

10.5.1 Integrating MERFISH and MALDI-imaging MS for enhanced spatial resolution

The integration of MERFISH and MALDI-imaging MS represents a powerful synergy that holds significant promise for pushing the boundaries of spatial metabolomics in plants. These techniques are highly complementary. MERFISH excels at spatially resolving hundreds to thousands of RNA transcripts at single-molecule resolution, generating a highly detailed map of gene expression patterns within intact tissue sections. In parallel, MALDI-imaging MS enables the untargeted, label-free visualization and quantification of a vast array of metabolites—from primary metabolites to specialized lipids and secondary compounds—directly from the same or adjacent tissue sections. The spatial resolution of MALDI-imaging MS has seen remarkable improvements, now reaching as fine as 5 μm in some applications, which begins to approach the single-cell level and allows for the mapping of metabolites at subcellular scales (Hansen and Lee, 2018). Recent technological advancements, including dual-modal workflows that combine different matrix application chemistries or integrate mid-infrared (MIR) microscopy guidance, are further expanding both the chemical coverage and the spatial accuracy of metabolite detection (Wei *et al.*, 2025; Yuan *et al.*, 2025).

Table 10.5. Spatial multi-omics for mapping metabolites in 3D plant architectures. Author's own table.

Key area	Core concept/methodology	Key technologies & techniques	Applications & insights	Citations
Technology integration	Combining transcriptomic and metabolomic imaging to correlate gene expression with metabolite production *in situ*.	MERFISH: multiplexed RNA imaging MALDI-imaging MS: label-free metabolite mapping Data co-registration: bioinformatics alignment	• Direct correlation of biosynthetic gene clusters with their metabolic products. • Holistic, multi-layered view of biochemical pathways in morphological context.	(Yu *et al.*, 2023; Zhang *et al.*, 2024a; Wei *et al.*, 2025; Yuan *et al.*, 2025)
Tissue-specific pathway resolution	Mapping metabolic and transcriptomic heterogeneity across different plant tissues and cell types to understand functional specialization.	High-resolution mass spectrometry imaging (MSI) (~5 μm), computational metabolomics	• Identification of tissue-specific metabolic specialties (e.g. floral metabolites in *Nicotiana*). • Validation of enzyme roles in specific tissues (e.g. flavonoid biosynthesis). • Mapping hormone gradients (e.g. auxin) governing development.	(Novák *et al.*, 2012; Li *et al.*, 2016; Dong *et al.*, 2020; Hurgobin and Lewsey, 2022)
Environmental response & regulation	Visualizing dynamic, spatially resolved metabolic reprogramming in response to biotic and abiotic stresses.	Time-course spatial omics, stress perturbation studies	• Mapping lipid/pigment redistribution under nutrient/herbicide stress. • Identifying tissue-specific defense compound accumulation. • Deciphering regulatory networks driving spatial metabolic adaptation.	(Sharma *et al.*, 2021; Katam *et al.*, 2022; Horn and Chapman, 2024; Kajrolkar, 2025)

Continued

Table 10.5. Continued

Key area	Core concept/methodology	Key technologies & techniques	Applications & insights	Citations
Applications in metabolic engineering	Using spatial maps to inform precision engineering strategies for improved crop traits and compound production.	Targeted engineering based on spatial data	• Engineering valuable compounds in specific accumulating tissues (e.g. glandular trichomes). • Enhancing stress resilience by targeting protective metabolite production sites. • Avoiding growth penalties by avoiding constitutive pathway expression.	(Sharma *et al*., 2021; Hurgobin and Lewsey, 2022; Barmukh *et al*., 2025)
Future potential & impact	Advancing toward a comprehensive 3D understanding of plant metabolism for fundamental and applied science.	Improved resolution/sensitivity, 3D reconstruction, integration with other omics	• Discovery of novel pathways and spatial biomarkers. • True 3D metabolic mapping of entire organs. • Systems-level understanding of plant development and adaptation.	(Zhang *et al*., 2024a; Barmukh *et al*., 2025)

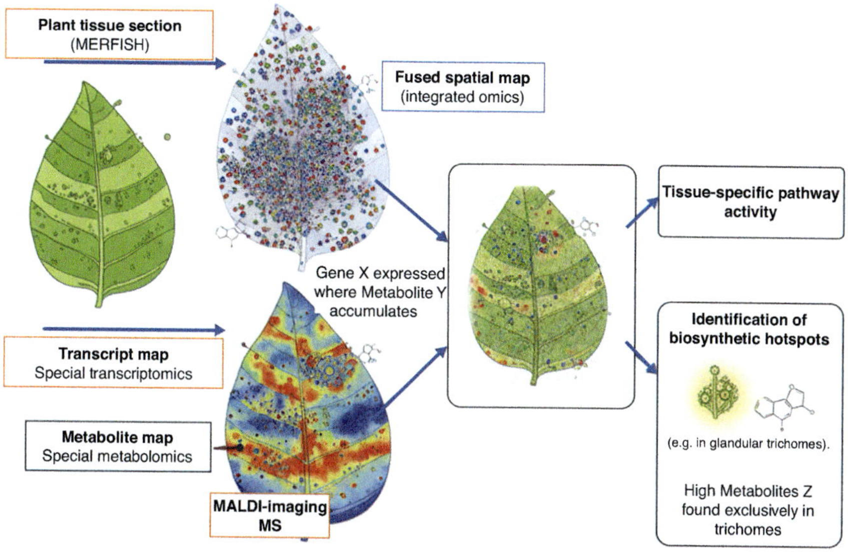

Fig. 10.4. Spatial multi-omics: Mapping metabolites in 3D plant architectures. (Author's own figure.)

The true power of integration lies in the co-registration of data from these two modalities. By aligning the high-throughput, multiplexed transcriptomic maps generated by MERFISH with the comprehensive metabolite landscapes revealed by MALDI-imaging MS, researchers can achieve a holistic, multi-layered view of biochemical pathways. This co-analysis facilitates the direct correlation of specific gene expression events with the spatial accumulation of their corresponding metabolic products. For instance, the expression of a biosynthetic gene cluster revealed by MERFISH can be directly overlaid with the distribution of its enzymatic products mapped by MALDI-imaging MS, providing compelling evidence for pathway activity in specific tissue domains. This integrative approach is critically dependent on sophisticated bioinformatics pipelines capable of aligning, normalizing, and interpreting these complex, multi-modal datasets to reveal the spatial relationships between the genome and the metabolome (Zhang *et al.*, 2024a; Wei *et al.*, 2025). Such a strategy moves beyond simple correlation to infer causal relationships within metabolic networks, directly within the morphological context of the plant.

10.5.2 Implications of tissue-specific pathway resolution in plant metabolomics

The ability to resolve metabolic pathways at the tissue-specific or even single-cell level is transformative for plant science, as it directly addresses the functional specialization of different plant organs and cell types. Plants are highly compartmentalized organisms, where different tissues—such as roots, leaves, vasculature, and reproductive structures—perform distinct physiological functions, a fact reflected in their unique gene expression and metabolite accumulation profiles (Novák *et al.*, 2012; Hurgobin and Lewsey, 2022). Spatial metabolomics has been instrumental in illuminating this diversity. A seminal study used computational metabolomics to illuminate the tissue-specific metabolic diversity in *Nicotiana attenuata*, revealing that floral organs, particularly anthers, possess highly specialized metabolomes rich in unique secondary metabolites crucial for pollination and defense (Li *et al.*, 2016).

By simultaneously mapping the spatial distribution of metabolites and the transcripts of their biosynthetic genes, researchers can infer gene function with high precision directly in the context of the plant's anatomy. This approach has successfully identified and validated the roles of key enzymes in specialized metabolic pathways. For example, high-mass-resolution spatial metabolite mapping has been used to pinpoint enzymes involved in flavonoid biosynthesis, with subsequent gene silencing experiments confirming their tissue-specific roles (Li *et al.*, 2016; Dong *et al.*, 2020). Furthermore, the tissue-specific profiling of plant hormones, such as auxin, has elucidated how precise gradients of these signaling molecules orchestrate complex developmental processes like organogenesis and tropic responses (Novák *et al.*, 2012).

This granular understanding of metabolic compartmentalization has profound implications. It enables a deeper, more mechanistic understanding of how plants coordinate growth, defense, and adaptation by allocating resources and activating pathways in specific cell types. From an applied perspective, tissue-specific pathway resolution provides a foundational knowledge base for targeted metabolic engineering. Instead of manipulating pathways constitutively throughout the plant, which can lead to growth penalties, scientists can now aim to engineer specific metabolic traits—such as enhanced nutrient content, improved stress resilience, or increased production of valuable compounds—in precisely defined tissues, thereby optimizing crop performance and yield (Sharma *et al.*, 2021; Hurgobin and Lewsey, 2022; Barmukh *et al.*, 2025).

10.5.3 Revealing novel insights into plant metabolic networks and environmental regulation

Spatial multi-omics approaches are uniquely positioned to uncover novel insights into the plasticity and regulation of plant metabolic networks in response to environmental cues. The spatial heterogeneity of metabolites is now recognized not as noise, but as a critical feature of plant adaptation, with distinct metabolic profiles emerging in different tissues under biotic and abiotic stresses (Katam *et al.*, 2022; Kajrolkar, 2025). MALDI-imaging MS has been pivotal in visualizing these dynamic responses. For instance, it has been deployed to map the redistribution of lipids and photosynthetic pigments in algal cells subjected to herbicide treatment or nutrient deprivation, revealing compartment-specific metabolic adaptations at the single-cell level (Katam *et al.*, 2022). In higher plants, spatial metabolomics has identified tissue-specific shifts in the abundance of defensive compounds, antioxidants, and osmolytes under drought, salinity, or pathogen attack, providing direct visual evidence of the mechanisms underpinning stress tolerance and local signaling events (Sharma *et al.*, 2021; Kajrolkar, 2025).

The integration of spatial transcriptomics via MERFISH with spatial metabolomics from MALDI-imaging MS is particularly powerful for deciphering the regulatory networks that coordinate these complex metabolic responses. This holistic perspective allows researchers to move from observing *where* a metabolic change occurs to understanding *how* it is regulated. By seeing which transcripts are co-localized with accumulating or depleting metabolites under stress conditions, it becomes possible to identify key regulatory genes and transcription factors that drive the spatial reprogramming of metabolism. This is essential for understanding how plants integrate external environmental signals with internal developmental programs to shape their growth, architecture, and ultimate survival (Horn and Chapman, 2024; Zhang *et al.*, 2024a; Barmukh *et al.*, 2025).

As these spatial multi-omics technologies continue to mature, with improvements in sensitivity, resolution, and multiplexing capacity, they are poised to drive the next wave of innovations in basic and applied plant biology. They will accelerate the discovery of novel metabolic pathways, identify spatial biomarkers for early stress detection, and refine strategies for crop improvement. Ultimately, by mapping the molecular geography of plants in 3D, we gain not just a snapshot, but a dynamic, systems-level understanding of plant life, from the single cell to the whole organism (Table 10.5; Barmukh *et al.*, 2025).

10.6 Quantum Computing for Omics Data Optimization in Medicinal Plants

The study of medicinal plants stands at the precipice of a computational revolution, driven by the emergence of quantum computing. This nascent technology promises to overcome one of the most significant bottlenecks in plant biology and biotechnology: the simulation and optimization of complex metabolic networks that are fundamentally intractable for classical computers. Medicinal plants synthesize a vast pharmacopeia of specialized metabolites—alkaloids, terpenoids, flavonoids—with immense therapeutic value. However, the biosynthetic pathways for these compounds often form intricate, highly regulated networks whose dynamic behavior eludes accurate computational modeling. Quantum computing, by leveraging the principles of quantum mechanics such as superposition and entanglement, offers a paradigm shift. It holds the potential to provide exponential speedups for specific problem classes, enabling researchers to unravel these complex networks, predict metabolic fluxes with unprecedented accuracy, and accelerate the discovery and engineered production of valuable plant-derived drugs (Table 10.6; Baiardi *et al.*, 2023; Zhang *et al.*, 2024b).

10.6.1 Challenge of metabolic network modeling in medicinal plants

The very characteristic that makes medicinal plants so valuable—their profound metabolic diversity—also presents a formidable computational challenge. These plants produce specialized metabolites through complex, branched biosynthetic pathways that are often tissue-specific, developmentally regulated, and responsive to environmental cues. Classical computational methods for modeling these networks, such as flux balance analysis (FBA) and kinetic modeling, are essential tools but face inherent limitations. As the scale and complexity of a network increase, the computational resources required to simulate its dynamics grow exponentially, a problem known as combinatorial explosion. This is exacerbated by the integration of multi-omics data (genomics, transcriptomics, proteomics, metabolomics), which adds layers of regulatory complexity and high dimensionality, pushing classical systems beyond their limits (Zhang *et al.*, 2024b; Rao and Liu, 2025). Consequently, traditional models often rely on simplifications and steady-state assumptions that fail to capture the full, dynamic, and stochastic nature of metabolic processes *in vivo*, leaving critical questions about pathway regulation, rate-limiting steps, and the emergence of metabolic phenotypes unanswered.

10.6.2 Quantum computing: A paradigm shift for omics data optimization

Quantum computing represents a fundamental departure from classical computing. While classical bits are either 0 or 1, quantum bits, or qubits, can exist in a superposition of both states simultaneously. This, combined with quantum entanglement, allows quantum processors to explore a vast number of potential solutions to a problem in parallel. For the analysis of omics data and metabolic networks, this translates to a potential quantum advantage in solving nondeterministic polynomial time (NP) hard optimization problems, which are central to systems biology. These include finding the global minimum in high-dimensional energy landscapes of molecular structures, optimizing metabolic fluxes across large networks, and simulating the quantum mechanical behavior of electrons in enzymatic reactions (Cordier *et al.*, 2022). Quantum algorithms are particularly well suited for multi-scale modeling, where different levels of biological organization—from an enzyme's active site to an entire metabolic pathway—must be integrated. In the context of medicinal plants, this means that quantum computing can provide a unified computational framework to bridge the gap between the quantum chemistry of a catalytic event and the systems-level behavior of a biosynthetic pathway, a task that is currently segmented and approximated across different classical software platforms (Zhang *et al.*, 2024b).

Table 10.6. Quantum computing for omics data optimization in medicinal plants. Author's own table.

Key area	Core concept/methodology	Quantum algorithms/ approaches	Potential applications in medicinal plants	Citations
Computational challenge	Classical simulation of complex, dynamic plant metabolic networks is intractable due to exponential complexity (combinatorial explosion).	N/A	• Modeling branched, regulated pathways for alkaloids, terpenoids, flavonoids. • Integrating multi-omics data into dynamic models.	(Zhang *et al.*, 2024b; Rao and Liu, 2025)
Quantum paradigm shift	Uses qubit superposition & entanglement to solve specific problems with exponential speedup potential.	Principles of quantum parallelism	• Solving NP-hard optimization problems in systems biology. • Multi-scale modeling from enzyme catalysis to pathway flux.	(Cordier *et al.*, 2022; Zhang *et al.*, 2024b)
Quantum algorithms	Specialized algorithms for simulating molecular systems and solving optimization problems.	VQE and QPE embedding strategies	• Elucidating reaction mechanisms of key enzymes (e.g. P450s). • Simulating electron behavior in catalytic sites. • Calculating energy barriers for complex cyclizations.	(Reiher *et al.*, 2017; Grimsley *et al.*, 2019; Cheng *et al.*, 2020)
Hybrid approaches & limitations	Combines quantum and classical computing to overcome current hardware constraints in the NISQ era.	Hybrid quantum-classical workflows, quantum circuit mapping	• Modeling logical regulatory networks controlling metabolism. • Limitations: qubit decoherence, noise, error rates, need for error correction.	(Baiardi *et al.*, 2023; Weidner *et al.*, 2023; Zhang *et al.*, 2024b)
Drug discovery & metabolic engineering	Applying quantum computing to predict, optimize, and design biological systems.	QML, *in silico* pathway simulation	• Predicting novel bioactive metabolites from omics data. • *In silico* optimization of metabolic engineering strategies. • Accelerated drug discovery via protein-ligand simulation.	(Pyrkov *et al.*, 2023; Sagingalieva *et al.*, 2023; Singh, 2024)
Future directions	Path to achieving practical quantum advantage in plant biotechnology.	Hardware advances, algorithm development, user-friendly software	• Scalable algorithms for large network simulation. • Increased qubit count and stability. • Accessible platforms for biologists.	(Cheng *et al.*, 2020; Baiardi *et al.*, 2023; Zhang *et al.*, 2024b)

10.6.3 Quantum algorithms for metabolic network simulation

Specific quantum algorithms have been developed that show great promise for direct application to biochemical systems. The variational quantum eigensolver (VQE) and quantum phase estimation (QPE) are two such algorithms designed for quantum chemistry simulations. They can calculate the electronic structures and energy levels of molecules with high accuracy, capturing electron correlation effects that are computationally prohibitive for classical methods such as density functional theory (DFT) on large molecules (Grimsley *et al.*, 2019; Cheng *et al.*, 2020). This capability is directly applicable to medicinal plant research. For instance, these algorithms can be used to elucidate the precise reaction mechanism of a key cytochrome P450 enzyme involved in the diversification of a terpenoid backbone, or to understand the energy barriers in a complex alkaloid cyclization reaction. Embedding strategies, where the quantum-mechanically complex core of a reaction (e.g. the active site) is simulated on a quantum processor while the surrounding protein environment is handled classically, make these simulations more feasible with current hardware and are particularly promising for modeling biosynthetic pathways (Cheng *et al.*, 2020). Early demonstrations have shown the viability of quantum computers in simulating fundamental biochemical processes like nitrogen fixation, paving the way for their application to the unique transformations found in plant specialized metabolism (Reiher *et al.*, 2017).

10.6.4 Hybrid quantum-classical approaches and current limitations

The current era of quantum computing is often termed the noisy intermediate-scale quantum (NISQ) era, characterized by processors with a limited number of qubits that are susceptible to decoherence and noise. Given these hardware constraints, the most practical and promising near-term strategies are hybrid quantum-classical approaches. These methods strategically partition a computational problem, offloading the most challenging sub-problems—those with exponential complexity on classical machines—to the quantum processor, while the rest of the computation is managed by powerful classical computers (Baiardi *et al.*, 2023; Zhang *et al.*, 2024b). A compelling example in systems biology is the modeling of logical regulatory networks. The dynamic analysis of these networks, which govern gene expression and subsequently metabolic activity, faces exponential complexity. This task can be efficiently mapped onto quantum circuits, significantly reducing computational overhead and enabling more realistic simulations of how metabolic networks in medicinal plants respond to genetic or environmental perturbations (Weidner *et al.*, 2023). Despite this progress, significant challenges remain. The accuracy of NISQ-era simulations is limited by quantum noise, necessitating ongoing research into error mitigation and, ultimately, fault-tolerant quantum computing with robust error correction. The development of scalable, resource-efficient quantum algorithms and their seamless integration into existing bioinformatics workflows for plant omics data are active and critical areas of investigation (Grimsley *et al.*, 2019; Cheng *et al.*, 2020).

10.6.5 Applications in drug discovery and metabolic engineering

The ultimate impact of quantum computing in medicinal plant science will be realized through its applications in drug discovery and metabolic engineering. By enabling accurate, dynamic simulations of entire metabolic pathways, quantum computing can transform the process of identifying and optimizing the production of bioactive compounds. It can predict the behavior of complex, non-linear biosynthetic routes, identify hidden regulatory nodes, and optimize the design of metabolic engineering strategies *in silico* before costly and time-consuming laboratory experiments are undertaken (Singh, 2024). Furthermore, the field of quantum-enhanced machine learning (QML) is emerging as a powerful tool. Hybrid quantum-classical neural networks have already demonstrated improved performance in predicting drug responses and molecular properties compared to purely classical models (Sagingalieva *et al.*, 2023). Applied to medicinal plants, QML models could sift through vast multi-omics datasets to predict novel bioactive metabolites, identify the genetic basis of high-yielding phenotypes, or design optimized cultivation strategies to enhance the production of target compounds. The integration of quantum computing with AI is thus creating a powerful new toolkit for personalized medicine, where therapeutic strategies could be designed based on a deep, quantum-simulated understanding of plant metabolite interactions with human biological targets (Pyrkov *et al.*, 2023).

The application of quantum computing to omics data optimization in medicinal plants is a frontier field, brimming with potential but also facing substantial research challenges. Key future directions include the development of more scalable and robust quantum algorithms specifically tailored for biological problems, such as simulating large-scale metabolic networks and protein–ligand interactions. Advances in quantum hardware, with increased qubit counts, improved coherence times, and lower error rates, are essential to achieve a clear and consistent quantum advantage for real-world problems. Furthermore, the development of user-friendly software platforms that allow plant biologists and bioinformaticians without quantum physics expertise to leverage these tools is crucial for widespread adoption (Baiardi *et al.*, 2023; Zhang *et al.*, 2024b). Success in this endeavor will hinge on deep, cross-disciplinary collaboration between quantum physicists, computational biologists, chemists, and plant scientists. As the hardware and algorithms mature, it is anticipated that quantum computing will transition from a theoretical curiosity to a practical tool, enabling the simulation of plant metabolic networks at a scale and resolution unattainable by classical computers. This will not only facilitate the accelerated discovery of new plant-derived drugs and the engineering of high-yielding plant strains but also lead to a more fundamental and holistic understanding of plant metabolism itself (Table 10.6; Grimsley *et al.*, 2019; Cheng *et al.*, 2020).

10.7 Case Study: Cannabinoid Engineering in Medicinal Plants via AI/CRISPR-omics

The field of plant biotechnology is undergoing a profound transformation, driven by the convergence of advanced technologies that are collectively reshaping how we approach the production of valuable specialized metabolites. In the realm of medicinal plants, particularly *Cannabis sativa*, this transformation is most evident in the quest to engineer tailor-made cannabinoids. The integration of AI, CRISPR-based genome editing, and multi-omics technologies—often referred to as "Ch10 technologies"—creates a powerful, synergistic platform to overcome the limitations of traditional breeding and extraction methods. This integrated approach enables the precise, efficient, and scalable production of specific cannabinoids, addressing critical challenges in therapeutic demand, regulatory compliance, and the production of rare analogs with unique pharmacological profiles (Table 10.7; Das *et al.*, 2024).

10.7.1 Promise and challenge of cannabinoid biosynthesis

Cannabinoids, a diverse class of terpenophenolic compounds primarily produced by *C. sativa*, exhibit a wide range of pharmacological properties, including analgesic, anti-inflammatory, antiepileptic, and neuroprotective effects. However, the natural production of these compounds is fraught with challenges. Yields of specific, therapeutically desirable cannabinoids are often low, and the plant's complex chemical profile includes the psychoactive tetrahydrocannabinol (THC), which is heavily regulated. Furthermore, over 150 unique cannabinoids have been identified, many of which are present only in trace amounts, making their isolation and study difficult. Traditional agricultural and extraction techniques are insufficient for reliably producing high yields of these minor cannabinoids or for creating novel analogs with optimized therapeutic indices. This creates a significant bottleneck for pharmaceutical development and underscores the need for precision engineering

Table 10.7. Case study: Cannabinoid engineering via AI/CRISPR-omics—an integrated workflow. Author's own table.

Technology/stage	Role in cannabinoid engineering	Specific tools/actions	Outcome/application	Citations
Multi-omics discovery	Provides a systems-level map of the cannabinoid biosynthetic pathway and its regulation.	Genomics, transcriptomics, metabolomics, proteomics	• Identifies genetic variation (SNPs, CNVs) in *THCAS/CBDAS*. • Pinpoints rate-limiting enzymes and regulatory nodes. • Reveals subcellular enzyme localization for flux optimization.	(Matchett-Oates *et al*., 2021; Yao *et al*., 2023; Borah *et al*., 2024; Yoosefzadeh Najafabadi and Torkamaneh, 2025)
AI-driven design	Interprets omics data to predict optimal genetic interventions and design novel components.	ML-guided sgRNA design (e.g. Cas-OFFinder), genomic prediction models, generative AI	• Designs high-efficiency, low off-target CRISPR guides. • Predicts cannabinoid profiles from genotype. • Generates novel cannabinoid analogs *in silico*.	(Ramzi *et al*., 2020; Hesami *et al*., 2021b; Bian and Xie, 2022; Yoosefzadeh Najafabadi and Torkamaneh, 2025)
CRISPR-based build	Executes precise genetic modifications to rewire the cannabinoid pathway as designed.	CRISPR knockout (e.g. *THCAS*), multiplex editing, synthetic biology	• Creates high-CBD, low-THC hemp varieties. • Enables production of rare cannabinoids (e.g. CBG, CBC). • Introduces heterologous genes for novel analogs.	(Hesami *et al*., 2021a; Zhang *et al*., 2021; Borah *et al*., 2024; Das *et al*., 2024)

Continued

Table 10.7. Continued

Technology/stage	Role in cannabinoid engineering	Specific tools/actions	Outcome/application	Citations
Integrated workflow & output	The synergistic, closed-loop DBTL cycle that combines all technologies for continuous optimization.	Iterative multi-omics feedback, AI model refinement	• Tailor-made cannabinoids: customized profiles for specific therapeutics. • Rare cannabinoid production: scalable synthesis of minor compounds. • Microbial biofactories: engineering yeast/bacteria for production.	(Hesami *et al.*, 2023; Borah *et al.*, 2024; Das *et al.*, 2024; Sun *et al.*, 2024)
Remaining challenges	Key technical and regulatory hurdles limiting widespread industrial application.	Improved tissue culture protocols, regulatory strategy, public engagement	• Transformation efficiency: low and genotype-dependent in cannabis. • Regulatory hurdles: complex global genetically modified organisms (GMO) regulations. • Public acceptance: addressing societal concerns over engineered plants.	(Zhang *et al.*, 2021; Mipeshwaree Devi *et al.*, 2023; Borah *et al.*, 2024; Das *et al.*, 2024)

of the cannabinoid biosynthetic pathway (Hesami *et al.*, 2023; Yoosefzadeh Najafabadi and Torkamaneh, 2025).

10.7.2 CRISPR-Cas genome editing: Precision tools for cannabinoid pathways

The advent of CRISPR-Cas genome editing technology has provided plant biotechnologists with a scalpel-like tool for precise metabolic engineering. In the context of cannabinoid production, CRISPR-Cas systems are being deployed to make targeted modifications to the key genes governing the biosynthetic pathway. Successful experiments have demonstrated the knockout of crucial genes such as *THCAS* (tetrahydrocannabinolic acid synthase), which diverts metabolic flux away from THC and toward the accumulation of non-psychoactive cannabidiol (CBD), a trait of immense medical and regulatory value (Zhang *et al.*, 2021; Das *et al.*, 2024). The feasibility of stable genetic transformation and gene editing in the historically recalcitrant cannabis plant was conclusively demonstrated with the CRISPR-Cas9-mediated knockout of the phytocene desaturase gene (*CsPDS1*), which produces an albino phenotype, serving as a clear visual marker for successful editing (Hesami *et al.*, 2021a; Zhang *et al.*, 2021).

Beyond single-gene knockouts, the potential of multiplex genome editing allows for the simultaneous targeting of multiple genes within a pathway. This capability is essential for the fine-tuning of complex metabolic networks or for the reconstruction of entire synthetic pathways to produce rare or novel cannabinoids that are not naturally abundant in the plant. Synthetic biology approaches, combined with CRISPR, can introduce heterologous genes or rewire regulatory circuits, vastly expanding the chemical diversity of cannabinoids that can be produced in plant systems (Borah *et al.*, 2024; Das *et al.*, 2024). Despite this progress, technical challenges such as low transformation efficiency, genotype-dependent responses, and the need for robust tissue culture regeneration protocols remain active areas of research and development (Hesami *et al.*, 2021a; Zhang *et al.*, 2021).

10.7.3 Omics technologies: Mapping and optimizing the cannabinoid network

The rational design of CRISPR editing strategies is critically dependent on a deep, systems-level understanding of the plant's molecular machinery, which is provided by multi-omics technologies. Genomics and pangenomics have been instrumental in characterizing the genetic diversity within *C. sativa*, revealing extensive copy number variations (CNVs) and single nucleotide polymorphisms (SNPs) in key biosynthetic genes like *THCAS* and *CBDAS*. This information is vital for informed guide RNA design and for predicting the functional consequences of genetic variations across different cultivars (Matchett-Oates *et al.*, 2021; Yoosefzadeh Najafabadi and Torkamaneh, 2025).

Transcriptomics and metabolomics provide a dynamic view of the pathway, identifying which genes are expressed and which metabolites accumulate under different developmental stages or environmental conditions. This helps pinpoint rate-limiting enzymes, identify novel candidate genes, and understand the regulatory mechanisms that control flux through the cannabinoid network (Borah *et al.*, 2024; Das *et al.*, 2024). Furthermore, proteomics and studies on subcellular compartmentalization reveal the precise localization of biosynthetic enzymes within the cell. Understanding this spatial organization is key to metabolic engineering, as targeting pathway enzymes to specific organelles, such as plastids or the endoplasmic reticulum, can enhance precursor availability and sequester toxic intermediates, thereby optimizing overall metabolic flux and final product yield (Yao *et al.*, 2023; Borah *et al.*, 2024). The integration of these disparate omics datasets creates a comprehensive map of the cannabinoid biosynthetic system, guiding the selection of the most impactful targets for genetic intervention.

10.7.4 AI: Accelerating design and optimization

The vast and complex datasets generated by omics technologies and CRISPR experiments necessitate advanced computational tools for interpretation and prediction. This is where AI and machine learning (ML) become force multipliers in the engineering cycle. AI-driven tools are now routinely used to enhance the precision of CRISPR workflows by designing optimal guide RNAs with high on-target efficiency and minimal off-target effects. Tools like Cas-OFFinder and ML-based ensemble strategies have been specifically applied to cannabis, improving the reliability and safety of genome edits in this valuable medicinal plant (Hesami *et al.*, 2021b; Yoosefzadeh Najafabadi and Torkamaneh, 2025).

Beyond guide RNA design, ML models are powerful tools for genomic prediction. Algorithms such as Random Forest and Bayesian models can analyze high-density genotyping data to predict complex traits, such as cannabinoid profiles, based on genetic markers. This allows breeders and engineers to select optimal parent lines or to identify key genetic loci for targeted editing, accelerating the development of cultivars with customized cannabinoid content (Yoosefzadeh Najafabadi and Torkamaneh, 2025). AI also plays a crucial role in integrating multi-omics data, identifying non-obvious regulatory nodes, and predicting metabolic fluxes *in silico*. More advanced applications include the use of deep learning and generative models for *de novo* molecular design, creating virtual libraries of novel cannabinoid analogs with predefined, desirable pharmacological properties before any synthesis is attempted in the lab (Ramzi *et al.*, 2020; Bian and Xie, 2022). By rapidly analyzing complex patterns that are intractable for human researchers, AI dramatically accelerates the "design-build-test-learn" cycle, enabling the rapid prototyping and optimization of engineered medicinal plants.

10.7.5 Synergy of Ch10 technologies: Integrated workflows for tailor-made cannabinoids

The true power of Ch10 technologies is realized not in isolation, but through their seamless integration into a cohesive engineering workflow. This synergistic pipeline begins with multi-omics discovery, where genomics, transcriptomics, and metabolomics are used to comprehensively map the cannabinoid pathway and identify key regulatory nodes and metabolic bottlenecks. These targets are then passed to an AI-driven design module, where ML algorithms predict the optimal CRISPR guide RNAs and model the potential outcomes of various genetic interventions. The selected designs are then built using CRISPR-Cas tools to create precisely edited plant lines. The resulting plants are then subjected to a new round of multi-omics analysis (the "test" phase), and the data is fed back into the AI models (the "learn" phase) to refine the next cycle of engineering, creating a closed-loop, iterative system for continuous improvement (Borah *et al.*, 2024; Das *et al.*, 2024; Yoosefzadeh Najafabadi and Torkamaneh, 2025).

This integrated approach is already yielding tangible results. It enables the development of industrial hemp varieties with high CBD and legally compliant, negligible THC levels. It facilitates the production of rare cannabinoids like cannabigerol (CBG) or cannabichromene (CBC) by redirecting metabolic flux. Furthermore, it opens the door to the creation of entirely novel cannabinoid analogs with improved bioavailability, efficacy, or safety profiles, and even to the engineering of water-soluble derivatives for enhanced pharmaceutical formulation (Deguchi *et al.*, 2020; Hesami *et al.*, 2023). This synergy is not limited to plants; the same technological convergence is being used to engineer microbial hosts like yeast and bacteria into efficient biofactories for cannabinoid production, offering an alternative, scalable, and controlled manufacturing platform (Hesami *et al.*, 2023; Sun *et al.*, 2024).

Despite the remarkable progress, the path to widespread industrial application is not without obstacles. Technical hurdles, particularly the low and genotype-dependent transformation and regeneration efficiency in cannabis, remain a significant bottleneck that requires the development of more robust and universal tissue culture protocols (Zhang *et al.*, 2021; Mipeshwaree Devi *et al.*, 2023). Furthermore, a complete understanding of the cannabinoid pathway and its regulation is still evolving, and metabolic bottlenecks or compensatory mechanisms can arise after genetic modification, necessitating further optimization. Beyond the lab, regulatory frameworks for genetically engineered medicinal plants are complex and vary globally, while public perception and acceptance pose additional challenges to commercialization (Borah *et al.*, 2024; Das *et al.*, 2024).

Future research must focus on closing the identified research gaps. Key open questions include: How can transformation protocols in cannabis be universally improved? What are the most effective strategies for integrating multi-omics and AI to build predictive digital twins of cannabinoid biosynthesis? And how can regulatory and societal barriers be proactively addressed? The continued maturation of Ch10 technologies—through more precise CRISPR systems, higher-throughput omics, and more sophisticated AI algorithms—will be crucial to answering these questions. In conclusion, the synergy of AI, CRISPR, and multi-omics is ushering in a new era for cannabinoid engineering in medicinal plants. This powerful technological triad moves the field from observation to predictive design, enabling the creation of tailor-made cannabinoids with precision and efficiency. By overcoming the limitations of traditional methods, this approach holds the promise of unlocking a new generation of plant-derived pharmaceuticals, expanding the therapeutic landscape, and ensuring a sustainable and standardized supply of these valuable medicinal compounds (Table 10.7).

10.8 Challenges and Future Perspectives in Frontier Technologies

The convergence of next-generation AI, synthetic biology, and multi-omics technologies is forging a new frontier in biotechnology, promising to revolutionize how we understand, design, and engineer biological systems. However, this rapid advancement is accompanied by a set of complex technical, computational, and translational challenges that must be addressed to fully realize its potential. The path forward will be paved by interdisciplinary collaboration, continued innovation in both hardware and software, and the development of robust ethical and regulatory frameworks (Table 10.8).

10.8.1 Deep learning for *de novo* biosynthetic pathway prediction

A significant challenge in deploying deep learning for *de novo* pathway design is the scarcity and inconsistent quality of large, annotated biological datasets. Incomplete or erroneous enzyme function annotations and reaction mechanisms severely limit the accuracy and generalizability of these predictive models (Xie *et al.*, 2024). Furthermore, the "black box" nature of complex transformer and graph-based models complicates biological validation, as a lack of interpretability erodes trust and hinders the extraction of actionable biological insights (Momeni and Ghorbani-Vaghei, 2024). Bridging the gap between *in silico* predictions and wet-lab validation also remains a bottleneck, as AI-generated pathways may be kinetically or thermodynamically infeasible within a living cellular context due to unknown regulatory constraints or toxicity.

Looking forward, the application of transfer learning and LLMs trained on expansive chemical and biological databases will significantly expand model capabilities and improve the accuracy of novel enzyme discovery (Xie *et al.*, 2024). The future lies in creating tight, hybrid AI-experimental feedback loops, where model predictions are rapidly tested and the resulting data is used to refine the AI, accelerating the iterative cycle of pathway optimization (Liao *et al.*, 2025b). Concurrently, major research thrust will focus on developing explainable AI (XAI) to make model decisions more transparent, thereby building trust and facilitating wider adoption in synthetic biology workflows.

10.8.2 Cell-free biosynthesis and omics-enabled prototyping

Cell-free systems excel at rapid prototyping but face considerable hurdles in scalability and cost-effectiveness, making their transition to industrial-scale production economically

Table 10.8. Challenges and future perspectives in frontier technologies. Author's own table.

Technology domain	Key challenges	Future perspectives & solutions	Cross-cutting themes	Citations
Deep learning for pathway prediction	• Data scarcity & poor annotation quality. • "Black box" interpretability issues. • *In silico* to *in vivo* translation gap.	• Transfer learning & LLMs • Explainable AI (XAI). • Tight AI-experimental feedback loops.	Data quality & standardization, model trust & adoption.	(Momeni and Ghorbani-Vaghei, 2024; Xie *et al*., 2024; Liao *et al*., 2025b)
Cell-free biosynthesis & prototyping	• Scalability & cost for industrial use. • Pathway assembly complexity *in vitro*. • Correlation with *in vivo* performance.	• High-throughput automation & microfluidics. • AI-guided experimental design. • Enhanced multi-omics integration.	Translational gap (lab to industry), cost-effectiveness.	(Jiang *et al*., 2018; Dudley *et al*., 2020; Karim *et al*., 2020; Jeung *et al*., 2025)
CRISPR-omics	• Delivery in recalcitrant plants. • Off-target effects. • Regulatory & epigenetic complexity. • Multi-omics data integration.	• Compact Cas systems. (Cas12/13) • Epigenome editors. • AI-guided target selection & network modeling.	Delivery efficiency, specificity & safety, data integration complexity.	(Cardi *et al*., 2023; Tang and Zhang, 2023; Liao *et al*., 2025b; Yang *et al*., 2025)
Spatial multi-omics	• Achieving single-cell/subcellular resolution. • Data complexity & lack of standardized pipelines. • Sample preparation for plant tissues.	• Higher-resolution MSI & MERFISH. • 3D tissue reconstruction. • On-tissue chemical derivatization.	Computational tool development, technical resolution limits.	(Hansen and Lee, 2018; Dueñas *et al*., 2019; Cao and Gao, 2022; Yuan *et al*., 2025)
Quantum computing for omics	• NISQ-era hardware limitations (noise, qubits). • Scarcity of tailored biological algorithms. • Data encoding & loading bottleneck.	• Hybrid quantum-classical approaches. • Algorithm innovation for biology. • Empirical validation of "quantum advantage."	Hardware limitations, algorithm-application gap.	(Tang *et al*., 2023; Alexeev *et al*., 2024; Zhang *et al*., 2024b)
Cross-cutting challenges & outlook	• Lack of interdisciplinary training. • Data standardization & open access. • Biosafety, ethics & regulatory frameworks. • Scaling to industrial viability.	• Fostering interdisciplinary collaboration. • Promoting open science & data sharing. • Developing proactive governance & cost-reduction strategies.	Interdisciplinary collaboration, responsible innovation, economic viability.	(Cardi *et al*., 2023; Momeni and Ghorbani-Vaghei, 2024)

challenging (Jeung *et al.*, 2025). The complexity of assembling stable, multi-enzyme pathways *in vitro* is another major obstacle, often hampered by enzyme instability, costly cofactor requirements, and an incomplete understanding of optimal reaction conditions (Dudley *et al.*, 2020). Perhaps most critically, metabolic fluxes and yields observed in a simplified cell-free environment do not always correlate directly with performance in the more complex milieu of a living cell, complicating the predictive power of the DBTL cycle.

Future progress will be driven by high-throughput automation using robotic and microfluidic platforms, which will enable massive parallelization of experiments, drastically reducing costs and increasing prototyping speed (Jiang *et al.*, 2018). The integration of AI and ML will be pivotal for guiding experimental design, intelligently optimizing enzyme combinations and reaction conditions to bridge the gap between cell-free and *in vivo* performance (Karim *et al.*, 2020). Furthermore, the incorporation of real-time multi-omics feedback into the cell-free systems workflow will provide a systems-level view, refining pathway design and dramatically accelerating the DBTL cycle.

10.8.3 CRISPR-omics: Editing regulatory networks via multi-omics feedback

The precision of CRISPR-based editing of regulatory networks is constrained by persistent challenges in delivery efficiency and off-target effects, particularly in non-model plants with complex and often polyploid genomes (Cardi *et al.*, 2023). The regulatory complexity itself is a hurdle, as editing promoters or enhancers is complicated by epigenetic modifications and chromatin architecture, dynamics that are not yet fully understood (Tang and Zhang, 2023). From a computational standpoint, the integration of multi-omics data (transcriptomic, epigenomic, proteomic) to inform precise CRISPRa/i targeting is both methodologically immature and computationally intensive.

The future of CRISPR-omics will be shaped by advanced editing platforms, including base and prime editors, as well as epigenome editors, which will allow for more precise and tunable regulation of gene expression without altering the underlying DNA sequence (Yang *et al.*, 2025). AI-guided target selection will become indispensable, using ML to identify the most impactful regulatory nodes and predict the systemic consequences of genetic perturbations (Liao *et al.*, 2025b). Coupling high-throughput CRISPR screens with multi-omics readouts will enable the functional mapping of regulatory networks at an unprecedented scale, providing the data needed to rationally rewire cellular machinery.

10.8.4 Spatial multi-omics: Mapping metabolites in 3D plant architectures

A primary technical challenge in spatial multi-omics is achieving true single-cell or subcellular resolution in dense and heterogeneous plant tissues, as current capabilities of MSI and spatial transcriptomics are often limited by sensitivity and resolution (Hansen and Lee, 2018). The data generated is inherently complex, and the integration of spatial coordinates with transcriptomic and metabolomic datatypes suffers from a lack of standardized bioinformatics pipelines despite requiring sophisticated computational tools (Cao and Gao, 2022). Furthermore, the physical properties of plant tissues, such as rigid cell walls and diverse secondary metabolites, complicate sample preparation and can hinder reproducibility.

Technological innovation will be key to the future of this field. Advances in instrumentation, such as higher-resolution MALDI-imaging MS, multiplexed techniques like MERFISH, and on-tissue chemical derivatization protocols, will progressively improve spatial resolution and expand metabolite coverage (Yuan *et al.*, 2025). A major goal is the move from 2D mapping to true 3D tissue reconstruction, combining serial sectioning with spatial omics and computational modeling to visualize metabolic pathways within the volumetric architecture of an organ (Dueñas *et al.*, 2019). These detailed maps will directly inform precision breeding and metabolic engineering strategies, enabling the targeted enhancement of valuable traits in crops.

10.8.5 Quantum computing for omics data optimization

The application of quantum computing to biological problems is currently constrained by significant hardware limitations. The NISQ era is defined by processors with limited qubit counts, short coherence times, and high error rates, which restrict their utility for large-scale omics data simulation (Zhang *et al.*, 2024b). There is also a scarcity of quantum algorithms specifically tailored for biological network optimization or omics data integration, and for many problems, highly optimized classical solvers remain fiercely competitive (Tang *et al.*, 2023). A fundamental bottleneck is the challenge of efficiently loading and encoding massive omics datasets into a quantum state for processing.

In the near term, the most promising path is the development of hybrid quantum-classical approaches, which strategically allocate computational tasks to exploit the nascent strengths of quantum processors for specific sub-problems while relying on classical systems for the rest (Alexeev *et al.*, 2024). Future progress hinges on algorithm innovation, particularly for combinatorial optimization, network inference, and QML, which could unlock new capabilities for simulating complex metabolic networks. Ultimately, rigorous empirical validation and benchmarking against classical methods will be essential to demonstrably achieve a "quantum advantage" in real-world omics applications.

10.8.6 Cross-cutting challenges and future outlook

Beyond the technology-specific hurdles, several cross-cutting challenges define the landscape. A critical bottleneck is the lack of interdisciplinary training, which is essential to bridge the deep expertise required across computer science, biology, chemistry, and engineering (Momeni and Ghorbani-Vaghei, 2024). The field is also hampered by a lack of data standardization and limited open access to high-quality datasets, which impedes model training, reproducibility, and collaborative progress. As these powerful technologies advance, robust biosafety, ethical, and regulatory frameworks must be developed in parallel to ensure their responsible deployment (Cardi *et al.*, 2023). Finally, translating proof-of-concept studies into economically viable, industrial-scale applications will require significant advances in automation, cost reduction, and the development of reliable biocontainment strategies. In conclusion, the convergence of AI, synthetic biology, and omics is poised to unlock a new era of biological engineering and discovery. Overcoming the current constellation of challenges will demand sustained innovation, a commitment to interdisciplinary collaboration and open science, and a proactive approach to governance. By navigating these complexities, we can harness these frontier technologies to address some of the most pressing challenges in health, agriculture, and environmental sustainability (Table 10.8).

10.9 Conclusion

The convergence of AI, synthetic biology, and multi-omics technologies marks a pivotal inflection point in medicinal plant biotechnology. This chapter has delineated how these frontier technologies are moving the field from observation and description to predictive design and rational engineering. Deep learning has transitioned from a pattern recognition tool to a creative partner in *de novo* pathway and enzyme design. Cell-free systems have dismantled the temporal and biological barriers of cellular hosts, enabling unprecedented speed and control in pathway prototyping. CRISPR-omics has evolved gene editing into a sophisticated discipline of network-level control, allowing for the nuanced fine-tuning of complex metabolic pathways. Spatial multi-omics has provided a long-missing dimension to our understanding, revealing the intricate molecular geography of plant tissues. On the horizon, quantum computing looms as a transformative force with the potential to tackle computational problems of a scale currently beyond our grasp. The integrated application of these technologies, as exemplified in the engineering of tailor-made cannabinoids, demonstrates a powerful new paradigm. It forges a closed-loop, data-driven workflow where multi-omics discovery informs AI-driven design, which in turn guides precise genetic interventions, with the results feeding back to refine the entire process. However, the path forward is not without its challenges, including issues of data quality, model interpretability, technical delivery in recalcitrant plants, hardware limitations, and the urgent need for interdisciplinary collaboration and robust regulatory frameworks. Despite these hurdles, the trajectory is clear. The continued maturation and integration of these frontier technologies promise to unlock the full therapeutic potential of the plant kingdom, enabling the sustainable and on-demand production of a new generation of plant-derived pharmaceuticals. This will not only accelerate drug discovery but also fundamentally deepen our understanding of plant metabolism itself, ultimately transforming the future of medicine, agriculture, and biotechnology.

Acknowledgments

The authors express their gratitude for the financial support received through the Distinguished High-Level Talents Research Grant from the Guizhou Science and Technology Corporation Platform Talents Fund (Grant nos.: [2017]5733-001 and CK-1130-002), high-level specialists project fund of Zunyi Medical University (1092/2000006)F-1092, the National Natural Science Foundation of China (82373981), and the support provided by Zunyi Medical University, China. Special appreciation is extended to all laboratory colleagues and research staff members for their valuable insights, constructive guidance, and assistance throughout this study.

References

Alamillo, J.M., López, C.M., Martínez Rivas, F.J., Torralbo, F., Bulut, M. *et al.* (2023) Clustered regularly interspaced short palindromic repeats/CRISPR-associated protein and hairy roots: A perfect match for gene functional analysis and crop improvement. *Current Opinion in Biotechnology* 79, 102876. DOI: 10.1016/j.copbio.2022.102876.

Alexeev, Y., Amsler, M., Barroca, M.A., Bassini, S., Battelle, T. *et al.* (2024) Quantum-centric supercomputing for materials science: A perspective on challenges and future directions. *Future Generation Computer Systems* 160, 666–710. DOI: 10.1016/j.future.2024.04.060.

Badia-i-Mompel, P., Wessels, L., Müller-Dott, S., Trimbour, R., Ramirez Flores, R.O. *et al.* (2023) Gene regulatory network inference in the era of single-cell multi-omics. *Nature Reviews. Genetics* 24(11), 739–754. DOI: 10.1038/s41576-023-00618-5.

Baiardi, A., Christandl, M. and Reiher, M. (2023) Quantum computing for molecular biology. *Chembiochem* 24(13), e202300120. DOI: 10.1002/cbic.202300120.

Bao, H., Zhao, J., Zhao, X., Zhao, C., Lu, X. *et al.* (2023) Prediction of plant secondary metabolic pathways using deep transfer learning. *BMC Bioinformatics* 24(1), 348. DOI: 10.1186/s12859-023-05485-9.

Barbadilla-Martínez, L., Klaassen, N., Van Steensel, B. and De Ridder, J. (2025) Predicting gene expression from DNA sequence using deep learning models. *Nature Reviews Genetics* 26(10), 666–680. DOI: 10.1038/s41576-025-00841-2.

Barmukh, R., Garg, V., Liu, H., Chitikineni, A., Xin, L. *et al.* (2025) Spatial omics for accelerating plant research and crop improvement. *Trends in Biotechnology* 43(8), 1904–1920. DOI: 10.1016/j.tibtech.2025.03.007.

Baysoy, A., Bai, Z., Satija, R. and Fan, R. (2023) The technological landscape and applications of single-cell multi-omics. *Nature Reviews Molecular Cell Biology* 24(10), 695–713. DOI: 10.1038/s41580-023-00615-w.

Bian, Y. and Xie, X.-Q. (2022) Artificial intelligent deep learning molecular generative modeling of scaffold-focused and cannabinoid CB2 target-specific small-molecule sublibraries. *Cells* 11(5), 915. DOI: 10.3390/cells11050915.

Bogart, J.W., Cabezas, M.D., Vögeli, B., Wong, D.A., Karim, A.S. *et al.* (2021) Cell-free exploration of the natural product chemical space. *ChemBioChem* 22(1), 84–91. DOI: 10.1002/cbic.202000452.

Borah, A., Singh, S., Chattopadhyay, R., Kaur, J. and Bari, V.K. (2024) Integration of CRISPR/Cas9 with multi-omics technologies to engineer secondary metabolite productions in medicinal plant: Challenges and prospects. *Functional & Integrative Genomics* 24(6), 207. DOI: 10.1007/s10142-024-01486-w.

Cao, Z.-J. and Gao, G. (2022) Multi-omics single-cell data integration and regulatory inference with graph-linked embedding. *Nature Biotechnology* 40(10), 1458–1466. DOI: 10.1038/s41587-022-01284-4.

Cardi, T., Murovec, J., Bakhsh, A., Boniecka, J., Bruegmann, T. *et al.* (2023) CRISPR/Cas-mediated plant genome editing: Outstanding challenges a decade after implementation. *Trends in Plant Science* 28(10), 1144–1165. DOI: 10.1016/j.tplants.2023.05.012.

Cheng, H.-P., Deumens, E., Freericks, J.K., Li, C. and Sanders, B.A. (2020) Application of quantum computing to biochemical systems: A look to the future. *Frontiers in Chemistry* 8, 587143. DOI: 10.3389/fchem.2020.587143.

Cordier, B.A., Sawaya, N.P.D., Guerreschi, G.G. and McWeeney, S.K. (2022) Biology and medicine in the landscape of quantum advantages. *Journal of the Royal Society, Interface* 19(196), 20220541. DOI: 10.1098/rsif.2022.0541.

Costello, Z. and Martin, H.G. (eds) (2018) A machine learning approach to predict metabolic pathway dynamics from time-series multiomics data. *NPJ Systems Biology and Applications* 4(1), 19. DOI: 10.1038/s41540-018-0054-3.

Das, S., Kwon, M. and Kim, J.-Y. (2024) Enhancement of specialized metabolites using CRISPR/Cas gene editing technology in medicinal plants. *Frontiers in Plant Science* 15, 1279738. DOI: 10.3389/fpls.2024.1279738.

Deguchi, M., Kane, S., Potlakayala, S., George, H., Proano, R. *et al.* (2020) Metabolic engineering strategies of industrial Hemp (*Cannabis sativa* L.): A brief review of the advances and challenges. *Frontiers in Plant Science* 11, 580621. DOI: 10.3389/fpls.2020.580621.

Dong, Y., Sonawane, P., Cohen, H., Polturak, G., Feldberg, L. *et al.* (2020) High mass resolution, spatial metabolite mapping enhances the current plant gene and pathway discovery toolbox. *The New Phytologist* 228(6), 1986–2002. DOI: 10.1111/nph.16809.

Dudley, Q.M., Karim, A.S. and Jewett, M.C. (2015) Cell-free metabolic engineering: Biomanufacturing beyond the cell. *Biotechnology Journal* 10(1), 69–82. DOI: 10.1002/biot.201400330.

Dudley, Q.M., Karim, A.S., Nash, C.J. and Jewett, M.C. (2020) *In vitro* prototyping of limonene biosynthesis using cell-free protein synthesis. *Metabolic Engineering* 61, 251–260. DOI: 10.1016/j.ymben.2020.05.006.

Dueñas, M.E., Larson, E.A. and Lee, Y.J. (2019) Toward mass spectrometry imaging in the metabolomics scale: Increasing metabolic coverage through multiple on-tissue chemical modifications. *Frontiers in Plant Science* 10, 860. DOI: 10.3389/fpls.2019.00860.

Gainza, P., Bunker, R.D., Townson, S.A. and Castle, J.C. (2025) Machine learning to predict *de novo* protein–protein interactions. *Trends in Biotechnology* 43(12), 3056–3070. DOI: 10.1016/j.tibtech.2025.04.013.

Gan, R., Cabezas, M.D., Pan, M., Zhang, H., Hu, G. *et al.* (2022) High-throughput regulatory part prototyping and analysis by cell-free protein synthesis and droplet microfluidics. *ACS Synthetic Biology* 11(6), 2108–2120. DOI: 10.1021/acssynbio.2c00050.

Goshisht, M.K. (2024) Machine learning and deep learning in synthetic biology: Key architectures, applications, and challenges. *ACS Omega* 9(9), 9921–9945. DOI: 10.1021/acsomega.3c05913.

Grimsley, H.R., Economou, S.E., Barnes, E. and Mayhall, N.J. (2019) An adaptive variational algorithm for exact molecular simulations on a quantum computer. *Nature Communications* 10(1), 3007. DOI: 10.1038/s41467-019-10988-2.

Hansen, R.L. and Lee, Y.J. (2018) High-spatial resolution mass spectrometry imaging: Toward single cell metabolomics in plant tissues. *The Chemical Record* 18(1), 65–77. DOI: 10.1002/tcr.201700027.

Hesami, M., Baiton, A., Alizadeh, M., Pepe, M., Torkamaneh, D. *et al.* (2021a) Advances and perspectives in tissue culture and genetic engineering of cannabis. *International Journal of Molecular Sciences* 22(11), 5671. DOI: 10.3390/ijms22115671.

Hesami, M., Yoosefzadeh Najafabadi, M., Adamek, K., Torkamaneh, D. and Jones, A.M.P. (2021b) Synergizing off-target predictions for *in silico* insights of CENH3 knockout in cannabis through CRISPR/Cas. *Molecules (Basel, Switzerland)* 26(7). DOI: 10.3390/molecules26072053.

Hesami, M., Pepe, M., Baiton, A. and Jones, A.M.P. (2023) Current status and future prospects in cannabinoid production through *in vitro* culture and synthetic biology. *Biotechnology Advances* 62, 108074. DOI: 10.1016/j.biotechadv.2022.108074.

Horn, P.J. and Chapman, K.D. (2024) Imaging plant metabolism *in situ*. *Journal of Experimental Botany* 75(6), 1654–1670. DOI: 10.1093/jxb/erad423.

Hu, J., Zhang, Y., Xie, J., Yuan, Z., Yin, Z. *et al.* (2025) Learning motif features and topological structure of molecules for metabolic pathway prediction. *Journal of Cheminformatics* 17(1), 56. DOI: 10.1186/s13321-025-00994-6.

Hurgobin, B. and Lewsey, M.G. (2022) Applications of cell- and tissue-specific omics to improve plant productivity. *Emerging Topics in Life Sciences* 6(2), 163–173. DOI: 10.1042/ETLS20210286.

Jamil, I.N., Remali, J., Azizan, K.A., Nor Muhammad, N.A., Arita, M. *et al.* (2020) Systematic multi-omics integration (MOI) approach in plant systems biology. *Frontiers in Plant Science* 11, 944. DOI: 10.3389/fpls.2020.00944.

Jeung, K., Kim, M., Jang, E., Shon, Y.J. and Jung, G.Y. (2025) Cell-free systems: A synthetic biology tool for rapid prototyping in metabolic engineering. *Biotechnology Advances* 79, 108522. DOI: 10.1016/j.biotechadv.2025.108522.

Ji, X., Liu, W.-Q., Cao, Z., Huang, S. and Li, J. (2025) Establishing a high-yield *Bacillus subtilis*-based cell-free protein synthesis system for *in vitro* prototyping and natural product biosynthesis. *ACS Synthetic Biology* 14(4), 1288–1297. DOI: 10.1021/acssynbio.5c00021.

Jiang, L., Zhao, J., Lian, J. and Xu, Z. (2018) Cell-free protein synthesis enabled rapid prototyping for metabolic engineering and synthetic biology. *Synthetic and Systems Biotechnology* 3(2), 90–96. DOI: 10.1016/j.synbio.2018.02.003.

Kajrolkar, A. (2025) Integrating multi-omics data for plant stress response: Current advances and future directions. *Premier Journal of Plant Biology* 3, 100012. DOI: 10.70389/PJPB.100012.

Karim, A.S. and Jewett, M.C. (2016) A cell-free framework for rapid biosynthetic pathway prototyping and enzyme discovery. *Metabolic Engineering* 36, 116–126. DOI: 10.1016/j.ymben.2016.03.002.

Karim, A.S., Dudley, Q.M., Juminaga, A., Yuan, Y., Crowe, S.A. *et al.* (2020) *In vitro* prototyping and rapid optimization of biosynthetic enzymes for cell design. *Nature Chemical Biology* 16(8), 912–919. DOI: 10.1038/s41589-020-0559-0.

Karim, A.S. and Jewett, M.C. (2018) Cell-free synthetic biology for pathway prototyping. In: Scrutton, N. (ed.) *Metabolic Engineering*, Vol. 608. Academic Press, Cambridge, Massachusetts, pp. 31–57.

Katam, R., Lin, C., Grant, K., Katam, C.S. and Chen, S. (2022) Advances in plant metabolomics and its applications in stress and single-cell biology. *International Journal of Molecular Sciences* 23(13), 6985. DOI: 10.3390/ijms23136985.

Kightlinger, W., Duncker, K.E., Ramesh, A., Thames, A.H., Natarajan, A. *et al.* (2019) A cell-free biosynthesis platform for modular construction of protein glycosylation pathways. *Nature Communications* 10(1), 5404. DOI: 10.1038/s41467-019-12024-9.

Kim, D., Tran, A., Kim, H.J., Lin, Y., Yang, J.Y.H. *et al.* (2023a) Gene regulatory network reconstruction: Harnessing the power of single-cell multi-omic data. *NPJ Systems Biology and Applications* 9(1), 51. DOI: 10.1038/s41540-023-00312-6.

Kim, G.B., Kim, J.Y., Lee, J.A., Norsigian, C.J., Palsson, B.O. *et al.* (2023b) Functional annotation of enzyme-encoding genes using deep learning with transformer layers. *Nature Communications* 14(1), 7370. DOI: 10.1038/s41467-023-43216-z.

Kim, T., Lee, S., Kwak, Y., Choi, M., Park, J. *et al.* (2024) READRetro: Natural product biosynthesis predicting with retrieval-augmented dual-view retrosynthesis. *New Phytologist* 243(6), 2512–2527. DOI: 10.1111/nph.20012.

Laohakunakorn, N., Grasemann, L., Lavickova, B., Michielin, G., Shahein, A. *et al.* (2020) Bottom-up construction of complex biomolecular systems with cell-free synthetic biology. *Frontiers in Bioengineering and Biotechnology* 8, 213. DOI: 10.3389/fbioe.2020.00213.

Li, D., Heiling, S., Baldwin, I.T. and Gaquerel, E. (2016) Illuminating a plant's tissue-specific metabolic diversity using computational metabolomics and information theory. *Proceedings of the National Academy of Sciences* 113(47), E7610–E7618. DOI: 10.1073/pnas.1610218113.

Li, R., Guo, S., Wang, D., Yang, T., Li, W. *et al.* (2025) Elucidation of the plant progesterone biosynthetic pathway and its application in a yeast cell factory. *Metabolic Engineering* 90, 197–208. DOI: 10.1016/j.ymben.2025.03.016.

Liao, C., Li, H., Zheng, H., Li, C., Liu, L. *et al.* (2025a) Optimization and application of the purified cell-free system. *ChemBioChem* 26(11), e202500176. DOI: 10.1002/cbic.202500176.

Liao, L., Xie, M., Zheng, X., Zhou, Z., Deng, Z. *et al.* (2025b) Molecular insights fast-tracked: AI in biosynthetic pathway research. *Natural Product Reports* 42(5), 911–936. DOI: 10.1039/d4np00003j.

Liu, Y., Yang, M., Deng, Y., Su, G., Enninful, A. *et al.* (2020) High-spatial-resolution multi-omics sequencing via deterministic barcoding in tissue. *Cell* 183(6), 1665–1681. DOI: 10.1016/j.cell.2020.10.026.

Madani, A., Krause, B., Greene, E.R., Subramanian, S., Mohr, B.P. *et al.* (2023) Large language models generate functional protein sequences across diverse families. *Nature Biotechnology* 41(8), 1099–1106. DOI: 10.1038/s41587-022-01618-2.

Matchett-Oates, L., Braich, S., Spangenberg, G.C., Rochfort, S. and Cogan, N.O.I. (2021) *In silico* analysis enabling informed design for genome editing in medicinal cannabis; gene families and variant characterisation. *PLoS ONE* 16(9), e0257413. DOI: 10.1371/journal.pone.0257413.

Mathema, V.B., Sen, P., Lamichhane, S., Orešič, M. and Khoomrung, S. (2023) Deep learning facilitates multi-data type analysis and predictive biomarker discovery in cancer precision medicine. *Computational and Structural Biotechnology Journal* 21, 1372–1382. DOI: 10.1016/j.csbj.2023.01.043.

Mipeshwaree Devi, A., Khedashwori Devi, K., Premi Devi, P., Lakshmipriyari Devi, M. and Das, S. (2023) Metabolic engineering of plant secondary metabolites: Prospects and its technological challenges. *Frontiers in Plant Science* 14, 1171154. DOI: 10.3389/fpls.2023.1171154.

Momeni, S. and Ghorbani-Vaghei, R. (2024) Copper immobilized on modified ldhs as a novel efficient catalytic system for three-component synthesis of Pyrano [2,3-*d*] pyrimidine and Pyrazolo [4′,3′:5,6] pyrano [2,3-*d*] pyrimidine derivatives. *ACS Omega* 9(9), 10332–10342. DOI: 10.1021/acsomega.3c07913.

Moore, S.J., Lai, H.-E., Chee, S.-M., Toh, M., Coode, S. *et al.* (2021) A *Streptomyces venezuelae* cell-free toolkit for synthetic biology. *ACS Synthetic Biology* 10(2), 402–411. DOI: 10.1021/acssynbio.0c00581.

Moret, M., Pachon Angona, I., Cotos, L., Yan, S., Atz, K. *et al.* (2023) Leveraging molecular structure and bioactivity with chemical language models for *de novo* drug design. *Nature Communications* 14(1), 114. DOI: 10.1038/s41467-022-35692-6.

Nana Teukam, Y.G., Zipoli, F., Laino, T., Criscuolo, E., Grisoni, F. *et al.* (2024) Integrating genetic algorithms and language models for enhanced enzyme design. *Briefings in Bioinformatics* 26(1), bbae675. DOI: 10.1093/bib/bbae675.

Nobori, T. (2025) Exploring the untapped potential of single-cell and spatial omics in plant biology. *The New Phytologist* 247(3), 1098–1116. DOI: 10.1111/nph.70220.

Novák, O., Hényková, E., Sairanen, I., Kowalczyk, M., Pospíšil, T. *et al.* (2012) Tissue-specific profiling of the *Arabidopsis thaliana* auxin metabolome. *The Plant Journal* 72(3), 523–536. DOI: 10.1111/j.1365-313X.2012.05085.x.

Olivecrona, M., Blaschke, T., Engkvist, O. and Chen, H. (2017) Molecular *de-novo* design through deep reinforcement learning. *Journal of Cheminformatics* 9(1), 48. DOI: 10.1186/s13321-017-0235-x.

Parthiban, S., Vijeesh, T., Gayathri, T., Shanmugaraj, B., Sharma, A. *et al.* (2023) Artificial intelligence-driven systems engineering for next-generation plant-derived biopharmaceuticals. *Frontiers in Plant Science* 14, 1252166. DOI: 10.3389/fpls.2023.1252166.

Poole, W., Kapasiawala, M., Roychoudhury, A., Haines, M., Freemont, P. *et al.* (2025) Metabolomics-informed coarse-grained model enables prediction of cell-free protein expression dynamics. *bioRxiv*. DOI: 10.1101/2025.06.20.660830.

Popova, M., Isayev, O. and Tropsha, A. (2018) Deep reinforcement learning for *de novo* drug design. *Science Advances* 4(7), eaap7885. DOI: 10.1126/sciadv.aap7885.

Pyrkov, A., Aliper, A., Bezrukov, D., Lin, Y.-C., Polykovskiy, D. *et al.* (2023) Quantum computing for near-term applications in generative chemistry and drug discovery. *Drug Discovery Today* 28(8), 103675. DOI: 10.1016/j.drudis.2023.103675.

Ramzi, A.B., Baharum, S.N., Bunawan, H. and Scrutton, N.S. (2020) Streamlining natural products biomanufacturing with omics and machine learning driven microbial engineering. *Frontiers in Bioengineering and Biotechnology* 8, 608918. DOI: 10.3389/fbioe.2020.608918.

Rao, X. and Liu, W. (2025) A guide to metabolic network modeling for plant biology. *Plants (Basel, Switzerland)* 14(3), 484. DOI: 10.3390/plants14030484.

Rao, A., Barkley, D., França, G.S. and Yanai, I. (2021) Exploring tissue architecture using spatial transcriptomics. *Nature* 596(7871), 211–220. DOI: 10.1038/s41586-021-03634-9.

Rasor, B.J., Yi, X., Brown, H., Alper, H.S. and Jewett, M.C. (2021) An integrated *in vivo*/*in vitro* framework to enhance cell-free biosynthesis with metabolically rewired yeast extracts. *Nature Communications* 12(1), 5139. DOI: 10.1038/s41467-021-25233-y.

Rasor, B.J., Chirania, P., Rybnicky, G.A., Giannone, R.J., Engle, N.L. *et al.* (2023) Mechanistic insights into cell-free gene expression through an integrated-omics analysis of extract processing methods. *ACS Synthetic Biology* 12(2), 405–418. DOI: 10.1021/acssynbio.2c00339.

Reiher, M., Wiebe, N., Svore, K.M., Wecker, D. and Troyer, M. (2017) Elucidating reaction mechanisms on quantum computers. *Proceedings of the National Academy of Sciences* 114(29), 7555–7560. DOI: 10.1073/pnas.1619152114.

Rice, A.J., Sword, T.T., Chengan, K., Mitchell, D.A., Mouncey, N.J. *et al.* (2025) Cell-free synthetic biology for natural product biosynthesis and discovery. *Chemical Society Reviews* 54(9), 4314–4352. DOI: 10.1039/d4cs01198h.

Ryu, G., Kim, G.B., Yu, T. and Lee, S.Y. (2023) Deep learning for metabolic pathway design. *Metabolic Engineering* 80, 130–141. DOI: 10.1016/j.ymben.2023.09.012.

Sagingalieva, A., Kordzanganeh, M., Kenbayev, N., Kosichkina, D., Tomashuk, T. *et al.* (2023) Hybrid quantum neural network for drug response prediction. *Cancers* 15(10), 2705. DOI: 10.3390/cancers15102705.

Sawasaki, T., Ogasawara, T., Morishita, R. and Endo, Y. (2002) A cell-free protein synthesis system for high-throughput proteomics. *Proceedings of the National Academy of Sciences* 99(23), 14652–14657. DOI: 10.1073/pnas.232580399.

Selvarajoo, K. and Maurer-Stroh, S. (2024) Towards multi-omics synthetic data integration. *Briefings in Bioinformatics* 25(3), bbae213. DOI: 10.1093/bib/bbae213.

Sharma, V., Gupta, P., Priscilla, K., SharanKumar, Hangargi, B. *et al.* (2021) Metabolomics intervention towards better understanding of plant traits. *Cells* 10(2), 346. DOI: 10.3390/cells10020346.

Singh, D. (2024) Quantum computing assays: Advancing drug metabolism studies and drug delivery design. *Drug Metabolism and Bioanalysis Letters* 17(3), 99–103. DOI: 10.2174/0118723128358210250219103853.

Subramanian, I., Verma, S., Kumar, S., Jere, A. and Anamika, K. (2020) Multi-omics data integration, interpretation, and its application. *Bioinformatics and Biology Insights* 14, 1177932219899051. DOI: 10.1177/1177932219899051.

Sun, X., Zhang, H., Jia, Y., Li, J. and Jia, M. (2024) CRISPR-Cas9-based genome-editing technologies in engineering bacteria for the production of plant-derived terpenoids. *Engineering Microbiology* 4(3), 100154. DOI: 10.1016/j.engmic.2024.100154.

Tang, X. and Zhang, Y. (2023) Beyond knockouts: Fine-tuning regulation of gene expression in plants with CRISPR-Cas-based promoter editing. *New Phytologist* 239(3), 868–874. DOI: 10.1111/nph.19020.

Tang, Z., Liu, X., Li, Z., Zhang, T., Yang, B. *et al.* (2023) SpaRx: Elucidate single-cell spatial heterogeneity of drug responses for personalized treatment. *Briefings in Bioinformatics* 24(6), bbad338. DOI: 10.1093/bib/bbad338.

Tinafar, A., Jaenes, K. and Pardee, K. (2019) Synthetic biology goes cell-free. *BMC Biology* 17(1), 64. DOI: 10.1186/s12915-019-0685-x.

Upadhyay, V., Boorla, V.S. and Maranas, C.D. (2023) Rank-ordering of known enzymes as starting points for re-engineering novel substrate activity using a convolutional neural network. *Metabolic Engineering* 78, 171–182. DOI: 10.1016/j.ymben.2023.06.001.

Valous, N.A., Popp, F., Zörnig, I., Jäger, D. and Charoentong, P. (2024) Graph machine learning for integrated multi-omics analysis. *British Journal of Cancer* 131(2), 205–211. DOI: 10.1038/s41416-024-02706-7.

Vandereyken, K., Sifrim, A., Thienpont, B. and Voet, T. (2023) Methods and applications for single-cell and spatial multi-omics. *Nature Reviews Genetics* 24(8), 494–515. DOI: 10.1038/s41576-023-00580-2.

Vögeli, B., Schulz, L., Garg, S., Tarasava, K., Clomburg, J.M. *et al.* (2022) Cell-free prototyping enables implementation of optimized reverse β-oxidation pathways in heterotrophic and autotrophic bacteria. *Nature Communications* 13(1), 3058. DOI: 10.1038/s41467-022-30571-6.

Wang, H., Fu, T., Du, Y., Gao, W., Huang, K. *et al.* (2023) Scientific discovery in the age of artificial intelligence. *Nature* 620(7972), 47–60. DOI: 10.1038/s41586-023-06221-2.

Wang, X., Quinn, D., Moody, T.S. and Huang, M. (2024) ALDELE: All-purpose deep learning toolkits for predicting the biocatalytic activities of enzymes. *Journal of Chemical Information and Modeling* 64(8), 3123–3139. DOI: 10.1021/acs.jcim.4c00058.

Wang, Y., Guo, M., Chen, X. and Ai, D. (2025) Screening of multi deep learning-based *de novo* molecular generation models and their application for specific target molecular generation. *Scientific Reports* 15(1), 4419. DOI: 10.1038/s41598-025-86840-z.

Wei, M., Wang, H., Zeng, Y., Ye, D., Wang, G. *et al.* (2025) Multi-modal spatial metabolomics with mid-infrared microscope guided mass spectrometry imaging. *Journal of Visualized Experiments* 221. DOI: 10.3791/68709.

Weidner, F.M., Schwab, J.D., Wölk, S., Rupprecht, F., Ikonomi, N. *et al.* (2023) Leveraging quantum computing for dynamic analyses of logical networks in systems biology. *Patterns* 4(3), 100705. DOI: 10.1016/j.patter.2023.100705.

Wu, Y. and Xie, L. (2025) AI-driven multi-omics integration for multi-scale predictive modeling of genotype-environment-phenotype relationships. *Computational and Structural Biotechnology Journal* 27, 265–277. DOI: 10.1016/j.csbj.2024.12.030.

Xie, X., Gui, L., Qiao, B., Wang, G., Huang, S. *et al.* (2024) Deep learning in template-free *de novo* biosynthetic pathway design of natural products. *Briefings in Bioinformatics* 25(6), bbae495. DOI: 10.1093/bib/bbae495.

Xu, X., Liu, Y., Du, G., Ledesma-Amaro, R. and Liu, L. (2020) Microbial chassis development for natural product biosynthesis. *Trends in Biotechnology* 38(7), 779–796. DOI: 10.1016/j.tibtech.2020.01.002.

Yang, L., Wang, H., Zhu, Z., Yang, Y., Xiong, Y. *et al.* (2025) Network pharmacology-driven sustainability: AI and multi-omics synergy for drug discovery in traditional Chinese medicine. *Pharmaceuticals* 18(7), 1074. DOI: 10.3390/ph18071074.

Yang, Z., Liu, J., Shah, H.A. and Feng, J. (2022) A novel hybrid framework for metabolic pathways prediction based on the graph attention network. *BMC Bioinformatics* 23(S5), 329. DOI: 10.1186/s12859-022-04856-y.

Yao, L., Wu, X., Jiang, X., Shan, M., Zhang, Z. *et al.* (2023) Subcellular compartmentalization in the biosynthesis and engineering of plant natural products. *Biotechnology Advances* 69, 108258. DOI: 10.1016/j.biotechadv.2023.108258.

Yeh, A.H.-W., Norn, C., Kipnis, Y., Tischer, D., Pellock, S.J. *et al.* (2023) *De novo* design of luciferases using deep learning. *Nature* 614(7949), 774–780. DOI: 10.1038/s41586-023-05696-3.

Yi, X., Rasor, B.J., Boadi, N., Louie, K., Northen, T.R. *et al.* (2023) Establishing a versatile toolkit of flux enhanced strains and cell extracts for pathway prototyping. *Metabolic Engineering* 80, 241–253. DOI: 10.1016/j.ymben.2023.10.008.

Yoosefzadeh Najafabadi, M. and Torkamaneh, D. (2025) Machine learning-enhanced multi-trait genomic prediction for optimizing cannabinoid profiles in cannabis. *The Plant Journal* 121(1), e17164. DOI: 10.1111/tpj.17164.

Yurchenko, A., Özkul, G., Van Riel, N.A.W., Van Hest, J.C.M. and De Greef, T.F.A. (2024) Mechanism-based and data-driven modeling in cell-free synthetic biology. *Chemical Communications (Cambridge, England)* 60(51), 6466–6475. DOI: 10.1039/d4cc01289e.

Yu, X., Liu, Z. and Sun, X. (2023) Single-cell and spatial multi-omics in the plant sciences: Technical advances, applications, and perspectives. *Plant Communications* 4(3), 100508. DOI: 10.1016/j.xplc.2022.100508.

Yuan, J., Li, X., Shen, X., Xiong, P., Zhu, N. *et al.* (2025) Comprehensive metabolite profiling in single-cell systems via dual-modal MALDI-mass spectrometry imaging. *Analytical Chemistry* 97(16), 8729–8737. DOI: 10.1021/acs.analchem.4c05480.

Zhang, H., Lu, K.H., Ebbini, M., Huang, P., Lu, H. *et al.* (2024a) Mass spectrometry imaging for spatially resolved multi-omics molecular mapping. *Npj Imaging* 2(1), 20. DOI: 10.1038/s44303-024-00025-3.

Zhang, M., Zhang, W. and Ma, X. (2024b) ST-SCSR: Identifying spatial domains in spatial transcriptomics data via structure correlation and self-representation. *Briefings in Bioinformatics* 25(5), bbae437. DOI: 10.1093/bib/bbae437.

Zhang, X., Xu, G., Cheng, C., Lei, L., Sun, J. *et al.* (2021) Establishment of an *Agrobacterium*-mediated genetic transformation and CRISPR/Cas9-mediated targeted mutagenesis in Hemp (*Cannabis sativa* L.). *Plant Biotechnology Journal* 19(10), 1979–1987. DOI: 10.1111/pbi.13611.

Zheng, S., Zeng, T., Li, C., Chen, B., Coley, C.W. *et al.* (2022) Deep learning driven biosynthetic pathways navigation for natural products with bioNavi-NP. *Nature Communications* 13(1), 3342. DOI: 10.1038/s41467-022-30970-9.

Index

Printed and bound by CPI Group (UK) Ltd, Croydon, CR0 4YY

16/09/2026

14963051-0002